Heat Transfer in Medicine and Biology

Analysis and Applications
Volume 2

Heat Transfer in Medicine and Biology

Analysis and Applications
Volume 2

Edited by

Avraham Shitzer

Technion — Israel Institute of Technology
Haifa, Israel

and

Robert C. Eberhart

University of Texas Health Science Center
Dallas, Texas

PLENUM PRESS · NEW YORK AND LONDON

Library of Congress Cataloging in Publication Data

Main entry under title:

Heat transfer in medicine and biology.

   Bibliography: p.
   Includes index.
   1. Body temperature. 2. Animal heat.3. Heat—Transmission. 4. Medical ther-
mography. I. Shitzer, Avraham, 1940–   . II. Eberhart, Robert C., 1937–
[DNLM: 1. Biomedical Engineering. 2. Body Temperature Regulation. 3. Energy
Transfer. QT 34 H437]
QP135.H37   1984   599′.01912   84-17698
ISBN 978-1-4684-8287-4   ISBN 978-1-4684-8285-0 (eBook)
DOI 10.1007/978-1-4684-8285-0

© 1985 Plenum Press, New York
Softcover reprint of the hardcover 1st edition 1985

A Division of Plenum Publishing Corporation
233 Spring Street, New York, N.Y. 10013

CONTRIBUTORS

R. F. Boehm Department of Mechanical and Industrial Engineering, University of Utah, Salt Lake City, Utah

Thomas C. Cetas Division of Radiation Oncology, University of Arizona, Tucson, Arizona

John C. Chato Department of Mechanical and Industrial Engineering, University of Illinois, Urbana, Illinois

Michael M. Chen Department of Mechanical and Industrial Engineering, University of Illinois, Urbana, Illinois

Robert M. Curtis Shiley, Inc., Irvine, California

Kenneth R. Diller Department of Mechanical Engineering, Biomedical Engineering Center, University of Texas, Austin, Texas

Robert C. Eberhart Department of Surgery, University of Texas Health Science Center, Dallas, Texas

L. M. Hanna Department of Bioengineering, University of Pennsylvania, Philadelphia, Pennsylvania

Linda J. Hayes Department of Aerospace Engineering and Engineering Mechanics, University of Texas, Austin, Texas

Rakesh K. Jain Department of Chemical Engineering, Carnegie–Mellon University, Pittsburgh, Pennsylvania

T. J. Love School of Aerospace, Mechanical, and Nuclear Engineering, University of Oklahoma, Norman, Oklahoma

John J. McGrath Bioengineering Transport Processes Laboratory, Michigan State University, East Lansing, Michigan

Robert W. Olsen Department of Surgery, University of Texas Health Science Center, Dallas, Texas

P. W. Scherer Department of Bioengineering, University of Pennsylvania, Philadelphia, Pennsylvania

Avraham Shitzer Department of Mechanical Engineering, Technion, Israel Institute of Technology, Haifa, Israel

George J. Trezek Department of Mechanical Engineering, University of California, Berkeley, California

A. J. Welch Department of Electrical and Computer Engineering and Biomedical Engineering Program, University of Texas, Austin, Texas

PREFACE TO VOLUME 2

This volume presents applications of heat transfer in medicine. In recent years this subject has received increased attention as many more medical applications, both in the hyper- and hypothermic ranges, have been developed. Among the subjects covered in this volume are the heating of body tissues and organs, electrosurgery, skin burns, preservation of tissues by freezing and the application of cryosurgery, heat and mass transfer in the respiratory system, heat transfer in teeth, thermography, and temperature measurement. Also included are two appendices, one presenting thermophysical properties of biological tissues and the other introducing the principles of numerical techniques in bioheat transfer.

As in Volume 1, each of the chapters in this volume is written by a leading authority in the field. The chapters all begin with a review of the state of the art, which is followed by a rigorous analytical exposition of the problem treated. Examples are given, wherever applicable, for the use of the results in actual situations.

For a quickly expanding field of science, we see here only the beginning of the application of heat transfer analysis in medicine. In the coming years we may expect more problems to be defined and analyzed and more fruitful collaboration between life scientists and physical scientists. It is our sincere hope that this book shall serve the purpose of providing the required foundation for this needed collaboration.

AVRAHAM SHITZER
ROBERT C. EBERHART

CONTENTS OF VOLUME 2

Part IV: APPLICATIONS OF HEAT TRANSFER IN MEDICINE

Chapter 16
ANALYSIS OF HEAT TRANSFER AND TEMPERATURE DISTRIBUTIONS IN TISSUES DURING LOCAL AND WHOLE-BODY HYPERTHERMIA
Rakesh K. Jain

Chapter 17
TEMPERATURE FIELDS AND LESION SIZES IN ELECTROSURGERY AND INDUCTION THERMOCOAGULATION
Avraham Shitzer

Chapter 18
ANALYSIS OF SKIN BURNS
Kenneth R. Diller

Chapter 23
HEAT AND WATER TRANSPORT IN THE HUMAN RESPIRATORY SYSTEM
P. W. Scherer and L. M. Hanna

Chapter 24
HEAT TRANSFER IN TEETH
R. F. Boehm

Chapter 25
ANALYSIS AND APPLICATION OF THERMOGRAPHY IN MEDICAL DIAGNOSIS
T. J. Love

Chapter 26
COMPUTER-AIDED TOMOGRAPHIC THERMOGRAPHY
Michael M. Chen

Part V: SELECTED TOPICS

CONTENTS OF VOLUME 1

Chapter 4
THERMOREGULATION IN PATHOLOGICAL STATES
J. M. Lipton

Chapter 5
THERMOREGULATION AND SLEEP
H. Craig Heller and Steven F. Glotzbach

Part II: THERMAL MODELING OF TISSUES

Chapter 6
HEAT GENERATION, STORAGE, AND TRANSPORT PROCESSES
Avraham Shitzer and Robert C. Eberhart

Chapter 7
THE TISSUE ENERGY BALANCE EQUATION
Michael M. Chen

Part III: ANALYTICAL AND BIOHEAT TRANSFER STUDIES

Chapter 8
MEASUREMENT OF THERMAL PROPERTIES OF BIOLOGICAL MATERIALS
John C. Chato

Chapter 9
ESTIMATION OF TISSUE BLOOD FLOW
H. Frederick Bowman

Chapter 10
GENERAL ANALYSIS OF THE BIOHEAT EQUATION
Avraham Shitzer

Chapter 11
GREEN'S FUNCTION FORMULATION OF THE BIOHEAT TRANSFER PROBLEM
Hans G. Klinger

NOMENCLATURE

Numbers in parentheses after the description refer to chapters and equations in which symbols are first used or are thoroughly defined—e.g., "(17—1)" refers to equation (1) of Chapter 17. Equations are not listed for some symbols in such general usage as to be familiar to all readers. Dimensions are given in terms of mass (M), moles, length (L), time (t), temperature (T), volts (V), and ohms. SI and cgs dimensions are given in text and tables, where appropriate. Symbols that appear infrequently or in one section only are not listed.

A	area, L^2
A	heat conductance parameter (A3—20), M/Lt^2T
A	constant in burn injury equation (17—1), t^{-1}
A	coefficient in general solution of bioheat equation (15—6)
A_i	parameters in endurance time equations (13—66, 67, 69); parameters in shivering thermogenesis equations (13—60, 62, 64); parameter in glycogen depletion equation (13—68); attenuation parameter for vasoconstrictor outflow (13—45)
A_{ij}	sensitivity coefficient of variable x_j with respect to x_i (28—2)
a	tube or capillary spacing (14—37), L
a	radius, thickness (9—12), L
a	species activity (20—2), $ML^3/$mole
B	amplitude coefficient (7—16b)
B	endurance time parameter (13—67)
B	magnetic field (16—16), $(M\,\text{ohm}/t)^{1/2}$
B	coefficient in general solution of bioheat equation (15—7)
B	heat convection parameter (A3—20), M/Lt^3T
BM	basal metabolic rate (22—41), M/t^3
Bi	Biot number (A1—Table 4)
b	tissue thickness (14—47) or blood vessel spacing (4.5—28), L
b	coefficient determining effect of species concentration on perfusion rate (13—46), $L^3/$mole
C	speed of sound (16—10), L/t
C	thermal equivalent O_2 consumption (12—71), M/L^3t^2
C	w_{min}/w_{max} (12—34)
C	thermal capacitance matrix in finite element formulation of bioheat equation (A3—33), M/Lt^2T
CI	cardiac index (22—30), L^2/t
C, c	species concentration, mole/L^3
C_s	shear velocity (8—37), L/t
C_1, C_2	coefficients of Planck radiation equation (A1—5)

c, c_p	heat capacity at constant pressure (6—3), L^2/Tt^2
D	diffusion coefficient (14—17), L^2/t
D	vasodilator outflow signal intensity (13—53), t^{-1}
D	heat generation and convection parameter (A3—20), M/Lt^3
D	coefficient in general solution of bioheat equation (15—8)
D, d	diameter, L
$D_{\tilde{T}c}$	weighting factor for vasodilator outflow (13—56)
d	thickness, L
E	blood flow weighting function (12—37)
ΔE	activation energy (17—1), ML^2/t^2
E_r	emissive power (8—2), M/t^3
F	vessel flow (6—9), L^3/t
F	radiation shape factor (6—14)
F	heat source function (10—15), LT
ΔF	net force (8—31), ML/t^2
FBF	forearm blood flow (3—2), L^3/t
F, f	skin heat flux (14—37), ML^2/t^3
F_c, F_w	weighting functions for thermoreceptor afferent error signals (13—41), $(Tt)^{-1}$
$\tilde{F}_2$	shivering thermogenesis parameter (13—62), t^{-1}
f	frequency (16—10), t^{-1}
f	probe heat generation rate function (9—12), $t^{-1/2}$
f	number of independent intensive properties (20—1)
f	heat convection and surface exchange matrix in finite element formulation (A3—33), ML^2/t^3
f_c	percentage of surface area through which convective exchange occurs (6—11)
f_{cl}	ratio of surface area of clothed body to that of nude body (6—25)
f_i	solution function for generalized bioheat equation (10—8)
f_w	surface wetting coefficient (6—20)
G, G_1, G_2	gain factors in shivering thermogenesis (13—62, 63)
G, g	generalized initial temperature distribution (10—7, 14)
Gz	Graetz number (A1—Table 4)
$G_{n,j}$	temperature weighting function (14—51), T
G_v	Green's function (11—8)
g	gravitation constant (A1—Table 4), L/t^2
$g_{n,0}$	temperature weighting function (14—52), T
H	heat generation rate (12—71), ML^2/t^3
HR	heart rate (3), t^{-1}
H, h	heat transfer coefficient (6—11), M/Tt^3
H_l	heat loss (3—1), ML^2/t^2
h	Planck constant = 6.625×10^{-34} W s^2 (19—1)
h	height, L
h_{fg}	latent heat of vaporization (6—17), L^2/t^2
h_r	radiative heat exchange coefficient (6—16), M/Tt^3
I	radiation intensity (18—17) M/t^3

I_{cl}	total thermal resistance offered by clothing ensemble (6—26), Tt^3/M
$I_{i(x)}$	modified Bessel function of the first kind, of order i (12—14)
J	surface-absorbed thermal radiant heat flux (18—16), M/t^3
$J_{i(x)}$	Bessel function of the first kind, of order i (12—15)
J_w	volume flux (20—31), L^3/t
j	current density (16—15), amperes/L^2
K	skin temperature gradient (18—12), T/L
K	kernel of finite integral transform (10—21)
$\mathbf{K}$	thermal conduction matrix in finite element formulation of bioheat equation (A3—33), ML^2/Tt^3
K_i, K_{ij}	thermal conductivity parameter between body layers (2.2—1), (22—34), ML/t^3T
$K_{i(x)}$	modified Bessel function of the second kind, of order i (14—36)
K_m	mass transfer coefficient (23—11), L/t
k	thermal conductivity (6—4), ML/Tt^3
k	mass transfer coefficient (14—17), L^3/t
k	Boltzmann constant = 1.38×10^{-23} W s/K (27—1)
k	coefficient in vasodilator equation (13—53); in glycogen depletion equation (13—71); in sweating rate equation (13—58)
k'	chemical reaction rate constant (13—32), mole/L^3t
k_{ij}	thermal conductivity tensor (11—56), ML/Tt^3
L	work load (13—66)
L	flow rate function (10—1), L^{-2}
L, l	length, L
L_f	latent heat of fusion (20—2), L^2/t^2
M	molecular weight (23—9), M/mole
M	kernel of finite integral transform (10—39)
$\dot{M}$	molar flux rate of species (13—16), mole/L^3t
M, N	number of capillaries in average cube in x, y directions (11—44), L^{-3}
M, m	Mass, M
M_{sh}	rate of shivering thermogenesis (13—59), $ML^2/t^{3\cdot}$
m	water vapor permeation coefficient of skin (6—17), t/L
m	ratio of electrode to tissue thermal inertia (17—10)
m	mass fraction (8—9)
m	concentration (20—4), mole/L^3
$\dot{m}$	mass flow rate (6—21), M/t (M/L^3t in other usage)
N	number of heat transfer units (12—33)
N	mass transfer rate (13—19), mole/L^2t
N_i	thermal conductance (12—46), L^3
Nu	Nusselt number (A1—Table 4)
n_i	molecular concentration (20—21), L^{-3}
n_i	outward unit vector normal to surface element
P	perimeter, L

P	reduced temperature function (10—14), T
P	power (8—31), ML^2/t^3
Pe	Peclet number (A1—Table 4)
P, P_0	power deposition in tissue (17—2), M/Lt^3
Pr	Prandtl number (A1—Table 4)
P_i, p_i	partial pressure of species i (6—17), M/Lt^2
p	fluid pressure (20—15), M/Lt^2
p	wave number (7—25)
p	number of phases (20—1)
p_w	water permeability (20—32), L^2t/M
Q	heat input (16—13), ML^2/t^2
$\dot{Q}$	heat generation rate, cooling rate, heat storage rate (6—1), ML^2/t^3
Q_{10}	temperature coefficient of metabolism (22—40)
q	heat generation rate per unit volume, M/Lt^3
q_i, q	heat flux (6—6), M/t^3
q_1	heat source strength per unit length (8—26), ML/t^3
R	electrical resistance (9—33), ohm
R	ratio of heat loss via coronary arteries to myocardial heat production (12—Fig. 36)
R, r	molar chemical reaction rate (13—17), mole/L^3t
R	stretching ratio for finite difference grid (18—44)
R, R_0	universal gas constant, 8.317 W s/K mol (17—1)
RQ	respiratory quotient, $\dot{V}_{CO_2}/\dot{V}_{O_2}$ (22—41)
R, r	radial coordinate; L or dimensionless
R_s	real part of shear acoustic impedance (8—35), M/L^2t
Re	Reynolds number (A1—Table 4)
r	tissue: blood solubility ratio (13—33)
r, R	resistance to heat exchange (22—Fig. 5), Tt^3/M
S	body surface area (22—29), L^2
S	signal intensity in the autonomic nervous system (13—46), $T^{0.25}$ (t, t^{-1} in other usage)
S, s	heat source (26—4), M/Lt^3
S, s	sweat generation rate (14—11), L^3/t, M/L^2t
Sc	Schmidt number (A1—Table 4)
Sh	Sherwood number (A1—Table 4)
S_1, S_2, S_3	contributions to shivering thermogenesis (13—59), ML^2/t^2
s	sensitivity of tissue damage to temperature (19—49)
T	temperature, T
ΔT	temperature increment, T
T_0	thermal parameter (12—13), T
t	time, t
t	thickness (22—6), L
t_c	characteristic time for heat conduction (11—4), t
t_f	endurance time (13—65), t
U	overall heat transfer coefficient (12—31), M/Tt^3
U, u	temperature difference (9—5), T

$U_{\text{slug},\phi}$	temperature increment, heated, perfused tissue (9—4), T
$U_{\text{slug},0}$	temperature increment, heated, unperfused tissue (9—4), T
V	voltage (9—33), V
V	temperature difference (12—13), T
V	volume, L^3
$\dot{V}, \dot{v}$	volumetric gas flow rate (13—38), L^3/t
$\bar{V}_w$	partial molar volume of water (20—42), L^3/mol
v	vapor
v	solution function for generalized bioheat equation (10—8), T
v, V	velocity, L/t
v	flow distribution function (28—24)
W	weight (6—12), M
W	total flow rate (22—1), L^3/t
w	specific humidity (6—21)
w_b	blood flow rate in tissue, per unit volume (6—6), t^{-1}
X	weighting factor for perfusion response to thermally induced vasoconstrictor outflow (13—46), $T^{-0.25}$
X	concentration, one compartment model (28—23), M/L^3
X, x	rectangular coordinate; L or dimensionless
X_s	imaginary component of shear acoustic impedance (8—36), $M/L^2 t$
x	mole fraction (20—19)
Y	concentration, two compartment model (28—24), M/L^3
Y, y	rectangular coordinate; L or dimensionless
$Y_i(x)$	Bessel function of the second kind, of order i (14—42)
Z, z	rectangular or axial cylindrical coordinate; L or dimensionless
Z_c	vasoconstriction factor (13—43)
z	body height (6—12), L
z	perfusion parameter (9—36)

GREEK SYMBOLS

α	thermal diffusivity (8—20), L^2/t
α	radiation absorptivity (A1—9)
α, α_r	radiation absorption coefficient (18—18), L^{-1}
α_a	acoustic absorption coefficient (8—34), L^{-1}
α_n	heat flux function (14—38), T/L
α_s	sweating coefficient (14—11), $M/L^2 Tt$
β	parameter for transient probe heating rate (9—12), $M/Lt^{2.5}$
β	perfusion parameter (A1—Table 4)
β	coefficients determining intensities of autonomic responses (13—46, 50, 51, 58, 60), t^{-1}
β	thermal coefficient of expansion $= \dfrac{1}{\rho}\left(\dfrac{\partial \rho}{\partial T}\right)_p$, T^{-1}
Γ	steady state heat generation rate (9—12), M/Lt^3

γ	heat generation parameter (A1—Table 4)
γ'	metabolic heat generation parameter (12—75)
γ, γ_i	vessel spacing parameters (14—48), L^{-1}
γ_j	coefficients determining intensities of autonomic responses (13—42, 45)
δ	thermal inertia parameter (17—25)
δ	depth of layer with varying temperature (26—18), L
δA	control element surface area (7—Fig. 2), L^2
δV	control element volume (7—Fig. 2), L
ε	dielectric constant (14—12), $(ohm)^{-1}$
ε	radiative emission coefficient (6—14)
ε	heat transfer effectiveness (12—25)
ζ	equivalent length parameter (14—40), L^{-1}
η	tube diameter to tube spacing ratio (14—31)
η	coefficients determining intensities of autonomic responses (13—53, 54)
η	heat transfer effectiveness parameter (16—3)
η	dimensionless radial distance (19—11)
Θ, θ	dimensionless or reduced temperature
θ	angle (7—Fig. 2)
θ	freezing point depression (20—3), T
κ	reaction rate constant, Arrhenius Equation (18—4), t^{-1}
κ	perfusion ratio (28—13)
Λ	thermal equilibration length coefficient (7—7), $M/Lt^2 T$
Λ	surface heat transfer parameter (10—13)
λ	ratio of tissue to blood heat capacity (9—8)
λ	wavelength (16—10), L
λ	depth of embedded heat source (26—21).
λ_i	area fraction of the ith vessel (7—22)
μ	viscosity (12), M/Lt
μ	chemical potential (20—5), ML^2/t^2 mol
μ_j	dynamic shear stiffness $j = 1$ (8—35), M/Lt^2; shear viscosity $j = 2$ (8—76), M/Lt
ν	kinematic viscosity (23—4), L^2/t
ν	dissociation constant (20—21)
ν	configuration parameter for general solution of bioheat equation (10—1)
ξ	dimensionless length (12—47)
Π	osmotic pressure (20—6), M/Lt^2
ρ	density (6—3), M/L^3
ρ	radiation reflectivity (A1—9)
σ	Stefan–Boltzmann constant $= 5.67 \times 10^{-8}$ W/m^2 K^4 (6—14)
σ	electrical conductivity (16—12), $(ohm)^{-1}$
σ	image radius (19—8), L
σ_i	spacing parameter (14—49)
σ_i	heat generation parameter (28—13)
τ	dimensionless time, Fourier number (A1—Table 4)

τ	radiation transmissivity (A1—9)
τ^*	dimensionless freezing time (21—43)
Φ, ϕ	dimensionless or reduced temperature
ϕ	efflux of a flow path (11—36), L^3/t
ϕ	solution osmotic coefficient (20—18)
ϕ	basis function in finite element formulation of bioheat equation (A3—28a)
χ	heat source distribution and perfusion parameter (26—24)
Ψ	equivalent tissue heat production (9—14), M/Lt^3
Ψ	combination of modified Bessel functions (15—11)
ψ_i	surface heat transfer parameter (10—3)
Ω	axial temperature distribution function (10—60), T
Ω	tissue damage function (17—1)
Ω	solution osmolality (20—3), M/L^3
ω	volume element (11—21), L^3
ω	frequency (17—4), t^{-1}

SUPERSCRIPTS

$'$	dimensionless or reduced variable
$'$	transient
+	dimensionless quantity
*	limit of discrete blood vessels
*	setpoint, reference
*	dimensionless quantity
(B)	bound
(F)	free
(i)	phase
$(i), (n)$	ith, nth iteration
(s)	steady state
(T)	total
(0)	first spectral component

SUBSCRIPTS

a	air
a	afferent
a, art	artery
amb	ambient
avl	available
A	alveolus, airway
B	body
B	surface
B, b	blood

br	brain
c	conduction
c	core
c	coolant
c, conv	convection
CO_2	carbon dioxide
chem	chemical reaction
cl	clothing
d	dentin
d, diff	diffusion
e	equilibrium
e	enamel
e	equivalent
e	electrical
e	evaporation
e, env	environment
e, ex	expired, exhaled
eff	effective
es	esophageal
f	fabric
f	fat
f	length of exposure
f, fr	frozen
fg	fluid to gas
G, g	glycogen
g	generation
h, hy	hypothalamus
h	heating
i	tissue element in finite difference schemes
i	initial
i	blood vessel generation
i	inspired
j	blood vessel generation
k	conduction
L	lung
l	lactic acid
l	liquid
M	mucus–air interface
M, m	node in finite difference mesh
m	metabolic
m	mean
m	muscle
m	tissue, intrinsic
max	maximum
mbf	myocardial blood flow
min	minimum
N	necrotic

O_2	oxygen
o	outer
p	probe, wave number
pc	phase change
r	radiation
r	resting, basal
r, re	rectal
ra	right atrial
ref	reference
res	respiratory
s	surface, skin
s	solute
s, sw	sweating
set	set point
sh	shivering
sk	skin
ss	steady state
st	storage
t	tissue
ty	tympanic
u	uniform
v	vein
v	volumetric
W, w	water
0	reference, ambient
0	initial
1	prior to occlusion, inner
2	following occlusion, outer
λ	wave length
∞	ambient

OVERLINES

$$\cdot \qquad \frac{d}{dt}$$

~	time-weighted function
¯	normalized parameter
¯	transformed function, averaged parameter
^	unit vector

UNDERLINES

matrix

OPERATORS

∇	grad
∇^2	div $\cdot$ grad
δ	Dirac delta function
ψ	$\nabla^2 - \dfrac{\partial}{\partial t}$
$\int$	integral
$\sum$	summation
$\prod$	product
$\langle \ \rangle$	spatial average

Part IV

APPLICATIONS OF HEAT
TRANSFER IN MEDICINE

ANALYSIS OF HEAT TRANSFER AND TEMPERATURE DISTRIBUTIONS IN TISSUES DURING LOCAL AND WHOLE-BODY HYPERTHERMIA

Rakesh K. Jain

1. INTRODUCTION

Fire will succeed when all other methods fail.—HIPPOCRATES

Heat in various forms has been exploited by mankind for therapeutic purposes since ancient times. The Egyptians (~3000 B.C.) were the first to use cautery against tumors and various nonmalignant diseases[1]. The Hindus (~2000 B.C.) used cautery to control surface lesions during the Aryan civilization.[2] The importance of therapeutic application of heat in the Greek civilization is reflected in the preceding aphorism attributed to Hippocrates (460–357 B.C.). He recommended cautery (with a red-hot iron) for small tumors and many other diseases.[3–5] The application of cautery using heated metals or lenses remained popular among the medical community until the middle of the nineteenth century, when more sophisticated methods for elevating local tissue temperatures became available (e.g., diathermy and ultrasound).

It was soon realized that heat may be lethal to tumors at moderately elevated temperatures (40–42°C). In 1866, Busch reported a cure for a histologically verified sarcoma of the face after an attack of erysipelas that induced fever. Busch suggested the possibility of heat being selectively lethal to neoplastic cells. Another 25 years later, Coley administered bacterial toxins in cancer patients, which resulted in fever and led to regression of advanced and inoperable cancers.[6,7] Coley's toxins led to sustained cures in some patients for up to 50 years.[8] However, uncertainty in the preparation and biological activity of the mixed bacterial toxins used by Coley caused the method to be abandoned, since it proved harmful to patients.

In the early twentieth century, a major development occurred in diathermy—the advent of short-wave (or radiofrequency) diathermy, which

Rakesh K. Jain • Department of Chemical Engineering, Carnegie–Mellon University, Pittsburgh, Pennsylvania 15213.

allowed both local and deep noninvasive heating. Stevenson,[9] and Rohdenburg and Prime[10] were among the first to point out the relationship between treatment times and temperature for animal tumors. In 1927, Wester-mark,[165] using a diathermy apparatus, demonstrated that heat was lethal to rat tumors in the temperature range 44–45°C if tumors were heated long enough to inhibit respiration and glycolysis. His results were confirmed in 1940 by three different groups of workers using various animal tumor models.[11–13] Later, sporadic attempts were made at treating human tumors by radiofrequency,[14] microwaves ($\lambda = 80$ cm),[15] and ultrasonic heat-ing.[16,17] Despite these extensive results and the development of new heating methods, the clinical application of hyperthermia remained limited, largely because of new developments and hope in cancer treatment using surgery, radiation, and chemotherapy, and also because of the difficulty of applying heat and measuring temperatures.

In recent years, Cavaliere[18] and his co-workers heated human tumors in the extremities by local perfusion with warm blood; they showed that heat alone can lead to total regression of melanomas and sarcomas and an increase in survival of patients. In 1969, Stehlin[19] indicated that heat and an anticancer drug (Melphalan) together not only cured the primary tumor, but also reduced the incidence of metastases from these tumors. The encouraging work of these two groups, coupled with the discouraging results by surgery, radiotherapy, and chemotherapy led to a new worldwide resurgence of interest in hyper-thermia. This renewed enthusiasm in cancer thermotherapy is reflected by explosive growth in the literature on this subject during the last five years.[20–27] Despite such a large effort to treat cancer by hyperthermia, taken collectively, the long-term survival results in patients from different centers have been less than impressive. The reasons why the full potential of hyperthermia, used either alone or in conjunction with other currently available methods for human cancer treatment (Table 1), has not been realized, are as follows: (1) the lack of data on the susceptibility of tumors to various thermal doses, as determined by the temperature and duration of heating, (2) the technical difficulties of monitoring the temperature of internal tumors and the heat transfer from the energy source to the tumor; (3) the lack of precise control of temperature distributions within tumors and the surrounding normal tissues during local or whole-body hyperthermia; (4) poor understanding of the

TABLE 1
Current Methods of Cancer Treatment

Methods[a]	Major problems
Surgery	Only the primary tumor mass and the surrounding tissue can be removed
Radiotherapy	Hypoxic cells, which represent a major fraction of the tumor, are resistant to radiotherapy
Chemotherapy	Most anticancer agents are also toxic to normal tissues
Immunotherapy	Only the residual tumor can be treated
Hyperthermia	Thermal dosimetry is poorly understood (see text)

[a] In most cases, a combination of two or more methods is used for cancer treatment.

biochemical, physiological, and immunological responses of normal and neoplastic tissues at elevated temperatures; and (5) the paucity of data on optimal sequencing of hyperthermia with other modalities of cancer treatment, e.g., to minimize the damage to normal tissues while maximizing damage to neoplastic tissue.

The principal objective of this chapter is to present various theoretical frameworks that can be used to estimate heat transfer from an external or internal source to a tissue and predict resulting temperature distributions in the normal and neoplastic tissues of various mammals during normothermia and hyperthermia. This information is important for improving tumor detection by thermography and designing heating protocols for hyperthermic treatment. Whereas the response of normal and neoplastic tissues to thermal stress depends on the absolute temperature obtained, the duration of that temperature, and the treatment history, there are many physical, physiological, biochemical, immunological, and structural factors that must be considered in evaluating the effectiveness of hyperthermia in cancer treatment. Since these factors have been discussed in depth by this author elsewhere,[28] this presentation will focus on heat transfer and temperature distribution. It is interesting to note that heat has many applications in medicine (Table 2), but the major emphasis here is on cancer treatment.

There are four major problems encountered when analyzing heat transfer in normal and neoplastic tissues during hyperthermia:

1. The exact description of convective heat transfer in tissues is mathematically intractable. In most cases, therefore, a simplified scalar term is used to describe heat transfer by blood.

2. Actual geometries of tumors and normal tissues are complex. While finite-element techniques can be used to solve the system equations for irregular geometries, most investigators have obtained numerical and analytical solutions for "simple" geometries.

3. The physiological parameters (i.e., blood flow and metabolic heat generation) and biophysical parameters (i.e., thermal, electrical, and acoustic

TABLE 2
Applications of Therapeutic Heat in Medicine[a]

Deep heating for the treatment of various musculoskeletal diseases
  (e.g., rheumatoid arthritis, osteoarthritis, fibrositis, and myositis)
Deep heating for many neuromuscular disorders
  (e.g., muscular dystrophy, progressive muscular atrophy)
Treatment of various eye disorders
  (e.g., iritis, postoperative uveitis)
Dental problems
  (e.g., swelling and trismus following extractions, toothache)
Elevating body temperatures following hypothermia surgery
Warming refrigerated blood stored at 4°C
Rapid thawing of frozen biomaterials and tissues
Cancer therapy using hyperthermia (40–50°C)

[a] From Refs. 138 and 139.

properties) are not available for most neoplastic tissues. In addition, these parameters vary during the course of the treatment.

4. Analytical expressions for the thermal energy absorbed in a tissue due to microwave, radiofrequency, and ultrasonic fields are not available for realistic geometries. Therefore, prediction of temperatures in the presence of these fields involves numerical solution of two problems—energy absorption and dissipation—each being complex by itself.

In the light of these problems, two approaches have been used by investigators in this area of research: distributed and lumped parameter approaches. While the former approach provides a more detailed picture of the temperature field, it may require considerable computational time and effort. The latter approach, at the cost of detailed spatial information, often provides adequate information on the average temperature distribution, with little computation effort. In this review, we will discuss both approaches and compare the numerical results with the data available in the literature. Wherever possible, outstanding problems in the prediction of temperature distribution will be identified and attention will be directed to other chapters giving detailed treatment. Finally, some directions for future research in hyperthermia-related areas will be pointed out.

2. DISTRIBUTED PARAMETER APPROACH

The temperature field in a tissue is determined by heat conduction and convection, metabolic heat generation, thermal energy transferred to the tissue from an external source or the surrounding tissue, and the tissue geometry. Thermal conduction is characterized by a thermal conductivity k at steady state and by a thermal diffusivity α in transient states. Thermal convection is characterized by the topology of the vascular bed and the blood flow rate, which is subject to thermoregulation.

2.1. The Bioheat Transfer Equation

The most common representation of the spatial and temporal distribution of temperature in living systems is the so-called "bioheat transfer" equation (subsequently referred to as the bioheat equation). It was first suggested by Pennes[29] in the following form:

$$\rho c \frac{\partial T}{\partial t} = \nabla(k \nabla T) + q_b + q_m \qquad (1)$$

Here, T is the tissue temperature, c is the tissue heat capacity, ρ is the tissue density, k is the tissue thermal conductivity, q_m is the rate of metabolic heat generation, and q_b is the rate of heat exchange with blood. The derivation of Eq. (1) is discussed in detail in Chaps. 6 and 7. The evaluation of q_m from the oxygen consumption of the tissue is discussed in Chap. 6. A discussion of the estimation of q_b as it pertains to tumors follows.

Within the vasculature of a tissue, blood flows in all directions, and the local direction of convection depends on the vascular morphology, which is always intricate. The situation is even more complex in tumors where the direction and magnitude of blood flow are not fixed because of the vascular growth and necrotic processes. In tumors, a venous capillary may behave as an arterial capillary at a different time. Therefore, the description of the local convective heat transfer term q_b in tissues would be a time-dependent vector, a problem that is enormously complex and has thus far proven mathematically intractable. In order to circumvent a mathematical description of the details and complexities of the microcirculation in a capillary bed, two approaches have been used by investigators in this area of research.

In the first approach, the convection term is replaced by a "diffusion type" of term, and heat transfer in tissues is described in terms of an effective thermal conductivity (k_{eff}):

$$\rho c \frac{\partial T}{\partial t} = \nabla(k_{eff} \nabla T) + q_m \tag{2}$$

Implicit in this equation is the assumption that due to a large vascular surface area, the capillary blood temperature equilibrates with the tissue temperature. The concept of effective thermal conductivity has been used by several investigators in thermal physiology.[30] Jain and Wei[31] have also used this concept to describe the distribution of a drug in tumors.

In the second approach, the convective heat transfer term q_b is replaced by the thermal energy brought in by the arterial blood minus the thermal energy carried away with the venous blood:

$$q_b = \eta w_b \rho_b c_b (T_a - T) \tag{3}$$

Here, b is the blood density, c_b is the blood heat capacity, T_a is the arterial blood temperature, and η is a measure of the effectiveness of heat transfer between the tissue and the venous blood ($0 \leq \eta \leq 1$); η is equal to 1 when venous blood is in complete equilibrium with tissue. It is shown in Chap. 7 that it is reasonable to expect that, due to slow blood flow, η will be close to 1. Although it is possible to introduce a value of η different from 1 and carry it through, it introduces no new insight and changes the numerical value of the blood flow rate slightly.

The bioheat equation, with both of these assumptions, has been solved for various tissue geometries, and initial and boundary conditions (see Chap. 12).

Due to scalar treatment of the convective heat transport by blood, the bioheat equation has come under serious criticism. In a recent New York Academy of Science conference,[26] the limitations of the bioheat equation were discussed. Various alternatives are presented and discussed in Chaps. 7, and 9–12. Considering tissue as porous media, Wulff[32] introduced the blood velocity vector w_b in the bioheat equation, but did not attempt to specify the circulation vector at the microscopic level.

A second criticism of the bioheat equation originates from the fact that it does not account for the countercurrent heat exchange in the larger blood vessels and the capillary bed. Mitchell and Myers,[33] Keller and Seiler,[34] and Weinbaum[35] have developed models to account for macroscopic and microscopic countercurrent heat exchange. While these models are conceptually more elegant than the distributed or lumped parameter treatments, they are impractical for most tissues where the vascular morphology is much more complex and the velocity vector changes its magnitude and direction randomly.

Klinger[36] (see Chap. 11) has demonstrated the use of the Green's function to obtain an exact analytical solution of the diffusion equation with convection terms, without making special assumptions concerning the velocity field. Again, the absence of a detailed knowledge of the convection field limits the usefulness of this approach.

One of the most significant improvements in the bioheat equation has been made by Chen and Holmes[37] (Chap. 7). These authors point out that in addition to the blood perfusion term, similar to Eq. (3), the blood flow in the microvasculature may have at least two contributions to heat transfer: a contribution proportional to the local blood velocity vector w_b and a contribution proportional to the temperature gradient, similar to the effective thermal conduction term in Eq. (2). Chen and Holmes also suggest that in some circumstances, these two additional contributions may be negligible compared to the perfusion term.

While these various improvements in the bioheat equation provide new insight into the heat transfer process in a capillary bed, the mathematical complexity makes their application to normal and neoplastic tissues difficult at best. Therefore, in this chapter, we will use the bioheat equation of Pennes to describe heat transfer and temperature distribution in tissues. In the next section, we will present values for the parameters incorporated in Pennes's model.

2.2. Parameter Values

2.2.1. Thermal Properties

A comprehensive tabulation of the thermal properties of tissues, including neoplastic tissues, reported up to 1980, can be found in Appendix 2. This compilation is an extension of previous efforts.[38,39]

Using a noninvasive probe technique, Jain et al.[40] have recently measured the thermal conductivity (3 mW/cm K) and thermal diffusivity (10^{-3} cm^2/sec) of a tumor of mammary origin—Walker 256 carcinoma. In this study, tumors weighing 2–11 g and having blood perfusion rates (w_b) of 1–6 hr^{-1} were used. While the effective thermal conductivity of these tumors decreased as they grew larger, no definite correlation was found between the true thermal conductivity of a tumor and its weight. When the tumor's blood flow rate was modified by inducing hypo- or hypervolemia, its effective thermal conductivity (as measured by the temperature rise in the heating

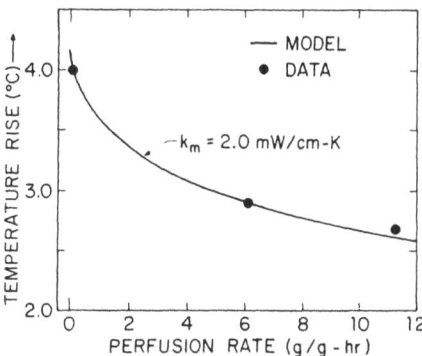

FIGURE 1
The effect of blood flow on the effective thermal conductivity of tumors. Temperature rise is inversely proportional to the effective thermal conductivity (see Ref. 40). The points represent data, and the solid line represents the solution of the bioheat transfer equation. By permission of the *J. Nat. Cancer Inst.*

probe embedded in the tumor) varied proportionally to the square root of the perfusion rate (Peclet number). But this result must be tempered by the finding that, in regions of varying perfusion, the prediction of perfusion from thermal properties is nonunique and perhaps inconclusive (cf. Chap. 12). Nevertheless, useful indications of trends in blood flow may be obtained by this method. Figure 1 suggests that in order to obtain a biologically significant increase in

TABLE 3
Thermal Conductivity of Various Animal and Human Tumors

Species	Tumor	k(mW/cmK)	Remarks
Rat	Walker 256 carcinoma	3.2 ± 0.9	*a*
Human	Breast		*b*
	Normal atrophic tissue	4.99 ± 0.04	
	Scirrhous carcinoma	3.97 ± 0.04	
	Mucinous (colloid) carcinoma	5.27 ± 0.41	
	Colon		
	Normal	5.56 ± 0.09	
	Metastatic colonic carcinoma	5.56 ± 0.12	
	Liver		
	Normal	5.72 ± 0.09	
	Metastatic colonic cancer	5.20 ± 0.08	
	Normal	5.08 ± 0.11	
	Metastatic pancreatic cancer	5.62 ± 0.21	
	Lung		
	Normal	5.18 ± 0.21	
	Squamous cell	6.66 ± 0.18	
	Pancreas		
	Normal	3.45 ± 0.05	
	Metastatic carcinoma	4.78 ± 0.39	
	Normal	4.68 ± 0.06	
	Metastatic gastric cancer	4.92 ± 0.54	
	Other		
	Acoustic Schwannoma	5.81 ± 0.17	

a *In vivo*; Jain *et al.* (1979), Ref. 40.
b *In vitro*; Bowman (1980), Ref. 44.

the effective thermal conductivity, a substantial increase in the blood flow rate is needed. In these experiments, thermal conductivity measured *in vitro* was found to be within 10% of the *in vivo* value.[41] Shah and Jain[42] have recently measured thermal conductivity, diffusivity, and perfusion rates of various animal tumors by implanting a probe noninvasively. Holmes and Chen[43] and Bowman[44] have also measured thermal properties of tumors using invasive-probe techniques (Table 3).

Until more data on the thermal properties of tumors are collected, an order of magnitude estimate of the thermal properties of tumors can be obtained using the correlation of Cooper and Trezek.[45] This correlation is a modification of the one developed by Poppendiek *et al.*[46] and relates the thermal properties of the tissue to its composition (water, protein, and fat). The need for more accurate measurements and predictions of thermal properties of tumors over a wide range of temperatures is urgent.

2.2.2. Blood Flow Rate

Blood flow rates and volumes of various organs and normal tissues of mammals (mouse, rat, hamster, dog, swine, rabbit, monkey, and man) were recently compiled by Bischoff,[47] Jain *et al.*,[48] and Geslowski and Jain.[49] Blood flow rates of various animal and human tumors during normothermia are given in Table 4. In general, the mean perfusion rates of tumors are less than those of normal tissues, with the exception of a canine lymphosarcoma. In addition, the average blood flow rate of an animal tumor decreases as it grows larger, with the exception of the data of Slotman *et al.*[50] on VX2 carcinoma in rabbits. This relationship has not been found to be valid in human tumors.[51]

After analyzing the perfusion rate w_b of various experimental tumors, Gullino[52] proposed the following simple relationship for a tumor with weight W in grams:

$$w_b = -0.667 \ (\log \ W) - a \quad (w_b \text{ in ml/g min}) \tag{4a}$$

where a, a constant is about 0.37 for W256 carcinoma.

Song *et al.*[53] obtained the following best-fitting curve for their data on blood flow rate of W256 as a function of weight, in the range 0.05–5.0 g

$$\log \ w_b = -0.1721 \ (\log \ W)^2 - 0.5382 \ (\log W) - 0.5255 \tag{4b}$$

Vaupel[54] found an exponential decay in blood flow rate of DS-carcinosarcoma in the range 3–13 g

$$\log \ w_b = -0.100 \ W - 0.315 \tag{4c}$$

It is well established that as a tumor grows larger, its center becomes necrotic and its average perfusion rate decreases (at least in animal tumors),

TABLE 4
Blood Flow Rates of Animal and Human Tumors

Tumor	Species	Blood flow (ml/g min)	Method	Reference
Hepatoma 5123	Rat	0.1–0.17	Direct collection of efferent blood	140a, b
Novikoff hepatoma	Rat	0.02–0.05	Direct collection of efferent blood	140a, b
Walker 256 carcinoma	Rat	0.03–0.1	Direct collection of efferent blood	140a, b
Walker 256 carcinoma	Rat	0.16–0.48	Radioactive microspheres	141
Sarcoma	Rat	0.04–0.21	^{133}Xe-clearance	142
Sarcoma	Rat	0.22–0.58	Plethysmography	142
DS-carcinosarcoma	Rat	0.07–0.32	Direct collection of efferent blood	54
Yoshida sarcoma	Rat	0.07	Uptake of ^{86}Rb	143
Nerve and brain tumors	Rat	0.44–0.79	Uptake of ^{14}C-antipyrine	144
Guerin carcinoma	Rat	0.20–0.21	Uptake of ^{86}Rb	145
DMBA-induced adenocarcinoma	Rat	0.025	^{133}Xe clearance	146
BA-1112 rhabdomyosarcoma	Rat		RBC velocity measurement	147
Melanoma	Hamster	0.60	^{131}I-antipyrine uptake	148
Cervical carcinoma	Hamster	0.22	^{133}Xe clearance	49
Sarcoma	Mouse	0.01–0.22	^{133}Xe clearnace	149
Sarcoma	Mouse	0.04–0.19	^{133}Xe clearance	150
Sarcoma	Mouse	0.07–0.14	^{133}Xe clearance	151
Mammary carcinoma	Mouse	0.01–0.17	^{133}Xe clearance	150
VX-2 carcinoma	Rabbit		^{85}Kr clearance	152
VX-2 carcinoma	Rabbit	0.24–1.13	Radioactive microspheres	50
Lymphosarcoma	Dog	0.63–3.4	Thermal dilution technique	60
Lymphoma	Human	0.34 ± 0.21	^{131}Xe clearance	51, 153
Anaplastic carcinoma	Human	0.15 ± 0.11	^{131}Xe clearance	51, 153
Differentiated tumors	Human	0.23 ± 0.15	^{131}Xe clearance	51, 153
Liver carcinoma	Human	0.12	^{131}Xe clearance	154

approximately according to Eqs. (4a)–(4c). However, only limited data and constitutive relationships are available for the distribution of blood flow in tumors.

Goldacre and Sylvan[55] have comprehensively reviewed the work done on the distribution of blood-borne dyes in the tumors. By changing the color of the systemic blood with lisammine green, the authors have isolated the poorly perfused region of several transplanted mouse tumors (mammary carcinoma, sarcoma 37, Ehrlich–Landschultz ascites) and Walker 256 carcinosarcoma in Wistar rats. These investigators did not propose a quantitative relationship for the spatial heterogeneities in blood flow. Using the data of Rowe-Jones,[56] Jain and Wei[31] proposed the following two-zone model for

W256 carcinosarcoma in rats:

$$w_b = \begin{cases} w_{bN} & \text{(in necrotic zone)} \\ 10 w_{bN} & \text{(in viable zone)} \end{cases} \tag{5a}$$

where w_{bN} is the perfusion rate in the necrotic zone. Recently, by using a thermal dilution technique,[57] we have found that blood flow rate can be nearly zero in some sections of tumors and around three times the average perfusion rate in other sections.

Shibata and MacLean[58] have used radioactive microspheres to study the relative perfusion rates of several human and animal tumors, although the radioactive microsphere technique is subject to large error in regions of very low flow. Their results on the distribution ratio of radioactivity between the periphery and the center of Walker carcinoma 256 (grown in rat thigh muscle) show that the ratio of radioactive counts between periphery and center ranges from 0.9 to 9.8, and it seems to have no correlation with tumor size. However, a definite decrease in the perfusion rate seems evident as the size of tumor increases.

Employing the bolus infusion, thermal dilution technique of Tuttle and Sadler,[59] Straw et al.[60] measured the regional perfusion rates of a canine

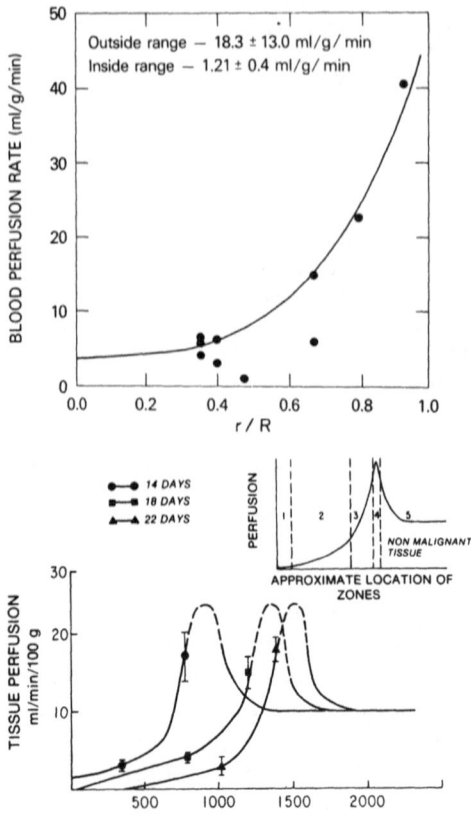

FIGURE 2

(a) Regional distribution of blood flow in a canine lymphosarcoma. (b) Regional distribution of blood flow in a two-dimensional ("sandwich") tumor. Adapted from Endrich et al. (see Ref. 147). By permission of the J. Nat. Cancer Inst.

lymphosarcoma. As shown in Fig. 2a, the perfusion rate is as high as 43.3 ml/g/min in the periphery and as low as 0.63 ml/g/min in the center. As shown in Chap. 9, this simple, lumped parameter thermodilution technique is subject to considerable error. Nevertheless, we have fit its data using the following parabolic relationship:

$$w_b = 1.2 + 15.1\ (r/R)^2 \tag{5b}$$

where r and R are the radial position in the tumor and tumor radius, respectively. In our previous works, we have used both Eqs. (5a) and (5b) in estimating temperature distributions in tumors, after normalizing these equations so that both lead to the same average perfusion rate, as given by Eq. (4a).[26,61]

Using a sandwich tumor preparation Endrich *et al.*[62] have recently studied the blood flow distribution in BA-1112 rhabdomyosarcoma in rats. They have divided the two-dimensional tumor tissue into five regions on the basis of perfusion rates (Fig. 2b). We are currently developing constitutive relationships to describe regional perfusion rates of VX2 carcinoma grown using a transparent chamber placed in a rabbit ear.[63–65]

2.2.3. Metabolic Heat Generation

Metabolic heat generation is, perhaps, the most elusive term in the bioheat equation; a detailed evaluation and model of this term is given in Chaps. 6 and 13. Prior to this new model, the most widely used approach in estimating this term has been to set q_m equal to oxygen consumption multiplied by the caloric value of oxygen.[66] Implicit in this approach is the assumption that oxygen, and not glucose, determines heat generation in a tissue; however, this assumption is tenuous in ischemic tissues (Chaps. 12 and 13). Oxygen consumption for various mammalian tissues can be found in the literature.[67,68] *In vivo* oxygen and glucose consumption and lactate production of various animal tumors are given in Table 5 and Refs. 69–74.

Gullino *et al.*[72–74] have shown that the oxygen consumption of tumors is linearly related to their perfusion rate

$$\dot{V}_{O_2} = -0.16 + 2.03\ w_b \tag{6}$$

TABLE 5
***In vivo* Oxygen and Glucose Consumption and Lactate Production of Rat Tumors**[a]

Tumor	Weight range (g)	Weight doubling time	mol/g hr		
			Oxygen	Glucose	Lactate
W256 carcinoma	2.0–9.8	33	0.24	0.54	0.38
Hepatocarcinoma 5123	3.8–7.0	196	0.42	0.31	0.20
Fibrosarcoma 4956	5.0–12.7	61	0.08	0.25	0.17
DS-carcinoma	3.2–13	?	0.13		

[a] From Refs. 54, 72–75.

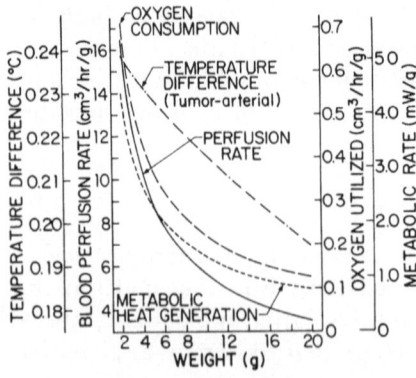

FIGURE 3
The effect of tumor weight on its oxygen consumption, blood flow rate, metabolic heat generation rate, and temperature rise due to metabolic heat generation (see Ref. 76). By permission of the *J. Therm. Biol.*

where $\dot{V}_{O_2}$ is in mmol/hr 100 g and w_b is in l/hr 100 g. Similarly, Vaupel[54,75] has shown that oxygen consumption by DS-carcinosarcoma decreases exponentially with tumor weight, in a fashion analogous to the perfusion rate

$$\log \dot{V}_{O_2} = -0.107 \ W - 1.469 \tag{7}$$

where $\dot{V}_{O^2}$ is in ml/g min. Vaupel[75] has also found maldistribution of oxygen concentration and O_2 consumption in tumors, similar to the local perfusion rate maldistribution.

Using Eq. (6), Sien and Jain[76] have developed the following constitutive relationship between the metabolic heat generation and the local blood supply:

$$q_m = 6 \times 10^4 \ w_b - 0.22 \qquad \text{mW/cm}^3 \tag{8}$$

This relationship includes the effect of the respiratory quotient ($\dot{V}_{CO_2}/\dot{V}_{O_2}$) in heat generation. In addition, this equation allows us to incorporate inhomogeneities in metabolic heat generation, analogous to the nonuniform perfusion rate.

Figure 3 shows the relationship among the tumor's blood flow rate (Eq. 4a), oxygen consumption rate (Eq. 6), and the metabolic heat generation rate (Eq. 8) of tumors as a function of their weight.

2.2.4. Tissue Geometry

Once all the model's parameter values are specified, the geometry of the model system must be defined. The situation for a tumor is more complex than for normal tissue. Figure 4 shows some of the model geometries that have been used to approximate thermal interactions between a tumor and the surrounding tissues. It is known that a tumor may infiltrate the surrounding tissue and assume a complex morphology. Because of the possibility of obtaining analytical solutions in simple geometries (i.e., cylinder, sphere), most work has been done for these cases. For more complex geometries, finite-difference or finite-element methods (Chaps. 12 and 18) are necessary to

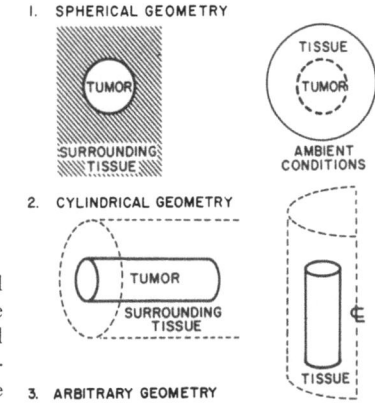

FIGURE 4
Typical model systems that approximate the physical and anatomical situation. While a tumor may have a more complex geometry, simple geometries (such as cylindrical or spherical) make it possible to obtain analytical solutions in most cases (see Ref. 26). By permission of the New York Academy of Sciences.

calculate the temperature field in the tumor and the surrounding normal tissues.

2.2.5. Boundary Conditions

In order to solve the bioheat equation for a given geometry, the boundary conditions must be specified. Within the tissue or organ, the heat flux and temperature at interfaces between regions of varying properties must be continuous. If the tissue or organ containing the tumor is exposed to the external environment, heat exchange with that environment, by conduction, convection, radiation, and evaporation must be accounted for.

Heat source terms must be added in either the bioheat equation or the boundary conditions, depending on the method used for inducing hyperthermia. (See Sec. 6 of this chapter and Chap. 27 for details of hyperthermia technology.) During surface heating by hot air, water, or a molten wax bath, heat transfer by conduction and convection to the overlying skin must be added to the boundary condition. During infrared or visible radiation-induced heating, a radiation term must be added to the skin boundary condition. During hyperthermic perfusion, the arterial temperature T_a must be set equal to the experimentally determined inlet blood temperature in the bioheat equation. During volume heating by ultrasound, microwave, or radiofrequency radiation, an additional term, describing the temporal and spatial distribution of absorbed energy must be added to the right-hand side of the bioheat equation. In a later section, we will discuss techniques for estimating the volume heating term.

3. LUMPED PARAMETER APPROACH

While the distributed parameter approach previously discussed describes the detailed temporal and spatial distribution of temperature in a tissue, the solution of system equations is often tedious and requires precise knowledge of tissue geometry, anisotropy, and orientation with respect to the surrounding tissues or heat source. Lumped parameter models overcome these problems at the cost of detailed information, which may not be needed in some cases of interest. In what follows, we discuss various lumped parameter models in brief and describe various heat transfer mechanisms that should be incorporated into such models.

3.1. Compartmental Approach

Lumped parameter models describing the mammalian thermal system have been developed for both specific organs and whole-body systems.[68,77-81] Analyses of the thermal interaction between a tumor and the host are more limited.[69,76,82-84] We will first discuss the whole-body model developed by Huckaba and Tam[68] and then our model for a tumor-bearing host.

The whole-body lumped parameter model of Huckaba and Tam is based on the distributed parameter models developed previously by Stolwijk,[80] Hardy *et al.*,[77] and Wissler.[85] The basic model is described in detail in Chap. 13. In this model, the body is divided into one spherical and a number of cylindrical segments: head, neck, upper trunk, lower trunk, upper arms, forearms, hands, fingers, thighs, legs, and feet. In the case of the extremities, single cylinders are used to represent corresponding pairs of segments together (Fig. 5a). Note that this symmetry assumption will not work for tumor-bearing organs.

Each segment is further divided into four subsections: skin, fat, muscle, and core (Fig. 5b); a total of 44 subsections are specified. Unlike the distributed parameter model, which is based on local satisfaction of the bioheat equation,

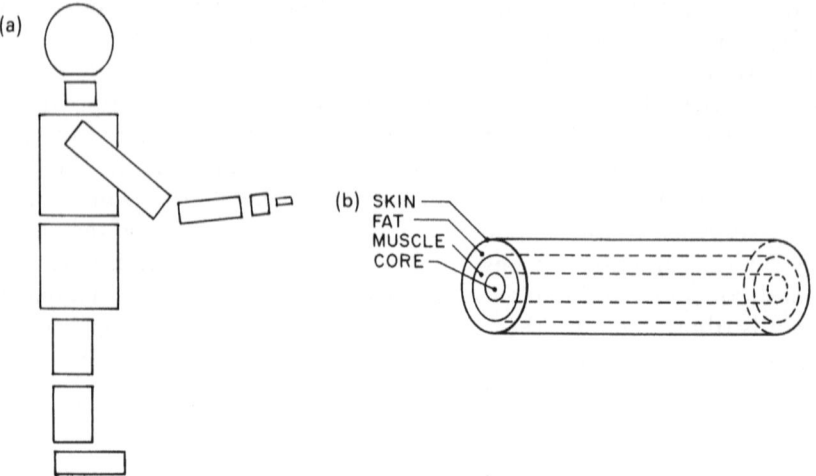

FIGURE 5

(a) A schematic diagram showing the geometrical arrangement of various segments of a human body for thermal-modeling purposes. Note that the presence of a tumor will require additional compartments. (b) Each segment of the mammalian body is divided into four concentric layers: core, muscle, fat, and skin.

each subsection is assumed to have a spatially uniform temperature, which is given by an unsteady-state energy-balance equation shown as follows:

$$
\begin{matrix}
\text{Net accumulation of} \\
\text{thermal energy in} \\
\text{each subsection}
\end{matrix}
=
\begin{matrix}
\text{metabolic} \\
\text{heat} \\
\text{production}
\end{matrix}
+
\begin{matrix}
\text{heat gained} \\
\text{by conduction} \\
\text{from interacting} \\
\text{subsection(s)}
\end{matrix}
$$

$$
\begin{matrix}
- \text{heat lost to the} \\
\text{perfusing blood}
\end{matrix}
\quad
\begin{matrix}
- \text{heat lost to the} \\
\text{environment}
\end{matrix}
\tag{9a}
$$

The detailed expression for each of the preceding terms is discussed in the next section.

In addition to the transient energy-balance equation for each of the 44 subsections, the following balance is written for the central blood pool:

$$
\begin{matrix}
\text{Net accumulation} \\
\text{of thermal energy} \\
\text{in blood pool}
\end{matrix}
=
\begin{matrix}
\text{energy brought} \\
\text{in with the} \\
\text{venous blood}
\end{matrix}
\quad
\begin{matrix}
\text{energy carried} \\
- \text{away with the} \\
\text{arterial blood}
\end{matrix}
\tag{9b}
$$

After substituting appropriate parameter values, these 45 equations are solved numerically to obtain transient temperature distributions in humans during hypo- and hyperthermia.[68] We have recently adapted and extended this model to predict temperature distributions in cancer patients during local, regional, and whole-body hyperthermia.[69a,b]

In our analyses,[76,82,84] a mammalian species is considered to be comprised of a tumor and of normal tissues, represented by compartments interconnected in an anatomical fashion. Because of our focus on interaction between the tumor and surrounding tissues, we lump all the normal tissues, except those next to the tumor, into a single compartment. This approach is illustrated in Fig. 6a for a rat carrying a subcutaneous tumor, where the model consists of the following seven compartments: the tumor, the surrounding normal tissue, the body (which represents the remaining normal tissues), the skin directly above the tumor, the skin above surrounding normal tissue, the rest of the skin, and the central blood pool. For predictions of intratumor temperature distributions, it is necessary to subdivide the neoplastic tissue and the skin above it into N equal compartments, where N is determined by the spatial resolution and precision desired in the computed temperature field (Fig. 6b). While dividing compartments further in this model increases the spatial resolution, its advantages are offset by the larger number of parameters necessary to formulate the system equations. In the limit of an infinite number of subcompartments, this model is equivalent to a distributed parameter model. The detailed schematic diagrams of various versions of our model are given elsewhere for the rat, rabbit, swine, dog, baboon, and humans.[86a,b,87]

Once the number of compartments has been specified, the analysis consists of applying unsteady-state energy-balance equations [cf. Eqs. (9a, 9b)] to each compartment. On substituting suitable numerical values for the various

(a)

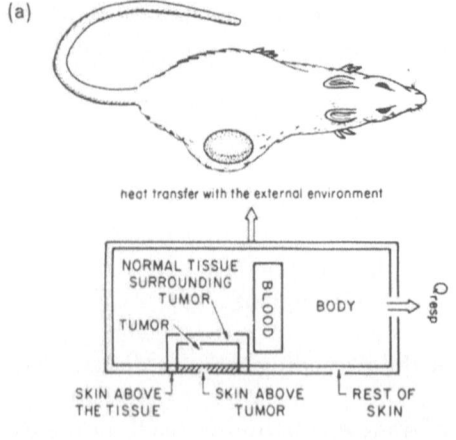

heat transfer with the external environment

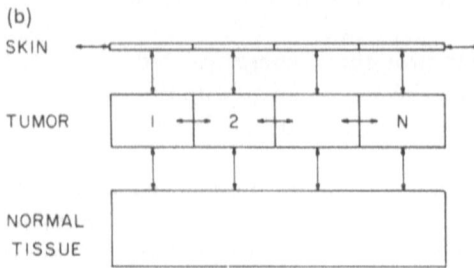

SKIN ABOVE — └ SKIN ABOVE   └ REST OF
THE TISSUE   TUMOR   SKIN

(b)

SKIN

TUMOR   1 ←→ 2 ←→ ←→ N

NORMAL
TISSUE

FIGURE 6

(a) Schematic of a compartmental model for the analysis of temperature distribution in a tumor-bearing mammal during hyperthermia. In this case, a rat is represented by seven compartments interconnected in anatomical fashion: tumor, skin above tumor, normal tissue surrounding the tumor, skin above the normal tissue, rest of body, rest of skin, and central blood pool. (b) For simulations of intratumor temperature gradients, it is necessary to subdivide the tumor and skin above it into N subcompartments. N is determined by the spatial resolution and precision desired at the calculated temperature distribution.

parameters and heat flux terms, the set of coupled, nonlinear ordinary differential equations is solved numerically.

3.2. Heat Transfer Mechanisms

Under normal physiological conditions, the following heat transfer terms should be incorporated in a lumped or distributed parameter model: metabolic heat generation, conduction and convection within the body, heat exchange with the environment by radiation, conduction and convection, heat loss from the skin by evaporation of sweat and water diffused across the skin and respiratory heat loss. Expressions for each of these terms are developed in Chap. 6 and in Refs. 69a and b, along with the parameter values; therefore, we will not discuss them here.

During hyperthermia, terms representing the heat input to a specific tissue or whole body must be added to the proper system equations. For example, during whole-body or local hyperthermia induced by a hot air/water/wax bath or infrared or visible radiation, a heat flux term is added to the skin surface area interacting with the source. During hyperthermia produced by radiofrequency currents, microwaves, or ultrasound, a heat flux term is added to the section of body being heated. During hyperthermia with blood perfusion, the afferent blood temperature is set at a desired value, and the efferent blood is circulated to the central blood pool or to the extracorporeal device used for heating the blood. Suitable numerical values for the various parameters involved in these heat flux terms are given elsewhere.[69a,b,76,82,83]

It remains to comment on the incorporation of thermoregulation in normal and neoplastic tissues in our model. The first reaction of a mammal exposed to high ambient temperature is to increase the blood flow to the skin, which in turn increases the heat flux from the skin to surroundings (see Ref. 88 and Chaps. 3 and 13). This type of physical thermoregulation is effective only when the skin temperature is higher than the ambient temperature. When the ambient temperature is higher than the body temperature, other means of cooling are needed (cf. Chaps. 3, 4 and 13). Table 6 summarizes the thermoregulatory mechanisms used by various homeotherms. There are two ways in which thermoregulation can be included in such a model: feedforward and feedback controls (cf. Chap. 2 and Refs. 68, 77, and 80). While the latter model is more sophisticated and realistic, we have used the former approach in our analysis because of its simplicity and because of our lack of understanding of the physiological feedback control system. In panting animals, we introduce thermoregulation by an increase in the respiration rate as a function

TABLE 6
Thermoregulatory Mechanisms in Homeotherms[a]

Animal	Rectal temperature Normal °C	Critical air temperature[b] Low °C	High °C	Temperature regulating mechanisms[c] Sweating	Shivering	Panting	Thermoneutrality zone[d] °C
Man	37	1	32	+	+	−	24–31
Camel	34–40			+	+	−	
Cat	39		36	−	+	+	24–27
Cattle, Brahman	38–39	1	32	−	−	+	10–27
Cattle, dairy	38–39		24	−	+	+	5–16
Dog	38–39	−80	42–58	−	+	+	18–25
Donkey	36–38			+	+	−	
Goat	38–39			−	+	+	20–26
Guinea pig	39	−15	32	−	−	+	30–31
Horse	38			+	−	−	
Monkey	37–39		40	+	+	−	27–30
Mouse, white	37	10	37	−	−	−	30–33
Rabbit	39	−29	32	−	+	+	28–32
Rat, white	37.5	−10	32	−	+	−	28–30
Seal	37	−30		−	−	−	−10–+30
Sheep	39		32	+	−	+	13–31
Swine	37–38		30	−	−	+	0–20
Chicken	41–42	−35	32	−	+	+	19–29
Pigeon	43	−85	42	−	+	+	20–30

[a] From Ref. 67.

[b] Critical air temperature: Air temperature at which the normal animal first begins to show a change in deep body temperature.

[c] Temperature regulating mechanisms: (+) present; (−) absent.

[d] Thermoneutrality zone: The range of air temperature at which the normal animal has the lowest metabolic rate.

TABLE 7
Effect of Hyperthermia on Tumor Blood Flow Rate

Tumor (host)	Measurement technique	Temp × time	Blood flow	Reference
W256 carcinoma (1.7–8.5g) (Sprague-Dawley rats)	Direct collection of tumor efferent blood	40–43°C × 1 hr	No significant change	119
		43°C × 1/2 hr	No significant change (18, 48, and 72 hr)	155
W256 carcinoma (0.3–5g) (Sprague-Dawley rats)	Radioactive micro-sphere method	43°C × 1 hr	No significant change	156, 157
		45°C × 1 hr	Significant decrease 3 hr after heating	157
Yoshida sarcoma (1–1.5 g) (Wistar rats)	The fractional distri-bution of [86]Rb	42°C × 1 hr	No change while heating; drop to zero in the next hour, recovered to control level 12 hr after heating	143
		42°C × 3 hr	No change (1 hr) pro-gressive decrease (2 hr)	143
DS-carcinosarcoma (3.0–5.0 g) (Sprague-Dawley rats)	Direct collection of tumor efferent blood	35–44°C × 20 min	Increase up to 39.5°C; decrease below con-trol at 44°C	75
Mammary adenocar-cinoma (0.3–0.6 g) (C3H mouse)	Hydrogen clearance method	Continuous increase from 32–45°C × 30–40 min	Increase up to 41°C; decrease after 42°C	158
Transplants of spon-taneous tumors (C3H mice)	Fractional distribu-tion of [86]Rb	37.5, 42, 44°C	No change	159
Ependymoblastoma (C57BL/6 mice)	[133]Xe clearance method	40, 42, 45°C × 15–75 min	Increased 30–45 min (40–42°C); rapid decrease after that; decreased to 50% of control in 15 min at 45°, 60 min at 42°C	128
BA-1112 Rhab-domyosarcoma (WAH-Rij rats)	[15]O clearance	41°C × 40 min; 42°C × 30 min	Decreased by 50%	160
BA-1112 Rhabdomy-osarcoma (trans-parent chambers in WAH-Rij rats)	Microphotography and histopathology	42·5°C × 3 hr	Permanent damage to microvasculature in 140 ± 60 min (with glucose, mic-onidazole or 5-TG)	161, 162
BA-1112 Rhabdomy-osarcoma (trans-parent chambers in WAH-Rij rats)	Red blood cell veloc-ity measured using a photodiode method	27–42°C	Continuous heating <35°C increased velocity; functional capillaries constant; decrease in velocity and number of functional capil-laries at 40°C; 40–42°C caused per-manent vascular damage	62
Squamous carcinoma (hamster cheek pouch)	Microphotography and histopathology	41–45°C × 30 min	Hemorrhage and stasis at 43°C or with intermittent heat applied (42°C at 1-hr intervals)	163, 164

TABLE 8

Effect of Hyperthermia on Oxygen and Glucose Consumption of Tumors (in vivo)

Tumor (host)	Method	Thermal	O_2 consump.	Glucose consump.	Ref.
W256 (2.3–13.2 g) Sprague-Dawley rats)	A-VO$_2$ difference	40–41.8°C ×1 or 3 hr	No significant change	No significant change	119, 115
DS-carcino-sarcoma (3–5 g) (Sprague-Dawley rats)	A-VO$_2$ difference	Stepwise increase (33–36°C) to 44°C, 20 min at each temp	Increase from 6.1 to 9.3 μl/g/min 0.28 (39.5°C); decrease to initial value at 42–44°C; no change in large tumors	Increase from 0.28 to 0.41 μl/g/min (39.5°C); decrease to 0.30 (42°C), to 0.22 (44°C); No change in large tumors	75

^{a} Several investigators have measured these consumption rates in tumor slices *in vitro.*[165,166]

of temperature and time. Similar increases are incorporated into the blood flow rates and metabolic heat generation rates of various organs, as reported in the literature.[68,77,80,89–91]

Unlike normal tissues, the physiological response of tumors is poorly understood. As shown in Table 7, most tumors exhibit an increase in their blood supply up to 40–41°C, and at higher temperatures, blood flow begins to decrease. The W256 carcinoma is an exception to this rule, since it shows no change in blood flow up to 45–46°C and then shows a marked impairment in its blood supply at higher temperatures. As shown in Table 8 the oxygen consumption of tumors at elevated temperatures follows essentially the same pattern as their blood supply.

With our current understanding of tumor thermoregulation, the following points must be borne in mind when modeling tumor thermal behavior at elevated temperatures:

(i) In the absence of any tumor data, the following assumption may be used in developing a mathematical model. Most tumors show a moderate increase in their blood supply up to 40–41°C, and then their supply may be impaired. In normal tissues that are heated artificially, the blood flow may increase up to 46°C and then may decrease at higher temperature and/or longer durations of heating. This differential physiological response may account for, in part, the selective lethal response to hyperthermia.[63]

(ii) Each tumor is different, and therefore mathematical generalizations about tumor thermal behavior are hard to make. In addition to the intertumor differences, heterogeneities within a tumor make the situation more complex for thermal modeling purposes.

4. THERMAL ENERGY ABSORBED DURING ULTRASOUND, MICROWAVE, AND RADIOFREQUENCY HEATING

In the past decade, the following three methods of heating a deep-seated tumor have received considerable attention: ultrasound, microwaves, and

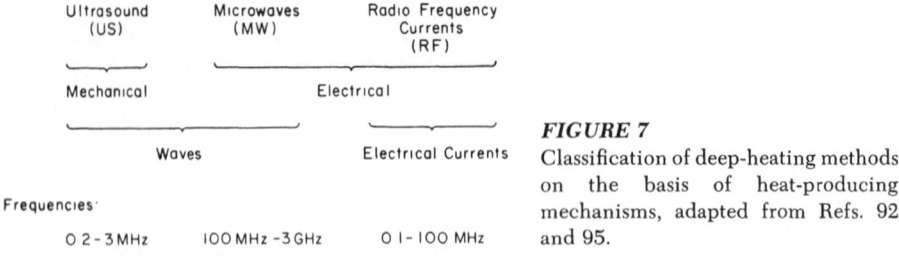

FIGURE 7

Classification of deep-heating methods on the basis of heat-producing mechanisms, adapted from Refs. 92 and 95.

radiofrequency (Fig. 7). Various aspects of applying these techniques, including the advantages and disadvantages of each, have been reviewed recently by many investigators.[28,92–95] In what follows, we will discuss the quantitative aspects of thermal energy generated in tissues while applying these techniques. This quantity is essential for computing the resulting temperature distributions.

4.1. Ultrasound

Ultrasonic heating is based on the absorption of high energy waves by the tissue. The mechanical energy carried by the longitudinal waves (i.e., the particles oscillate in the direction of wave propagation) is converted into thermal energy by frictional lossses. Depending on the need and the instrumentation, the beam may be focused or unfocused, stationary or translocating. In addition, more than one beam may be used simultaneously or in a predetermined sequence to increase heat deposition (cf. Chap. 8 and Refs. 96–98).

The wavelength λ and frequency f of the beam are related to the speed of propagation C in the medium by the following equation:

$$\lambda f = C \tag{10}$$

The velocity of sound in most tissues (except bone) is approximately equal to that in water, about 1500 m/sec (see Table 9). Air, on the other hand,

TABLE 9
Velocity of Ultrasound in Animal and Human Tissues[a]

Tissue	Velocity (m/s)
Muscle	1585 ± 20
Liver	1590
Spleen	1555
Kidney	1560
Brain	1540
Fat	1440
Bone, skull	3360

[a] From Ref. 92.

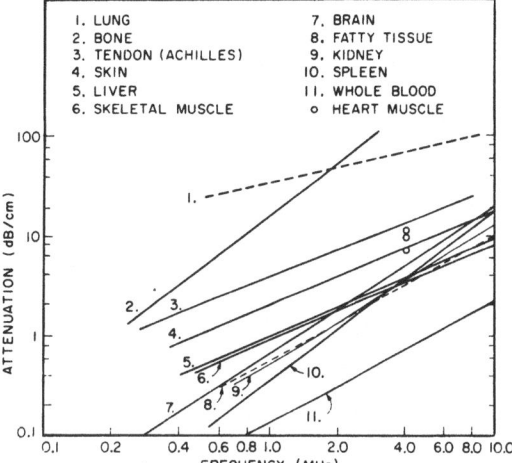

FIGURE 8
Ultrasound absorption coefficients for human tissue. The dashed line represents canine data. The data were compiled by Goss *et al.* (see Ref. 206) and plotted by Hahn *et al.* (see Ref. 207). Reproduced by permission of the New York Academy of Sciences.

does not allow the transmission of ultrasound from the source to the tissue. Therefore, ultrasound cannot be used to heat tissues that contain even minute amounts of air, such as the chest cavity.

The physical property that determines the absorption of ultrasound energy in a tissue is the attentuation (or absorption) coefficient α of the tissue. As a plane-parallel beam penetrates into a homogeneous tissue, its intensity I decreases exponentially with distance x as a result of absorption

$$I = I_0 \exp(-\alpha x) \tag{11}$$

Values of α as a function of the wave frequency for various normal tissues are shown in Fig. 8. Note that the value of α is largest for bone, smallest for fat, with intermediate values for muscle, suggesting a similar trend in energy absorption by these tissues.

While Eq. (11) has been used by many investigators to compute temperature distributions in tissues, we must be aware of the following constraints while using it.

(i) As a beam travels away from the transducer, its cross section increases linearly with the distance from the source. The divergence of the beam is inversely proportional to its frequency. Note that the depth of penetration ($D = 1/\alpha$) is proportional to the frequency. These two opposite requirements restrict the operational range between 0.2 and 3 MHz, with an optimum around 1 MHz.

(ii) Due to constructive and destructive interference of waves arriving at a point near the source, the intensity of the beam goes through relative maxima and minima in space. These interference patterns are quite complex in the "near" field of the transducer.

(iii) Heterogeneities in the transducer may lead to cross-sectional variation in the intensity, even in the "far" field of the transducer.

(iv) Near bones or metal implants, the longitudinal waves turn into transverse (or shear) waves, which are absorbed more rapidly, leading to "hot spots." In addition, reflections by the bone may create complex standing wave patterns. Recently, Chan *et al.*[99] have computed the energy distribution resulting from the change of longitudinal waves into shear waves.

(v) In the case of a focused beam, the intensity distribution in the focal region is Gaussian, the dimensions of which are difficult to predict theoretically.

In light of these constraints, investigators have either arbitrarily assumed Eq. (11) to be valid[100–103] or fitted the measured intensity distribution with a set of empirical relationships.[103–105]

4.2. Microwaves

Microwave diathermy, similar to ultrasound, is based on the absorption of high-energy electromagnetic waves by the tissue. The energy carried by these waves is converted into thermal energy by dielectric and resistive losses. Similar to ultrasonic waves, the product of the frequency f and wavelength λ of microwaves is equal to the speed of these waves C in the medium. (In air and vacuum, $C = 3.0 \times 10^8$ m sec^{-1}.) While frequencies between 10 MHz to 100 GHz can be used in principle, the Federal Communications Commission allows the use of 13.56, 27.12, 40.68, and 915 MHz and 2.45, 5.8, and 22.15 GHz frequencies for industrial, medical, and scientific purposes in the United States. Outside the United States, the 433-MHz frequency is also allowed. Use of unassigned frequencies requires careful screening, so that the radiated energy is less than 15 μV/m at a distance of 1,000 ft from the applicator.[94] In addition, there is a limit on occupational exposure of 10 mW/cm power averaged over any 0.1-hr period for the 10 MHz–100 GHz range.

Similar to ultrasound, the strength of the electrical field E, resulting from the absorption of plane-parallel microwaves in a homogeneous tissue, decreases exponentially according to Eq. (11).[11] The absorption coefficient α is related to the dielectric constant ε, the electrical conductivity σ (mho/cm), and the wavelength in air λ (cm), as follows:

$$\alpha = \frac{2\pi}{\lambda} \sqrt{2\varepsilon} \left[\left(1 + \frac{60\lambda\sigma}{\varepsilon} \right)^{1/2} - 1 \right]^{1/2} \tag{12}$$

The absorbed power density Q, resulting from the electrical field E, is given by

$$Q = \frac{\sigma}{2} E^2 \tag{13}$$

Since both σ and ε are functions of the wave frequency (Figs. 9a and b), Q will depend on the frequency. Note that both ε and σ vary with the

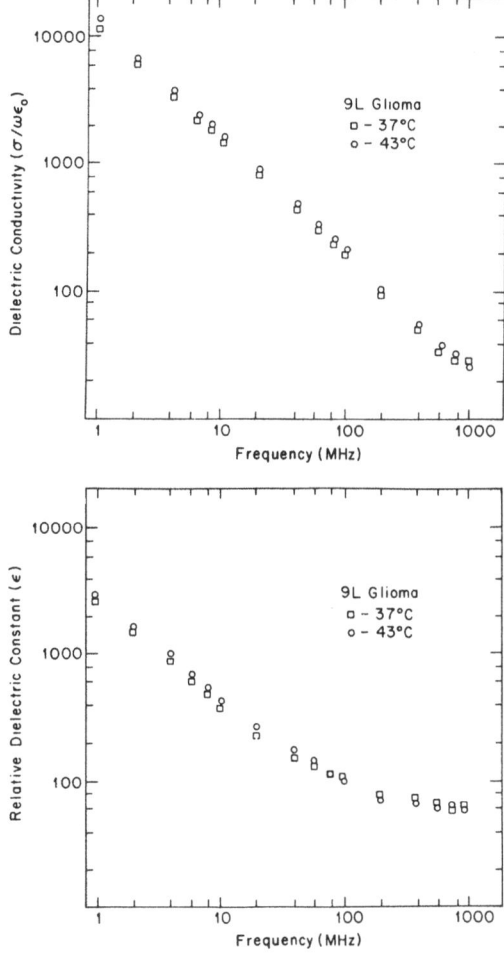

FIGURE 9
(a) Reduced electrical conductivity $\sigma/\omega\varepsilon_0$, of tumor tissues as a function of frequency. (b) Relative dielectric constants of tumor tissues and muscle as a function of frequency.[209]

temperature according to the following relationships:

$$\frac{\Delta\varepsilon}{\varepsilon} \approx -0.05\%/°C$$

$$\frac{\Delta\sigma}{\sigma} \approx \ \ 2.0\%/°C \tag{14}$$

While evalution of the heating pattern is relatively straightforward for a homogeneous tissue exposed to "far-field" radiation, the following factors make it difficult to compute heating patterns for heterogeneous media with realistic geometries.

(i) The microwaves are transmitted through, absorbed by, and reflected at biological interfaces as a function of the tissue size and geometry, tissue composition and properties, wave frequency, and source design. Absorption is high and depth of penetration ($D = 1/\alpha$) is low in tissues of high water

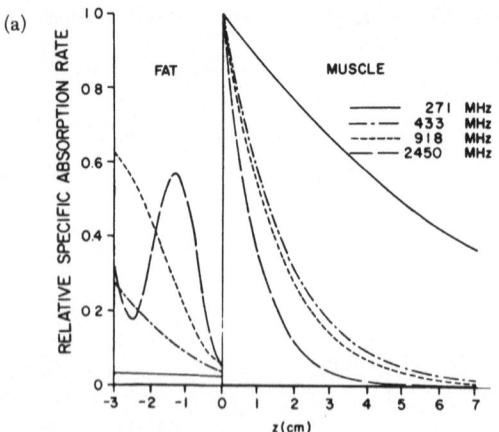

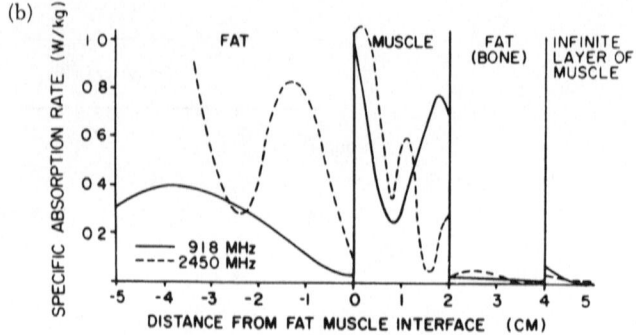

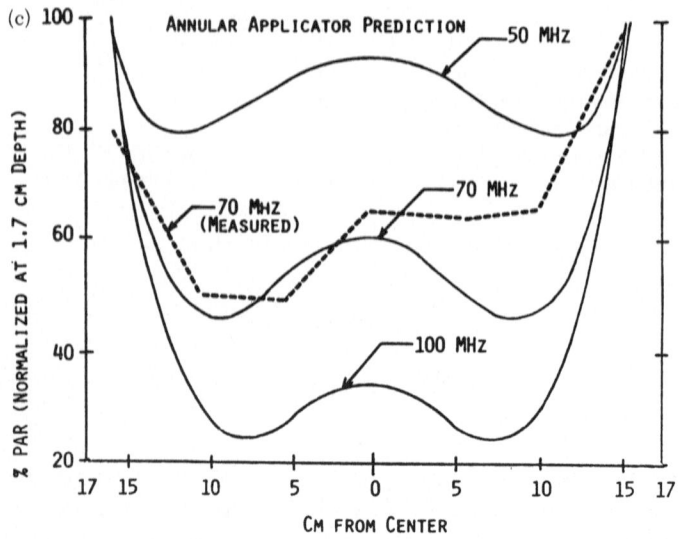

FIGURE 10

(a) Relative absorbed power density patterns in a two-layer model (fat–muscle) exposed to a plane-wave microwave source (see Ref. 171). By permission of IEEE. (b) Relative absorbed power density patterns in a four-layer model (fat–muscle–bone–muscle) exposed to a plane-wave microwave source (see Ref. 171). By permission of IEEE. (c) Relative power absorption rates in a homogeneous thorax model exposed to an annular array applicator at three frequencies: 50, 70, and 100 MHz. The broken line shows a typical measured pattern at 70 MHz (see Ref. 110). By permission of the *J. Nat. Cancer Inst.*

TABLE 10
Theoretical Analysis of Power Absorbed by Tissues Exposed to Plane Microwaves

Tissue model	Reference
Plane tissues layers (semi-infinite slab)	
Isotropic	
2-layer model (fat–muscle)	167–169
3-layer model (skin–fat–muscle)	92, 170
4-layer model (fat–muscle–bone–muscle)	171
Anisotropic	
3-layer model (skin–fat–muscle)	172
Cylindrical tissue layers	
Homogeneous tissue	173[a]
3-layer model (fat–muscle–bone)	174, 175
Spherical tissue layers	
Homogeneous sphere	171, 176–181[b]
2-layer model (fat–muscle)	171
3-layer model (skin–fat–muscle)	182
4-layer model (brain–skull–fat–skin)	183
5-layer model (brain–CSF–dura–bone/fat–skin)	181
6-layer model (brain–CSF–dura–bone–fat–skin)	184–187
Prolate spheriod and ellipsoid models	
Homogeneous spheroid	188–192[b,c]
Homogeneous ellipsoid	193
Long cylinders of arbitrary cross section	
Homogeneous tissue	194[d]
3-layer model (skin–fat–muscle)	195[e]
Finite biological bodies	
Finite planar model	196
Isolated section of body	197, 198
Part-body and multibody analysis	189
Block model of man	199–201

[a] Geometrical-optics method.
[b] Radiofrequency (1–20 MHz); solution obtained by combination of quasi-static electric and magnetic solutions.
[c] Extended-boundary condition method.
[d] Integral equations and moment method technique.
[e] Finite-element method.

content (e.g., muscle, brain, internal organs, skin), and the opposite is true in tissues of low water content (e.g., fat, bone). In addition, interference between the transmitting and reflecting waves coming from an interface separating tissues of different complex dielectric constant, $\varepsilon^* = \varepsilon - i\sigma/\omega\varepsilon_0$ may create standing waves, resulting in hot spots. Figures 10a,b show heating patterns generated in two-layer (fat–muscle) and four-layer (fat–muscle–bone–muscle) models of tissue exposed to plane waves. Analytical and numerical results for plane-wave absorption by tissues of various geometries have been obtained by many investigators as summarized in Table 10. Although "non contact" applicators are widely used to induce hyperthermia, "direct contact" applicators are being developed, because they reduce unwarranted reflection from the skin and make their use safer for both the patient and operator. The complex "near-field" heating patterns generated by the contact applicators, however, have hindred progress in their design. Investigators have, therefore, taken two approaches to optimize applicator design: measure experimentally,

TABLE 11
Sources Other Than a Plane Wave

Tissue model	Source	Reference
Plane tissue layers (semi-infinite slab)		
2-layer model (fat–muscle)	Dielectric loaded dipole-corner reflector applicator	202
	Direct contact rectangular aperture	203
Cylindrical tissue layers		
3-layer model (fat–muscle–bone)	Direct contact aperture	204
Spherical tissue layer		
1- and 5-layer model (brain–CSF–dura–bone/fat–skin)	Direct contact applicator	181
6-layer model (brain–CSF–dura–bone–fat–skin)	Loop or dipole antennas	205

the heating patterns in a test material[107] or evaluate the heat deposition numerically, using various novel solution techniques. Table 11 summarizes the attempts of the latter for various types of contact apertures.

(ii) In order to heat large volumes of the body (e.g., the chest cavity), or a deep-seated tumor, many investigators have used multiple applicators placed in a specified geometry.[108,109] Recently, Turner and Kumar[110] have obtained approximate numerical solutions for such applicators by replacing the original aperture field with an array of point-source dipole radiators. Shown in Fig. 10c is the result of such a computation for a homogeneous tissue cylinder heated by an annular array applicator. Note that the analysis is in qualitative agreement with the data.

(iii) In order to overcome the poor penetration problem associated with microwaves in the GHz range, several investigators place needle-like microwave applicators inside the tumor or in an adjacent cavity. Insertion of radio-active needles is a common practice in radiotherapy. Although the heating patterns around these probes are being studied experimentally by many investigators,[111,112] theoretical attempts to predict the patterns have not been reported extensively.

4.3. Radiofrequency Heating

Heating by radiofrequency currents† can be achieved by coupling the electromagnetic energy to the tissue either capacitively (also referred to as dielectrically) or inductively. These techniques can be applied either invasively or noninvasively. In principle, electrical currents with frequencies greater than 10 kHz (to avoid muscle contractions associated with electric shock) may be used for radiofrequency heating; due to FCC regulations, 13.56- and 27.12-MHz frequencies are used most commonly.

---

† Also referred to as high-frequency or ultrashortwave diathermy. These names *incorrectly* suggest that waves are involved in heating the tissue.

In the capacitive heating mode, high-frequency currents are passsed through the tissue volume via appropriately placed electrodes. The current density in the tissue can be controlled by varying either the interelectrode distance or the electrode geometry. Conduction currents lead to ohmic (or resistive) heating, and displacement currents lead to dielectric heating. Therefore, from the knowledge of the electromagnetic properties (ε and σ), wave frequency, and geometry, it is possible to calculate heat generation in the tissue. When the ohmic heating dominates, (i.e., when $\sigma > \omega\varepsilon$), the heat generated Q is given by

$$Q = j^2/\sigma \quad \text{W/cm}^2 \tag{15}$$

where j is the current density (A/cm^3).

In the inductive heating mode, RF currents are passed through a well-insulated, flexible heavy wire arranged in regular coils. Alternating current passing through the coil generates an oscillating magnetic field, which in turn produces "eddy" currents in the tissue volume. The current density j is proportional to the rate of change of the magnetic field dB/dt and to the tissue conductivity σ. Therefore, the heat generated Q is given by

$$Q = \frac{\sigma}{2}\left(\frac{dB}{dt}\right)^2 \quad \text{W/cm}^2 \tag{16}$$

Since σ for fat and skin is an order of magnitude less than for the muscle, inductive heating, unlike capacitive heating, permits heating the deeper tissues more easily. In addition, skin cooling by air or water may be used in both modes of heating to heat the deeper tissue more effectively.

Similar to microwaves, RF heating may be produced by one or many applicators surrounding the tissue volume. In addition, the probes may be implanted inside the tumor or in a cavity adjacent to it. Unlike microwaves, however, not much effort has been put into computing the heat distribution in tissues.

Guy *et al.*[113] have recently computed specific heat deposition rates in a two-layer (fat–muscle) model exposed to a flat "pancake" induction coil. These investigators found heat deposition to be nonuniform radially (toroidal in shape) and exhibiting a maximum at the muscle–fat interface.

Better methods are now needed to predict more accurately the power distribution in normal and neoplastic tissues in order to exploit these novel methods of heating.

5. TEMPERATURE DISTRIBUTIONS DURING NORMOTHERMIA

Knowledge of the temperature distribution during normothermia is needed both for cancer detection and to provide a basis for thermal dosimetry. Thermographic methods have been developed to measure surface temperature patterns (cf. Chap. 25 and Refs. 114 and 115). For tissues at depth, other methods are required (cf. Chaps. 26 and 27).

If the tumor were considered infinite and system parameters were constant, the temperature of the tumor T would be equal to the arterial blood temperature T_a plus the temperature rise due to heat deposition or metabolic generation T_m: $T = T_a + T_m$, where $T_m = q_m/p_b c_b w_b$. Shown in Fig. 3 is the value of T_m as a function of tumor weight.[76] It is of interest to note that T_m is less than 0.25°C and does not change significantly with tumor weight.

For tumors of finite size that interact thermally with the surroundings at temperature T_s, the calculated temperature profiles are shown in Fig. 11a. The results indicate that the temperature profile throughout a small tumor is lower than that in a larger tumor for a fixed, uniform blood flow rate. This result is not surprising if we realize that the temperature profile is determined by both the total heat generation rate (which is proportional to the tumor volume) and the total rate of heat loss to the surroundings (which is proportional to the outer surface area). Consequently, for a fixed w_b, q_m, and T_s, a higher ratio of heat generation to heat rejection rates in large tumors leads to higher temperatures when compared to small ones.[61]

Figure 11b shows the temperature distribution in human mammary epitheliomas, measured by Gautherie and co-workers.[116] The data suggest that the temperature is a maximum in the center of these tumors and are in qualitative agreement with our analysis (Fig. 11a). For quantitative com-

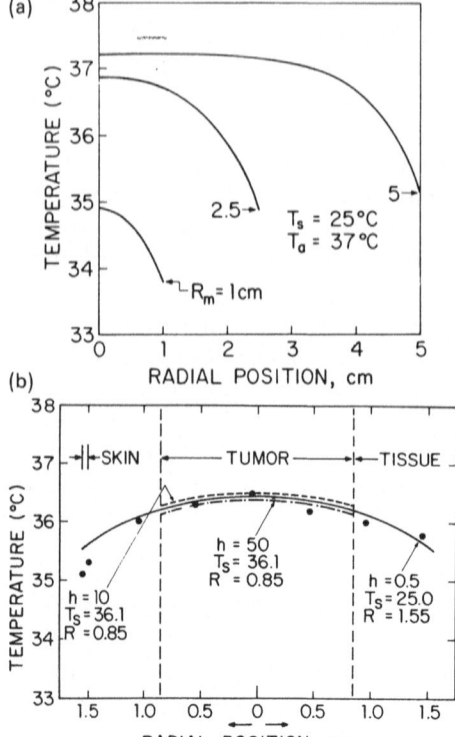

FIGURE 11

(a) Effect of tumor radius on the temperature distribution in uniformly perfused tumors of various radii. For an infinite tumor, initial temperature would be uniform at 37.2°C ($T = T_a + T_m$). Note that blood perfusion rate is assumed independent of tumor weight in these simulations, adapted from Jain, (Ref. 130). (b) Comparison of model predictions with the data on temperature distributions in a human mammary epithelioma and the surrounding tissue. Solid line refers to the first approach, and dotted [(- - -) and (-·-·)] refer to the second approach discussed in the text. Adapted from Jain (Ref. 130).

parison, some additional assumptions must be made about the thermal symmetry of the system. In reality, the tumor is located between the skin, exposed to the ambient air at 25°C, and the chest at 36–37°C. Hence, the temperature of surroundings T_s is not uniform. However, it is possible to set bounds on the temperature distributions in the tumor by considering two limiting cases: (1) the tumor and the surrounding tissue can be considered as a sphere of radius $R = 1.55$ cm, exposed to ambient air temperature (25°C); or (2) the tumor alone ($R = 0.85$ cm) is surrounded by a tissue at 36.1°C. The results of computation using these two approaches are compared with data in Fig. 11b.

The effect of nonuniformities in the perfusion rate on the temperature profile is shown in Fig. 12 for a tumor of radius $R = 1.5$ cm. The results are similar for tumors of different radii. Because of its large blood flow rate and higher rate of metabolism, the temperature in the periphery of a necrotic tumor is higher than in a tumor that has a uniform blood flow rate, although both tumors have the same average perfusion rate. Inhomogeneities in blood flow rate, therefore, tend to make the temperature profile flat. These results are in qualitative agreement with the data of Gautherie.[117] Our current efforts are directed toward improving this analysis by considering thermal asymmetry of the tumors.

6. TEMPERATURE DISTRIBUTIONS DURING HYPERTHERMIA

Currently available "physical" methods of producing hyperthermia can be divided into two categories: surface heating and volume heating.

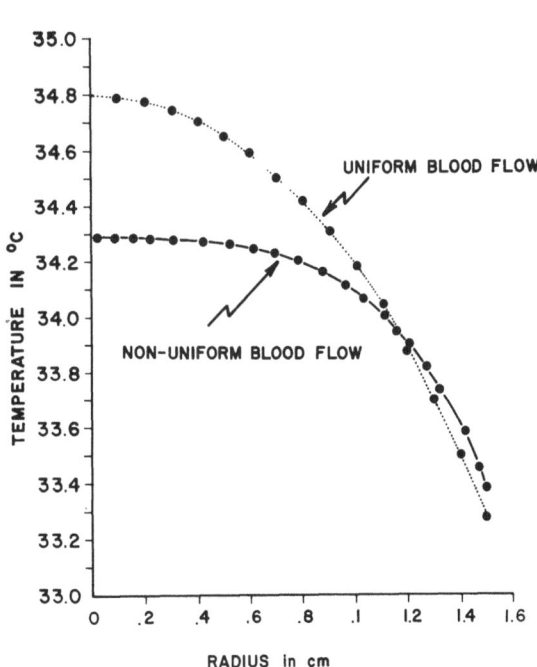

FIGURE 12
Effect of nonuniformity in perfusion rate on temperature distribution in a 1.5-cm (radius) tumor. Note that nonuniformity in these calculations is considered parabolic (see Ref. 26). By permission of the New York Academy of Sciences.

In surface heating, the temperature is increased using one or more of the following methods: immersion in hot water bath or hot air incubator, exposure to visible or infrared radiation, or direct contact with an interface at a higher temperature. Heat transfer takes place from the skin to the underlying tissue by conduction and convection. In volume heating, the temperature of underlying tissue is brought to an elevated temperature by either perfusing the tissue with preheated blood or exposing it to ultrasound, microwave, radiofrequency currents or bacterial toxin injection.

Depending on the surface area or volume heated, hyperthermia can be divided into three categories: (1) local, (2) regional, and (3) whole body. In local hyperthermia, the tumor mass and (hopefully) a minimum of the surrounding tissue are heated; in regional hyperthermia, usually the limb containing the tumor mass is heated; and in whole-body hyperthermia (WBH), the temperature of the host is brought to an elevated temperature.

No matter which of these methods is used to induce hyperthermia, it is essential to monitor the resulting temperature distributions in the normal and neoplastic tissues continuously and to control the power input and position of heating source, so that the temperatures are maintained in the desired range for an optimal time. This type of control can be achieved either manually or automatically and is incorporated in most commerical or custom hyperthermia systems used currently.

The objective of the following section is to discuss the quantitative aspects of heat transfer and temperature distributions during hyperthermia.

6.1. Hot Water Bath and Moist Hot Air Incubator

These two are the simplest methods for surface heating *in vitro* and *in vivo* and have been used to induce localized, regional, and whole-body hyperthermia in animals and humans. The principal mode of heat transfer in these methods is conduction.

For the past ten years, several research groups in Europe have been using a hyperthermia cabin designed by Siemens in which hot air (50–60°C) is circulated to maintain the patient's temperature around 41–42°C. In order to increase the initial rate of heating, this unit has provisions to install a 27-MHz RF heating unit and a 433-MHz MW applicator. The cabin is made from a transparent polymeric material in order to allow observation of the patient.[118]

We have recently used the lumped parameter model to simulate the temperature distribution in a tumor-bearing rat during whole-body and local hyperthermia, induced by a water bath. Figure 13 shows the temperature data obtained by Gullino *et al.*[119] in W256 carcinoma (2.5 g) during whole-body hyperthermia. In these experiments, the tumor-bearing Sprague-Dawley female rat (200 g), anesthetized with urethan (1 mg/kg) was immersed vertically, tail down into a well-stirred, heated water bath (10 cm^3, 41°C) until the water reached its mandible. As shown in Fig. 13, the computed values adequately reproduce those measured experimentally: a rapid elevation of 4–5°C during the first 5 min and a slower increment of 4–5°C for the remaining 20 min; the time required to reach 41°C was within the experimental range of 20–30 min. The normal tissues followed the same time course.[76]

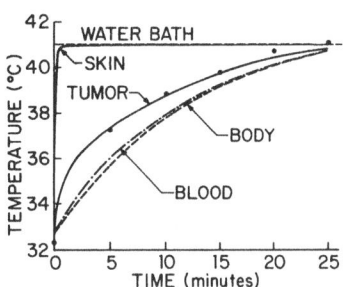

FIGURE 13

Temperature distributions in W256 carcinoma (2.5 g) and normal tissues of a Sprague-Dawley female rat (200 g) during whole-body hyperthermia induced by immersing the rat in a constant temperature water bath (41°C). Points represent the data points and lines represent the model predictions (see Ref. 76). By permission of the *J. Therm. Biol.*

Figure 14 shows the temperature data of Dickson and Suzanger[120] in Yoshida sarcoma transplanted in the foot of Wistar rats (200 g) during localized hyperthermia. In these experiments, the rat, anesthetized with pentobarbital sodium (Nembutal) (2 ml/kg), was placed over a water bath (15 cm³, 42.7 ± 0.05°C), and the tumor-bearing foot was inserted into the water bath through a 10-cm diam, padded opening to a depth that permitted complete submersion of the tumor. As shown in Fig. 14, the computed values agree closely with those measured experimentally: a rapid elevation of 6°C during the first 2–3 min, followed by a slow asymptotic convergence to the water bath temperature. The temperature of the normal tissues followed a smooth time course to reach 40°C in about 120 min.[83]

Stolwijk[80] has used the whole-body model to simulate the temperature distributions in normal humans when heat is deposited in the trunk or core via a water bath. In these computations, heat loss by sweat evaporation is set equal to zero (see Fig. 15). Huckaba and Tam[68] have also analyzed the temperature rise in various organs of a human subject when placed in a hot incubator. Recently, we have improved the models developed by Stolwijk and Huckaba *et al.* to predict temperature distributions in cancer patients.[69a,b] As seen in Fig. 16, our model reproduces the transient data of Huckaba *et al.* adequately.

In order to exploit whole-body hyperthermia effectively, it is essential to measure as many thermal, physiological, and biochemical parameters as possible without harming the patient and to incorporate this information into future analyses.

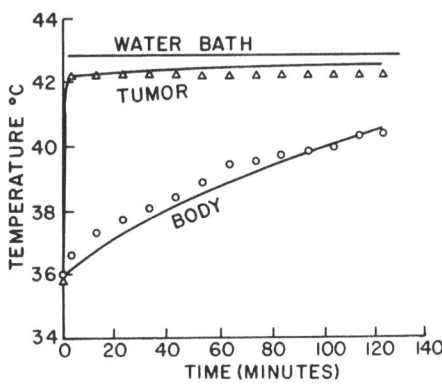

FIGURE 14

Temperature distributions in Yoshida sarcoma and normal tissues of a Wistar rat (200 g) during localized hyperthermia induced by immersing the tumor-bearing foot of the rat in a constant temperature water bath (42.7°C). Points represent the data points and lines represent the model predictions (see Ref. 83). By permission of the *J. Med. Phys.*

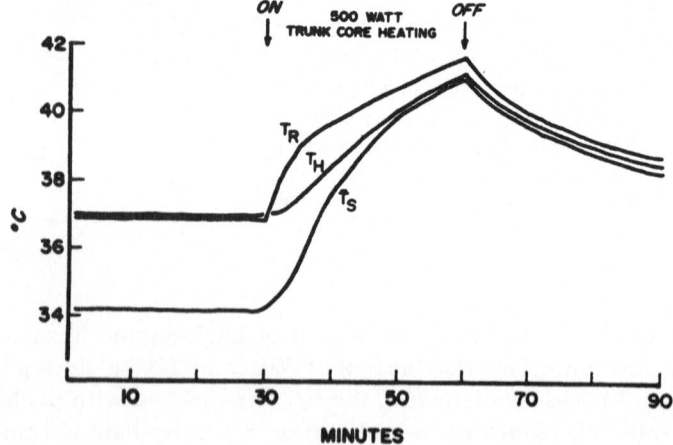

FIGURE 15

Simulation of the effect of depositing 500 W of heat into the trunk core of a man for a period of 30 min. Sweat is not allowed to evaporate. Shown are the trunk core temperature (T_R), brain temperature (T_H), and mean skin temperature ($\bar{T}_s$). From Ref 80, by permission of the New York Academy of Sciences.

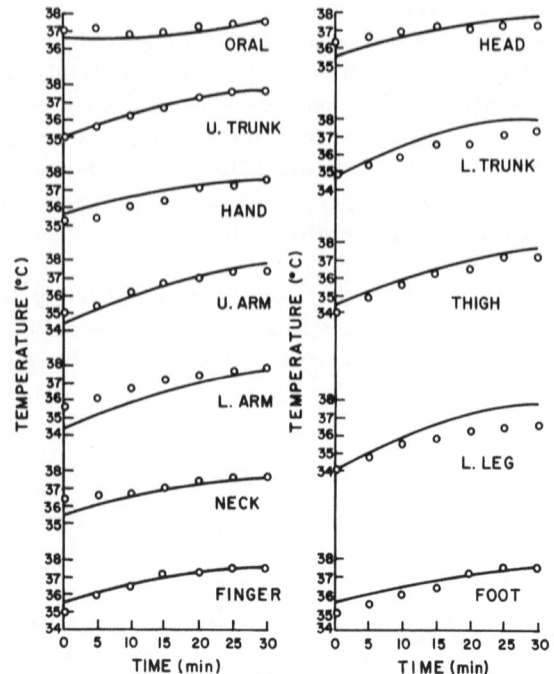

FIGURE 16

Comparison of computed (lines) and measured (dots) temperatures of the various segments of a human subject undergoing whole-body heating (see Ref. 69a). By permission of the American Institute of Physics.

6.2. Space Suit and Blanket Techniques

These two techniques are also based on conduction of heat from a hot fluid (usually air or water) to the skin, except that the patient is not in direct contact with the fluid. Containment of fluid in a suit or thermal blankets makes the procedure more manageable and convenient.[121] The lumped parameter models can be modified to predict temperature distributions during the use of these methods.[69a,b] However, the imprecision of these methods in regions of large temperature gradient, such as the cutaneous layers, through which this heat exchange must occur, suggests that the distributed parameter approaches may provide more satisfactory results; such results are given in Chap. 14.

6.3. Infrared Radiation

Every object with a temperature above absolute zero emits infrared radiation. The rate of emission is proportional to the fourth power of absolute temperature. By exposing the whole body or part of it to a controlled thermal radiation source, the temperature of the skin may be elevated, with heat transferred to the underlying tissues by convection and conduction; see Stoner[122] for the details of source design and applications. Thermal-radiative exchange is discussed in Chaps. 25, 27, and Appendix 1.

We have recently measured temperature distributions inside a tumor during localized hyperthermia induced by infrared radiation.[82] Figure 17 shows our data on intratumor temperature distributions measured by two thermistors placed 1 cm apart in W256 during localized hyperthermia. In these experiments, a specified portion of skin above the tumor was heated by a diathermic lamp. As shown in Fig. 17, the computed values agreed adequately with the data: a rapid elevation of 2°C during the initial 10 min and a slower increment for the remainder of the experiment in the thermistor close to the heated skin area. The initial temperature rise measured further away by the second thermistor was less than 1°C in 20 min. A temperature difference of 2.2°C between two points in the tumor persisted after 20 min. The variation in temperature with position by thermal-radiative exchange should be kept in mind while heating patients for cancer treatment, especially when the precise location of a tumor is not known.

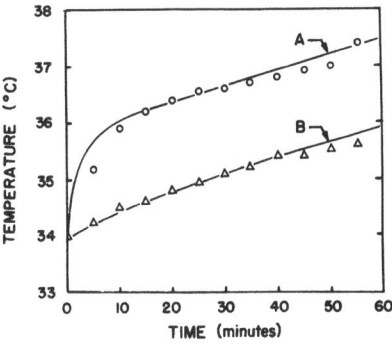

FIGURE 17
Comparison between model simulations and experimental data on intratumor temperature distributions during localized hyperhermia. Points represent the experimental data measured by thermistors *A* and *B* placed 1 cm apart in the tumor, and solid lines represent model simulations. Note the intratumor temperature gradient of 2.2°C/cm after 20 min (see Ref. 82). By permission of the *J. Therm. Biol.*

6.4. Pettigrew Technique (Hot Wax Bath)

An effective and simple method for producing whole-body hyperthermia was introduced by Henderson and Pettigrew.[123] In this method, a patient is enclosed in a sealed covering of polyethylene to eliminate evaporative heat losses and then immersed in a bath of molten paraffin wax having a melting point of 46°C. Heat is transferred from the wax to the patient's skin primarily by conduction; the wax subsequently begins to solidify at the polyethylene envelope surrounding the patient. As a result, an increasingly thick layer of solid wax begins to form, which is in equilibrium with the skin temperature on one side and at its melting point (46°C) on the other side. The skin temperature rises rapidly in the beginning and then approaches steady state more slowly, since the thickening solid layer decreases the temperature gradient at the skin surface. The wax layer also acts as an insulator to minimize heat loss by conduction and convection from the patient. This method is also beneficial for those with decubitus ulcers (pressure sores), which is a common problem in cancer patients who are lying on their back for a long time. In this application, since the density of the molten wax is slightly higher than that of the patient, the buoyancy force provides an ideal support.

In order to increase the initial rate of heating, some investigators ventilate their patients with a mixture of humidified hot air (usually a 50% mixture of helium and oxygen). Others use a microwave diathermy apparatus (433 MHz) to increase the tumor temperature.[124]

Recently, Law and Pettigrew[125] have developed a simple mathematical model to predict the core and skin temperature during hyperthermia using Pettigrew's technique (Fig. 18). While this is in qualitative agreement with their data, it underpredicts the thickness of solid wax and as a result, leads to discrepancies in heat transfer to the patient. Nevertheless, this model can

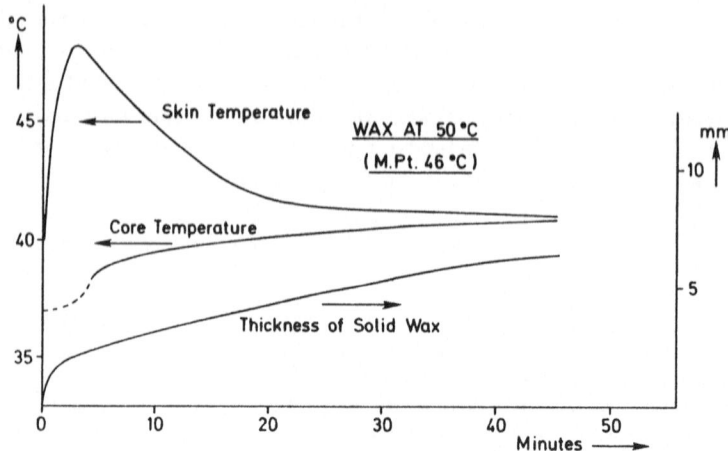

FIGURE 18

Computed skin and core temperatures in a patient undergoing whole-body hyperthermia via the hot wax bath technique. Note that the molten wax solidifies with time (see Ref. 125). By permission of the New York Academy of Sciences.

be easily improved to provide more detailed and accurate information on the temperature distribution in a patient, as shown in Refs. 69a,b.

6.5. Regional and Systemic Perfusion with Heated Blood

While the technology for isolated perfusion of an organ or a limb has existed for a long time, heating limbs containing tumors started only in 1967;[18,19,126,127] this led to significant improvements in the long-term survival of patients. In this technique, blood is heated to approximately 43°C in a heat exchanger and circulated through the limb to heat the muscle to 40–42°C by convection. There are two major thermal problems in this technique: (1) loss of heat through the skin, which is compensated for by the use of infrared lamps or warm rubber blankets; and more importantly, (2) exchange of heat, via the systemic circulation, with surrounding tissues; this latter problem makes selective heating difficult with regional perfusion. Additional problems are related to enhanced coagulation in the extracorporeal equipment at elevated temperature and management of bleeding with regional anti-coagulation.

6.6 Implantation of a Heat Source and Electrocoagulation

Direct implantation of a heat source in the tumor has been used by Sutton[128] in the treatment of malignant gliomas of the brain. The procedure showed promising results for early gliomas prior to widespread invasion of normal brain. Strauss[129] also electrocoagulated ($T \geqslant 70°C$) a large number of tumors by bringing a resistive heat source in contact with the tumors. Strauss obtained permanent cures in advanced patients, which he attributed to the stimulation of the host's immune response.

We have recently obtained analytical and numerical solutions for transient and steady-state temperature distributions in and around the heating probe for both uniformly and nonuniformly perfused tumors (Fig. 19). These solutions are valid for several types of heating functions; for example, a step input, a single or a series of pulses, or a transient input of the form $a + b\sqrt{t}$.[26,61,84,130]

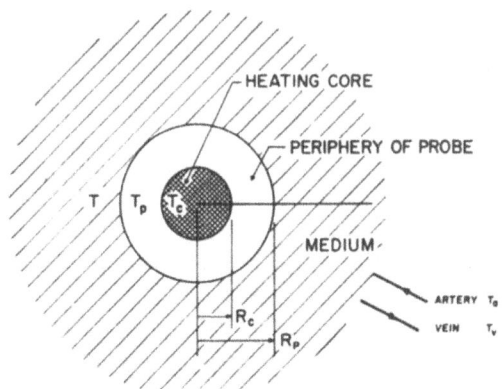

FIGURE 19
Schematic diagram of the tumor (medium) containing a spherical heating probe in its center. This model is applicable for estimating temperature distributions in tumors during localized hyperthermia. Adapted from Jain, Ref. 61.

(a)

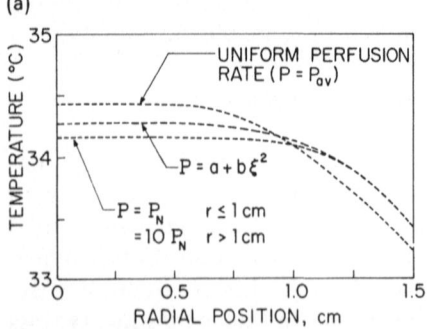

(b)

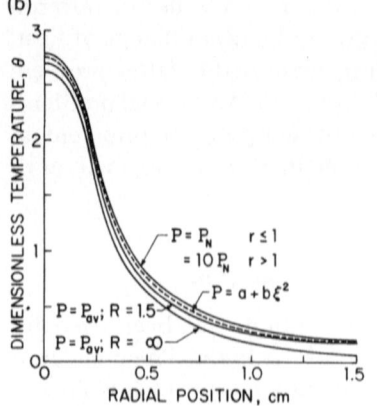

FIGURE 20
(a) Effect of nonuniformities in perfusion rate, P, on the temperature distribution in a 1.5 cm (radius) tumor grown around a 0.5 cm (radius) heating probe shown in Fig. 19. Note that the heater is off. Adapted from Jain, Ref. 61. (b) Effect of nonuniformities in perfusion rate on steady-state temperature rise in the probe and tumor as a result of step heat input. The probe radius is 0.5 cm, and the tumor radius is 1.5 cm. Adapted from Jain, Ref. 61.

Shown in Fig. 20a is the temperature profile in and around a 1.5 cm radius tumor before the onset of heating. Note that, as expected, the temperature profile within the probe is flat. Temperature in the center of a necrotic tumor is lower when compared to a viable tumor, although both tumors have the same average perfusion rate. These results are analogous to the case when the tumor is not grown around a probe (Fig. 12).

Shown in Fig. 20b is the steady-state temperature profile for the same probe and tumor resulting from a step heat input. Since the temperature response is linear with respect to the magnitude of the step input Q, the temperature is nondimensionalized in the following manner:

$$\theta = \frac{T - T_0}{QR_p^2/3k} \tag{17}$$

Here, T_0 is the steady-state temperature before the onset of heating (Fig. 20a), T is the steady-state temperature during heating, and R_p is the radius of heating probe. As expected, because of a poor convective heat transfer, the temperature rise in a necrotic tumor is higher than that in a viable tumor. Temperature rise in an infinite and uniformly perfused tumor is the lowest (Fig. 20b).

Analytical methods for predicting temperature profiles induced by electrosurgical methods are treated in further detail in Chap. 17.

6.7. Pyrogenic Agents and Bacterial Toxins

Coley[6] obtained remarkable success in controlling advanced disease in many patients by injecting mixed bacterial toxins that induced fever. However, due to poor biological characterization, many investigators have been unable to reproduce Coley's work. To this author's knowledge, no attempt has been made to model temperature distributions during induced fever.

6.8. Ultrasound

Although ultrasound has received special attention in inducing hyperthermia for cancer treatment, there is a paucity of data and analyses of the resulting temperature distributions in tissues. The first attempts to compute temperature distributions during ultrasonic therapy are attributed to Guttner[100] and Schwan *et al.*[101] Schwan and co-workers calculated temperature distributions in a two-layer (fat-muscle) model exposed to plane-parallel (unfocused) waves. Guttner carried out the analysis for a three-layer (fat–muscle–bone) model. Both authors neglected convection and conduction in their analyses and calculated temperature rise on the basis of the local absorption of ultrasonic energy given by Eq. (11).

Chan *et al.*[102] modified the analysis of Guttner by including conduction and convection in their layered model. In 1974, these authors[99] presented a detailed analysis of the reflection and transmission of ultrasonic waves as well as conversion of longitudinal waves into shear waves at the tissue boundaries. These authors also studied the role of the angle of incidence on the relative heating pattern; however, no attempt was made to compute the resulting temperatures. Recently, ter Haar[103] has repeated these calculations for an unfocused beam impinging on a three-layer model in order to study the sensitivity of temperature distributions to the wave frequency, incident energy (I_0, in Eq. 11), skin temperature (which can be controlled by a water bath), tissue composition, and blood flow rate. In one set of computations, ter Haar has also used the measured intensity distribution from an ultrasonic source to compute temperature distributions in a layered model. Her results are in qualitative agreement with data she obtained on the hind leg of an anesthetized pig.

One of the first attempts to simulate temperature elevations caused by a stationary focused beam is due to Robinson and Lele.[104] Although the focal region of their beam was a prolate spheroid with a Gaussian intensity distribution, these authors assumed the source to be either spherical or cylindrical in shape in their computations. In addition, the role of blood flow was analyzed by using an effective thermal conductivity term, similar to that in Eq. (2). Despite many assumptions made in their analysis, these authors obtained adequate agreement between their analysis and data on a cat's brain.

Recently, Parker and Lele[105] have calculated the temperature distribution in a cat's brain heated with one or more moving, focused beams of ultrasound. These authors have assumed that the insonation creates a temperature "forcing function" along its axis, with highest temperatures occurring in the focal plane. Using the bioheat equation, they obtained a steady-state temperature profile that reproduced their data adequately.

While some investigators have reported temperature distributions in tumors due to unfocused and focused beams,[96,97] theoretical analyses of these data have not been carried out.

6.9. Microwaves

Although considerable effort has been put into the analysis of energy absorbed by model tissues during microwave diathermy, only limited studies have been done on the measurement or prediction of resulting temperature distributions *in vivo*.

The efforts to monitor temperature distributions in tissues exposed to electromagnetic fields by thermocouples and thermistors have been limited for the following reasons: (1) They scatter incident electromagnetic radiation into the surrounding tissue, causing a perturbation of the EM field and modifications in the thermal pattern. (2) The incident electromagnetic field induces an electric field in the probe material and as a result, leads to ohmic heating. (3) The current induced in the connecting wires of a probe interferes with the signal from the probe. Investigators using microwaves and radio-frequency heating have avoided some of these problems by: (1) turning off the field for a short time for periodic temperature measurements; (2) orienting the probe and connecting leads properly, so that the interference is minimum; (3) using a probe as small as possible; and (4) using nonconducting probe and leads (see Table 12). Electromagnetic interference in temperature measurement and its control is discussed in detail in Chap. 27.

Recently, as a result of developments in optical fiber technology, several new probes have been developed and tested that are electrically isolated and nonperturbing to the EM field. Table 12 compares four probes that are either available commercially or under development. While these probes differ from each other in several ways, they all modify the intensity, wavelength, polarization, emission, or reflection of incident light as a function of the probes' temperature. This change in the optical data is conducted through optical

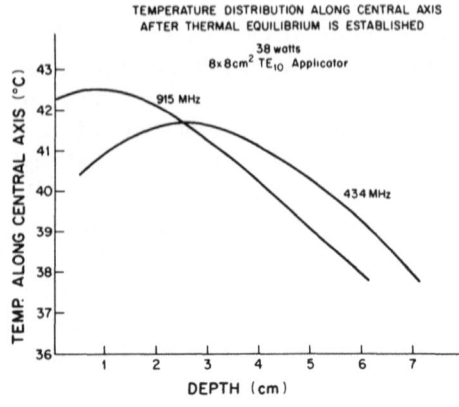

FIGURE 21
Steady-state temperatures measured as a function of tissue depth in pigs exposed to 434 and 915 MHz microwaves. Note that 434 MHz is more effective for heating at depths greater than 2 cm. From Ref. 132, by permission of the New York Academy of Sciences.

TABLE 12

Temperature Probes for Measurement in Electromagnetic Fields[a]

	Thermistor	Liquid crystal[b]	Birefringent crystal (lithium tantalate)[c]	Semiconductor (gallium arsenide)[c]	Phosphors[d]
Leads	Carbon impregnated plastic	Fiber optic	Fiber optic	Fiber optic	Fiber optic
Temperature range °C	0–250	35–50	18–49	20–50	9–250
Resolution °C	0.05	0.1–0.5	0.1	0.1–0.2	0.025
Size mm	0.5–2	1–2	1	0.25	—
Advantages	Relatively accurate, reproducible and stable	Available commercially	Available commercially, the greatest sensitivity in the required temperature range	Available commercially, minimum transmission loss in the GaAs range; relatively small	Available commercially, accurate, low cost, interchangeable, small probe
Disadvantages	Use of high-resistance leads requires sensitive electronics and short lead length; probes are large	Instability, hysteresis; probes are large	Source instabilities; probes are large	Source instabilities	Sophisticated optics and electronics

[a] Cf. Chapter 27.
[b] Ramal Inc., P.O. Box 275, Sandy, UT 84070.
[c] Yellow Spring Instrument Co., Yellow Springs, OH.
[d] Luxtron Corp., 2916 Scovill Blvd, Santa Clara, CA 95060

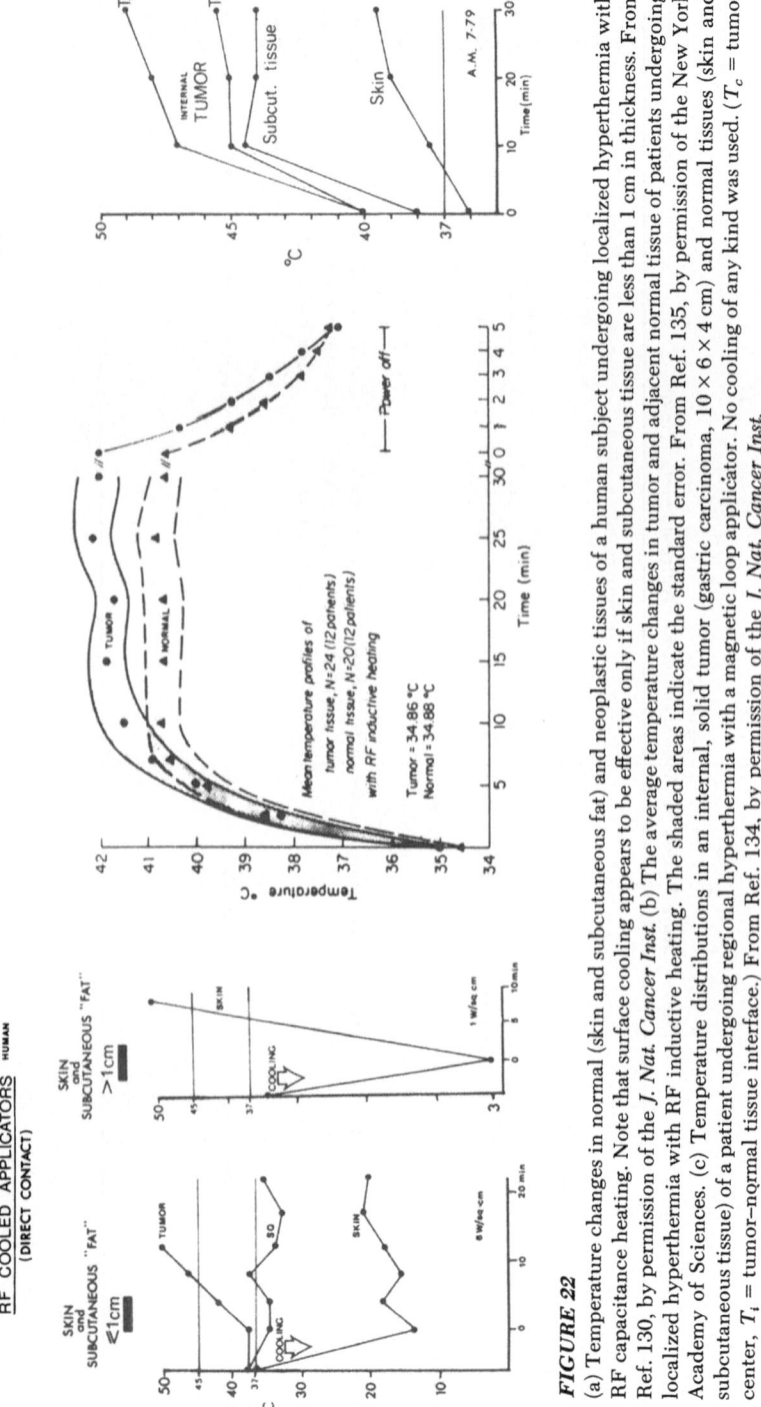

FIGURE 22

(a) Temperature changes in normal (skin and subcutaneous fat) and neoplastic tissues of a human subject undergoing localized hyperthermia with RF capacitance heating. Note that surface cooling appears to be effective only if skin and subcutaneous tissue are less than 1 cm in thickness. From Ref. 130, by permission of the *J. Nat. Cancer Inst.* (b) The average temperature changes in tumor and adjacent normal tissue of patients undergoing localized hyperthermia with RF inductive heating. The shaded areas indicate the standard error. From Ref. 135, by permission of the New York Academy of Sciences. (c) Temperature distributions in an internal, solid tumor (gastric carcinoma, $10 \times 6 \times 4$ cm) and normal tissues (skin and subcutaneous tissue) of a patient undergoing regional hyperthermia with a magnetic loop applicator. No cooling of any kind was used. (T_c = tumor center, T_i = tumor-normal tissue interface.) From Ref. 134, by permission of the *J. Nat. Cancer Inst.*

fibers to an optoelectronic processor to obtain an analog or digital signal for recording the temperature. Chap. 27 also discusses this new technology in detail.

Despite these new developments, only limited data have been reported in the literature on the detailed temperature distributions during MW diathermy.[131] Figure 21 shows the steady-state temperature distribution in a pig exposed to microwaves. The 915 MHz is more effective in heating at less than a 2 cm depth, while 430 MHz is effective at greater depths. The depth of heating can be increased by skin cooling.[132] Assuming exponential decay in the electromagnetic field (cf. Eq. 11), Chan *et al.*[102] have calculated the temperature distribution in a three-layer (fat–muscle–bone) model of human thigh. To account for thermoregulation of the tissue, these authors introduced a nonlinear, convective heat transfer term with a delay function into the bioheat equation. Their results are in qualitative agreement with their data obtained for 915-MHz microwaves. Kritikos and Schwan[133] have also calculated the temperature rise in a spherical region simulating a hot spot in the central region of a human head. However, the assumption that the energy absorbed is uniform in a "spherical" active region embedded in an infinite region is unrealistic and makes this analysis of limited use in designing heating protocols. Trembly *et al.*[112] have recently calculated temperature distributions around an implanted antenna in a homogeneous tissue.

Needless to say, more effort is now needed in analyzing the temperature distributions in tissues exposed to EM radiation.

6.10. Radiofrequency Heating

Similar to MW diathermy, analyses of heat transfer and temperature distributions during RF heating have been limited. Temperature measurements *in vivo* have been hindered by the problems listed in the previous section. Figs. 22a,b show the results on two groups of the temperature distributions in tumors and normal tissues exposed to capacitive (13 MHz) and inductive (27 MHz) heating, respectively.[134,135] Note that despite surface cooling, the high specific absorption by the subcutaneous fat limits the use of capacitive heating to "nonobese" patients.

The recently developed induction loop applicator, in which the patient can be placed, allows heating of a deep-seated tumor (Fig. 22c). While limited theoretical results are available for any of these devices, the experimental results on the temperature distribution look promising, especially for large tumors (greater than 5 cm diam).

7. SUMMARY AND RECOMMENDATIONS

The objectives of this chapter are to present various theoretical frameworks that may be used to predict temperature distributions in normal and neoplastic tissues during hyperthermia and to describe the thermal energy

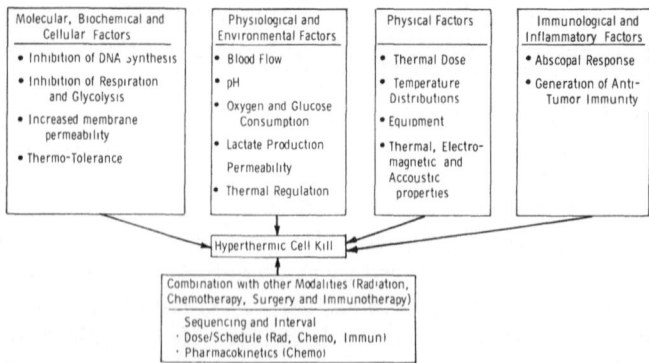

FIGURE 23

Various physical, physiological, biochemical, and immunological factors that determine the effectiveness of hyperthermia, alone, or in combination with other modalities, in cancer treatment.

deposition modalities. The various biological, physiological, and immuno-logical factors (Fig. 23) have been mentioned only briefly (for details, see Ref. 28). In addition, the role of hyperthermia as an adjuvant to radiotherry or chemotherapy was not discussed (see Refs. 26, 136, 137). The emphasis here has been on the physical factors involved in hyperthermia. To this end, two model approaches, distributed and lumped parameter methods, were presented for calculating the temperature distribution between the normal and neoplastic tissues of various mammals. Evaluation of thermal energy deposition during US, MW, and RF heating is discussed in depth. The experimental and theoretical results for the temperature distributions were summarized according to the method used for heating. Several important and not well-understood problems were pointed out at various places in the text in the hope of stimulating interest in this multidisciplinary research.

Some heat transfer related areas that deserve further attention, in my view, are listed here.

1. While the bioheat equation, as it stands, appears to give adequate results in several applications, a precise description of heat transfer in tissues remains a tedious but challenging problem.

2. As pointed out, data on the thermal, electrical and acoustic properties, intratumor blood flow rates, and metabolism of tumors are limited. These parameters should be measured accurately, over a wide range of temperatures, and at various stages of tumor growth and regression.

3. It is essential to provide accurate measurements of intratissue tem-perature distributions during normothermia and hyperthermia. New developments in noninvasive thermometric techniques are needed in order to optimally exploit hyperthermia.

4. Physiological studies in animals and humans are needed in order to understand thermoregulation during hyperthermia. This information should then be incorporated into the mathematical models of the type described in this chapter.

5. Analysis of heat transfer and temperature distributions during US, MW, and RF diathermy is imperative for the proper use of these techniques. Such efforts in the past have been limited at best.

6. Quantitative comparisons of various methods of inducing hyperthermia should be made on the basis of temperatures reached in various parts of normal and neoplastic tissues. This type of analysis will help in deciding the "best" modality of heating a given tumor and help in designing feedback control mechanisms and heating protocols for maintaining desired temperatures in normal and neoplastic tissues.

Our current research is directed toward some of these areas.

ACKNOWLEDGMENT. This work was supported in part by research grants from the National Science Foundation (ENG-78-22814 and ENG-78-25432) and the American Cancer Society (PDT-150), and by a Career Development Award from the National Cancer Institute (CA-00643). The author is grateful to Dr. Sudhir Shah for his many helpful comments, to Daniel Solomon and Robert Peloso for their help in the literature search, and to Maria Strati for typing this manuscript.

This material has appeared in a similar form as "Bioheat Transfer: Mathematical Models of Thermal Systems," Chapter 2 in *Hyperthermia in Cancer Therapy* edited by F. K. Storm. The author is thankful to G. K. Hall and Co. for their permission to use the published material.

REFERENCES

1. Breasted, J. H., The Edwin Smith Surgical Papyrus (Chicago: 1930).
2. Jee, B. S., *A Short History of Aryan Medical Sciences* (London, 1896).
3. Fienus, T., *De cauteriis libri quinqui* (Louvain, 1598).
4. Severin, M. A., *De la medicine efficae* (Geneva: 1668).
5. Wolff, J., *Die Lehre von der Krebskrankheit* (Jena: 1907).
6. Coley, W. B., The treatment of malignant tumor by repeated inoculations of erysipelas with a report of original cases, *Am. J. Med. Sci.* **105**, 487–511, 1893.
7. Coley, W. B., Late results of the treatment of inoperable sarcoma by the mixed toxins of erysipelas and bacillins prodigiosus, *Am. J. Med. Sci.* **131**, 375, 1906.
8. Nauts, H. C., Fowler, G. A., and Bogarko, F. H., A review of the influence of bacterial infection and bacterial products (Coley's toxins) on malignant tumors in man, *Acta Med. Scand.* **145**:suppl. 276, 1–103, 1953.
9. Stevenson, H. N., The effect of heat upon tumor tissue, *J. Cancer Res.* **4**, 54, 1919.
10. Rohdenburg, G. L., and Prime, F., The effect of combined radiation and heat in neoplasms, *Arch. Surg. (Chicago)* **2**, 116, 1921.
11. Crile, G., Jr., Heat as an adjuvant to the treatment of cancer: Experimental studies, *Clin. Annals, Cleveland* **28**, 577–582, 1940.
12. Johnson, H. J., The action of short radiowaves on tissues. III. A comparison of the thermal sensitivities of transplantable tumors *in vivo* and *in vitro*. *Am. J. Cancer* **38**, 533–550, 1940.
13. Overgaard, K., and Okkels, H., Uber den Einfluss der Warmebehandlung auf Woods Sarkom, *Strahlentherapie* **68**, 39, 1940.
14. Schliephake, E., *Les ondes courtes en biologie*, Paris, 1938.
15. Denier, A., Les ondes hertziennes ultracourtes de 80 cm., *J. Radiol. Electrol.* **20**, 193, 1936.
16. Horvath, J., Ultraschallwirkung beim menschlichen Sarkom. *Strahlentherapie* **75**, 119–125, 1944.

17. Horvath, J., New possibilities in the treatment of malignant tumors by ultrasonic waves, *Dtsch Med. Wochenschr.* **27–28**, 392, 1947.

18. Cavaliere, R., Ciocatto, E. C., Giovannela, B. C., Heidelberger, C., Johnson, R. O., Martotini, M., Mondovi, B., Moricca, G., and Rossi-Fanelli, A., Selective heat sensitivity of cancer cells: Biochemical and clinical studies, *Cancer* **20**, 1351–1381, 1967.

19. Stehlin, J. S., Hyperthermic perfusion with chemotherapy for cancer of the extremities, *Surg. Gynecol. Obst.* **129**, 305–308, 1969.

20. Robinson, J. E., and Wizenberg, M. J., eds. *Proc. 1st Int. Symp., Cancer Therapy by Hyperthermia and Radiation* (Bethesda, MD: American College of Radiology, 1976).

21. Lett, J. T., and Adler, H., eds., *Advances in Radiation Biology*, vol. 6 (New York: Academic Press, 1976).

22. Rossi-Fanelli, A., Cavaliere, R., Mondovi, B., and Moricca, G., eds., *Selective Heat Sensitivity of Cancer Cells* (New York: Springer-Verlag, 1977). Published as vol. 59 of *Recent Results in Cancer Research.*

23. Streffer, C., ed., *Proc. 2d Int. Symp. Cancer Therapy by Hyperthermia and Radiation* (Baltimore, MD: Urban and Schwarzenberg, 1978).

24. Caldwell, W. E., and Durand, R. E., eds. *Proc. Conf. Clin., Prospects of Hypoxic Cell Sensitizers and Hyperthermia* (Madison: University of Wisconsin Press, 1978).

25. Midler, J. W. ed., Proc. conf. on hyperthermia in cancer treatment. *Cancer Res.* **39**, 2231–2340, 1979.

26. Jain, R. K., and Gullino, P. M., eds., Thermal characteristics of tumors: Applications in detection and treatment, *Ann. N. Y. Acad. Sci.* **335**, 1980.

27. Dewey, W. C., and Dethlefsen, L., eds., *Proceedings 3rd International Conference on Cancer Therapy by Hyperthermia, Drugs, and Radiation*, National Cancer Institute, Monograph Vol. 63, 1980.

28. Jain, R. K., Bioheat transfer: mathematical models of thermal systems, in *Hyperthermia in Cancer Therapy*, Storm, F. K., ed. (Boston: G. K. Hall and Co., in press).

29. Pennes, H. H., Analysis of tissue and arterial blood temperatures in the resting human forearm, *J. Appl. Physiol.* **1**, 93, 1948.

30. Trezek, G. J., and Jewett, D. L., Nodal network of transient temperature fields from cooling sources in anesthetized brain, *IEEE Trans. Biomed. Eng.* **4**, 281, 1970.

31. Jain, R. K., and Wei, J., Dynamics of drug transport in solid tumors: a distributed parameter model, *J. Bioeng.* **1**, 313–330, 1977.

32. Wulff, W., The energy conservation equation for the living tissue, *IEEE Trans. Biomed. Eng.* **21**, 494, 1974.

33. Mitchell, J. W., and Myers, G. E., An analytical model of the countercurrent heat exchange phenomenon, *Biophys. J.* **8**, 897, 1968.

34. Keller, K. H., and Seiler, L., An analysis of peripheral heat transfer in man, *J. Appl. Physiol.* **30**, 779–786, 1971.

35. Weinbaum, S., General discussions, *Ann. N. Y. Acad. Sci.* **335**, 173–175, 1980.

36. Klinger, H. G., Heat transfer in perfused biological tissue, I: General theory, *Bull. Math. Biol.* **36**, 403, 1974.

37. Chen, M. M., and Holmes, K. R., Microvascular contributions in tissue heat transfer, *Ann. N. Y. Acad. Sci.* **335**, 137–150, 1980.

38. Chato, J. C., Heat transfer in bioengineering, in *Advanced Heat Transfer*, Chao, B. T. ed. (Urbana: University of Illinois Press, p. 395, 1969).

39. Bowman, H. F., Cravalho, E. G., and Woods, M., Theory, measurements and applications of thermal properties of biomaterials, *Ann. Rev. Biophys. Bioeng.* **4**, 43–80, 1975.

40. Jain, R. K., Grantham, F. H., and Gullino, P. M., Blood flow and heat transfer in Walker 256 mammary carcinoma, *J. Nat. Cancer Inst.* **62**, 927–933, 1979.

41. Jain, R. K., and Gullino, P. M., Analysis of transient temperature distribution in a perfused medium due to a spherical heat source with application to heat transfer in tumors: Homogeneous and perfused medium, *Chem. Eng. Comm.* **4**, 95–118, 1980.

42. Shah, S., and Jain, R. K., Modification of blood flow in W256 carcinoma by hyperglycemia and hyperthermia: a thermal probe method, *Proc. Am. Assoc. Cancer Res.* **22**, 60, 1981.

43. Holmes, K. R., and Chen, M. M., Local thermal conductivity of *para*-7 fibrosarcoma in hamster, *Adv. Bioeng. (ASME)* **4**, 147–149, 1979.

44. Bowman, H. F., Heat transfer mechanisms and thermal dosimetry, paper presented at the *3d Int. Symp. Cancer Therapy Hyperthermia, Drugs, and Radiation*, Fort Collins, CO, June 22–26, 1980.

45. Cooper, T. E., and Trezek, G. J., Correlation of thermal properties of some human tissues with water content, *J. Aerospace Med.* **42**, 24–28, 1971.

46. Poppendiek, H. F., Randall, R., Breeden, J. A., Chambers, J. E., and Murphy, J. R., Thermal conductivity measurements and predictions for biological fluids and tissues, *Cryobiology* **3**, 318, 1967.

47. Bischoff, K., Some fundamental considerations of the applications of pharmacokinetics to cancer chemotherapy, *Cancer Chemother. Rep.* **59**, 777–793, 1975.

48. Jain, R. K., Weissbrod, J., and Wei, J., Mass transfer in tumors: Characterization and applications in chemotherapy, *Adv. Cancer Res.* **33**, 251–311, 1980.

49. Geslowski, L. E., and Jain, R. K., Physiologically based pharmacokinetics: Principles and applications, *J. Pharm. Sci.* **72**, 1103–1127, 1983.

50. Slotman, G. J., Swaminathan, A. P., Casey, K. F., and Rush, B. F., Jr., Quantitative changes in tumor blood flow with expanding tumor mass in the VX2 carcinoma, *Proc. Am. Assoc. Cancer Res.* **21**, 51, 1980 (abstract no. 203).

51. Mantyla, M. J., Kuikka, J., and Rekonen, A., Regional blood flow in human tumours with special reference to the effect of radiotherapy, *Br. J. Radiol.* **49**, 335–338, 1976.

52. Gullino, P. M., *In vitro* perfusion of tumors, in J. C. Norman, Folkman, J., Hardison, W. G., et al, eds., *Organ Perfusion and Preservation* (New York: Appleton-Century-Crofts, 1968), pp. 877–898.

53. Song, C. W., Kanz, M. S., Rhee, J. G., and Levitt, S. H., Effect of hyperthermia on vascular function in normal and neoplastic tissues *in vivo*, *Ann. N. Y. Acad. Sci.* **335**, 35–47, 1980.

54. Vaupel, P., Interrelationship between mean arterial blood pressure, blood flow and vascular resistance in solid tumor tissue of DS-carcino-sarcoma, *Experientia* **31**, 587, 1975.

55. Goldacre, R. J., and Sylven, B., A rapid method of studying tumour blood supply using systemic dyes, *Nature (London)* **14**, 63, 1959.

56. Rowe-Jones, D. C., The penetration of cytotoxins in malignant tumours, *Br. J. Cancer* **22**, 156–162, 1968.

57. Gullino, P. M., Jain, R. K., and Grantham, F. H., Temperature gradients and local perfusion in a mammary carcinoma, *J. Nat. Cancer Inst.* **68**, 519–533, 1982.

58. Shibata, H. R., and MacLean, L. D., Blood flow to tumors, *Prog. Clin. Cancer* **11**, 33–47, 1966.

59. Tuttle, E. P., and Sadler, J. S., Measurement of renal tissue fluid turnover rates by thermal washout technique, *Hypertension* **13**, 3, 1964.

60. Straw, J. A., Hart, M. M., Klubes, P., Zaharko, D. S., and Dedrick, R. L., Distribution of anticancer agents in spontaneous animal tumors, I. Regional blood flow and methotrexate distribution in canine lymphosarcoma, *J. Nat. Cancer Inst.* **52**, 1327–1331, 1974.

61. Jain, R. K., Effect of inhomogeneities and finite boundaries on temperature distributions in a perfused medium with application to tumours, *J. Biomech. Eng., Trans. ASME* **100**, 235–241, 1978.

62. Endrich, B., Zweifach, B. W., Reinhold, H. S., and Intaglietta, M., Quantitative studies of microcirculatory function in malignant tissue: Influence of temperature on microvascular hemodynamics during the early growth of the BA 1112 rat sarcoma, *Int. J. Rad. Oncol. Biol. Physics* **5**, 2021–2030, 1979.

63. Dudar, T. E., and Jain, R. K., Microcirculatory changes during hyperthermia in normal and neoplastic tissues, *Cancer Res.* **44**, 605–612, 1984.

64. Nugent, L. J., and Jain, R. K., Diffusional transport and permeability of macromolecules in normal and neoplastic tissues, *Cancer Res.* **44**, 238–244, 1984.

65. Gerlowski, L. E., and Jain, R. K., 1982 (unpublished results).

66. Gemmill, C. L., and Brobeck, J. R., Energy exchange, in *Medical Physiology*, edited by Mountcastle, V. B. (St. Louis: C. V. Mosby, 1968), pp. 485–488.

67. Altman, P., and Ditmer, D., *Biology Data Book* (Bethesda, MD: Fed. Am. Soc. Exptl., 1954).

68. Huckaba, C. E., and Tam, H. S., Modeling of human thermal system, in *Advances in Biomedical Engineering*, part I, ed. D. O. Cooney, (New York: Marcel Dekker, 1980), pp. 1–58.

69. (a) Volpe, B. T., and Jain, R. K., Temperature distributions and thermal responses in humans, I. Simulations of various modes of whole-body hyperthermia in normal subjects, *Med. Phys.* **9**, 506–513, 1982.

69. (b) Volpe, B. T., and Jain, R. K., Temperature distributions and thermal responses in humans, II. Simulations of whole-body, regional and localized hyperthermia in cancer patients, *Am. Inst. Chem. Eng. Symp. Ser.* **79**, 116–123, 1983.

70. Gullino, P. M. *In vivo* utilization of oxygen and glucose by neoplastic tissue, in *Oxygen Transport to Tissue*, J. Groste, D. Reneau, G. Thews, eds., (New York: Plenum, 1976), pp. 521–536.

71. Vaupel, P., Astemgaswechsel und Glucosestoffwechsel von Implantationstumoren (DS-Carcinosarkom) *in vivo* (Mainz: Akademie der Wissenshaften und der Literatur, 1974).

72. Gullino, P. M., Grantham, F. H., and Courtney, A. H., Utilization of oxygen by transplanted tumors *in vivo*, *Cancer Res.* **27**, 1020–1030, 1967.

73. Gullino, P. M., Grantham, F. H., Courtney, A. H., Glucose consumption by transplanted tumors *in vivo*, *Cancer Res.* **27**, 1031–1040 1967.

74. Gullino, P. M., Grantham, F. H., Courtney, A. H., and Losonczy, I., Relationship between oxygen and glucose consumption by transplanted tumors *in vivo*, *Cancer Res.* **27**, 1041–1052, 1967.

75. Vaupel, P., Frinak, S., Mueller-Klieser, W., and Bicher, H. I., Impact of localized hyperthermia on the cellular microenvironment in solid tumors, *Proc. 3d Int. Symp. Cancer Therapy Hyperthermia, Drugs, and Radiation* June 22–26, 1980.

76. Sien, H. P., and Jain, R. K., Temperature distributions in normal and neoplastic tissues during hyperthermia: lumped parameter analysis, *J. Therm. Biol.* **4**, 157–164, 1979.

77. Hardy, J. D., Gagge, A. P., and Stolwijk, J. A. J., *Physiological and Behavioral Thermoregulation* (Springfield, IL: Charles C. Thomas, 1970).

78. Fan, L. T., Hsu, F. T., and Hwang, C. L., A review of mathematical models of the human thermal system, *IEEE Trans. Biomed. Eng.* **BME-18**, 218, 1971.

79. Hwang, C. L., and Konz, S. A., Engineering models of the human thermoregulatory system—a review, *IEEE Trans. Biomed. Eng.* **BME-24**, 309–315, 1977.

80. Stolwijk, J. A. J., Mathematical models of thermal regulation, *Ann. N. Y. Acad. Sci.* **335**, 98–106, 1980.

81. Shitzer, A., Mathematical models of thermoregulation and heat transfer in mammals, *NASA Tech. Mem.* **X62**, 172, 1974.

82. Sien, H. P., and Jain, R. K., Intratumor temperature distributions during hyperthermia, *J. Therm. Biol.* **5**, 127–130, 1980.

83. Chrysanthopoulos, G., and Jain, R. K., Thermal interactions between normal and neoplastic tissues in the rat, rabbit, swine, and dog during hyperthermia, *J. Med. Phys.* **7**, 529–539, 1980.

84. Jain, R. K., Heat transfer in tumors: Characterization and applications to thermography and hyperthermia, in *Advances in Biomedical Engineering*, Cooney, D. O., ed., part 1 (New York: Marcel Dekker, 1980).

85. Wissler, E. H., Steady-state temperature distribution in man, *J. Appl. Physiol.* **16**, 734, 1961.

86. (a) Sien, H. P., Dynamics of temperature distributions in normal and neoplastic tissues during hyperthermia (M. S. thesis, Columbia University, New York, 1978). (b) Chrysanthopoulas, G., Thermal interactions between normal and neoplastic tissues in mammalian systems during hyperthermia (M. S. thesis, Columbia University, New York, 1979).

87. Volpe, B. T., Analysis of temperature distributions and thermal response in cancer patients during hyperthermia (M. S. thesis, Carnegie–Mellon University, Pittsburgh, 1981).

88. Cunningham, D. J., An evaluation of heat transfer through skin in the human extremity, in *Physiological and Behavioral Thermoregulation* (Springfield, IL: Charles C. Thomas, 1970).

89. Thauer, R., Circulatory adjustments to climatic requirements, in *Handbook of Physiology*, vol. III, Hamilton, W. F., and Dow, P., eds. (Washington, DC: Am. Physiol. Soc., 1965), pp. 1921–1966.

90. Esmay, M. L., *Principles of Animal Environment* (Westport, CT: AVI Publishing, 1969), pp. 77–80.
91. Hales, J. R. S., Physiological responses to heat, in MTP Int. Rev. Ser, *Physiology* series 1, vol. 7, Robertshaw, D., ed. (London: Butterworths), pp. 107–162, 1970.
92. Schwan, H. P., Biophysics of diathermy, in *Therapeutic Heat and Cold*, Licht, S., ed., (New Haven, CT: E. Licht Publishers, 1965), pp. 63–125.
93. Har-Kedar, I., and Bleehan, N. J., Experimental and clinical aspects of hyperthermia applied to the treatment of cancer with special reference to the role of ultrasonic and microwave heating, *Adv. Rad. Biol.* 6, 229–266, 1976.
94. Dobson, J., Equipment for local and regional hyperthermia, report prepared by WSA, Inc. San Diego, CA, for the Radiotherapy Development Branch, DCT, NCI, NIH, Feb. 1980.
95. Hunt, J. W., Application of microwave, ultrasound, and radiofrequency heating *in vivo*, *Proc. 3d Int. Symp. Cancer Therapy Hyperthermia, Drugs, and Radiation*, June 22–26, 1980.
96. Marmor, J. B., Pounds, D., Hahn, N., and Hahn, G. M., Treating spontaneous tumors in dogs and cats by ultrasound-induced hyperthermia, *Int. J. Rad. Oncol. Biol. Phys.* 4, 967–973, 1978.
97. Lele, P. P., A transient thermal pulse technique for measurement of tissue thermal diffusivity *in vivo*, *Ann. N. Y. Acad. Sci.* 335, 83–85, 1980.
98. Pounds, D. D., personal communication, 1980.
99. Chan, A. K., Siglemann, R. A., and Guy, A. W., Calculations of therapeutic heat generated by ultrasound in fat–muscle–bone layers, *IEEE Trans. Biomed. Eng.* BME-21, 280–284, 1974.
100. Guttner, W., Ultraschall in Menschlichen Korper, *Acustica* 4, 547–554, 1954.
101. Schwan, H. P., Carstensen, E. L., and Li, K., Electric and ultrasonic deep-heating diathermy, *Electronics*, Mar. 1954, pp. 172–175.
102. Chan, A. K., Sigelmann, R. A., Guy, A. W., and Lehman, J. F., Calculation by the method of finite differences of the temperature distribution in layered tissues, *IEEE Trans. Biomed. Eng.* BME-20, 86–90, 1973.
103. Haar, ter G. R., Computed temperature profiles in tissues resulting from ultrasonic irradiation, abstr. no. 39, *3d Int. Symp. Cancer Therapy Hyperthermia, Drugs, and Radiation*, June 22–26, 1980.
104. Robinson, T. C., and Lele, P. P., An analysis of lesion development in the brain and in plastics by high-intensity focused ultrasound at low megaheartz frequencies, *J. Acoust. Soc. Am.* 51, 1333–1351, 1972.
105. Parker, K., and Lele, P. P., discussion, *Ann. N. Y. Acad. Sci.* 335, 64, 1980.
106. Bladel, van J., *Electromagnetic Fields* (New York: McGraw-Hill, 1964).
107. Kantor, G., New types of microwave diathermy applicators: comparison of performance with conventional types, *Proc. Symp. Biol. Effects and Meas. of Radiofrequency/Microwaves*, Hazzard, D. G., ed. (DHEW pub. no. FDA-77-8026, July 1977).
108. Holt, J. A. G., The use of VHF radiowaves in cancer therapy, *Aust. Radiol.* 19, 223–241, 1975.
109. Turner, P. F., Deep heating of cylindrical or elliptical tissue masses, *Proc. 3d. Int. Symp. Cancer Therapy Hyperthermia, Drugs, and Radiation*, June 22–26, 1980.
110. Turner, P., and Kumar, L., Computer solution for applicator heating patterns, *Proc. 3d Int. Symp. 'Cancer Therapy Hyperthermia, Drugs, and Radiation*, June 22–26, 1980.
111. Arcangeli, G., Cividalli, A., Creton, G., Nervi, C., Biological rationale for an optional scheduling of heat and ionizing radiation: Clinical results on neck node metastases, *Proc. 3d Int. Symp. Cancer Therapy Hyperthermia, Drugs, and Radiation*, June 22–26, 1980, pp. 54.
112. Trembly, B. S., Strohbein, J. W., de Seiges, D. C., and Douple, E. B., Hyperthermia induced by an array of invasive microwave antennas, *Proc. 3d Int. Symp. Cancer Therapy Hyperthermia, Drugs, and Radiation*, June 22–26, 1980, pp. 74.
113. Guy, A. W., Lehman, J. F., and Stonebridge, J. B., Therapeutic applications of electromagnetic power, *Proc. IEEE* 62, 55–75, 1974.
114. Amalric, R., Spitalier, J. M., Giraud, D., and Altschuler, C., Thermography in diagnosis of breast diseases, *Bibl. Radiol.* 6, 65–76, 1975.

115. Gershon-Cohen, J., Berger, S. M., Haberman, J. D., and Brueschke, E. E., Advances in thermography and mammography, *Ann. N. Y. Acad. Sci.* **121**, 283–300, 1964.

116. Gautherie, M., Bourjal, P., Quenneville, Y., and Gros, C., Puissance thermogène des epithéliomas mammaries, I. Determination par thermométrie intratumorale et thermographie infrarouge cutanée, *Rev. Eur. Etud. Clin. Biol.* **17**, 776–781, 1972.

117. Gautherie, M., Thermopathology of breast cancer: Measurement and analysis of *in vivo* temperature and blood flow, *Ann. N. Y. Acad. Sci.* **335**, 383–415, 1980.

118. Reinhold, H. S., van der Zee, J., Faithfull, N. S., and van Rhoon, G. C., Utilization of the Siemens unit techniques, *Proc. 3d Int. Symp. Cancer Therapy Hyperthermia, Drugs, and Radiation*, June 22–26, 1980, pp. 114.

119. Gullino, P. M., Yi, P. N., and Grantham, F. H., Relationship between temperature and blood supply or consumption of oxygen and glucose by rat mammary carcinomas, *J. Nat. Cancer Inst.* **60**, 835–847, 1978.

120. Dickson, J. A., Suzanger, M., *In vitro-vivo* studies on the susceptibility of the solid Yoshida sarcoma to drugs and hyperthermia, *Cancer Res.* **34**, 1263–1274, 1974.

121. Herman, T. S., Zukoski, C. F., and Anderson, R. M., Whole-body hyperthermia via blanket technique, *Proc. 3d Int. Symp. Cancer Therapy Hyperthermia, Drugs, and Radiation*, June 22–26, 1980, p. 112.

122. Stoner, E. K., Luminous and infrared heating, in *Therapeutic Heat and Cold*, 2d ed., Licht, S., ed. (New Haven, CT: E. Licht Publishers, 1965), pp. 252–265.

123. Henderson, M. A., and Pettigrew, R., Induction of controlled hyperthermia in treatment of cancer, *Lancet* **1**, 1275–1277, 1971.

124. Levin, W., Wasserman, H., and Blair, R. M., Tumor temperature augmentation utilizing 433-MHz microwaves in patients undergoing whole-body hyperthermia, *Proc. 3d Int. Symp. Cancer Therapy Hyperthermia, Drugs, and Radiation*, June 22–26, 1980, p. 20.

125. Law, H. T., and Pettigrew, R. T., Heat transfer in whole-body hyperthermia, *Ann. N. Y. Acad. Sci.* **335**, 298–310, 1980.

126. Cavaliere, R., Moricca, G., DiFillippo, F., Caputo, A., Heat transfer problems during local perfusion in cancer treatment, *Ann. N. Y. Acad. Sci.* **335**, 311–326, 1980.

127. Stehlin, J. S., Hyperthermic perfusion for melanoma of the extremities: Experience with 65 patients, 1967 to 1979, *Ann. N. Y. Acad. Sci.* **335**, 352–355, 1980.

128. Sutton, C. H., Discussion., *Ann. N. Y. Acad. Sci.* **335**, 45–47, 1980.

129. Strauss, A. A., *Immunologic Resistance to Carcinoma Produced by Electrocoagulation, Based on Fifty-Seven Years of Experimental and Clinical Results* (Springfield, IL: Charles C. Thomas, 1969).

130. Jain, R. K., Transient temperature distributions in an infinite-perfused medium due to a time-dependent, spherical heat source, *J. Biomech. Eng., Trans. ASME* **101**, 82–86, 1979.

131. Sandhu, T. S., Kowal, H., and Johnson, R. J., The development of hyperthermia microwave generators and thermometry, *Int. J. Rad. Oncol. Biol. Phys.* **1**: suppl. 100, 1976.

132. Subjeck, J., Sciandra, J., Johnson, R., Drechsel, R., and Kowal, H., Cell survival dependence on heating method, *Proc. 3d Int. Symp. Cancer Therapy Hyperthermia, Drugs, and Radiation*, June 22–26, 1980, pp. 46.

133. Kritikos, H. N., and Schwan, H. P., Potential temperature rise induced by electromagnetic field in brain tissues, *IEEE Trans. Biomed. Eng.* **26**, 29–34, 1979.

134. Strom, F. K., Morton, D. L., Kaiser, L., Harrison, W. H., Elliott, R. S., Weisenberger, T., Parker, R. G., and Haskell, C. M., Clinical local hyperthermia by radiofrequency: A review, *Proc. 3d Int. Symp. Cancer Therapy Hyperthermia, Drugs, and Radiation*, June 22–26, 1980, pp. 108.

135. Hahn, E. W., and Kim, J. H., Clinical observations on the selectve heating of cutaneous tumors with the radio-frequency inductive method, *Ann. N. Y. Acad. Sci.* **335**, 347–355, 1980.

136. Overgaard, J., Influence of sequence and interval on the biological response to combined hyperthermia and radiation, *Proc. 3d Int. Symp. Cancer Therapy Hyperthermia, Drugs, and Radiation*, June 20–22, 1980, pp. 105–108.

137. Hahn, G. M., Studies on drug-hyperthermia interaction, *Proc. 3d Int. Symp. Cancer Therapy Hyperthermia, Drugs, and Radiation* June 22–26, 1980.

138. Licht, S., *Therapeutic Heat and Cold*, 2d ed. (Baltimore: Waverly Press, 1965).

139. Greene, J., Microwave diathermy: The invisible healer (HEW pub. no. FDA-79-8085, Feb. 1979).

140. (a) Gullino, P. M., and Grantham, F. H., Studies on the exchange of fluids between host and tumor, I. A method growing "tissue-isolated" tumors in laboratory animals, *J. Nat. Cancer Inst.* **27**, 679–693, 1961. (b) Gullino, P. M., and Grantham, F. H., Studies on the exchange of fluids between host and other tumor, II. The blood flow of hepatomas and other tumors in rats and mice, *J. Nat. Cancer Inst.* **27**, 1465–1491, 1961.

141. Song, C. W., Payne, J. T., and Levitt, S. H., Vascularity and blood flow in X-irradiated Walker carcinoma 256 of rats, *Radiology* **104**, 693–697, 1972.

142. Kjartansson, I. E., Tumour circulation: An experimental study in the rat with a comparison of different methods for estimation of tumour blood flow, *Acta Chir. Scand. Suppl.* **471**, 1–74, 1976.

143. Dickson, J. A., Calderwood, S. K., Temperature range and selective sensitivity of tumors to hyperthermia: A critical review, *Ann. N. Y. Acad. Sci.* **335**, 180–205, 1980.

144. Allen, N., Goldman, H., Gordon, W. A., and Clendenon, N. R., Topographic blood flow in experimental nervous system tumors and surrounding tissues, *Trans. Am. Neurol. Assoc.* **100**, 157, 1975.

145. Takacs, L., Debreczeni, L. A., and Farsang, C., Circulation in rats with Guerin carcinoma, *J. Appl. Physiol.* **38**, 696, 1975.

146. Moller, U., and Bojsen, J., Temperature and blood flow measurements in and around 12-dimethylbenz(a) anthracene-induced tumours and Walker 256 Carcinomas in rats, *Cancer Res.* **35**, 3116–3121, 1975.

147. Endrich, B., Intaglietta, M., Reinhold, H. S., and Gross, J. F., Tissue perfusion inhomogeneity during early tumor growth in rats, *J. Nat. Cancer Inst.* **62**, 387–395, 1979.

148. Rogers, W., Tissue and blood flow in transplantable tumors of the mouse and hamster, *Diss. Abstr. Int. B* **28**, 5185, 1968.

149. Robert, J., Martin, J., and Burg, C., Evolution de la vascularisation d'une tumeur isolgue solide de la souris au cours de sa croissance, *Strahlentherapie* **133**, 621, 1967.

150. Peterson, H.-I., Appelgren, K. L., Rudenstam, C.-M., and Lewis, D. H., Studies on the circulation of experimental tumours, I, Effect of induced fibrinolysis and antifibrinolysis on capillary blood flow and the capillary transport function of two experimenal tumours in the mouse, *Eur. J. Cancer* **5**, 91, 1969.

151. Kallman, R. F., de Nardo, G. L., and Stasch, M. J., Blood flow in irradiated mouse sarcoma as determined by the clearance of Xenon-133, *Cancer Res.* **32**, 483, 1972.

152. Gump, F. E., and White, R. L., Determination of regional tumor blood flow by krypton-85, *Cancer (Philadelphia)* **21**, 871–875, 1969.

153. Mantyla, M. J., Regional blood flow in human tumors, *Cancer Res.* **39**, 2304–2306, 1979.

154. Plengvanmit, U., Suwanik, R., Chearanai, O., Intrasupt, S., Sutayavanich, S., Kalayasiri, C., and Viranuvatti, V., Regional hepatic blood flow studied by intrahepatic injection of xenon in normals and in patients with primary carcinoma of the liver, with particular reference to the effect of hepatic artery ligation, *Aust. N.Z.J. Med.* **1**, 44, 1972.

155. Gullino, P. M., Influence of blood supply on thermal properties and metabolism of mammary carcinomas., *Ann. N. Y. Acad. Sci.* **335**, 1–21, 1980.

156. Song, C. W., Effect of hyperthermia on vascular functions of normal tissues and experimental tumours, *J. Nat. Cancer Inst.* **60**, 711–713, 1978.

157. Song, C. W., *Proc. 3d Int. Symp. Cancer Therapy by Hyperthermia, Drugs, and Radiation*, June 22–26, 1980.

158. Bicher, H. I., Sandhu, T. S., Vaupel, P., Hetzel, F. W., Physiological mechanisms of action of localized microwave hyperthermia, *Proc. 3d Int. Symp. Cancer Therapy Hyperthermia, Drugs, and Radiation*, June 20–22, 1980, p. 118.

159. Robinson, J. E., McCulloch, D., McCready, W. A., Blood perfusion of murine tumor at normal and hyperthermal temperatures, *Proc. 3d Int. Symp. Cancer Therapy by Hyperthermia, Drugs, and Radiation*, June 20–22, 1980, p. 77.

160. Nussbaum, G. H., Emami, B., Tenhaken, R. K., Hahn, N., and Hughes, W. L., Changes in tumor blood flow following hyperthermia, *Int. J. Rad. Oncol. Biol. Physics*, in press.

161. Reinhold, H. S., Blachiewicz, B., Berg-Blok, A., Decrease in tumor microcirculation during hyperthermia, *Proc. 2d Int. Symp. Cancer Therapy Hyperthermia and Radiation* (Essen: Urban and Schwarzenberg, 1978), pp. 231–232.

162. Reinhold, H. S., and van den Berg-Blok, A., Enhancement of thermal damage to sandwich tumours by additional treatment, *Proc. 3d Int. Symp. Cancer Therapy Hyperthermia, Drugs, and Radiation,* June 20–22, 1980, p. 96.

163. Eddy, H. A., Alterations in tumor microvasculature during hyperthermia, *Radiology* 137, 515–522, 1980.

164. Eddy, H. A., Sutherland, R. M., Chielewski, R., Tumor microvascular response: Hyperthermia, drug, radiation combinations, *Proc. 3d Int. Symp. Cancer Therapy Hyperthermia, Drugs, and Radiation,* 1980.

165. Westermark, N., The effect of heat upon rat tumors, *Scan. Arch. Physiol.* 52, 257–322, 1927.

166. Dickson, J. A., The effects of hyperthermia in animal tumour systems, *Rec. Results Cancer Res.* 59, 43–111, 1977.

167. Schwan, H. P., Heating of fat–muscle layers by electromagnetic and ultrasonic diathermy, *Proc. AIEE* 72, 483–487, 1953.

168. Schwan, H. P., and Piersol, G. M., The absorption of electromagnetic energy in body tissues, *Am. J. Phys. Med.* 33, 371, 1954.

169. Schwan, H. P., and Piersol, G. M., The absorption of electromagnetic energy in body tissues, I. Biophysical aspects, II. Physiological aspects, *Am. J. Phys. Med.* 33, 371–404, 1954; 34, 425–448, 1955.

170. Schwan, H. P., and Li, K., Hazards due to total body irradiation by radar, *Proc. IRE* 44, 1572–1581, 1956.

171. Johnson, C. C., and Guy, A. W., Nonionizing electromagnetic wave effects in biological materials and systems, *Proc. IEEE* 60, 692–717, 1972.

172. Johnson, C. C., et al., Electromagnetic power absorption in anisotropic tissue media, *IEEE Trans. Microw. Theor. Tech.* MTT-23, 52932, 1975-a.

173. Massoudi, H., Durney, C. H., and Johnson, C. C., A geometrical optics and an exact solution for internal fields in the energy absorption by a cylindrical model of man irradiated by an electromagnetic plane wave, *Radio Sci.* 14, 35–42, 1979.

174. Ho, H., Guy, A. W., Sigelmann, R. A., and Lehman, J. F., Electromagnetic heating patterns in circular cylindrical models of human tissue, *Proc. 8th Int. Conf. Med. & Biol. Eng.,* July 1969.

175. Ho, H. S., Dose rate distribution in triple-layered dielectric cylinder with irregular cross-section irradiated by planewave sources, *J. Microw. Power* 10, 421–431, 1975.

176. Anne, A., Scattering and absorption of microwaves by dissipative dielectric objects: the biological significance and hazard to mankind (Ph.D. diss., University of Pennsylvania, Philadelphia, 1963).

177. Kritikos, H. N., and Schwan, H. P., Hot spots generated in conducting spheres by EM waves and biological implications, *IEEE Trans. Biomed. Eng.* BME-19, 53–58, 1972.

178. Kritikos, H. N., and Schwan, H. P., The distribution of heating potential inside lossy spheres, *IEEE Trans. Biomed. Eng.* BME-22, 457–463, 1975.

179. Lin, J. C., Guy, A. W., and Johnson, C. C., Power deposition in a spherical model of man exposed to 1–20 MHz electromagnetic fields, *IEEE Trans. Microw. Theor. Tech.* MTT-21, 791–797, 1973.

180. Lin, J. C., *et al.,* Microwave selective brain heating, *J. Microw. Power* 8, 275–286, 1973.

181. Ho, H. S., Contrast of dose distribution in phantom leads due to aperture and planewave sources, *Ann. N. Y. Acad. Sci.* 247, 454–472, 1975.

182. Hand, J. W., Microwave heating patterns in simple tissue models, *Phys. Med. Biol.* 22, 981–987, 1977.

183. Kritikos, H. N., and Schwan, H. P., Formation of hot spots in multilayer spheres, *IEEE Trans. Biomed. Eng.* BME-23, 168–172, 1976.

184. Shapiro, A. R., Lutomirski, R. F., and Yura, H. T., Induced heating within a cranial structure irradiated by an electromagnetic plane wave, *IEEE Trans. Microw. Theor. Tech.* MTT-19, 187–196, 1971.

185. Joines, W. T., and Spiegel, R. J., Resonance absorption of microwaves by the human skull, *IEEE Trans. Biomed. Eng.* **BME-21**, 46–48, 1974.

186. Weil, C. M., Absorption characteristics of multilayered sphere models exposed to UHF/Microwave radiation, *IEEE Trans. Biomed. Eng.* **BME-22**, 468–476, 1975.

187. Neuder, S. M., *et al.*, Microwave power density absorption in a spherical multilayered model of the heat, in *Biological Effects of Electromagnetic Waves*, vol. 2, Johnson, C. C., and Shore, M. L., eds. (Washington, D.C.: HEW pub. no. FDA-77-8011, Dec. 1976), pp. 199–210.

188. Johnson, C. C., Durney, C. H., and Massoudi, H., Long-wavelength electromagnetic power absorption in prolate spheroidal models of man and animals, *IEEE Trans. Microw. Theor. Tech.* **MTT-23**, 739–747, 1975.

189. Gandhi, O. P., Hagerman, M. J., and D'Andrea, J. A., Partbody and multibody effects on absorption of radiofrequency electromagnetic energy by animals and by models of man, *Radio Sci.* **14**, 15–21, 1979.

190. Wu, C., and Lin, J. C., Absorption and scattering of electromagnetic waves by prolate spheroidal model of biological structures, in *Proc. 3d Int. Symp. Cancer Therapy Hyperthermia, Drugs, and Radiation*, June 20–22, 1980, pp. 142–145.

191. Barber, P. W., Electromagnetic power deposition in prolate spheroidal models of man and animals at resonance, *IEEE Trans. Biomed. Eng.* **BME-24**, 513–521, 1977.

192. Rowlinson, G. J., and Barber, P. W., Absorption of higher frequency RF energy in biological models: Calculations based on geometrical optics, *Radio Sci.* **14**, 43–50, 1979.

193. Massoudi, H., Durney, C. H., and Johnson, C. C., Long wavelength electromagnetic power absorption in ellipsoidal models of man and animals, *IEEE Trans. Microw. Theor. Tech.* **MTT-25**, 47–52, 1977.

194. Wu, T. K., and Tsai, L. L., Electromagnetic fields induced inside arbitrary cylinders of biological tissues, *IEEE Trans. Microw. Theor. Tech.* **MTT-25**, 61–65, 1977.

195. Neuder, S. M., and Meijer, P. H. E., Finite-element variational calculus approach to the determination of electromagnetic fields in irregular geometry, in *Biological Effects of Electromagnetic Waves*, vol. 2, Johnson, C. C., and Shore, M. L., eds. (Washington, D.C.: HEW pub. no. FDA-77-8011, Dec. 1976), pp. 193–198.

196. Livesay, D. E., and Chen, K., Electromagnetic fields induced inside arbitrary shaped biological bodies, *IEEE Trans. Microw. Theor. Tech.* **MTT-22**, 1273–1280, 1974.

197. Guru, B. S., and Chen, K. M., Experimental and theoretical studies on electromagnetic fields induced inside finite biological bodies, *IEEE Trans. Microw. Theor. Tech.* **MTT-24**, 433–440, 1976.

198. Rukspollmuang, S., and Chen, K. M., Heating of spherical versus realistic models of human and infrahuman heads by electromagnetic waves, *Radio Sci.* **14**, 51–62, 1979.

199. Chen, K. M., and Guru, B. S., Internal EM field and absorbed power density in human torsos induced by 1–500 MHz EM waves, *IEEE Trans. Microw. Theor. Tech.* **MTT-25**, 746–756, 1977.

200. Chen, K. M., and Guru, B. S., Induced EM fields inside human bodies irradiated by EM waves of up to 500 MHz, *J. Microwave Power* **12**, 173–183, 1977.

201. Hagerman, M. J., and Gandhi, O. P., Numerical calculation of electromagnetic energy deposition in models of man with grounding and reflector effects, *Radio Sci.* **14**, 23–29, 1979.

202. Guy, A. W., and Lehman, J. F., On the determination of an optimum microwave diathermy frequency for a direct contact applicator, *IEEE Trans. Biomed. Eng.* **BME-13**, 76–87, 1966.

203. Guy, A. W., Electromagnetic fields and relative heating patterns due to a rectangular aperture source in direct contact with bilayered biological tissue, *IEEE Trans. Microwave Theory Tech.* **MTT-19**, 214–223, 1971.

204. Ho, H. S., Guy, A. W., Sigelmann, R. A., and Lehman, J. F., Microwave heating of simulated human limbs by aperture sources, *IEEE Trans. Microwave Theory Tech.* **MTT-19**, 224–231, 1971.

205. Hizal, A., and Baykal, Y. K., Heat potential distribution in an inhomogeneous spherical model of a cranial structure exposed to microwaves due to loop or dipole antennas, *IEEE Trans. Microwave Theory Tech.* **MTT-26**, 607–612, 1978.

206. Goss, S. A., Johnston, R. L., and Dunn, F., Comprehensive compilation of empirical ultrasonic properties of mammalian tissues. *J. Acoust. Soc. Am.* **64**(2), 423–457, August 1978.
207. Hahn, G. M., Kernahan, P., Martinez, A., Pounds, D., and Prionas, S., Some heat transfer problems associated with heating by ultrasound, microwaves, or radiofrequency, *Ann. N. Y. Acad. Sci.* **335**, 327–346, 1980.
208. Schepps, J. L., and Foster, K. R., The UHF and microwave dielectric properties of normal and tumor tissue, *Phys. Med. Biol.* **25**, 1149–1159, 1980.
209. Peloso, R., Tuma, D. T., and Jain, R. K., Dielectric properties of solid tumors during normothermia and hyperthermia, *IEEE Trans. Biomed. Eng.*, in press.

Chapter 17

TEMPERATURE FIELDS AND LESION SIZES IN ELECTROSURGERY AND INDUCTION THERMOCOAGULATION

Avraham Shitzer

1. INTRODUCTION

Several ingenious methods are currently used for tissue removal and destruction, including techniques based on mechanical, chemical, and physical principles. The usual mechanical method involves a cutting knife—the "scalpel." The chemical method is based on the introduction of suitable chemical agents, systemically or locally, to achieve the desired effect. An example is thrombolysis by streptokinase infusion. The physical method, on the other hand, entails a variety of techniques, such as nonionizing radiation, ultrasonic waves, laser beam irradiation, cryosurgery, and electrosurgery. The choice of the method to be applied depends on the nature of the problem and the experience of the surgeon. The end effector of these techniques is the same: the input or removal of heat. This chapter focuses on the use of tissue heating due to electricity and radiofrequency radiation as a lesion-producing agent. Following a physical description of pertinent methods of tissue heating, the production of lesions in tissue by these methods will be quantified by the bioheat transfer analysis.

When energy is deposited in the tissue faster than it can be dissipated, the inevitable result is tissue heating, manifested as an elevation of temperature. Prolonged periods of elevated temperature may cause irreversible damage to the tissue. Controlled deposition of heat in the tissue, particularly in small, well-defined regions, is an intricate but necessary manipulation to avoid unnecessary damage to normal tissue. It may be achieved by various means, e.g., application of heated probes, passage of electrical current through the tissue, and exposure to radiofrequency radiation. The analysis of the latter two methods is the subject of this chapter.

Avraham Shitzer • Department of Mechanical Engineering, Technion, Israel Institute of Technology, Haifa, Israel.

Electrical current may be applied to the tissue in either the direct or the alternating methods. Direct current, when passing through the tissue, causes Joule heating, but also leads to the electrolytic dissociation of the tissue fluids, both of which eventually cause cell death.[1,2] The passage of high-frequency alternating current, on the other hand, causes Joule heating of the tissue as the primary phenomenon. Some heat is transferred to tissue from the electric arc between the active electrode and the tissue. No nerve or muscle stimulation occurs when the proper frequency range is employed.[3] The ac- or dc-induced high tissue temperature causes protein denaturation, blood and tissue coagulation, and sometimes even evaporation of the fluids with the subsequent destruction of the tissue.[1,4–6]

These methods of electrosurgery may be classified according to the mode of application and the power involved, as follows:

(a) *Electrocautery.* This technique involves scarring or burning tissues by means of heat. In its medical application, tissue heating is obtained by application of a resistively heated probe. Current does not flow in the tissues. Electrocautery is typically employed to destroy solitary tumors. Point cautery is also an effective hemostatic technique. The healing of electrocautery crusts is slow, but scarring is minimal and tends to reduce with time.[7]

(b) *Electrosection or electrotomy.* This technique is employed primarily for cutting tissues. The cutting action is achieved by intense heating of the tissue, which denatures connective tissue and also creates intercellular steam that causes the tissue's disruption. When high frequency, undamped current is applied, considerable hemorrhage may be caused. This problem may be alleviated by applying a moderately damped waveform as in the Bovie machine. Such an application results in cutting accompanied by hemostasis and dehydration.[8] Electrosection is always applied in the biterminal mode, i.e., using both an active and a dispersive (indifferent) electrode. The active electrode is equipped with a suitable conducting small tip, sometimes in the shape of a thin wire, held by an insulated handle. The dispersive electrode is usually made of a large conducting plate, which is brought into contact with remote, high-conductivity tissue region.

(c) *Electrocoagulation.* This technique is essentially similar to electrosection. It is applied in the biterminal mode and achieves blood and tissue coagulation, and eventual charring, by the deposition of high-energy fluxes in the tissue. The typical application of electrocoagulation, in contrast with electrosection, does not involve moving the active electrode. This technique therefore requires preplanning of the details of the heat application. Due to the biterminal mode, the flow of the electrical current in the tissue tends to follow a preferred path, an additional factor that should be taken into consideration.

Some investigators believe that electrocoagulation possesses specific beneficial advantages in carcinoma treatment. They believe it enhances the inherent natural immunological system of the tumor-bearing patient.[8] The coagulated and subsequently necrosed cancer cells are absorbed slowly and are regarded by the host as an antigen, which may elicit a specific antibody response. Electrocoagulation produces, they claim, a demonstrable immunologic effect that increases host resistance and may cause rejection of subsequent tumor challenges. However, the claim is not conclusive, since no definitive proof was presented in Ref. 8.

(d) *Electrodesiccation and electrofulguration.* Both of these techniques are used synonymously to indicate a monoterminal application.[7] The electrical capacitance of the tissue is employed to deposit energy without any preferred direction. Both of these techniques require a high-frequency, highly damped waveform. Both generally produce the most dependable hemostasis, the greatest precision in application, and the surest results.[8] Electrodesiccation usually involves inserting a needle electrode into the tissue. Electrofulguration relies on a spark action at the tissue surface and is generally applied to create superficial lesions by dehydration resulting from the spark. The main destructive effects of desiccation are the shriveling of the cells, their histological death, and thrombosis of smaller blood vessels. Histological death of the cells and thrombosis of the smaller blood vessels occur with little hemorrhage. Healing takes place with crust formation, mild exudation, and sloughing of the crust in 7–21 days. Even with extensive electrodesiccation, scars tend to be minimal and improve with time. Typical applications of electrodesiccation and electrofulguration are coagulation of sebaceous and epidermoid cysts, destruction of malignant tumors, destruction of skin tags and benign superficial lesions, e.g., keratoses and virus warts.[7]

Electrosurgical effects are produced by diathermy machines that operate at frequencies in the region 1–3 MHz and produce total output power in the range of 0.5–5 W.[10] The diathermy machine is essentially a vacuum tube or solid state oscillator. Vacuum tube oscillators generate an electric arc in the gap between internal metal plates; the Bovie machine is a typical embodiment. The frequency spectrum of the resulting potential waveform has been empirically adjusted to produce the desired arcing between the active electrode tip and the adjacent tissue. Solid-state oscillators provide arc-generating waveforms without the spark gap of the vacuum tube oscillator. The solid-state oscillators produce better controlled waveforms for cutting, coagulation, or the other applications previously mentioned. Diathermy machines employ a capacitance coupling for the biterminal and inductive coupling for monoterminal applications.

The electrodes employed vary in shape and size but typically measure a few millimeters in diameter and thickness. Selecting the design and application parameters, e.g., the duration of application, size and shape of the electrode, power, waveform, and voltage applied, remains more an art than a science at present. The medical practitioner wishing to apply this technique is therefore presented with empirical methods for proper selection and application of the apparatus and operating parameters.

Another method of producing thermal lesions is by induction thermocoagulation. In this technique, a region of the body into which a so-called "electroseed" has been implanted is exposed to a radiofrequency (rf) electromagnetic field. The electroseed is heated by absorption of a portion of the electromagnetic energy. This energy is transferred to the tissue in contact with the electroseed and causes tissue destruction by thermocoagulation. The created lesion is subsequently replaced by fibrous tissue to produce a scar. Within a few weeks, almost the entire lesion has been reduced by this process.[9,10] The "thermoseed" is not reduced by the process.

A point of concern in this method is the possible migration of the implanted thermoseed to undesired locations. However, one study demonstrated that the thermoseed was anchored firmly by the surrounding coagulated tissue and no displacement took place.[10] However it was also noted in this study that induction thermocoagulation caused secondary lesions. These lesions were caused by infarcts of adjoining tissue regions, whose blood supply was cut off by coagulated blood vessels.

Induction thermocoagulation employs the ferromagnetic skin effect. Accordingly, most of the induced current flow and heating are concentrated near the surface of the electroseed, with an exponential falloff toward the center. In early applications, 430 stainless steel was used.[9] When exposed to rf radiation, this material may attain excessively high temperatures that may be difficult to control. Consequently, new materials were developed to solve this problem.[10]

The basic idea is to develop a low–Curie point alloy that would "saturate" at a temperature compatible with tissue thermocoagulation. At this saturation temperature, a magnetic to nonmagnetic transformation occurs and absorption of energy by the electroseed ceases. Among the alloys that exhibit this characteristic and are also compatible with the tissue environment are nickel alloys with additions of either chromium, copper, or palladium. These low–Curie point alloys were termed thermoseeds.[10]

Thermoseeds have been employed in neurosurgery for the treatment of movement disorders (Parkinson's disease) and intractable pain. The thermoseed, typically in the shape of a small cylinder, is inserted into the tissue and hopefully causes minimal damage along the insertion track. After about a two-week recovery, exposure to rf radiation commences. Treatment is performed at a number of sittings, thus incrementing the extent of the thermal damage.

Table 1 summarizes some typical data that characterize the electrosurgical and thermoseed techniques. A comprehensive review of the state of the art of thermal "knives" is given in.[11] Operating parameters for electrosurgical apparatus are given in.[12]

2. ANALYSIS

The remainder of this chapter is devoted to the analysis and quantification of thermal "lesions" produced by the aforementioned methods. In all of these methods, a central question is the prediction of the irreversible damage and histological death of the tissue. A quick reference to the literature will reveal the variety of approaches used and the prevailing uncertainty. For instance, Overmyer et al. assume the threshold of tissue damage to be in the "neighborhood" of 45°C.[13] Aronow[14] and Cooper and Groff[15] consider 55°C as the threshold for irreversible tissue damage. These two conflicting figures

TABLE 1
Typical Applications and Operating Parameters of Electrosurgical Modalities

Technique	Applications	Typical operating parameters					Typical electrode size, shape, material
		Waveform	Electrodes	Voltage (V)	Power (W)	Frequency (MHz)	
Electrolysis	Desiccation of hair follicles, creation of small vascular lesions	dc	2	22–25	1/2–2 mA		2 mm diam., cylindrical
Electrocautery	Solitary verrucae, pyogenic granuloma, superficial benign tumors	dc or ac	1	6	2–10	0.525	0.025–2 mm diam. needle or wire, cylindrical, nichrome or platinum
Electrosection or electrotomy	Production of surgical lesions	Continuous or moderately damped ac	2	30–350 R.M.S.	15–150	2–10	
Electrocoagulation	Ablation of sebaceous and epidermoid cysts	Damped ac	2				
Electrodesiccation or electrofulguration	Keratoses, virus warts, destruction of malignant tumors	Highly damped ac	1	10,000–20,000			
Radiofrequency induction heating	Creation of brain lesions, movement disorders intractable pain	ac			5,000	0.610	1–1.6 mm diam., 3–10 mm long, cylindrical 430 stainless steel or nickel–palladium alloy

probably represent an oversimplified approach to the problem. On the other hand, Kach and Incropera[16] and Erez and Shitzer[17] adopted the tissue destruction model suggested by Henriques.[18] It should be noted that all of these approaches are apparently gross simplifications of the problem, inasmuch as details of subcellular damage mechanisms are ignored. Of the approaches just cited, the one suggested by Henriques appears better founded. According to Henriques, the thermal destruction of the tissue can be approximated by a rate process. He assumed that the kinetics of the destruction process in the living tissue is similar to any other physical and/or chemical process. Thus, the possible occurrence of the process depends on the total energy deposition in the involved tissue. If this energy content is less than a critical value, known as the activation energy, the process can not take place. If the energy content is greater than this value, the process will proceed.

Based on observations, Henriques assumed three possible mechanisms for tissue thermal injury: (1) thermal coagulation (denaturation) of proteins; (2) other possible alterations in metabolic processes; and (3) nonprotein, thermally induced alterations in the chemical characteristics of cells, e.g., changes in intracellular ion concentration. Henriques proposed a combined, single-rate process of the Arrhenius type to represent tissue thermal injury

$$\frac{d\Omega}{dt} = A \exp\left[-\frac{E}{R(T + 273)}\right] \tag{1}$$

where Ω represents an arbitrary tissue damage function, A is a numerical coefficient, E is the energy of activation, R is the universal gas constant, and T is tissue temperature in degress Celsius. Equation (1) can be integrated with respect to time to yield

$$\Omega = A \int_0^t \exp\left[-\frac{E}{R(T + 273)}\right] dt \tag{2}$$

When fitted to data obtained in his laboratory and assigning a value of $\Omega = 1$ for irreversibly destroyed tissue, Henriques found the values of $A = 3.1 \times 10^{98}$ sec^{-1} and $E = 150,000$ cal/mole with $R = 2$ cal/mole K.

These values are probably not general, and variations may be expected for different tissues. Moreover, tissue thermal damage may not be a single-rate process. Furthermore, incineration and ablation of tissue are not included in this model. Nevertheless, and in view of the lack of additional data, both Eq. (2) and the values of A and E derived by Henriques may be assumed as first approximations to predict tissue destruction by thermal means.

Several attempts have been made to analyze and quantify the effects of electrosurgery on tissue destruction.[13-23] In certain cases, the investigators were interested in only the development of the electrical field.[22-25] In other studies, both the electrical and thermal fields were considered.[13,16,17]

Honig analyzed the electrical field as applied to cutting (electrosection).[22] He described the cutting mechanism as a "physical rupturing of the tissue" due to "expansion of the evolving steam bubbles." Using the power density imparted to the tissue by the electric current as the main parameter, Honig estimated tissue temperature rise, assuming no boiling to occur. He concluded that tissue destruction due to elevated temperature alone would occur at, and around, the wire tip, provided it was moved fast enough. Little or no damage would occur at short distances away from the cutting tip.

Overmyer and co-workers concerned themselves with temperature distributions at the dispersive electrode site,[13] as undesired burns have been reported in tissue regions that are in contact with the dispersive electrode. Overmyer *et al.* studied this problem analytically and experimentally, employing both human subjects and surrogate materials. The investigators first calculated the electric potential distribution that developed in the tissue. From this information, the current density distribution in the tissue was obtained. The current density was subsequently assumed to represent a volumetric heating term in the bioheat equation. These investigators employed a numerical solution technique to show that "temperature gradients are most severe near the perimeter of the electrode." They concluded by stating that their predicted temperature fields are similar to those observed in human subjects.

In the following sections, more detailed analyses of three specific electrosurgical applications are presented.

2.1. Electrocoagulation

This analysis is applicable to electrosurgery with a monoterminal electrode. Aronow was among the first to analyze the temperature field in tissues through which a high frequency electrical current is flowing.[14] Subsequent works were presented by Van den Berg *et al.*[19] and Rutkin *et al.*[20] None of these works considered the effects of blood flow or the effect of the rf heating on the transient temperature field. These factors were considered by Cooper and Gengler,[21] who presented the results of a numerical solution on the effects of a small spherical electrode embedded in the tissue.

In 1972, Drabkin modeled the active electrode as a small sphere embedded in tissue and assumed a spherically symmetric homogeneous tissue mass with constant thermophysical properties.[26,27] The effects of blood flow on heat transport were not considered. The analytical solution obtained under these assumptions is apparently in error due to the incorrect employment of a mathematical transformation.[26]

Erez and Shitzer re-analyzed this problem, considering the effects of blood perfusion and metabolic heat generation.[17] The energy balance in the tissue was assumed to be described by the bioheat equation. Other assumptions included the following:

(a) The tissue is homogeneous and isotropic, with constant thermophysical and physiological properties. This assumption is made due to the lack of more detailed data and to facilitate a first-order analysis.

(b) The coagulating electrode is spherical, completely embedded, and in intimate contact with the tissue. There is no interfacial thermal resistance, and arcing in a gas phase region between electrode and tissue is ignored. Angular symmetry is assumed such that heat flows in the radial direction only.

(c) The tissue and electrode are at a uniform and equal temperature prior to the power application. This assumption would hold for internal tissues and may cause a slight deviation in results when superficial layers are analyzed.

(d) Blood perfusion rate is unaffected by the coagulation process. This severe and somewhat restrictive assumption implies that blood continues to perfuse the coagulated tissue, but does not collect in the extravascular space, and that neither blood nor blood vessels are damaged by the elevated temperature. In reality, some of the blood vessels would be heat denatured, thus cutting off perfusion to the dependent tissue region. A model of this somewhat idealized situation would require a division of the problem into two regions: one with no blood perfusion, and no convective cooling effects, and a non-coagulated region in which blood perfusion cooling is still in effect. These two regions would be governed by two different diffusion equations similar to those pertaining to cryosurgery.[29] However, the dividing line between the two regions may not be a clear cut one and would be rather difficult to define analytically. Based on these considerations and to facilitate a first analytical approximation of the problem, it was decided to assume uniform blood perfusion throughout the tissue, except for a very narrow region adjacent to the electrode, which does not alter the governing equation.

(e) Metabolic heat generation rate is constant and uniform and is unaffected by the coagulating process. As it turns out, the heat generation term may be omitted from the analysis without causing any significant errors.

(f) No boiling of tissue fluids occurs during the coagulation process. This simplifying assumption may not be valid, depending on the temperature attained. However, it was found difficult to predict the exact temperature and region in which boiling would occur. Consequently, it was decided to omit this phenomenon from the analysis.

(g) The tissue electrical impedance is modeled as a pure resistance, with no effects of frequency on resistivity.

(h) The thermal conductivity of the coagulating electrode is much larger than that of the tissue, $k_0 \gg k$.

(i) The heat of reaction for the hemostatic process is not included in the present analysis. This quantity of heat may affect the thermal behavior of the process and also cause the problem to become mathematically nonlinear, with an effect similar to that of latent heat of solidification (cf. Ref. 29).

Based on these assumptions, the monoterminal coagulation process is represented by the following model: the active spherical electrode is completely enveloped by an infinite tissue. The electric current spreads radially through the tissue; the current densities, at equal distances from the active electrode, are equal. The model geometry is shown schematically in Fig. 1. The problem is divided into two time domains: (1) the tissue heating phase, occurring for as long as the electrical current passes through the tissue, and (2) the tissue cooling phase, which commences as soon as the flow of the electrical current has ceased. These two phases are analyzed separately.

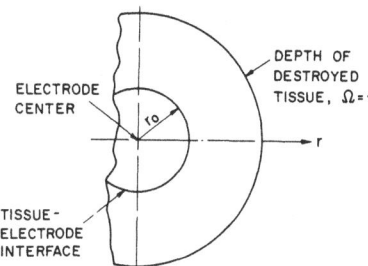

FIGURE 1
Schematic diagram of the system geometry (see Ref. 17).

2.1.1. Heating Phase

Based on the foregoing assumptions, the problem may be stated mathematically as follows:*

$$\frac{\partial T_h}{\partial t} = \frac{\alpha}{r^2} \frac{\partial}{\partial r} \left[r^2 \frac{\partial T_h}{\partial r} \right] + \frac{\rho_b c_b w_b}{\rho c} (T_a - T_h) + \frac{q}{\rho c} \tag{3}$$

where q represents all the evenly distributed heat sources in the tissue per unit volume and is a linear combination, in the present case, of two terms:

1. q_m, which is the metabolic heat generation rate per unit volume of tissue
2. P_0, which is the Joule heat produced per unit volume of tissue by the electrical current passing through the tissue. P_0 is given by[16]

$$P_0 = \frac{Pr_0}{8\pi r^4} (1 + \cos 2\omega t) \tag{4}$$

Recall that the frequency range (ω) in which coagulation equipment is operated is 1–3 MHz. Consequently, the rate of change of the periodic term in Eq. (4), when compared to the thermal inertia of the tissue, is so much larger that an average value might be assumed without causing any significant error. Accordingly, following Drabkins's analysis, Eq. (4) becomes

$$P_0 = \frac{Pr_0}{4\pi r^4} \tag{5}$$

With Eq. (5) substituted into it, Eq. (3) becomes

$$\frac{\partial T_h}{\partial t} = \frac{\alpha}{r^2} \frac{\partial}{\partial r} \left(r^2 \frac{\partial T_h}{\partial r} \right) + \frac{\rho_b c_b w_b}{\rho c} (T_a - T_h) + \frac{q_m}{\rho c_p} + \frac{Pr_0}{4\pi \rho c r^4} \tag{6}$$

Initial and boundary conditions for this model are based on the assumption suggested by Trezek and Cooper,[30] who presume that tissue is in thermal equilibrium as long as it has not been perturbed by an external factor.

* See Nomenclature.

Accordingly,

$$T_h(r, 0) = T_a + \frac{q_m}{\rho_b c_b w_b} \tag{7}$$

and

$$T_h(\infty, t) = T_a + \frac{q_m}{\rho_b c_b w_b} \tag{8}$$

The remaining boundary condition at the interface between the coagulating electrode and the tissue is governed by the thermal inertia of the electrode. Assuming the thermal conductivity of the electrode to be much larger than that of the tissue, foregoing assumption (h), the following expression is obtained:

$$\rho_0 c_0 \frac{4\pi r_0^3}{3} \frac{\partial T_h}{\partial t}(r_0, t) = 4\pi r_0^2 k \frac{\partial T_h}{\partial r}(r_0, t) \tag{9}$$

Equation (9) is a statement of the energy balance at the electrode–tissue interface assuming the electrode to be at a uniform temperature equal to that of the immediately adjacent tissue, assumption (b).

Introducing the following dimensionless parameters

$$x = \frac{r}{r_0}; \qquad \tau = \frac{\alpha t}{r_0^2}; \qquad \beta = \left(\frac{\rho_b c_b w_b r_0^2}{k}\right)^{1/2} r_0$$

$$m = \frac{\rho_0 c_0}{\rho c}; \qquad \theta = \frac{4\pi r_0 k}{P}\left[(T_a - T) + \frac{q_m}{\rho_b c_b w_b}\right] \tag{10}$$

yields the set of dimensionless equations

$$\frac{\partial \theta_h}{\partial \tau} = \frac{\partial^2 \theta_h}{\partial x^2} + \frac{2}{x}\frac{\partial \theta_h}{\partial x} + \beta^2 \theta_h + \frac{1}{x^4} \tag{11}$$

$$\theta_h(x, 0) = 0 \tag{12}$$

$$\theta_h(\infty, \tau) = 0 \tag{13}$$

$$\frac{m}{3}\frac{\partial \theta_h}{\partial \tau}(1, \tau) = \frac{\partial \theta_h}{\partial x}(1, \tau) \tag{14}$$

Equations (11)–(14) were solved by applying a Laplace transformation and a second transformation given by

$$U(x) = x\bar{\theta}_h(x, s) \tag{15}$$

to yield

$$\frac{dU}{dx^2} - (s + \beta^2)U = -\frac{1}{2sx^3} \tag{16}$$

$$U(\infty) = 0 \tag{17}$$

$$3\frac{dU}{dx}(1) - (ms + 3)U(1) = 0 \tag{18}$$

The solution of Eqs. (16)–(18) is given by

$$\bar{\theta}_h(x, s) = \frac{1}{2sx}\int_{u=1}^{\infty}\left\{\frac{1}{\nu}[\exp(-\xi\nu) - \exp(-\lambda\nu)]\right.$$
$$\left. + \frac{6}{ms + 3(\nu + 1)}\exp(-\lambda\nu)\right\}\frac{du}{u^3} \tag{19}$$

where

$$\nu = (s + \beta^2)^{1/2} \tag{20}$$

$$\xi = |u - x| \tag{21}$$

$$\lambda = u + x - 2 \tag{22}$$

Performing the inverse transformation indicated in Eq. (19) yields the solution to the heating phase

$$\theta_h(x, \tau) = \frac{1}{2x}\int_{u=1}^{\infty}\left\{\frac{1}{2\beta}\left[\exp(-\beta\xi)\,\mathrm{erfc}\,(\xi/2\sqrt{\tau} - \beta\sqrt{\tau})\right.\right.$$

$$\left. - \exp(\beta\xi)\,\mathrm{erfc}\left(\frac{\xi}{2\sqrt{\tau}} + \beta\sqrt{\tau}\right)\right]$$

$$+ \frac{1 + \beta}{1 - \beta}\exp(\beta\lambda)\,\mathrm{erfc}\left(\frac{\lambda}{2\sqrt{\tau}} + \beta\sqrt{\tau}\right)$$

$$\left. + \frac{1 - \beta}{1 + \beta}\exp(-\beta\lambda)\,\mathrm{erfc}\left(\frac{\lambda}{2\sqrt{\tau}} - \beta\sqrt{\tau}\right)\right\}$$

$$+ \frac{1}{\gamma}\left\{\frac{\delta - \gamma}{1 + \gamma - \delta}\exp[(\delta + \gamma)\lambda - 2\delta\tau(1 + \gamma - \delta)]\right.$$

$$\times \mathrm{erfc}\left[\frac{\lambda}{2\sqrt{\tau}} + (\delta - \gamma)\sqrt{\tau}\right]\frac{\delta + \gamma}{1 - \gamma - \delta}\exp[-(\delta + \gamma)\lambda$$

$$\left. + 2\delta\tau(1 + \gamma + \delta)]\,\mathrm{erfc}\left[\frac{\lambda}{2\sqrt{\tau}} - (\delta + \gamma)\sqrt{\tau}\right]\right\}\frac{du}{u^3} \tag{23}$$

where

$$\gamma = (\delta^2 - 2\delta + \beta^2)^{1/2} \tag{24}$$

$$\delta = \frac{3}{2m} \tag{25}$$

The steady-state tissue temperature for the heating phase would be obtained a long time after application of the coagulating process, i.e., $\tau \to \infty$. The temperature distribution for this case might be directly obtained from the transformed solution by employing the following property of the Laplace transformation

$$\lim_{\tau \to \infty} F(x, \tau) = \lim_{s \to 0} s\bar{F}(x, s) \tag{26}$$

where $\bar{F}(x, s)$ denotes the Laplace transformation of $F(x, \tau)$. Applying Eq. (26) to Eq. (19) gives

$$\theta_h^s(x) = \frac{1}{2x\beta} \int_{u=1}^{\infty} \exp(-\beta\xi) - g(\beta) \exp(-\beta\lambda) \frac{du}{u^3} \tag{27}$$

where

$$g(\beta) = \begin{cases} \dfrac{1 + \beta}{1 - \beta}; & m \text{ finite} \\ 1; & m \text{ infinite} \end{cases} \tag{28}$$

Of particular interest to the present analysis is the case when there is no blood perfusion in the tissue. This case would correspond to the *in vitro* steady-state condition or, more generally, to the solution of a general heat conduction equation (e.g., Ref. 31). The expression describing this condition is obtained by taking the limit of Eq. (19) when $\beta \to 0$

$$\theta_h^s(x) = \left. \begin{cases} \dfrac{1}{x}\left(1 - \dfrac{1}{2x}\right); & m \text{ finite} \\ \dfrac{1}{2x}\left(1 - \dfrac{1}{x}\right); & m \text{ infinite} \end{cases} \right\} \quad \text{no blood perfusion} \tag{29}$$

from which the maximum tissue temperature due to the coagulating process is calculated

$$\theta_{h,\max}^s = \left. \begin{cases} \frac{1}{2} \text{ at } x = 1; \ m \text{ finite} \\ \frac{1}{8} \text{ at } x = 2; \ m \text{ infinite} \end{cases} \right\}; \quad \text{no blood perfusion} \tag{30}$$

These steady-state results may be verified by directly solving the steady-state heat conduction problem without the heat generation or blood perfusion terms, i.e., $q_m \rightarrow 0$ and $\dot{m}_b \rightarrow 0$.

2.1.2. Cooling Phase

The governing equation for the cooling phase differs from Eq. (11) in that the term representing the heating due to the passage of electrical current through the tissue, i.e., $1/x^4$, is omitted. The initial condition for this phase corresponds to the steady state achieved at the end of the heating process, which is obtained after a lengthy period of application. Consequently,

$$\theta_c(x, 0) = \theta_h^s(x) \tag{31}$$

where $\theta_h^s(x)$ is given by Eq. (27). Boundary conditions for this phase are given in Eqs. (13) and (14). Substituting the following transformation into the governing equation,

$$V(x, \tau) = \theta_h^s(x) - \theta_c(x, \tau) \tag{32}$$

yields

$$\frac{\partial V}{\partial \tau} = \frac{\partial^2 V}{\partial x^2} + \frac{2}{x}\frac{\partial V}{\partial x} + \beta^2 V + \frac{1}{x^4} \tag{33}$$

with the initial and boundary conditions

$$V(x, 0) = 0 \tag{34}$$

$$V(\infty, \tau) = 0 \tag{35}$$

$$\frac{m}{3}\frac{\partial V}{\partial \tau}(1, \tau) = \frac{\partial V}{\partial x}(1, \tau) \tag{36}$$

The solution of Eqs. (33)–(36) is given by Eq. (23); The expression for $\theta_c(x, t)$ may be readily obtained from Eqs. (23) and (31).

The effects of the various parameters on the tissue temperature distribution and extent of destruction due to the coagulating process can now be studied. Physical and physiological values of the parameters employed are listed in Table 2. Numerical evaluation of Eq. (23) was performed on a CDC Cyber computer. Integration was done by employing a quadratic numerical integration technique based on the Hermite–Laguerre polynomials, the degree of which is less than or equal to 36. Convergence problems were encountered for values of β greater than five. The absolute value of the error due to the numerical integration for the range of $\beta < 2$ and $x < 15$ was always less than 10^{-5}.

Numerical results indicate that the metabolic heat generation term has a negligible effect on tissue temperature and extent of destruction during

TABLE 2
Tissue and Coagulating Electrode Data[a]

(a) Tissue physical and physiological parameters	
Tissue specific heat, c_p (kJ/kg °C)	3.72
Blood specific heat, c_p (kJ/kg °C)	3.64
Tissue thermal conductivity, k (W/m °C)	0.46
Tissue thermal diffusivity, (m²/s × 10⁷)	1.1
Arterial temperature, T_a (°C)	37
Metabolic heat generation, $\dot{q}_m$ (W/m³)	240
Tissue density, ρ (kg/m³)	1050
Tissue electrical resistivity, ρ_r (Ω cm)	200

(b) Ranges of values of various parameters employed	
Parameter	Range
Electrode radius, r_0 (mm)	0.5–2.0
Dimensionless time, $\tau = \alpha t/r_0^2$	0–100
Blood perfusion rate, $\rho_b w_b$ (kg/m³s)	0–10
Dimensionless blood perfusion, $\beta = \rho_b c_b w_b r_0^2/k$	0–1.2
Dimensionless electrode thermal inertia, $m = \rho_0 c_0/\rho_c$	0–1.0
Electrical power, P (W)	0.3–5
Ratio of metabolic heat to heat transported by blood perfusion, $q_m/w_b c_b$ (°C)	0–0.006

[a] From Ref. 17.

coagulation. This result is in accordance with Kach and Incropera,[16] and the metabolic heat generation term is therefore omitted from the discussion.

First, the *in vitro* behavior of the model is studied. This case, represented by setting blood perfusion equal to zero, indicates what might be expected to occur in a tissue undergoing coagulation while blood flow to it is otherwise occluded, as with vessel cross clamping.

Temperature variations during the heating phase in the "dead" tissue at three different depths are shown in Fig. 2. This figure is plotted for a coagulating electrode having an infinite thermal inertia, i.e., $m \to \infty$; also plotted in this figure are the results obtained by Drabkin.[26] As is seen, tissue temperatures increase faster in regions close to the tissue–electrode interface for the first period of heating. After a certain time, $\tau = 2$ in this case, tissue temperatures

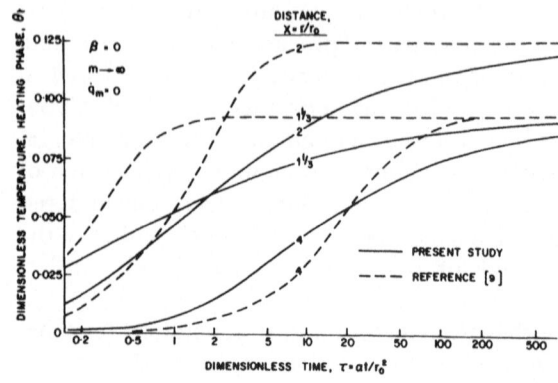

FIGURE 2

In vitro temperature variations in the tissue during the heating phase for an electrode with infinite thermal inertia (see Ref. 17).

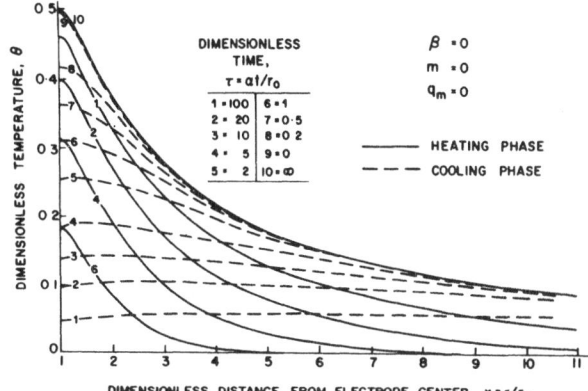

FIGURE 3

In vitro temperature distributions in the tissue during the heating and cooling off phases for an electrode with zero thermal inertia (see Ref. 17).

at larger distances, $x = 2$ in this case, increase faster and attain higher values. The results plotted by Drabkin do not coincide with the present ones and do not tend to equality for large values of τ. This discrepancy is apparently due to the incorrect mathematical transformation employed in Ref. 26, as discussed in the foregoing.

Tissue heating and cooling for an electrode with no thermal inertia is shown in Fig. 3. It is seen that tissue temperatures increase quite rapidly in regions adjacent to the electrode and may reach some 35% of the maximum value at $\tau = 1$, which implies about 10 sec for a 1-mm radius electrode. As time progresses, temperatures continue to increase until the maximum is reached. During the cooling phase, temperatures at distances $x > 9$ remain almost unaffected for the first period, $\tau < 10$, while a significant decrease is occurring next to the electrode. At $\tau = 20$, which implies about 200 sec for a 1-mm radius electrode, the temperature of the tissue is essentially uniform but continues to decrease gradually.

In Fig. 4, the effect of the electrode thermal inertia on the nonperfused tissue temperature distribution is examined. It is noted that for a range of values, i.e., $m = 0$–100, very little effect on tissue temperature is noticeable. The three curves shown almost coincide, and all differ quite substantially from the curve drawn for $m \to \infty$. Comparison of these two groups of curves reveals the advantage of using an electrode with as high a thermal inertia as possible for tissue destruction. This is due to the more moderate and evenly distributed temperatures that would be obtained for this case. However, as indicated in Table 2, the practical range of values of m for the commonly employed electrodes made of stainless steel or gold is $m = 0.5$–1. Thus, the phenomenon of excessive temperatures in regions adjacent to the electrode may not be avoided by a proper selection of the coagulation electrode alone. Furthermore, as a good first approximation, the electrode may be regarded for all practical purposes as having zero thermal inertia.

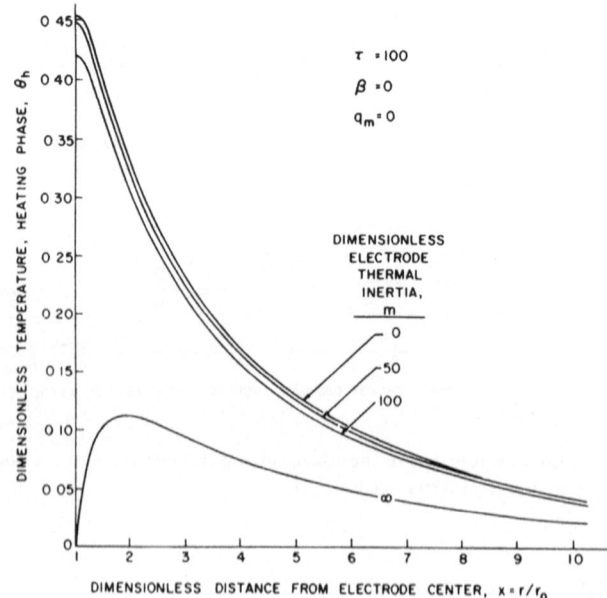

FIGURE 4
Effect of electrode thermal inertia on the *in vitro* temperature distributions in the tissue during the heating phase (see Ref. 17).

Attention is now turned to the behavior of the tissue in which blood perfusion exists. Figure 5 shows the effects of various dimensionless blood perfusion rates on the temperature distribution in the tissue during the heating phase. Curves are plotted for an electrode having an infinite thermal inertia and for two times, $\tau = 1$ and $\tau = 100$. It is seen that tissue temperature elevations are drastically suppressed by the cooling effect of blood flow through the tissue. Furthermore, an almost steady-state temperature distribution is obtained much faster, the higher the value of the blood perfusion parameter. For a normal tissue and typical electrode dimensions, the value of the blood perfusion parameter is of the order of $\beta = 0.5$. For this value,

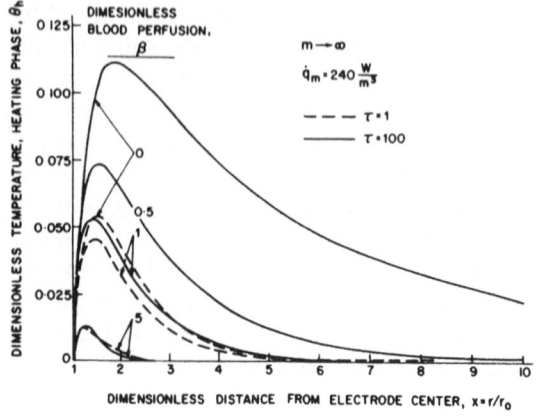

FIGURE 5
Effect of blood perfusion parameter on the tissue temperature distributions at $\tau = 1$ and $\tau = 100$ for an electrode with infinite thermal inertia (see Ref. 17).

the steady state is attained at about $\tau = 5$, with the maximum temperature being only about 60% of the maximum value for a nonperfused tissue. To illuminate the results of this study and to qualify them for the clinician, the extent of damage caused in the tissue is now calculated. The thermal damage is assumed to be a single rate process, Eq. (2). This function is divided into three ranges[18]:

1. $\Omega < 0.53$—tissue damage is reversible, i.e., healing will occur in due time.
2. $0.53 < \Omega < 1.0$—some tissue damage is irreversible.
3. $\Omega < 1.0$—irreversible damage occurs for all tissue reaching the threshold temperature.

Equation (2) was integrated employing the same numerical technique as indicated in the foregoing. The computation turned out to be quite time consuming, with about 0.5 CPUs required for each double integration. Calculation was carried out until the value $\Omega = 1$ was attained. Tissue regions for which this value is obtained are regarded as destroyed irreversibly, and the smallest depth for which $\Omega = 1$ is termed "depth of destruction."

Figure 6 shows the effect of the blood perfusion parameter on the depth of destruction as a function of electrode power. It is seen that blood flow effects are quite pronounced; the higher the perfusion rate, the smaller the depth of destruction. For a richly perfused tissue, e.g., liver for which $\beta \approx 1$, the power required to produce a 3-mm-deep lesion after 50 sec of application is about five times larger than that required to produce the same damage in epidermal tissue for which $\beta = 0.1$. Further results indicate that for an identical case, a 200-sec application will result in a maximum depth of destruction of about 4 mm for a richly perfused tissue. This maximum depth would be obtained at an electrical power of 15 W; the tissue then becomes insensitive to further increases in power. No maximum depth of destruction is observed for a poorly perfused tissue after 200 sec of application; the damaged region continues to grow with an increase in the electrical power.

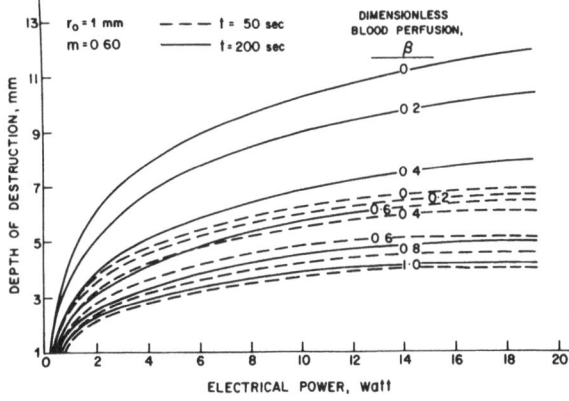

FIGURE 6
Depth of tissue destruction ($\Omega = 1$) vs. electrode power and blood perfusion rate for a 1-mm radius gold electrode ($m = 0.60$) (see Ref. 17).

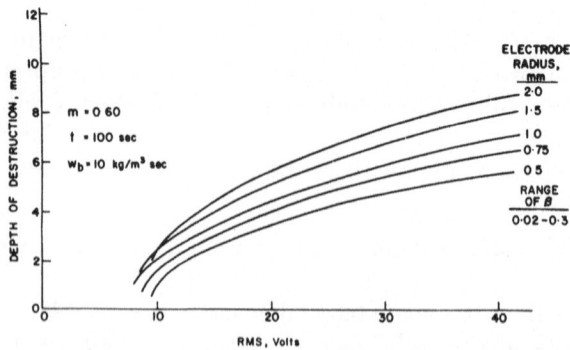

FIGURE 7

Depth of tissue destruction ($\Omega = 1$) vs. RMS voltage and electrode radius for a gold electrode ($m = 0.60$) (see Ref. 17).

It should be noted that as the electrical power of the electrode is increased, tissue temperature levels are also elevated. Thus, if a 4 mm depth of destruction is required for epidermal tissue and a 1-mm radius gold electrode is applied, 2.5 W would cause a 160°C temperature at the tissue–electrode interface after 50 sec of application. If the time of application is increased to 200 sec, the power required is reduced to 1.1 W, and the maximum temperature is only 106°C. This observation leads to the conclusion that it would be more advantageous to employ less power, applied for a longer duration, as is also suggested by Drabkin.[26] Another practical way of achieving a similar result is to apply pulses of short duration. This method, however, is not included in the present analysis. In addition, when the desired depth of destruction is smaller, e.g., 2–3 mm, even lower power levels would have to be employed, with a consequent lowered tissue temperature.

The depth of destruction is plotted as a function of the electrode radius and the root mean square (RMS) voltage applied to the electrode. Results are shown in Fig. 7 for an electrode made of gold and a moderately richly perfused tissue. It is seen that the depth of destruction at a given RMS voltage is inversely proportional to the electrode radius. The effects of the voltage applied versus time duration to achieve a certain degree of tissue damage are

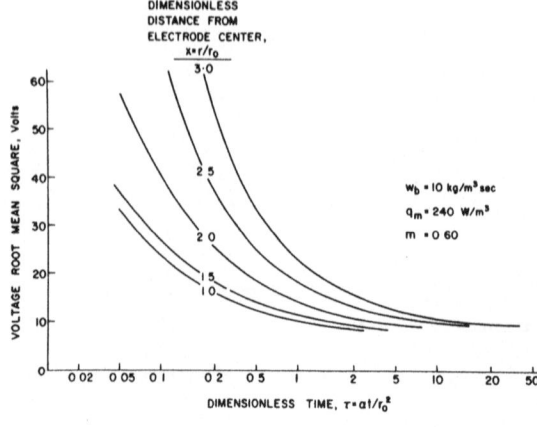

FIGURE 8

Variation of the depth of destruction vs. voltage and duration of application for a richly perfused tissue and a gold electrode ($m = 0.60$) (see Ref. 17).

shown in Fig. 8. It is again noted that the higher the voltage, the shorter the application time required to achieve the same depth of destruction. However, as time progresses, the effects of increasing the voltage diminish, and at about 15 V, it would be possible to destroy the tissue to depths of $x = 3$ or less by extending the time of application.

Based on the results of this study, the following observations may be made:

1. Blood perfusion through the tissue plays a major role in determining tissue temperature and, consequently, the depth of tissue destruction that is achieved. Also, due to the sensitivity of the results to this parameter, a fairly good knowledge of the magnitude of blood perfusion is required in order to quantitate the coagulation process.

2. Effects of metabolic heat generation on tissue temperature and extent of destruction are negligible and might be omitted from the analysis for all practical purposes.

3. The thermal inertia of the coagulating electrode influences both tissue temperature levels and the depth of destruction. However, the electrodes employed in practice possess a dimensionless thermal inertia of up to about $m = 1$. This small value permits the assumption of zero thermal inertia of the electrodes for all practical purposes.

4. The depth of tissue destruction is almost directly proportional to the electrical power and duration of application for a given electrode. However, excessive tissue temperatures might be obtained, and a suitable power reduction with prolonged time of application is advisable to achieve a desired depth of destruction.

5. For the tissue destruction problem, charring of tissue layers in regions next to the electrode may not be avoidable. The extent of charring may be limited by appropriately reducing the voltage applied to the electrode and prolonging the time of application.

The analysis presented should be regarded as a first approximation to the rather complex coagulation process occurring in biological tissues during the passage of alternating current. Several important factors—e.g., skin effects from alternating current applications, skin surface heat loss, boiling of tissue fluids, liberation of heat of reaction, contraction of denatured tissue, coagulation of blood vessels and consequently the blocking of blood perfusion, preferential heat transfer in tissue planes, and plane separation, etc.—are not considered in the present analysis. Nevertheless, some basic conclusions pertaining to the behavior of the tissue undergoing coagulation are drawn. These conclusions should serve as guidelines for applications of the technique. Further extensions and improvements of this model would have to be based on quantification of some of the experimental observations previously noted.

2.2. Induction Thermocoagulation

Induction thermocoagulation has been used for years to create brain lesions.[9,10,32–39] Most of the early analyses of the temperature fields and extent of damage were rather crude and failed to consider many of the parameters involved. Many of these studies considered the power deposition problem with little or no regard to the temperature field.[22–25] In 1974, Kach and Incropera presented a more complete analysis of this process; this section is based on their work.[16] These investigators made the following assumptions:

(a) The tissue is homogeneous and isotropic.

(b) The small cylindrical thermoseed is completely embedded in the tissue, which is of infinite extent.

(c) Transport of energy from the cylinder occurs in the radial direction only. No consideration is given to edge effects.

(d) The tissue and the electrode are at a uniform and equal temperature prior to the exposure to the RF field.

(e) Blood flow effects are presented mathematically as a conduction process. Accordingly, a conductivity increment was added to the so-called "normal" thermal conductivity to create the effective thermal conductivity that accounts for both conduction and perfusion effects in the tissue. When the tissue is destroyed, blood perfusion is assumed to cease, and only the "normal" thermal conductivity remains. This assumption, although mathematically convenient, was shown to be inaccurate and should, therefore, be regarded as only a crude approximation to the problem.[40]

(f) Metabolic heat generation is neglected.

Stated mathematically the problem is

$$\rho c \frac{\partial T}{\partial t} = \frac{1}{r} \frac{\partial}{\partial r}\left(k_{eff} r \frac{\partial T}{\partial r}\right) \tag{37}$$

with the initial and boundary conditions

$$T(r, 0) = T_0 \tag{38}$$

$$T(\infty, t) = T_0 \tag{39}$$

$$T(r_0, 0 \leq t \leq t_h) = F(t)$$

$$\frac{\partial T}{\partial r}(r_0, t > t_h) = 0 \tag{40}$$

where $F(t)$ is the temperature at the thermoseed surface described by

$$F(t) = T_0 + \frac{T_1 - T_0}{t_1}, \qquad 0 < t \leq t_1$$

$$F(t) = T_1, \qquad t_1 < t \leq t_h \tag{41}$$

with $T_0 = 37°C$, $t_1 = 15$ sec, which is the warm-up time required to achieve the steady-state implant temperature, and T_1 is the steady-state temperature at the thermoseed surface that was allowed to vary between 60–90°C.

Solution to the preceding set of equations was obtained numerically using a time interval of 0.5 sec and a radial increment of 0.2 mm. Temperature

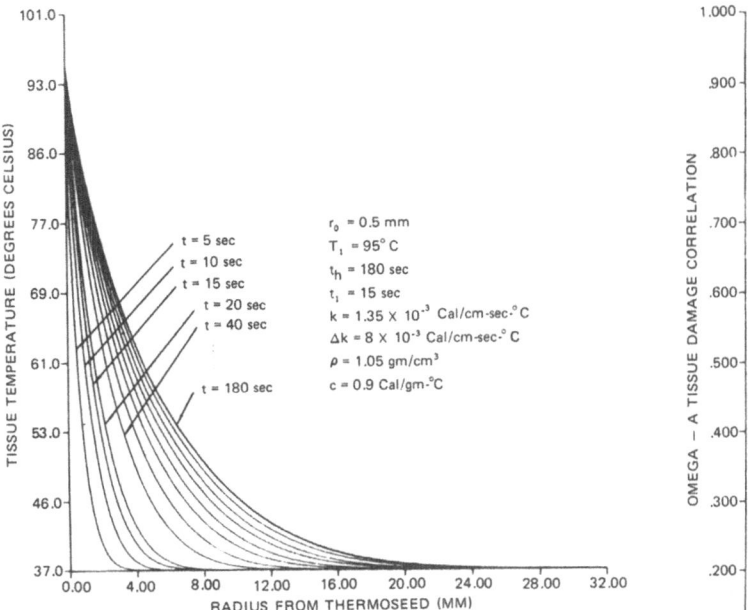

FIGURE 9

Temperature distribution in the tissue at various times during induction heating. From Ref. 16 with permission.

distributions in the tissue during induction heating $(0 < t \le t_h)$ and subsequent cooling $(t > t_h)$ are shown in Figs. 9 and 10, respectively, for representative thermoseed operating conditions $(r_0 = 0.5$ mm, $t_h = 180$ sec, $T_1 = 95°C)$. Figure 9 reveals the nature and the extent of the temperature buildup

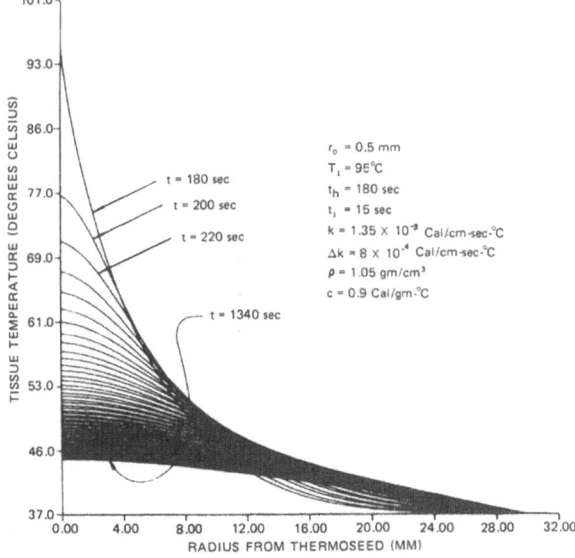

FIGURE 10

Temperature distribution in the tissue at various times following cessation of induction heating. From Ref. 16 with permission.

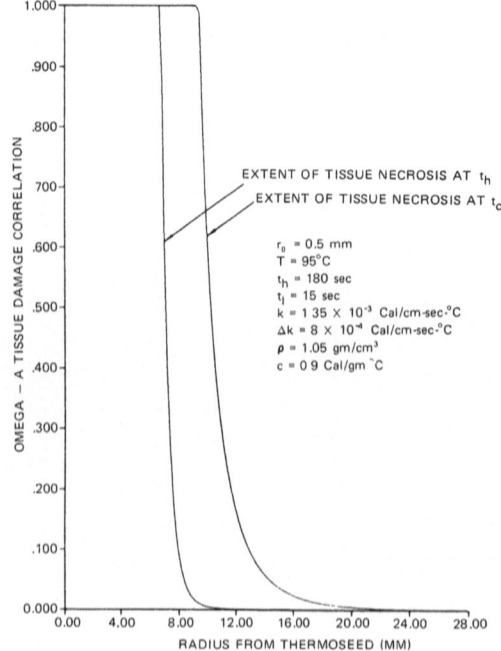

OMEGA – A TISSUE DAMAGE CORRELATION

RADIUS FROM THERMOSEED (MM)

EXTENT OF TISSUE NECROSIS AT t_h

EXTENT OF TISSUE NECROSIS AT t_c

r_0 = 0.5 mm
T = 95°C
t_h = 180 sec
t_l = 15 sec
k = 1 35 × 10^{-3} Cal/cm-sec-°C
Δk = 8 × 10^{-4} Cal/cm-sec-°C
ρ = 1.05 gm/cm³
c = 0 9 Cal/gm °C

FIGURE 11

The spatial extent of tissue damage incurred by the thermocoagulation process. From Ref. 16 with permission.

that occurs during induction heating. It is evident that for t_h = 180 sec, tissue thermal alterations will be experienced at points as far as 25 mm from the thermoseed surface and that the temperature change will be appreciable (> 7°C) up to a distance of 10 mm. The tissue cooling characteristics, Fig. 10, indicate that the time required for all of the tissue to cool below 44°C is in excess of 1300 sec. This seems excessive, perhaps due to the omission of evaporation and radiative cooling. This cooling time $t_c - t_h$ will increase with increasing t_h and T_1. Moreover, during cooling, the region that experiences elevated temperatures continues to increase. From knowledge of the temperature distribution, the extent of damage may be predicted by the method due to Henriques,[18] which was previously discussed.

The damage obtained from the temperature histories of Figs. 9 and 10 is presented in Fig. 11. The two curves pertain to the tissue damage that exists at the conclusion of heating and to the final damage figure (which occurs when the tissue temperature has dropped below 44°C everywhere). Note that although the largest contribution to complete tissue destruction ($\Omega \geq 1$) is made during the heating period, a significant contribution is also made during the subsequent cooling period. The size of the lesion is increased by approximately 2.5 mm as a result of additional tissue damage sustained during cooling. Note also that although the zones of irreversible ($\Omega \geq 1$) and reversible ($\Omega <$ 0.53) tissue damage are significant, the extent of the incomplete damage region ($0.53 \leq \Omega < 1.0$) is comparatively small (approximately 0.8 mm for the specified thermoseed conditions). Hence, the model predicts that only a thin layer separates the region of necrosis from tissue that experiences no damage.

In Fig. 12, the extent of tissue damage is plotted against heating time t_h for representative values of r_0, t_1, and T_1. The dashed line indicates the radius

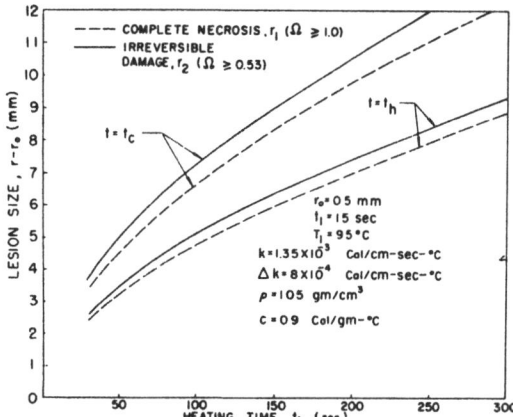

FIGURE 12

Lesion size as a function of thermoseed heating time. From Ref. 16 with permission.

r_1 of a cylindrical surface that encloses necrotised tissue ($\Omega \geqslant 1$), and the solid line provides the radius r_2 of a cylindrical surface enclosing tissue that has experienced some irreversible damage ($\Omega \geqslant 0.53$). Results are presented for the time t_h corresponding to the cessation of heating and for the time t_c beyond which no further damage occurs. As expected, the lesion size increases with t_h. The amount of tissue that experiences some reversible damage, but not complete necrosis, is included in the annular region defined by $r_2 - r_1$. It also increases with t_h, varying at $t = t_c$ from approximately 0.03 mm for a t_h of 30 sec to approximately 1 mm for a t_h of 300 sec. In addition, the amount of tissue sustaining thermal damage during the cooling transient also increases with increasing t_h. The variation of tissue damage with the steady-state thermoseed temperature T_1 is shown in Fig. 13. In addition to increasing the lesion size, an increase in T_1 will significantly increase the extent of tissue damage incurred during cooling. An increase in T_1 will also slightly increase the value of $r_2 - r_1$.

The preceding results provide an indication of the effect of thermoseed operating parameters on lesion size. However, due to the absence of compatible experimental data, it is presently impossible to determine the accuracy of the computed results. The major sources of error appear to be the assumption

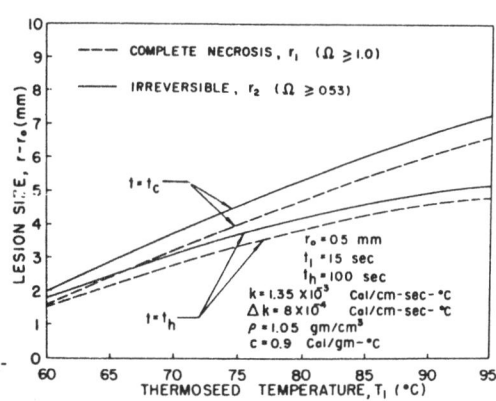

FIGURE 13

Lesion size as a function of thermoseed temperature. From Ref. 16 with permission.

of one-dimensional heat transfer, the uncertainty associated with the knowledge of the thermal conductivities, and the omission of evaporative cooling and reradiation in the vicinity of the thermoseed. Of these, the assumption of the one-dimensional conditions is thought to contribute the least error. However, little is known about the precise effect of perfusion on energy transfer in tumors and the conductivity of tissues that have undergone coagulation. Calculations reveal that a ±100% change in the value of k_{eff} will alter the lesion size by approximately ±25% (lesion size increases with increasing k_{eff}). Refinements in the model will await the acquisition of appropriate data for comparison with the present results.

2.3. Resistance Heating (Electrocautery)

Electrocautery, whether applied to superficial or deeper layers of the tissue, is used extensively to treat medical problems in a number of specialties,[7] including dermatology, neurology, otolaryngology, and gynecology. The principle of operation is simple: electric current is passed through a resistive element applicator. As a result of Joule heating, there being no alternative path, this heat is dissipated to the surrounding tissue, causing coagulation.

A number of probe shapes are commonly used, including wire tips or loops, flat disk tips, and cylindrical probes. Several studies of these embedded probes indicate the need for improved temperature measurement techniques and analysis.[41–43] Both these problems were addressed by Cooper and Groff, whose work forms the basis for this section.[15]

The heating probe considered by these investigators was a cylinder made of 4-mm glass tubing wrapped tightly with resistance wire over a length of 40 mm. The wire was used as both a resistor and a resistance thermometer. Probe surface temperature was measured by placing the resistance wire in one arm of a Wheatstone bridge. The thermal behavior of this probe was studied based on the following assumptions:

(a) The tissue is homogeneous, of constant thermal properties, and infinite in extent.
(b) The tissue is perfused isotropically and uniformly by capillary blood flow. The bioheat equation is assumed to describe the heat balance of the tissue.
(c) Metabolic heat generation is neglected.
(d) The heat flow from the probe to the tissue occurs in the radial direction only. This assumption was subsequently removed, and an analysis of a two-dimensional problem was performed.

Mathematically, the problem is stated as follows:

$$\frac{k}{r} \frac{\partial [r(\partial T/\partial r)]}{\partial r} - \rho_b c_b w_b (T - T_0) = pc \frac{\partial T}{\partial t} \tag{42}$$

with the boundary and initial conditions

$$T = T_p \quad \text{at} \quad r = r_0 \tag{43}$$

$$T \to T_0 \quad \text{as} \quad r \to \infty \tag{44}$$

$$T = T_0 \quad \text{at} \quad t = 0, \quad r > r_0 \tag{45}$$

Equation (42) and its boundary and initial conditions may be normalized by introducing the following set of quantities:

$$R = \frac{r}{r_0} \quad \theta = \frac{T - T_0}{T_p - T_0} \quad \tau = \frac{\alpha t}{r_0^2} \quad \beta^2 = \frac{\rho_b c_b w_b r_0^2}{k} \tag{46}$$

to yield

$$\frac{1}{R} \frac{\partial[R(\partial\theta/\partial R)]}{\partial R} - \beta^2 \theta = \frac{\partial\theta}{\partial\tau} \tag{47}$$

with boundary and initial conditions

$$\theta = 1 \quad \text{at} \quad R = 1 \tag{48}$$

$$\theta \to 0 \quad \text{as} \quad R \to \infty \tag{49}$$

$$\theta = 0 \quad \text{at} \quad \tau = 0, R > 1 \tag{50}$$

Equation (47) was solved using the Laplace transformation, and the following solution was obtained:

$$\theta = \frac{K_0(\sqrt{\beta R^2})}{K_0(\sqrt{\beta})} - \frac{2}{\pi} \int_0^\infty \frac{e^{-\tau u^2} C_0(u, Ru) \, du}{(u + \beta^2/u)[J_0^2(u) + Y_0^2(u)]} \tag{51}$$

where

$$C_0(u, R) = J_0(u)Y_0(uR) - J_0(uR)Y_0(u) \tag{52}$$

J_0, Y_0, and K_0 are Bessel functions, and u is a dummy variable of integration. Values of θ vs. R, τ, and β were generated using Simpson's rule on Eq. (51). Additional results can be found in Ref. 44.

Figure 14 depicts the effect of blood flow on the location of a particular isotherm corresponding to a nondimensional temperature of $\theta_c = 0.50$. Note that blood flow may have a strong influence on the temperature field. Of particular interest is the influence that blood flow would have on the field surrounding a typical surgical probe of 1-m diameter or less. As an illustration, assume that a 1-mm probe is to be used in brain tissue. Brain tissue has the following properties: thermal conductivity, 1.26×10^{-3} cal/cm sec °C; density, 1.05 g/cm^3; specific heat, 0.88 cal/g °C; thermal diffusivity, 1.36×10^{-3} cm^2/sec; and blood flow rate, 8×10^{-3} g/cm^3 sec. Using these values, β^2 is calculated to be 0.016. From Fig. 14, it is noted that for this value of β^2, the influence of the blood flow is minimal, at least for short times. As can be seen by examining the definition of the blood flow parameter β^2, Eq. (46), large probes will experience much larger blood flow effects than small probes, since β^2 varies as the probe radius squared. To more realistically model the behavior of actual probes, a two-dimensional model was employed. An actual

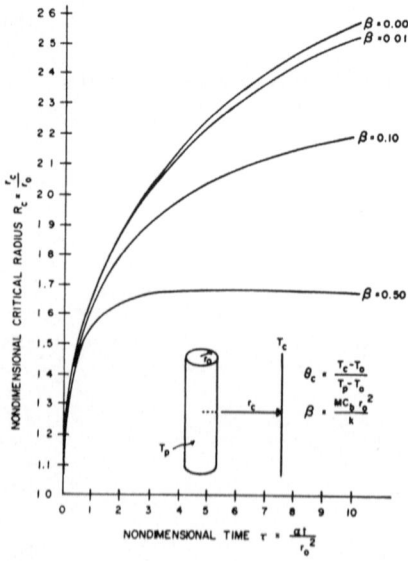

FIGURE 14

Influence of blood flow on the radial movement of the $\theta_c = 0.50$ isotherm (see Ref. 15).

surgical probe usually consists of a constant-temperature section of finite length attached to an unheated cylindrical stem. If the length-to-diameter ratio of the heated section is small, end effects will destroy the uni-dimensionality of the heat flow process, and the problem will become two dimensional. The prototype probe under consideration had a heated section with a length-to-diameter ratio of ten. To study the two-dimensional transient temperature field produced by this probe, a digital computer program with the code name TRUMP[45] was used. Unfortunately, blood flow effects were not considered in the two-dimensional study.

To study experimentally the temperature field around a heated surgical probe, Cooper and Groff imbedded the probe in a 0.3% agar–99.7% water medium. Temperatures were measured by the liquid crystal technique.[46–48] Experimentally determined values of the radius corresponding to the 29.9°C

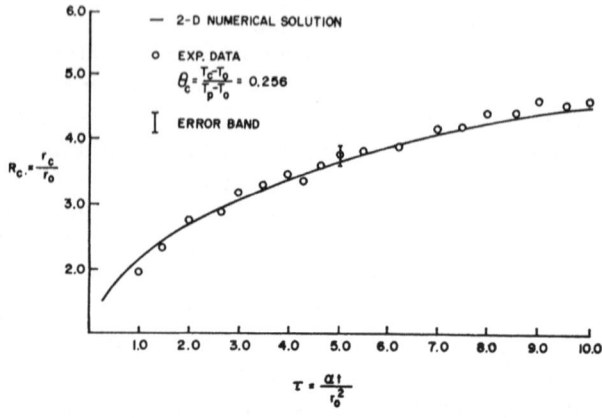

FIGURE 15

Comparison of experimentally determined midplane values of the critical radius R_c vs. time with results obtained from the two-dimensional numerical solution (see Ref. 15).

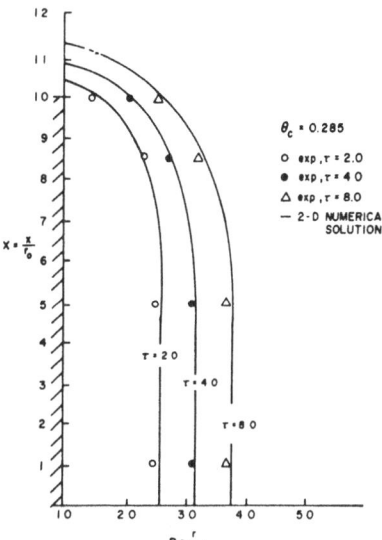

FIGURE 16

Comparison of experimentally determined axial values of the critical radius R_c vs. time with results obtained from the two-dimensional numerical solution (see Ref. 15).

isotherm R_c vs. nondimensional time were compared with the numerical results generated with TRUMP. Figures 15 and 16 show typical comparisons between experiment and theory. In all cases studied, agreement was well within the estimated experimental uncertainty of 10%, indicating that theoretical predictions of the temperature field produced by this resistance probe may be usefully employed in this situation.

3. SUMMARY

The analyses presented in this chapter demonstrate the variety of electrosurgical procedures that may be treated analytically to provide the surgeon with a damage prediction. Obviously, much work needs to be done before the analysis becomes a generally useful and reliable tool. Among the factors to be considered in the future are:

(a) Comparison of the analytical results with experimentally measured values. This imperative requirement presents a number of difficulties, e.g., the need to invade the treated tissue, the possible distortion of the electrical field, etc. However, ultimate validation of any analytical tool is dependent on the ability to demonstrate its conformity to the experimental results. Thus, much work will have to be done on phantom and animal models before the analysis of electrosurgical techniques gains confidence.

(b) More accurate modeling of both the electrical and the thermal fields is needed. Up to now, both fields have been approximated by first-order models. These models assume homogeneous, isotropic tissue models, with[13,15,17,21] or without[19,20,26] consideration of the effects of blood perfusion. In order to improve the predictive power of these models, both the thermophysical and the dielectric properties have to be considered in detail.

Furthermore, the effects of blood vessels, the larger ones in particular, should be modeled more carefully. In this respect, the question of what actually happens to the blood flow distribution in the larger region following local hemostasis should be addressed.

(c) Formulation of a damage function to more accurately describe the biochemical processes of thermal denaturation and destruction is required. The methods currently used, e.g., a damage threshold temperature,[13,15] a single-rate process,[18] or even an entirely subjective method based on touching the skin,[6] attest to the fact that much additional work is needed to identify and quantify the actual process involved. In the present situation, it seems advisable to use the single-rate process, as proposed by Henriques,[18] until more accurate and reliable models are developed.

(d) Consideration of the thermal conditions at the probe–tissue interface is required. In certain cases, due to excessively high temperatures, steam bubbles may form and tissue charring may also result. Both these conditions modify the behavior of the tissue electrically and thermally and introduce nonlinearities into the modeling equations. These nonlinearities should be considered specifically.

In addition, tissues tend to separate along planes defined by fascia, the thin membranes that envelop muscle and other tissue bundles. This acts to thermally separate the tissue in an anisotropic fashion. Also, thermal denaturation and coagulation of connective tissue tend to shorten it, in some cases causing adjacent tissues to be in tension and perhaps to tear. These effects should also be considered in the analysis.

(e) Development of an analysis of pulsed electrosurgery seems warranted. It appears that preplanned pulsed, rather than continuous, application of either the rf current or the electromagnetic induction field, offers the promise of better control over the produced lesion. The distinction made here is between the aforementioned damped waveform, which is a pulsed mode of application, and an arbitrarily discontinued operation. It appears that the pulsed mode of operation may offer the advantage of avoiding tissue charring and steam formation and should therefore be analyzed systematically.

(f) Development of a spatial placement function to facilitate the destruction of irregularly shaped tumors and tissue regions. As is apparent from the foregoing, all current models assume regular geometries, e.g., cylindrical[15,16] and spherical.[17,26] Evidently, most tumors assume irregular shapes that should be treated as such. Thus, the analysis should consider the question of optimal placement of a number of electrosurgical probes and their shapes (thermal activity). These should be considered along with operating details to achieve the optimal destruction of the tissue region compatible with preservation of normal tissue.

In spite of these present inadequacies, it appears that mathematical modeling offers certain advantages in quantifying the effects of electrosurgery on the tissue as well as providing a basis for improved equipment design. With further development and experimental verification, it may also become a useful tool in the hands of the practicing surgeon.

REFERENCES

1. Glasser, O., *Medical Physics*, vols. 1 and 2 (Chicago: The Yearbook Publishers, 1950).
2. Nightingale, A., *Physics and Electronics in Physical Medicine* (London: G. Bell & Sons, 1959).
3. Ray, E. D., ed., *Medical Engineering* (Chicago: The Year Book Publishers, 1974), pp. 1048–1053.
4. Kelly, H. A., and Ward, G. E., *Electrosurgery* (Philadelphia: W. B. Saunders & Co., 1959).
5. Krusen, F. H., Kottke, F. J., and Ellwood, P. M., *Handbook of Physical Medicine and Rehabilitation*, 2d ed. (Philadelphia: W. B. Saunders & Co., 1971).
6. Rook, A., Wilkinson, B. B., and Elbing, F. J. G., eds., *Textbook of Dermatology*, 2d ed. (Oxford: Rockwell Scientific Publications, 1972).
7. Otto, J. F., ed., *Principles of Minor Electrosurgery* (Liebel-Florsheim, 1957).
8. Strauss, A. A., *Immunologic Resistance to Carcinoma Produced by Electrocoagulation* (Springfield, IL.: Charles C. Thomas, 1969).
9. Burton, C. V., Mozley, J. M., Walker, A. E., and Braitman, H. E., Induction thermocoagulation of the brain: a new neurological tool, *IEEE Trans. Biomed. Eng.* **BME-13**, 114–120, 1966.
10. Burton, C. V., Hill, M., and Walker, A. E., The RF thermoseed—a thermally self-regulating implant for the production of brain lesions, *IEEE Trans. Biomed. Eng.* **BME-18**, 104–109, 1971.
11. Glover, J. L., Bendick, P. J., and Link, W. J., The use of thermal knives in surgery: Electrosurgery, lasers, plasma scalpel, *Current Prob. Surg.* **15(1)**, 1–78, 1978.
12. McLean, A. J., Characteristics of adequate electrosurgical current, *Am. J. Surg.* **18**, 417–441, 1932.
13. Overmyer, K. M., Pearce, J. A., and DeWitt, D. P., Measurement of temperature distributions at electrosurgical dispersive electrode sites, ASME paper no. 77-WA/HT-47, 1977.
14. Aronow, S., The use of RF power in making lesions in the brain, *J. Neurosurg.* **17**, 431–438, 1960.
15. Cooper, T. E., and Groff, J. P., Thermal mapping via liquid crystals of the temperative field near a heated surgical probe, *ASME Trans. J. Heat Transfer* **95**, 250–256, 1973.
16. Kach, E., and Incropera, F. P., Induction thermocoagulation: Thermal response and lesion size, *IEEE Trans. Biomed. Eng.* **BME-21**, 8–12, 1974.
17. Erez, A., and Shitzer, A., Controlled destruction and temperature distributions in biological tissues subjected to monoactive electrocoagulation, *ASME Trans. J. Biomech. Eng.* **102**, 42–49, 1980.
18. Henriques, F. C., Studies of thermal injury, V. Predictability of thermally induced rate processes leading to epidermal injury, *Arch. Path.* **43**, 489–502, 1947.
19. Van den Berg, J., and Van Manen, J., Graded coagulation of brain tissue, *Acta Physiol. Pharmacol. Nederlandica* **10**, 353–377, 1962.
20. Rutkin, B. B., and Barish, E. Z., Localized thermal distributions in the brain, *Proc. 17th Ann. Conf. Eng. Med. and Biol.* **6**, 14, 1964.
21. Cooper, T. E., and Gengler, P. L., Heat transfer analysis of a radio frequency probe, *Proc. 24th Ann. Conf. Eng. Med. and Biol.* **13**, 216, 1971.
22. Honig, W. M., The mechanism of cutting in electrosurgery, *IEEE Trans. Biomed. Eng.* **BME-22**, 58–62, 1975.
23. Merry, A. G., Hale, R., and Zervos, N. T., Induction thermocoagulation—a seed power study, *IEEE Trans. Biomed. Eng.* **BME-20**, 302–303, 1973.
24. Lin, J. C., Comments on "Induction thermocoagulation—A seed power study," *IEEE Trans. Biomed. Eng.* **BME-21**, 419, 1974.
25. Lin, J. C., Induction thermocoagulation of the brain—quantitation of absorbed power, *IEEE Trans. Biomed. Eng.* **BME-22**, 542–546, 1975.
26. Drabkin, R. L., Analysis of tissue temperature in monoactive electrocoagulation, *Biomed. Eng. (N. Y.)* **7(8)**, 80–84, 1972.
27. Drabkin, R. L., Electrocoagulation of the sclera, *Biomed. Eng. (N. Y.)* **8(2)**, 76–79, 1975.
28. Shitzer, A., Studies of bioheat transfer in mammals, in *Topics in Transport Phenomena*, Gutfinger, C., ed. (New York: Halsted Press, 1975), pp. 211–343.

29. Rubinsky, B., and Shitzer, A., Analysis of Stefan-like problem in a biological tissue around a cryosurgical probe, *ASME J. Heat Transfer* **98**, 514–519, 1976.

30. Cooper, T. E., and Trezek, G. J., Mathematical predictions of cryogenic lesions, in *Cryogenics in Surg.*, H. Von Leden and W. G. Cahan, eds. (Flushing, NY: Medical Examination Publishing Co., 1971).

31. Carslaw, H. S., and Jaeger, J. C., *Conduction of Heat in Solids*, 2d ed. (Boston: Oxford Press, 1959).

32. Dusser de Barenne, J. B., Method of laminar coagulation of the cerebral cortex, *Yale J. Biol. Med.* **10**, 573, 1938.

33. Walker, A. E., and Silver, M. L., Histopathology of thermocoagulation of the cerebral cortex, *J. Neuropath. Exp. Neurol.* **6**, 311–322, 1947.

34. Burton, C. V., Walker, A. E., Adamkiewcz, J. J., Mozley, J. M., and Dillon, E. T., High-frequency thermal induction lesions of the brain, *J. Nerv. Men. Dis.* **136**, 298–301, 1963.

35. Burton, C. V., Walker, A. F., RF telethermocoagulation, *J. Am. Med. Ass.* **197**, 700–704, 1966.

36. Riechert, T., and Gabriel, E., A new surgical method of producing localized tissues lesions by induction heating, *Dtsch. Med. Wochenschr. (Eng. lang. ed.)* **7**, 357–359, 1967.

37. Riechert, T., Forester, Ch. F., and Krainick, J. U., The technique of induction heating in stereotactic surgery, *Top. Prob. Psychiat. Neurol.* **10**, 154–159, 1970.

38. Burton, C. V., Conference on RF neuromagnetics—summary of proceedings, *IEEE Trans. Biomed. Eng.* **BME-18**, 242–245, 1971.

39. Riechert, T., and Krainick, J. U., Application of inductive coagulation to produce reversible nerve tissue damage, in *Special Topics in Stereotaxis*, Umbach, W., ed. (Stuttgart: Hippokrates, 1971), pp. 121–129.

40. Eberhart, R. C., Shitzer, A., and Hernandez, E. J., Thermal dilution methods: Estimation of tissue blood flow and metabolism, *Ann. N. Y. Acad. Sci.* **335**, 107–130, 1980.

41. Carpenter, M., and Whittier, J. R., Study of methods for producing experimental lesions of the central nervous system with special reference to stereotaxic technique, *J. Comp. Neurol.* **97**, 73–117, 1952.

42. Gildenberg, P. L., Studies in stereoencephalotomy X, *Confinia Neurol.* **20**, 53–65, 1960.

43. Watkins, W. S., Heat gains in brain during electrocoagulative lesions, *J. Neurosurg.* **23**, 319–328, 1965.

44. Groff, J. P., The design and analysis of a resistively heated surgical probe (M.S. thesis, Naval Postgraduate School, Monterey, CA, 1971).

45. Edwards, A. L., TRUMP: a computer program for transient and steady-state temperature distributions in multidimensional systems, Lawrence Radiation Laboratory, report UCRL-14754, rev. 11, 1969.

46. Fergason, J. L., Liquid crystals, *Scientific Am.* **211**(2), 76–86, 1964.

47. Fergason, J. L., Liquid crystals in nondestructive testing, *Appl. Topics* **7**(9), 1729–1737, 1968.

48. Fergason, J. L., Experiments with cholesteric liquid crystals, *Am. J. Phys.* **38**(4), 425–428, 1970.

Chapter 18

ANALYSIS OF SKIN BURNS

Kenneth R. Diller

1. INTRODUCTION

The burn injury constitutes one of the most commonly encountered types of trauma. The average number of burns reported annually in the United States is in excess of 2,000,000,[1] which represents a source of morbidity and mortality of major concern. About 50% of these burns or scalds are severe enough to require medical attention and restrict physical activity; 25% of the burns necessitate confinement to a bed.[2] Burns resulting from fires and explosions are the most frequent causes of fatal accidents among children and the elderly. Accidents occurring in the home are responsible for more than three-fourths of these deaths.[3] The concentration of burns among the younger portion of the population results in a disproportionately large social cost in comparison with other serious pathologies, such as heart disease, cancer, and stroke, which tend to be contracted at a more advanced age following a useful, productive life. From these statistics, it is obvious that burns constitute a major injury entity to modern society.

Burns typically are caused by one of many types of interaction with high temperature sources, including exposure to flames, contact with a hot solid or liquid, inhalation of a hot vapor, or electrical energy dissipation. In the present discussion, we shall be concerned with only thermal sources exterior to the body resulting in heating the skin and underlying tissues.

The magnitude of a burn injury may vary from a minor first-degree wound to the most severe form of injury to which a person is liable. A major burn evokes a myriad of physiological changes involving both local and systemic responses. Unlike other wounds in which closure can be completed within hours or days, a serious burn requires a prolonged time for removal of dead eschar before it can be closed. The continued presence of necrosed dead tissue produces further injury and elicits additional systemic complications. The complex disease process presents a tremendous challenge in understanding the mechanisms of healing and in developing effective therapeutic modalities.

Kenneth R. Diller • Department of Mechanical Engineering, Biomedical Engineering Center, University of Texas, Austin, Texas 78712. The research reported herein was sponsored in part by a grant from the National Institutes of Health, No. GM-22693.

2. PHYSIOLOGY OF SKIN

A proper description of the burn wound process requires a basic understanding of the physiology of the tissue involved. In both size and weight,
skin constitutes the largest organ in the body; in an average adult, the skin
surface area is about $1.7 \, m^2$, and in weight, it represents about 15% of the
total for the entire body.[4] The functions of skin are vital to life; they include:
(1) protection of underlying tissues from physical, chemical, and thermal
trauma; (2) thermal regulation by sweating, heat conduction (insulation), and
control of blood flow to a profuse plexus of minute surface vessels; (3)
impermeability to both tissue fluids and environmental chemicals; and (4)
sensory perception of touch, pain, and temperature. In addition, skin is the
organ by which we are presented to the world. Although skin may normally
be a source of visual beauty, if scarred and deformed by a burn, skin can
represent a serious physical and social handicap.

Anatomically, the skin consists of a stratified structure with three parallel
layers: an outer, thin epidermis; a dense, noncellular connective tissue called
the dermis or corium; and a thick, subcutaneous fatty tissue.[5] A cross-sectional
anatomy of the normal skin is shown in Fig. 1. Although the skin is a single
organ, it is highly diversified phsyiologically and anatomically according to
different locations on the body. These structural variations result in differing
sensitivities to thermal insult and susceptibility to severe burn injury.

Although large regional variations in skin anatomy exist, the general
structural features are constant. A brief review of the anatomical organization
of skin will be helpful in subsequent discussions of the physiological response
to thermal trauma.

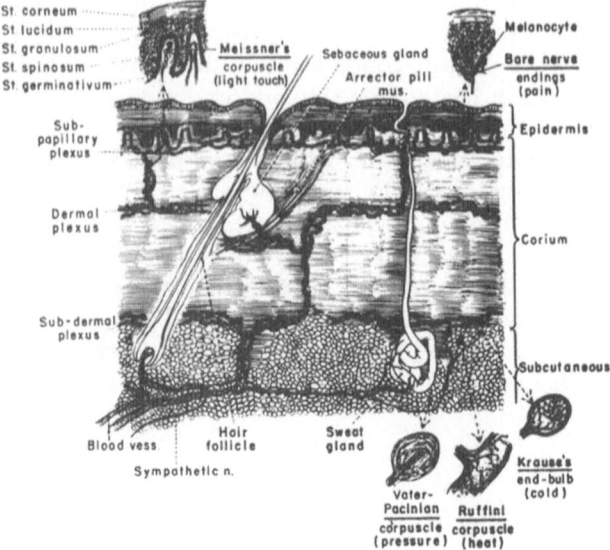

FIGURE 1
Normal cross-sectional anatomy of the skin. From Ref. 4 with permission.

The epidermis is a stratified, highly cellular membrane measuring 0.06–0.8 mm in thickness. It contacts the corium substrate via a three-dimensional network of irregular interpapillary convolutions that afford a large surface area between the two layers and provide much of the resistance of normal skin to tangential stress. Since the epidermis is devoid of blood vessels, lymphatics, and connective tissue, it is parasitic to the underlying dermis, which can furnish nourishment from its rich capillary plexuses.

The innermost stratum of the epidermis is the basal layer that contacts the neurovascular supply (Fig. 2). Cells are generated in this layer and then gradually extend toward the skin surface, forming first the prickle layer, which is characterized by interlocking cell wall projections that aid the skin's ability to withstand shearing forces. The next layer is the stratum granulosum, which has a large, transferable water content (70%) that is involved in water retention and heat regulation.[4] These cells eventually die and form the outermost, dead, waterproof layer of fibrous keratin protein, i.e., the stratum corneum.[6] The keratin normally contains about only 15% water; immersion into a water bath increases this percentage, resulting in softening and wrinkling of the skin, whereas drying produces chapping and calluses.

The underlying corium is 20 to 30 times thicker than the epidermis and contains the vascular, nervous, lymphatic, and supporting structures of the

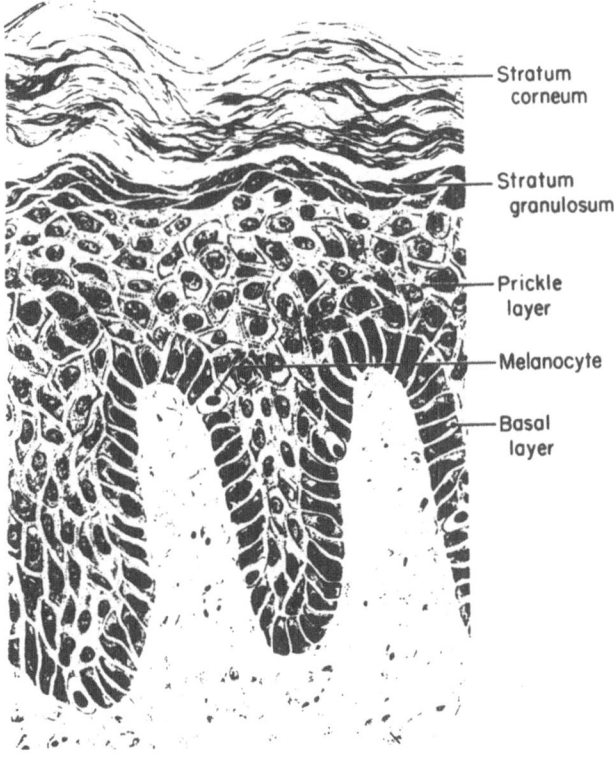

Stratum
corneum

Stratum
granulosum

Prickle
layer

Melanocyte

Basal
layer

FIGURE 2
Stratified structure of the epidermis. From Ref. 5 with permission.

skin. Three basic anatomical components are present in the corium: cells, matrix, and fibers, of which collagen fibers are the overwhelmingly predominant element. An important type of cell in the corium is the mastocyte, which contains considerable quantities of histamine and heparin and which also produces the mucopolysaccharides, hyaluronic and chondroitin sulfuric acids that form the outerfibrillary ground matrix of the dermis. Damage to this cell releases granules of histamine and heparin, initiating the localized inflammatory response.

The papillary layer is a superficial layer of the corium, lying adjacent to the epidermis. It consists of numerous small, vascularized, and highly sensitive protuberances called papillae, which rise perpendicularly from its surface. The papillae are minute conical eminences with rounded or blunted extremities that are received into corresponding pits of the undersurface of the epidermis. Each papilla has many very small and closely interlaced bundles of finely fibrillated tissue, with a few elastic fibers and a single blood capillary loop (Fig. 1).

The blood supply to the skin accounts for nutritional supply, cellular and humoral defenses, and a major portion of the thermal regulatory function. The vasculature is highlighted by the presence of a rich network of branching arterial and venous arcades that are interconnected in a regular pattern.[7] Small arteries entering the subcutaneous tissue form a system of long arterioles about 50 μm in diameter as the subdermal plexus. Adjacent links of the arteriolar system branch off to form the dermal plexus, which in turn forms a mesh of much finer arcuate capillaries, the subpapillary plexus (Figure 1). The capillaries combine into an even more complex venular mesh.

An important and unique aspect of the skin circulation is the presence of a large number of direct shunts from arteries to veins, called arteriovenous (AV) anastomoses. Activation of these AV shunts enables nearly all blood flow to be diverted past the capillaries to increase the effective thermal insulation between the subcutaneous tissue and the environment. On the other hand, this complex and vast vascular bed can accept up to 20% of the total cardiac output when the subpapillary plexus becomes engorged with blood. The small blood vessels of the skin have extensive sympathetic innervation and are highly responsive to neurogenic constrictor and dilator influences arising from various forms of local and systemic stress. This responsiveness is known to play an important role in the human thermoregulatory mechanism.

The subcutaneous tissue comprises yet another stratum of connective tissue, which specializes in the production of fat. In contrast to the corium, where the fibrocyte is the key cell-forming collagen, in the subcutaneous tissue, the lipocyte manufactures and stores large quantities of fat. Individual and anatomical differences in this fat deposition are very large and can play a major role in determining the degree of injury for a severe thermal insult.

3. PHYSIOLOGICAL ASPECTS OF THE BURN INJURY

A thermal burn occurs as a result of an elevation in tissue temperature above a threshold value for a finite period of time. The values of both the absolute temperature and the exposure time are crucial in determining the extent of injury. Temperature and time are not independent parameters in effecting a burn; rather, clinical data indicate that a nonlinear coupling exists that fixes the severity of trauma. In general, the transient tissue temperature integrated over the time of exposure must be considered in creating a thermal lesion. The larger the value of this coupled integral, the greater is the potential for injury.

The most frequent cause of skin burns is associated with the surface application of a heat source, whether by a conduction, convection, or radiation mode or a combination thereof. In the case of radiation, the source term may have some absorption distribution function within the tissue. Since heat transport from the surface to interior regions is limited by the effective thermal resistance of the tissue, a three-dimensional temperature field is established in which significant gradients may exist. Consequently, the temperature history in the affected area is nonuniform, and regions of graded injury develop with the most acute involvement closest to the source.

The microscopic physiological response to a thermal insult can be primarily measured in terms of the injury to the microvascular bed. The initial microvascular response to injury is characterized by complex changes in blood flow, coagulation process, and the permeability of the endothelial barriers. Increased vascular permeability is one of the most important features of this response, since it results in substantial losses of fluid and electrolyte, which lead to widespread pathophysiological changes. The physiological response of the microcirculation to elevated temperatures comprises a singularly significant element of the total burn injury.

The response of the microcirculation to thermal injury may vary from a slight increase in permeability of the endothelium up to complete stasis and necrosis of tissue in the burned area. It has been well established, for example, that burns cause a direct impairment of microcirculatory perfusion and that viability of the affected tissue is dependent on the patency of the microvascular blood vessels.[8–20] In general, the extent to which circulation is altered is a direct function of the severity of the trauma.

In a first-degree burn, vasodilatation is the only major change that occurs, resulting in the familiar reddening of the tissue. A second-degree burn is characterized by capillary damage resulting in tissue edema and bleb formation. Increased vascular permeability can occur by the opening of gaps in the endothelium or, in more severe cases, by direct damage to the endothelial cells.[17,21,22] The cells may become swollen, with many large vacuoles present or with a loss of internal structure, and the distorted cells may project into the lumen, partially occluding the vessel. The permeability of the vasculature in the burned area rapidly increases, allowing loss of molecules of up to 125,000 D.[23] This vascular response to trauma appears to be a result of both direct thermal injury and chemical mediators.[19,24–26] This phenomenon, molecular sieving, results in the sequestering of plasma proteins in the extravascular and extracellular spaces where, by osmotic action, they enhance the edematous effect.[27,28] Concentrations of water and albumin in the affected area reach a peak within 30 min following trauma.[27–30] The alteration of the vasular endothelium allows an overall fluid loss and drop in plasma volume. The decreased plasma volume is a prime factor in causing shock in untreated burned patients.

A third-degree burn causes cessation of blood flow in the microcirculation and leads to eventual necrosis of the tissue. In severe burns that functionally alter the circulation deep in the tissue, a large volume of extravascular fluid may collect beneath the wound before visible swelling occurs at the surface. Thus, the extensive fluid loss associated with third-degree burns is a consequence of injury to tissue beneath and surrounding the area of full-thickness skin destruction.[21]

Changes in blood rheology are noted consistently during the postburn period[31]; blood viscosity rises significantly and remains elevated for several days.[32] A marked increase in platelet aggregation occurs immediately in the injured area.[33] With microscopic observation, these aggregates can be clearly seen adhering to vessel walls, breaking off and flowing with the blood stream, and re-adhering further downstream. In some cases, the aggregates partially or totally occlude the vessels. Platelet adhesiveness is increased, and the circulating platelet count also increases and remains elevated for three weeks.[32]

Other factors that vary on a microscopic scale can influence the tissue response to a given heat stress. These include the water content of skin, pigmentation, regional differences in epithelial structure and thickness, presence of hair and surface oil or contaminants, as well as the instantaneous state of the local capillary blood flow.

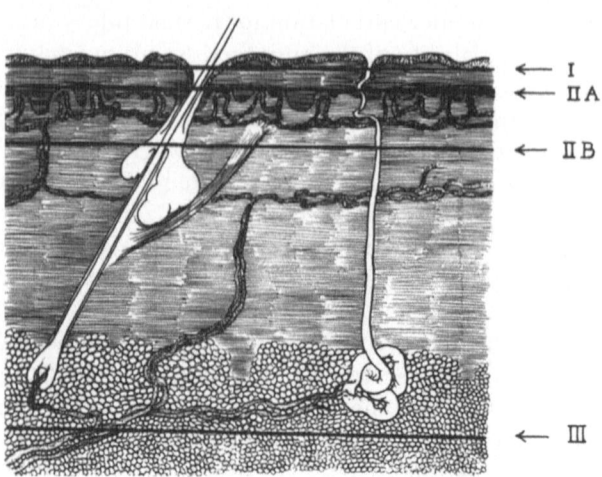

FIGURE 3
Depth of burn injury in skin: (I) Superficial burn; involves only the subcorneal layer. (IIA) Superficial partial thickness; involves some but not all of the basal layer. Healing should occur spontaneously, perhaps with some depigmentation due to loss of melanocytes. (IIB) Deep partial thickness; complete loss of the basal layer. Potential for spontaneous healing lies only in viable epidermal elements present in skin appendages. (III) Full thickness; all epidermal elements are destroyed, and spontaneous healing is impossible other than from the margins of the wound. From Ref. 4 with permission.

The macroscopic categorization of burn injury has been described in detail by Krizek, Robson, and Wray.[4] A first-degree burn is characterized in appearance by a general erythema caused by vasodilation of subpapillary vessels in the affected area. The response is localized; systemic effects are negligible. There are only minimal effects, other than slight edema and irritation of the nerve endings in layers deeper than the stratum corneum (Fig. 3). The first-degree injury involves only a temporary discomfort, with no permanent scarring or skin discoloration. Although denaturation of the outermost layer of dead keratin cells occurs, sometimes accompanied by slight peeling, healing proceeds uninhibited normally, with migration of cells to the epithelial surface.

Second-degree, or partial-thickness, burns can be classified clinically according to two subcategories, based on the penetration depth of the injured zone into the skin. Accordingly, both superficial and deep second-degree burns can be identified.[4] In general, a partial-thickness burn can be considered as one where the injury results in a loss of less than the full thickness of skin. For a superficial wound, a significant fraction of the basal cells of the corium are not destroyed, enabling the healing process to proceed in a normal pattern. The permeability of the endothelial cell lining of the subpapillary plexus is greatly increased, resulting in an accumulation of edema fluid with blistering. The stratum corneum forms a waterproof covering over the wound that prevents bacterial influx. Rupture of the blister may result in a weeping, open wound, increasing the susceptibility to infection and permitting a much greater evaporative loss with attendant requirements for metabolic expenditure of energy.[34] Because a majority of the basal cells are not injured during the burn, epithelial regeneration is normally prompt and complete without scarring.

In the deep second-degree burn, much of the basal cell layer of the corneum is lost, although certain viable epidermal elements, such as hair follicles and glands, may remain. The vascular injury is typified by widespread stasis and destruction of endothelial cells in the subpapillary plexus. Blistering is not widespread; however, an eschar of plasma and necrotic cells forms over the wound. Evaporative fluid loss through the water-permeable eschar is very high. In addition, destruction of the water barrier produces an easy route for bacterial invasion; the resulting wound sepsis is a major clinical deterrent to the healing process. If infection can be limited, this type of wound has the potential of spontaneously regenerating the epithelium. The wound becomes resurfaced from undamaged epithelial cells in the hair follicles and the margins of the injured area. However, the new epithelium is thinner than normal, lacks secretions necessary to lubricate the surface, has a diminished sensory capacity, and will be lighter than the surrounding tissue due to permanent loss of melanocytes. Also, the failure to redevelop interpapillary ridges results in an extended susceptibility to tangential stress loading.[35]

In a third-degree, or full-thickness, burn all epidermal elements and supporting dermal structures are destroyed.[4] Since the local blood vessels are obliterated, there can be no vascular response within the immediate injury area; rather, the celluar and fluid responses that typify inflammation are confined to the periphery of the affected region. The overlying eschar has no active nervous sensitivity and is highly permeable to both water and bacteria. Due to this physiological state, there exists no possibility for sponteneous healing of the wound. Resurfacing occurs only from the margins of the wound or by application of a skin graft.

A fourth-degree burn can occur, associated with incineration of tissue. The injury extends through the subcutaneous layer to involve the fascia, muscle, periosteum, and bone. The epithelial healing process is not greatly different from that for a third-degree burn, except for the greater complications associated with injury to the underlying tissues.

The preceding discussion has focused on the local response to burn injury. However, in major burns, systemic effects derived from local physiological phenomena play an important role in patient progression. Indeed, the function of all organ systems will eventually be altered due to the effects of a major burn. Although some changes may be related directly to the stress of the injury and the endogenous inflammatory mediators released locally into the circulatory system, changes are due for the most part to the altered functional capacity of the skin.[4] These changes are summarized briefly.

The primary and immediate response to a major burn is shock resulting from fluid loss.[36] Cardiac output drops in response to fluid loss and, perhaps, due to release of myocardial depressant factors.[37] A severe insult to the respiratory system can occur, even though no direct burning has occurred.[38] Both direct inhalation of combustion products and edema formation secondary to altered pulmonary capillary permeability act to impair alveolar gas diffusion.[39,40] The metabolic rate of a patient with a major burn is accelerated. For example, the surface evaporation of 5 liters of water daily is not uncommon and requires the expenditure of 2,880 kcal to compensate for the evaporative heat loss.[41] Evaporative cooling of the body leads to a further increase in metabolism by shivering.[42] Nutritional defects,[43-48] gastroduodenal mucosa ulceration,[49] and altered immune function[50-52] are additional complications of the burn wound.

4. DETERMINATION OF BURN INJURY FROM THE TEMPERATURE–TIME HISTORY

The production of thermal burns is dependent on both the elevation of tissue temperature and the duration of exposure to thermal stress. However, serious misconceptions have developed concerning the interaction between these two parameters in producing a burn. An example is the rule of a critical thermal load (CTL), defined as the total energy delivered in a given exposure that is necessary to produce a specific degree of injury.[53] The CTL is determined as the integral of the transient heat flux over the time of exposure to a given source. The CTL approach assumes that the thermal injury is a function of only total cumulative dosage, so that equal doses produce equal injury. Stoll has demonstrated clearly, to the contrary, that a large amount of energy delivered over an extended period of time may produce no injury whatsoever, whereas the same dose delivered nearly instantaneously may destroy the skin.[53] Experimental measurements by Stoll demonstrate that the integral of heat flux over time does not uniquely determine the threshold of injury. As shown in Fig. 4, human skin was irradiated at various intensity levels for the minimum time required to produce a blister. The fact that the product of irradiance and time is not constant for all values of irradiance suggests that the CTL concept is false.

Moritz and Henriques first demonstrated on pigs and humans what was to become the classical inverse relationship between the temperature and time required to produce a graded degree of thermal injury.[54] Their data, which are plotted in Fig. 5, show that a burn wound of a standard threshold severity can be produced by progressively decreasing temperatures as the duration of the thermal insult is logarithmically increased. Moritz and Henriques also conducted studies where pressure was applied to the skin during the burn, with the intention of reducing the continuous blood flow during the period of elevated temperature.[54] Within the sensitivity of the experimental technique, there was no detectable influence of blood flow in the superficial dermal capillaries on the vulnerability of the epidermis to thermal injury. These results were expected in light of the relative thinness of the epidermal layer and the normal absence of blood perfusion therein. Thus, the investigators reasoned that the subepidermal temperature should be controlled by diffusion of heat from the surface.

Subsequent investigators have confirmed the same type of temperature–time response curve as presented by Moritz and Henriques. Stoll and her co-workers conducted numerous detailed experiments on the interaction of burn time t, and temperature T, to cause specified injury levels in a human skin model.[55,56] Ross and Diller obtained similar results on a microvascular preparation in the hamster cheek pouch by using very precise temperature measurement and control techniques.[57] All of these data can be fit to a simple

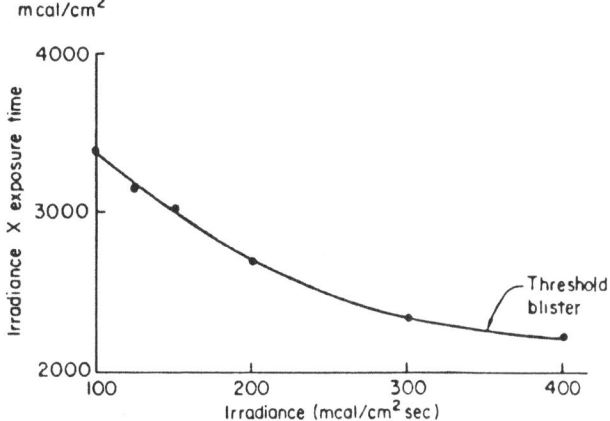

FIGURE 4

Product of irradiance and exposure time as a function of irradiance requisite to produce a threshold blister. From Ref. 53 with permission.

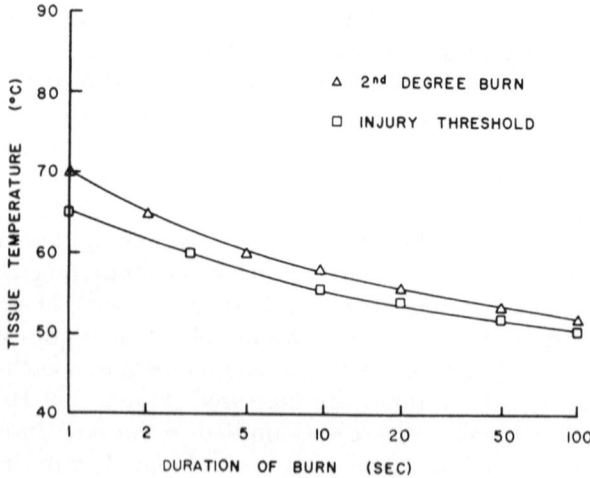

FIGURE 5

Inverse relationship between temperature and time necessary to produce a graded degree of thermal injury. From Ref. 54 with permission.

exponential function of the form

$$t = t_0 \exp\left[(T_0 - T)\right] \tag{1}$$

or, in terms of dimensionless time τ, and temperature ϕ

$$\tau = A \exp\left(-\phi\right) \tag{2}$$

where the reference time and temperature, t_0 and T_0, respectively, reflect the relative intensity of the injury.[57]

The consistency of data among numerous experimental protocols indicates that it should be possible to predict the extent of a burn wound using an appropriate analytical model. To be effective, such a model should incorporate parameters to account for thermal boundary conditions during a burn, constitutive and physiological properties of skin, and a criterion to define the thresholds for the levels of injury.

Henriques and Moritz were the first to propose a successful analytical model for thermal injury to skin.[58] They considered boundary conditions for conduction, convection, and radiation sources, coupled with conduction of heat within the skin. The transient temperature distribution is described in terms of the standard one-dimensional heat conduction equation (cf. Nomenclature)

$$\frac{\partial T(x, t)}{\partial t} = \frac{k}{\rho c} \frac{\partial^2 T(x, t)}{\partial x^2} \tag{3}$$

Subject to these boundary conditions, analytic solutions for Eq. (3) are obtained that describe temperatures in the affected tissue as a function of time and position. As part of the thermal injury model, Henriques devised a damage function that has subsequently been used quite widely. With this injury function, the cumulative damage incurred during a burn can be predicted. Thus, the simulation of a thermal burn in this model has two requisite steps; first, the transient temperature field must be determined for the boundary value problem of interest, and second, the thermal data must be applied to the evaluation of a damage rate function.

In deriving a damage function for burns, Henriques assumed that the governing biochemical processes could be depicted in terms of an Arrhenius relationship

$$\kappa = A \exp\left(-\frac{E}{RT}\right) \tag{4}$$

A is described as the frequency factor, κ as a second-order reaction rate constant, E is the activation energy for the reaction, and R is the universal gas constant. In his analysis, the term Ω was used to denote an arbitrary degree of tissue injury, and the rate of production of injury (the damage rate function) was given by

$$\frac{d\Omega(x, t)}{dt} = A \exp\left[-\frac{E}{RT(x, t)}\right] \tag{5}$$

The total injury at any point in the tissue is obtained by integrating the damage rate function over the entire burn period, using the corresponding local transient temperature history[59]

$$\Omega(x) = A \int_0^t \exp\left[-\frac{E}{RT(x, t)}\right] dt \tag{6}$$

The Ω function was quantified by Henriques to identify various injury thresholds. A value of $\Omega = 0.53$ was used to define the minimum conditions to obtain irreversible epidermal injury, whereas at $\Omega = 1.0$, complete transepidermal necrosis occurred. Thus, based on an evaluation of the local temperature history, and by applying Eq. (6), an estimate of the severity of burn injury can be made as a function of skin depth.

The boundary value problem in the model of Henriques and Moritz was based on several simplifying assumptions; the effects of local blood perfusion were neglected, and skin layer thickness and thermophysical properties were assumed to be constant.[58] Two specific experimental cases were considered: (1) a heat source was applied to the skin, and heat conduction alone caused a step increase to a constant burn temperature; and (2) the entire animal was subjected to an air environment of known convective and radiative properties. In both cases, the surface heat flux q_B due to a burning source at temperature T_B and heat transfer coefficient h, was given by

$$q_B = h[T_B - T(0, t)] \tag{7}$$

with a uniform initial skin temperature T_s

$$T(x, 0) = T_s \tag{8}$$

The analytical solution for this problem is well-known.[60] The temperature at a depth L below the skin surface is given by

$$\frac{T_B - T(L, t)}{T_B - T_s} = \text{erf}\left(\frac{\gamma}{\sqrt{t}}\right)\left\{\exp\left[\frac{hL}{k}\left(1 + \frac{hLt}{4\gamma^2 k}\right)\right]\right\}\left\{1 - \text{erf}\left[\frac{\gamma}{\sqrt{t}}\left(1 + \frac{hL}{2\gamma^2 k}\right)\right]\right\} \tag{9}$$

where

$$\gamma = \frac{L}{2(k/\rho c)^{1/2}} = \frac{L}{2\sqrt{\alpha}} \tag{10}$$

and erf denotes the error function. In the case of direct application of the burning source to the skin surface, the effective value of the heat transfer coefficient becomes infinitely large. When the source is applied to the skin very rapidly, effecting an instantaneous temperature increment, Eq. (9) reduces to

$$\frac{T_B - T(L, t)}{T_B - T_s} = \text{erf}\left(\frac{\gamma}{\sqrt{t}}\right) \tag{11}$$

Shortly following the pioneering work of Henriques and Moritz, Buettner performed a more comprehensive mathematical simulation of both the burn and postburn processes to identify the transient temperature fields that determine injury.[61] He addressed a broader scope of boundary conditions that approximated several commonly encountered conditions that may lead to a burn. His approach was similar to that of Henriques and Moritz, i.e., to solve Eq. (3) for specific boundary conditions and to apply a damage rate function to the resulting temperature field. Buettner assumed material constants to be uniform in time and space and adopted a one-dimensional, semi-infinite geometry. Comparison of his results with experimental data obtained for human subjects confirmed his analysis for some specific cases.[62] Several of the cases solved by Buettner will be reviewed.

In case 1, Buettner considers a time-invariant heat flux incident on the skin surface. The initial conditions assume a linear temperature profile at the skin surface

$$T(0, x) = T_{s,0} + Kx \tag{12}$$

where $T_{s,0}$ is the initial surface temperature of the skin, and K is constant, describing the gradient. The skin is isothermal for $K = 0$. The surface heat flux is described by

$$\frac{\partial T(t, 0)}{\partial x} = \frac{-q_e}{k} \tag{13}$$

where q_e includes the sum total of environmental heat exchanges attributable to all modes of transport. The solution of this problem is

$$T = \left[\left(\frac{q_e}{k}\right) + K\right]\left\{\frac{2\sqrt{\alpha t}}{\sqrt{\pi}}\exp\left(\frac{-x^2}{4\alpha t}\right) - x\left[1 - \text{erf}\left(\frac{x}{2\sqrt{\alpha t}}\right)\right]\right\} + T_{s,0} + Kx \quad (14)$$

At the surface, $T(0, t) = T_B$, so that

$$T_B - T_{B,0} = \left[\left(\frac{q_e}{k}\right) + K\right]\frac{2\sqrt{\alpha}\,t}{\sqrt{\pi}} \quad (15)$$

Figures 6 and 7 show skin temperature plotted as a function of position and time as calculated from Eq. (14).

In case 2, the initial temperature is constant, i.e., $K = 0$, and a penetrating radiation heat flux I is assumed in addition to a nonpenetrating flux J. Thus, the boundary condition is

$$\frac{\partial T(0, t)}{\partial x} = \frac{-J}{k} \quad (16)$$

and the internal temperature field is described by

$$\frac{\partial T(x, t)}{\partial x} = \alpha\frac{\partial^2 T}{\partial x^2} + \frac{I\alpha_r\,e^{-\alpha_r x}}{\rho c} \quad (17)$$

where α_r is the radiation absorption coefficient. For $t < 0$, $I = J = 0$, and for $t > 0$, the magnitudes of I and J are restricted by $I > 0$ and $J > 0$. The second term on the right side of Eq. (17) represents an internal energy source due to penetrating radiation, the magnitude of which decreases exponentially from the surface.

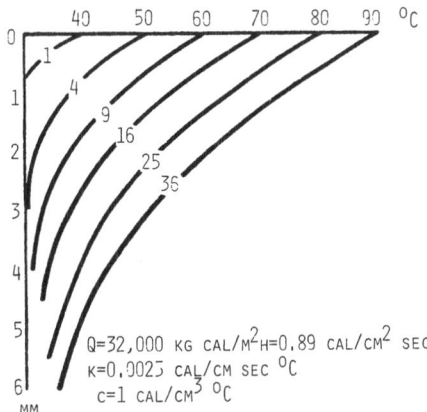

FIGURE 6
Calculated temperature profiles in skin for a nonpenetrating constant radiant heat flux of 0.89 cal/cm² sec, $k = 0.0025$ cal/cm sec °C, $\rho c = 1$ cal/cm °C. From Ref. 61 with permission.

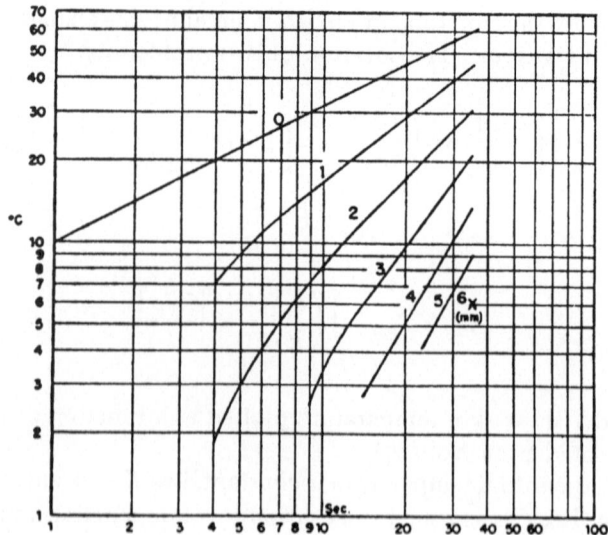

FIGURE 7

Calculated temperature time histories at various skin depths for a nonpenetrating constant radiant heat flux of 0.044 cal/cm^2 sec, $k = 0.0024$ cal/cm sec °C, $\rho c = 1$ cal cm^2/°C. From Ref. 61 with permission.

The solution of this problem is given by

$$T = \frac{I}{2k\alpha_r} \left\{ \exp\left(\alpha_r^2 t - \alpha_r x\right)\left[1 - \mathrm{erf}\left(\alpha_r\sqrt{\alpha t} - \frac{x}{2\sqrt{\alpha t}}\right)\right]\right.$$

$$\left. + \exp\left(\alpha_r^2 t + \alpha_r x\right)\left[1 - \mathrm{erf}\left(\alpha_r\sqrt{\alpha t} + \frac{x}{2\sqrt{\alpha t}}\right)\right]\right\}$$

$$- \frac{I}{2k\alpha_r} \exp\left(-\alpha_r x\right) + \frac{2(I + J)\sqrt{\alpha t}}{k\sqrt{\pi}}$$

$$\times \exp\left(\frac{-x^2}{4\alpha t}\right) - \frac{x(I + J)}{k}\left[1 - \mathrm{erf}\left(\frac{x}{2\sqrt{\alpha t}}\right)\right] + T_s \qquad (18)$$

This equation reduces to an expression for the transient surface temperature by setting $x = 0$

$$T_0 - T_s = \frac{I}{k\alpha_r} \exp\left(\alpha_r^2 t\right)1 - \mathrm{erf}\left(\alpha_r\sqrt{\alpha t}\right) + \frac{2(I + J)t}{\sqrt{\pi}k\rho c} - \frac{I}{2k\alpha_r} \qquad (19)$$

and we obtain the result in Eq. (11). If the radiation is nonpenetrating, $\alpha_r = \infty$. Equation (19) with $\alpha_r = 10$ desdribes solar irradiation, whereas for an infrared source, α_r exceeds 100. Transient temperature profiles at various skin depths, based on these results, are shown in Fig. 8.

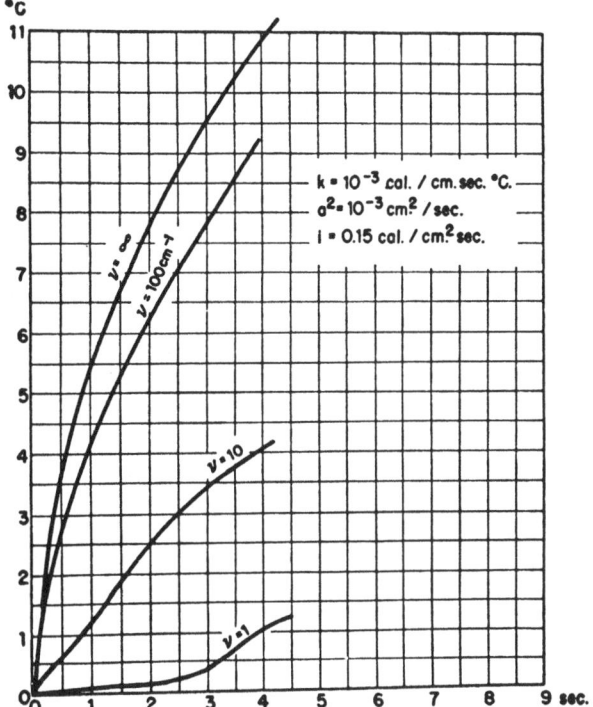

FIGURE 8

Calculated increase in skin temperature due to a penetrating radiant heat flux of 0.15 cal/cm² sec for different values of the radiation absorption coefficient $\alpha_r = \nu$. $k = 10^{-3}$ cal/cm sec °C, $\alpha = 10^{-3}$ cm²/sec. From Ref. 61 with permission.

Next, in case 3, a convective heating is considered. The initial conditions are given by the Newtonian relationship in Eq. (7), and the boundary heat flux is described by

$$\frac{\partial T(0, t)}{\partial x} = \frac{h}{k}(T_e - T_0) \tag{20}$$

The solution of Eq. 3 subject to these conditions is

$$\frac{T_e - T}{T_e - T_s} = \mathrm{erf}\left(\frac{x}{2\sqrt{\alpha t}}\right) + \exp\left(\frac{h^2 \alpha t}{k^2} + \frac{hx}{k}\right)\left[1 - \mathrm{erf}\left(\frac{x}{2\sqrt{\alpha t}} + \frac{h}{k}\sqrt{\alpha t}\right)\right] \tag{21}$$

At the surface, the temperature is given by

$$\frac{T_e - T_0}{T_e - T_s} = \exp\left(\frac{h^2}{k^2}\alpha t\right)\left[1 - \mathrm{erf}\left(\frac{h}{k}\sqrt{\alpha t}\right)\right] \tag{22}$$

This case corresponds to that studied by Henriques and Moritz[58] as given in Eq. (9). Temperature profiles for direct contact heating are plotted in Fig. 9.

In case 4, the heat flux is assumed to pass through an outer protective layer of clothing; $x = 0$ is still set at the surface of the skin. The temperature of the exposed surface of the garment at $x = -d$ is assumed to increase step wise to equal that of a hot environment,

$$T(-d, t) = T_e \tag{23}$$

The heat flux from the protective layer to the skin can be expressed as

$$\frac{\partial T(0, t)}{\partial x} = \frac{T_e - T_0}{d} G(\xi^2) \tag{24}$$

where

$$G(\xi^2) = 1 - 2 \sum_{n=1}^{\infty} (-1)^{n-1} \exp(-n^2 \xi^2) \tag{25}$$

$$\xi^2 = \frac{\alpha_{cl} \pi t}{d} \tag{26}$$

The subscript cl refers to properties of the clothing.

Values of G were considered for three domains:

(a) For small times, $\xi^2 < 0.2$, and G is zero, indicating no heat has reached the skin.

(b) For $0.2 < \xi^2 < 1.0$, the heat flux impinging on the skin increases rapidly. The flux is approximated by the relationship

$$\frac{\partial T(0, t)}{\partial x} = 1.2 \frac{k_{cl}}{k} \frac{T_e - T_0}{d} \sin[M(\xi^2 - 0.2)] \tag{27}$$

where $M = 0.872 \text{ sec}^{-1}$.

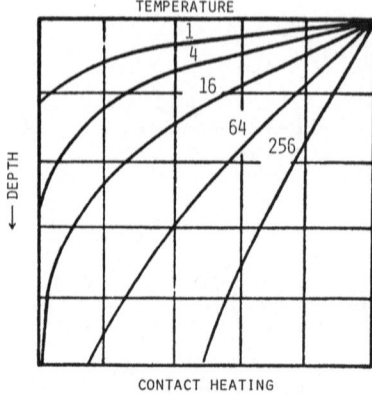

FIGURE 9
Calculated transient temperature profiles in skin following application of a warm contacting body to produce a constant surface temperature. $\alpha = 10^{-3} \text{ cm}^2/\text{sec}$. From Ref. 61 with permission.

(c) For $\xi^2 > 1.0$, G approaches a steady state, so that

$$\frac{\partial T(0, t)}{\partial x} = 1.2 \frac{k_{cl}}{k} \frac{T_e - T_0}{d} \tag{28}$$

The transient skin surface temperature during the second phase can be approximately determined by Eq. (3) for the boundary condition in Eq. (27). Buettner's analytical solution for this problem is

$$T(0, t) = 1.2 k_{cl} \frac{T_e - T_0}{d} (k_{cl} \rho_{cl} c_{cl} M)^{1/2} \left\{ \sin \left[M(\xi^2 - 0.2) - \frac{\pi}{4} \right] + \frac{1}{\sqrt{2}} \right\} + T_0 \tag{29}$$

which is plotted in Fig. 10 for $T_e - T_0 = 1000$ K.

As a final case, Buettner also considered the changes in skin temperature that occur subsequent to removing the heat source. This postheating period of time can play an important role in determining the cumulative injury of a burn, since temperatures will remain above the threshold value until sufficient heat can be transferred away from the site of the lesion.[63] In his final analysis, Buettner assumed that heat was dissipated from the skin by conduction to only the subcutaneous tissue, with convection from the clothing layer to the environment being negligibly small. As a first approximation, skin temperature was assumed to decrease linearly at the end of the heating process with distance from the skin surface. This situation corresponded to the cases of nonpenetrating heating and contact with a hot surface. Furthermore, beyond a depth $x = d$, a constant temperature, equal to the initial value, as assumed to hold. In this case, the boundary condition for cooling became

$$t = 0, \quad x \leq d, \quad T(0, x) = T_{s,0} + T_s \left(1 - \frac{x}{d} \right) \tag{30}$$

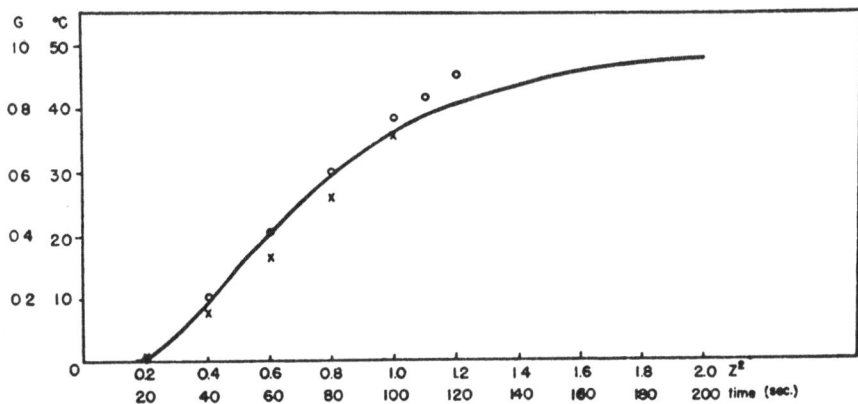

FIGURE 10

Increase with time of heat flow G and skin temperature under a protective clothing layer of 0.99 cm thickness. x indicates estimated skin temperature increase. 0 indicates values of °C vs. $\sin M(\xi^2 - 0.2)$, $\alpha = 10^{-3}$ cm^2/sec, $\Delta T = 10^{3}$°C. From Ref. 61 with permission.

The heat conduction, Eq. (3), was solved for this boundary condition using Smirnov's technique,[64] so that

$$T(x, t) = T_{s,0} + T_s \sum_{n=1}^{\infty} \frac{e^{-u(2n+1)^2}}{(2n + 1)^2} \tag{31}$$

$$u = \frac{\alpha \pi^2 t}{4d^2} \tag{32}$$

Under these conditions, the temperature decays for all positions according to an exponential function depending on t and the cutoff depth d.

In the case of exposure to a high-intensity source for a very short duration, such as might occur in an explosion, Buettner used a different temperature distribution function at the beginning of cooling, described by

$$t = 0, \quad x < x_0, \quad T(x, 0) = T_{0,s} + T_1 \tag{33}$$

$$t = 0, \quad x > x_0, \quad T(x, 0) = T_{0,s} \tag{34}$$

Here, the temperature elevation above the initial skin temperature is assumed to be uniform within a very thin surface layer of thickness x_0, below which there are no propagated thermal effects. The solution of Eq. (3) for these boundary conditions yields

$$T(x, t) = T_1 \left[\text{erf}\left(\frac{x + x_0}{2\sqrt{\alpha t}} \right) - \text{erf}\left(\frac{x - x_0}{2\sqrt{\alpha t}} \right) \right] + T_{0,s} \tag{35}$$

Equation (35) is plotted in Fig. 11, showing the temperature distribution within the skin at various times following the sudden heating of a very thin surface layer. It is clearly seen that significant elevations in temperature can remain long after termination of the thermal insult.

Buettner applied the damage rate model as derived by Henriques, Eq. (5), to his temperature–time data to obtain values for burning effects. The expression used was

$$\Omega = 10^{98} \int_0^t \exp\left[-\frac{75,000}{T + 273} \right] dt \tag{36}$$

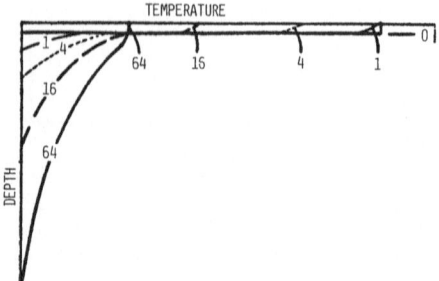

FIGURE 11

Calculated transient temperature profiles in skin following a sudden and short heating of a very thin surface layer. From Ref. 61 with permission.

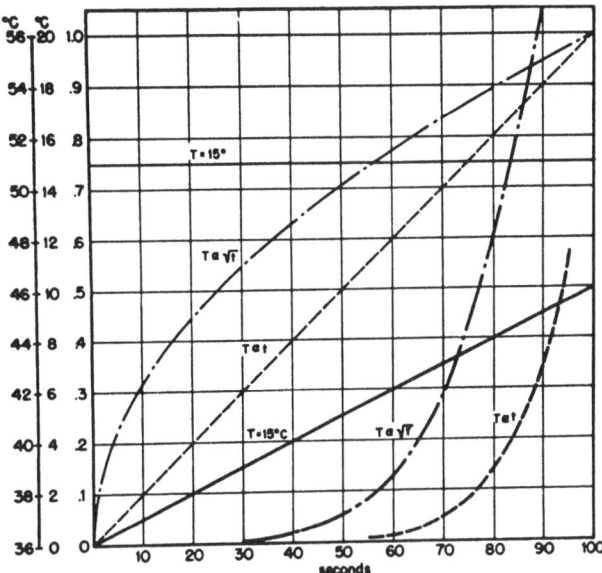

FIGURE 12
Calculated values of the burn integral Ω for specific temperature–time histories, corresponding to constant, square root, and linear temperature functions. From Ref. 61 with permission.

where T is temperature in °C. The burn integral is plotted in Fig. 12 for various transient temperature histories. Irreversible injury is assumed to occur when a certain threshold value of Ω is reached.

Stoll has made an extensive study of the relationship between tissue damage and the physical and physiological parameters that characterize the burn process, with particular emphasis on the development and evaluation of protective garments for fire hazards.[65,66] Her approach involved both experimental and analytical investigations. The measured temperature–time history of the skin was correlated with neurological sensations and observed thermal injury effects. The injury process was simulated by an analytical model to predict transient tissue temperatures; the temperature profiles were applied to a damage rate model based on Henriques's equation.

Transient skin temperature histories were measured at the surface and also at the dermal–epidermal interface, which was assumed to be 8×10^{-5} m deep.[55] Temperature–time profiles during the sequential heating and cooling phases at two different levels of irradiation are presented in Fig. 13. The subsurface temperature lags and does not reach the extreme value achieved at the surface location. It is apparent that the maximum temperature and the rate of rise are both greater for the high-intensity source. It follows that during the initial period of cooling, the tissue remains well above the minimum injury temperature that was defined by Stoll to be 44°C.[55]

An analytical temperature prediction model was developed by Weaver and Stoll to evaluate this type of experimental data.[67] The model assumed a nonpenetrating, constant radial heat flux of rectangular pulse shape with a

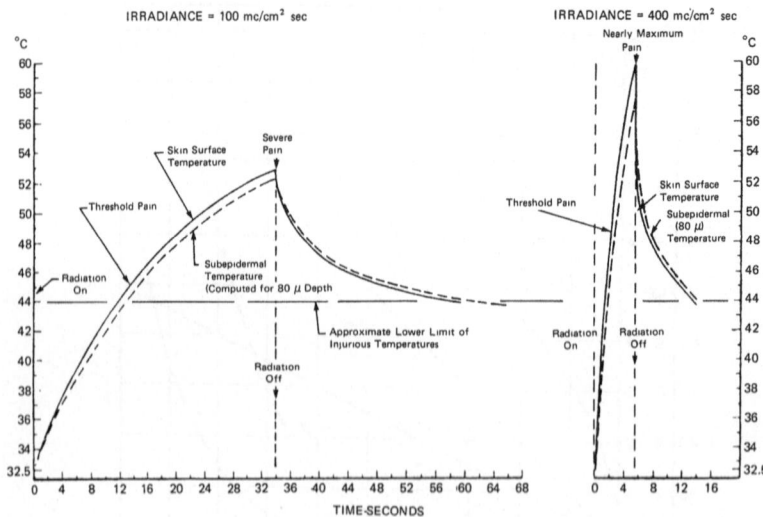

FIGURE 13
Measured skin surface and subepidermal temperature–time histories during and after radiation at two levels of intensity. From Ref. 65 with permission.

duration t_q. In this application, the solution was considered in two time domains: a heating phase for $0 < t < t_q$ and a cooling phase for $t_q < t$. Under these conditions, the solution of Eq. (3), (12), and (13) is given by

$$T - T_0 = \frac{q}{k}\left\{ \frac{2\sqrt{\alpha t}\,\exp\left(-x^2/4\alpha t\right)}{\sqrt{\pi}} - x\left[1 - \operatorname{erf}\left(\frac{x}{2\sqrt{\alpha t}}\right)\right] \right.$$

$$\left. - 2\frac{[\alpha(t - t_q)]^{1/2}\,\exp\left[-x^2/4\alpha(t - t_q)\right]}{\sqrt{\pi}} - x\left[1 - \operatorname{erf}\left(\frac{x}{2[\alpha(t - t_q)]^{1/2}}\right)\right] \right\}$$

$$(37)$$

This expression has a major advantage over previous models, in that both heating and cooling effects are included. Equation (37) was evaluated for the skin surface and a depth of 80×10^{-6} m corresponding approximately to the thickness of the epidermal layer. Irradiations of 100 mC/cm^2 sec for 34 sec[67] and 400 mC/cm^2 sec for 6 sec were considered.[68]

Based on experimental data, such as that shown in Fig. 13, Stoll and Chianta have described threshold temperatures for various thermal injury processes. Accordingly, for thermal injury associated with temperatures of 44°C or higher, injurious conditions would not be attained in the low-energy burn for the initial 12 sec, then would persist during the remaining 22 sec of heating and the first 24 sec of cooling. That the duration of supra-injurious conditions should be longer during cooling than heating is not surprising since the thermal driving potential is often much smaller after the heat source is removed. Thus, a primary rationale for postburn cooling therapy is to lower tissue temperatures below the threshold injury level by providing an effective heat sink at the skin surface. For the second case of a larger irradiation flux of shorter duration, the ratio of times at injurious temperatures during cooling in comparison to heating was found to be even larger, approaching a value of 2 (see Fig. 13 and ref. 55).

The temperature–time data just described, as obtained either experimentally or analytically, afford the possibility of calculating values for the damage integral. However, since the slopes of the heating and cooling curves are not linear, the computational procedures become long and tedious. Therefore, Stoll developed a procedure for performing the calculations on a digital computer.[63] The resultant damage rates are shown as a function of temperature in Fig. 14 for combined radiative and conductive heating. Note that two different damage rate constants appear as a function of the absolute value of the temperature. Subsequently, the damage rates are integrated over time to obtain a value for the integral. Graphical representations of the damage integrals computed for the two radiative heating experiments in Fig. 14 are shown in Fig. 15. In comparing these two curves, it is seen that only 10% of the injury occurred during cooling for the low-intensity source, but more than a third of the total injury occurred during cooling for the high-intensity source. These data clearly indicate the potential importance of the cool-down portion of the thermal insult process.

The analytical solutions for burn models discussed up to this point have proved useful in providing the capability for predicting the temperature fields produced by a few thermal insult regimens.[69] However, the solutions are subject to critical limitations that restrict both their overall domain of applicability and their accuracy when applied to any single protocol. For example,

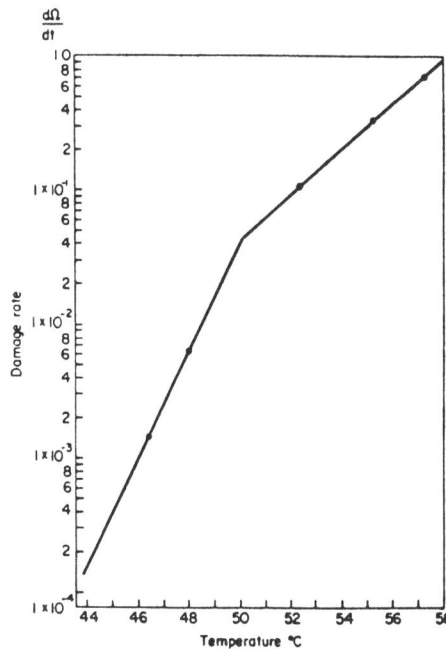

FIGURE 14
Temperature dependence of the damage rate function. From Ref. 63 with permission.

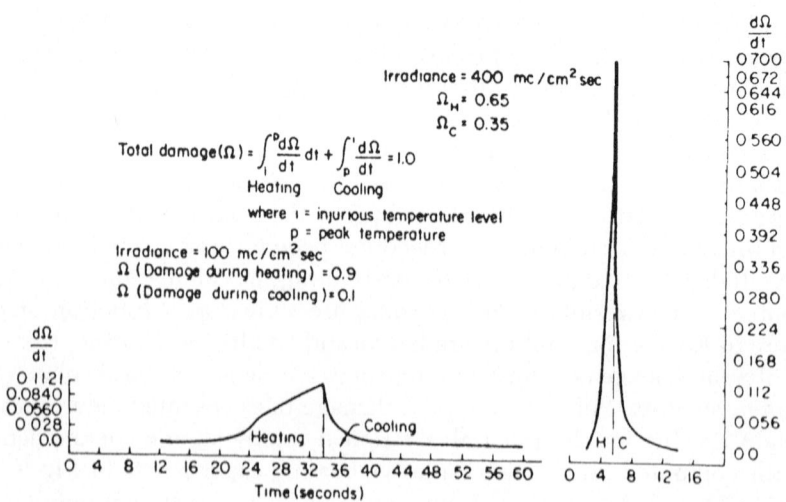

FIGURE 15

Damage integral functions computed from Fig. 13. From Ref. 63 with permission.

they contain no capability for including the effects of tissue blood flow and metabolic heat generation. Also, nonlinear initial temperature distributions, morphological and thermal variations in tissue properties, transient boundary conditions, and realistic representations of tissue geometry and anatomy have not been included. The use of numerical approximation techniques made possible by the availability of large and efficient digital computers has allowed more comprehensive computations to be performed in recent years.

The first application of digital analysis for the study of burn injury was by Stoll in 1960.[63] In this work, she developed an algorithm to evaluate the experimental data in Figure 13 and to calculate values for the damage integral. The skin surface temperature and time were measured throughout the entire heating and cooling cycle. The temperature of the skin at 80×10^{-6} m deep in the skin, representing the volar surface of the forearm where the epidermis separates from subepidermal layers at blistering,[59] was calculated analytically using the measured surface values. As previously noted, the computer was used only to calculate the slope of the damage rate curve and the area under the temperature–time curve to determine the damage integral.

Weaver also used the same approach to calculate temperature–time histories and damage integrals.[70] The analytical model in Eq. (37) was combined with experimental data to predict varying degrees of injury severity, such as the onset of pain and blistering (Fig. 16).[66] These data can be transposed to show the toleration time of skin for a given temperature rise, with injury varying between none and total destruction (Fig. 17).

In the 1970s, there arose both the need for, and capability of, models simulating more complex conditions of burn injury, including, for example, the effects of blood perfusion, metabolism, and temporal and multi-dimensional spatial variations in heating, such as might be produced by a laser. The solution of such problems is beyond the realm of possibility for

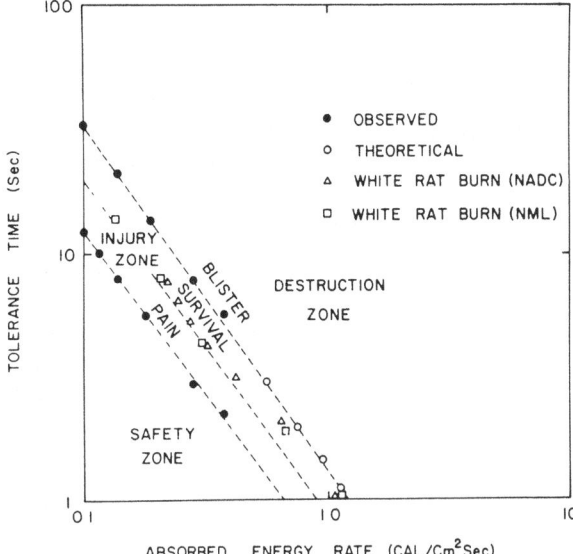

FIGURE 16
Threshold human tolerance for specified levels of burn injury and pain as a function of thermal energy flux. From Ref. 70 with permission.

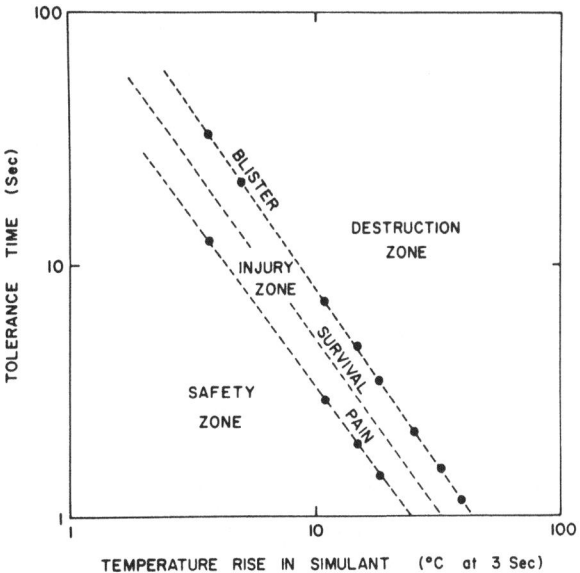

FIGURE 17
Human threshold tolerance for specified levels of burn injury and pain as a function of skin temperature rise. From Ref. 70 with permission.

analytical methods, necessitating the application of appropriate numerical approximation procedures.

One of the earliest applications of the digital computer to this end was by Mainster *et al.*[71] Their objective was to derive a numerical model for treating transient heat conduction problems involving heat sources or sinks with complicated spatial or temporal behavior and differential thermodynamic properties in systems having a layered morphology. Since the primary application was for laser irradiation, a cylindrical geometry was chosen that specifies temperature as a function of r, z, and t. The standard statement of the conduction equation for this application is

$$\rho(z)c(z)\frac{\partial T}{\partial t} = a(r,\,z,\,t) + \frac{k}{R}\frac{\partial T}{\partial R} + \frac{\partial}{\partial R}\left\{K\frac{\partial T}{\partial R}\right\} + \frac{\partial}{\partial Z}\left\{K\frac{\partial T}{\partial Z}\right\} \qquad (38)$$

where a represents the source term.

Mainster *et al.* formulated Eq. (38) in terms of a set of finite difference equations to be solved for a matrix of grid points using a Peaceman–Rachford procedure. In order to accurately simulate this type of system, the grid network must be large enough to approximate boundary conditions far from the source, and it must be fine enough to give high resolution in regions of interest including the effect of exponential absorption. To meet this requirement, Mainster *et al.* devised a nonuniform network with fine meshes in the central region of greatest interest, which was stretched to a coarse mesh size near extremities where the temperature gradients are smaller, to permit approximation of boundary conditions by a relatively small number of grid points. The resulting finite difference grid is shown in Fig. 18. In their notation, the indices for the coordinates z, r, and t are, respectively, i, j, and k.

The difference formulations of the terms in Eq. (38) are as follows:

$$\rho c\frac{\partial T}{\partial t}\bigg|_{i,j,k+1/2} \doteq \frac{2\rho_i c_i}{\Delta t_k}(T_{i,j,k+1/2} - T_{i,j,k}), \qquad j = 0, 1, 2, \ldots, N \qquad (39)$$

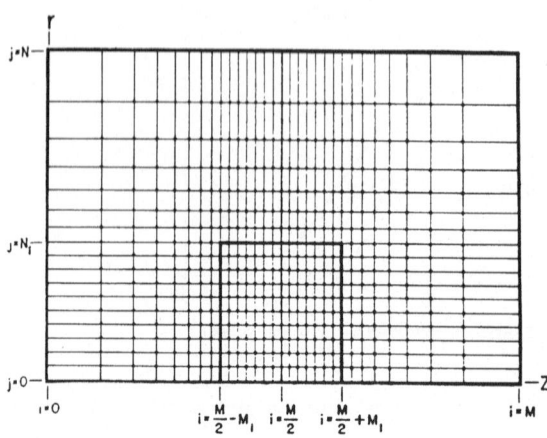

FIGURE 18
A nonlinear, two-dimensional finite difference grid for modeling temperature fields created by a distributed heat source. From Ref. 71 with permission.

$$\Delta t_k \doteq t_{k+1} - t_k \tag{40}$$

$$\frac{k}{r}\left(\frac{\partial T}{\partial r}\right)\bigg|_{i,j,k} \doteq \frac{k_j}{r_j}\left(\frac{T_{i,j+1,k} - T_{i,j-1,k}}{r_{j+1} - r_{j-1}}\right), \qquad j = 1, 2, \ldots, N-1 \tag{41a}$$

$$\doteq \frac{\partial}{\partial r}\left(k\frac{\partial T}{\partial r}\right)\bigg|_{i,0,k} \qquad i = 0 \tag{41b}$$

$$\frac{\partial}{\partial r}\left(k\frac{\partial T}{\partial r}\right)_{i,j,k} \doteq \frac{2k_i}{r_{j+1} - r_{j-1}}\left(\frac{T_{i,j+1,k} - T_{i,j,k}}{r_{j+1} - r_j} - \frac{T_{i,j,k} - T_{i,j-1,k}}{r_j - r_{j-1}}\right),$$

$$j = 1, 2, \ldots, N-1 \tag{42a}$$

$$\doteq \frac{k_i}{r_1^2}\left(T_{i,1,k} - T_{i,0,k}\right), \qquad j = 0 \tag{42b}$$

$$\frac{\partial}{\partial z}\left(k\frac{\partial T}{\partial z}\right)_{i,j,k} = \frac{2k_i}{z_{i+1} - z_{i-1}}\left(\frac{T_{i+1,j,k} - T_{i,j,k}}{z_{i+1} - z_i} - \frac{T_{i,j,k} - T_{i-1,j,k}}{z_i - z_{i-1}}\right)$$

$$+ \frac{k_{i+1} - k_{i-1}}{z_{i+1} - z_{i-1}}\frac{T_{i+1,j,k} - T_{i-1,j,k}}{z_{i+1} - z_{i-1}},$$

$$j = 0, 1, 2, \ldots, N \tag{43}$$

On the nonuniform grid, the distances between adjacent grid points for these equations are defined by

$$z_{i+1} - z_i = \begin{cases} \Delta z(R_1)\left[\left(\dfrac{M}{2}\right) - M_1 - i\right], & 0 \le i \le \left(\dfrac{M}{2}\right) - M_1 - 1 \tag{44a} \\[2ex] \Delta z, & \left(\dfrac{M}{2}\right) - M_1 \le i \le \left(\dfrac{M}{2}\right) + M_1 + 1 \tag{44b} \\[2ex] \Delta z(R_1)\left[i - \left(\dfrac{M}{2}\right) - M_1 + 1\right], & \left(\dfrac{M}{2}\right) + M_1 \le i \le M-1 \tag{44c} \end{cases}$$

$$r_{j+1} - r_j = \begin{cases} \Delta r, & 0 \le j \le N_1 - 1 \tag{45a} \\ \Delta r(R_2)^{j+1-N_1}, & N_i \le j \le N-1 \tag{45b} \end{cases}$$

where R_1 and R_2 are constant axial- and radial-stretching ratios, respectively, for the difference network. The z and r dimensions of the source term are defined as M_1 and N_1, respectively.

The computational accuracy of the model was tested on a simplified problem for which an analytical solution is known. A constant, uniform cylindrical heat source of 2.1×10^{-3} m diameter and 0.6×10^{-3} m length

(Fig. 19) was specified. The power density of the source is 10 cal/sec cm^3. Both the source and medium have identical thermodynamic properties: $k = 0.0014$ cal/°C cm, $c = 1$ cal/gm °C, and $\rho = 1$ g/cm^3. The initial uniform temperature is 0°C, and the outermost grid points are maintained at 0°C.

An analytical solution for this boundary value problem is obtained from Carslaw and Jaeger.[72]

$$T = \frac{a}{4}\left(\frac{pc}{\pi k^3}\right)^{1/2} \int_0^t \frac{dt'}{(t-t')^{3/2}} \int_{-0.03}^{0.03} \exp\left[\frac{-(z')^2 \rho c}{4(t-t')k}\right] dz'$$

$$\times \int_0^{0.105} \exp\left[\frac{-(r')^2}{4(t-t')}\frac{\rho c}{k}\right] r' \, dr' \tag{46}$$

Equation (46) was evaluated numerically and compared with the finite difference solution for various time-stepping and mesh size parameters. The values used are shown in Table 1, and the solutions for the individual cases are presented in Table 2.

Comparison of cases 1–4 indicates that it is not necessary to use extremely fine time steps to obtain accurate solutions for constant source terms. There was no significant difference in results for the cases reported, for which the Fourier number varied between 3.5×10^{-2} and 1.75; the authors reported obtaining excellent results for Fourier numbers as high as 3×10^4. In addition, the data show that solution accuracy is not affected by the size of the initial time step, provided it is not too large and that use of a nonuniform grid,

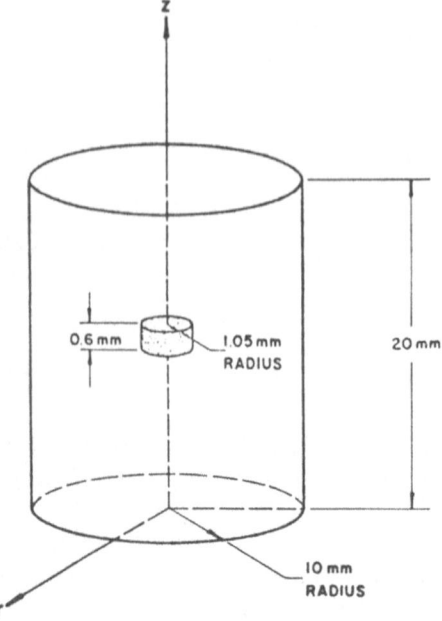

FIGURE 19
Model for analytical solutions for temperature fields with a uniform cylindrical heat source. From Ref. 71 with permission.

TABLE 1
Parameters for Finite Difference Solutions[a]

Case	R_1	R_2	M	N	Δ_z (mm)	Δ_r (mm)	Sequence of time steps executed (sec)
I	1	1	100	100	0.20	0.10	0.001, 0.01, 0.1, 1, 10, 100, 1000, ...
II	1	1	100	100	0.20	0.10	0.005, 0.01, 0.02, 0.05, 0.1, 0.2, 0.5, ...
III	1	1	100	100	0.20	0.10	0.05, 0.1, 0.2, 0.5, 1, 2, 5, ...
IV	1	1	100	100	0.20	0.10	0.5, 1, 2, 5, 10, 20, 50, ...
V	1	1	32	32	0.60	0.30	0.005, 0.01, 0.02, 0.05, 0.1, 0.2, 0.5, ...
VI	1.1	1.06	100	50	0.05	0.10	0.005, 0.01, 0.02, 0.05, 0.1, 0.2, 0.5, ...

[a] From Ref. 71 with permission.

within the tested limits, does not compromise the accuracy. Further evaluation of the program for a rapidly varying source term indicated the need to choose difference parameter values for which the temperature function remains well behaved.

The finite difference approach developed by Mainster *et al.* proved to be quite useful and has been applied to the analysis of numerous biological heating problems, including retinal irradiation.[73–77] However, a primary

TABLE 2
Finite Difference Solutions for Transient Temperature Problem Obtained Using the
Parameters Defined in Table 1[a]

Time steps executed (sec)	Case I (°C)	Case II (°C)	Case III (°C)	Case IV (°C)	Case V (°C)	Case VI (°C)	Analytic solution (°C)
1×10^{-3}	0.0096						0.0100
2×10^{-3}							0.0200
5×10^{-3}		0.0498			0.0499	0.0498	0.0500
1×10^{-2}	0.100	0.100			0.0996	0.100	0.100
2×10^{-2}		0.200			0.198	0.200	0.200
5×10^{-2}		0.495	0.494		0.490	0.498	0.498
1×10^{-1}	0.970	0.972	0.971		0.963	0.978	0.979
2×10^{-1}		1.85	1.85		1.85	1.84	1.83
5×10^{-1}		3.94	3.94	4.00	4.16	3.86	3.84
1×10^{0}	6.47	6.29	6.29	6.32	6.99	6.15	6.12
2×10^{0}		9.04	9.04	9.03	10.3	8.87	8.90
5×10^{0}		12.5	12.5	12.5	14.1	12.3	12.1
1×10^{1}	15.1	14.4	14.4	14.4	16.1	14.2	14.1
2×10^{1}		16.0	16.0	16.0	17.7	15.7	15.7
5×10^{1}		17.4	17.4	17.1	19.1	17.2	17.1
1×10^{2}	18.3	18.1	18.1	18.1	19.8	17.9	17.8
2×10^{2}		18.5	18.5	18.5	20.3	18.4	18.3
5×10^{2}		18.7	18.7	18.7	20.4	18.7	18.8
1×10^{3}	18.7	18.7	18.7	18.7	20.4	18.7	19.0

[a] From Ref. 71 with permission.

drawback to the Mainster *et al.* model for application in living tissues is that no term is included to account for the thermal effects of local blood flow. Subsequent investigators have incorporated the blood flow effect into computer models.[77]

Based on Mainster's approach, Takata *et al.*[78] have developed a more detailed computer model for the laser irradiation of skin. A sequence of calculations was carried out to develop a laser-induced transient temperature distribution and the resulting damage factor. First, the geometry of the skin was laid out as a nonlinear rectangular grid similar to that used by Mainster *et al.*[71] in order to quantify experimental absorption. Next, transient temperature profiles, created by a radially symmetric laser beam at normal incidence were computed using the Peaceman–Rachford technique as previously discussed. Finally, thermal injury was predicted from the temperature data by using a Henriques-type damage integral.

The heat conduction equation was modified by Takata *et al.* to include the effects of blood flow and water phase change from liquid to vapor. With the addition of these terms, Eq. (38) becomes

$$\rho c \frac{\partial T}{\partial t} = a(r, z, t) + \frac{k}{r} \frac{\partial T}{\partial r} + \frac{\partial}{\partial r} \left(k \frac{\partial T}{\partial r} \right) + \frac{\partial}{\partial z} \left(k \frac{\partial T}{\partial z} \right)$$

$$- w_b(z)(T_x - T_0) - H \tag{47}$$

The position-dependent convective term $w_b(z)(T_x - T_0)$ is a Fick-type relation, as described in Chapter 6; H is the latent heat loss with change of phase. The laser energy deposition function was assumed to obey Beer's law; the attenuation along the z axis is described by

$$a(r, z, t) = a(r, 0, t) \exp \left[- \int_0^z \alpha_r(z) \, dz \right] \tag{48}$$

where $\alpha_r(z)$ is a radiation absorption coefficient that is dependent on media properties.

Convective heat loss from the skin is described by

$$-k \frac{\partial T(r, 0, t)}{\partial z} = h_e [T(r, 0, t) - T_e] \tag{49}$$

where the value of the heat transfer coefficient h_e is dependent on whether a sweat layer is present. Formation of a steam blister changes the thermal resistance for heat flux to underlying tissues by introducing an insulating film of water vapor. By neglecting the heat capacity of the water vapor, the heat flux across the blister can be described by

$$-k \frac{\partial T}{\partial z} \bigg|_{z_b - \delta} = -k \frac{\partial T}{\partial z} \bigg|_{z_b + \delta} = h_b \left(T \bigg|_{z_b - \delta} - T \bigg|_{z_b + \delta} \right) \tag{50}$$

where z_b is the depth at which the blister formed, δ is a small increment in the z dimension, and h_b is the heat transfer coefficient across the blister.

A finite difference solution for the transient temperature history of the skin was obtained that was then used to predict the extent and severity of the burn injury. Takata *et al.* used a damage integral in the standard format for this calculation[78]

$$\Omega(r, z) = \int_0^\infty A_1 \exp\left[-A_2 t(r, z, t)\right] dt \tag{51}$$

where A_1 and A_2 were taken to be discontinuous functions of temperature that provided acceptable correlations with experimental data.

$$\left.\begin{array}{l} A_1 = 4.322 \times 10^{64}\ \text{sec}^{-1} \\ A_2 = 5.0 \times 10^4\ \text{K} \end{array}\right\} \quad 317 \leqslant T \leqslant 323\ \text{K} \tag{52a}$$

$$\left.\begin{array}{l} A_1 = 9.389 \times 10^{104}\ \text{sec}^{-1} \\ A_2 = 8.0 \times 10^4\ \text{K} \end{array}\right\} \quad 323\ \text{K} \leqslant T \leqslant 333\ \text{K} \tag{52b}$$

The integral for Ω in Eq. (51) was calculated for each grid point of interest.

Criteria for classifying burn severity were defined according to the numerical value of the parameter Ω or the peak temperature reached at a specific location in the skin. These criteria are shown in Table 3. Data are presented showing the radial and axial extents of irreversible damage produced by irradiation with a laser. Parameters of the irradiation included wavelength, power, beam radius, and pulse duration. A sample plot of the predicted transient temperature profiles induced by a ruby laser are shown in Fig. 20. A comparison of predicted depth and radius of injury vs. pulse duration for a CO_2 laser are shown in Fig. 21.

Quite recently, Palla[80] has presented a predictive burn model for scald injury, with the objective of providing a more thorough understanding of thermal injury from heated fluids as a prerequisite for the development of improved product standards. The bioheat equation was written, including blood perfusion and metabolic heat generation effects, but without convective interaction with major vascular components, to describe the transient tissue temperature for a step change in the environmental temperature. Forced

TABLE 3
Burn Injury Criteria[a]

First-degree burns	$\Omega = 0.1$ (assumed)
Second-degree burns	$\Omega = 1.0$
Third-degree burns	$\Omega = 10{,}000$
Fourth-degree burns	131°C at base of epidermis
Fifth-degree burns	400°C at exterior surface of epidermis

[a] From Ref. 78 with permission.

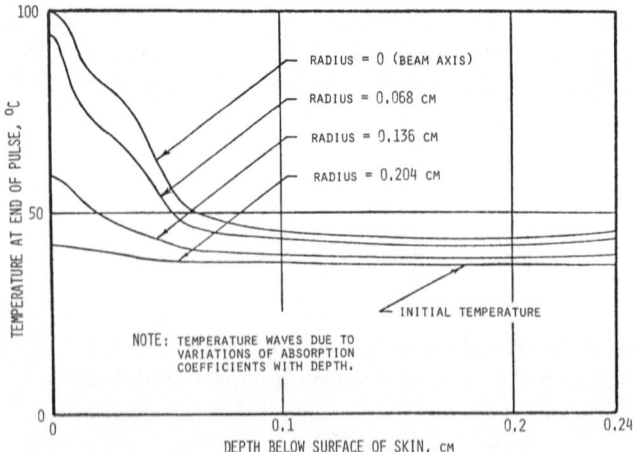

FIGURE 20

Temperature profiles predicted by finite element model at the completion of a ruby laser pulse to skin. Pulse energy = 1.74 J, pulse duration = 5×10^{-4} sec, beam radius = 0.17 cm at $1/e^2$ point (see Ref. 78).

convective heat transport between the skin and a hot, surrounding fluid was assumed during heating, and a combination of evaporation and low-velocity convection was assumed during cooling. A nonlinear initial temperature distribution was assumed, decreasing from a core maximum to a point approaching a selected ambient value. The tissue was modeled by a one-dimensional nonlinear finite difference grid, with the node spacing increasing

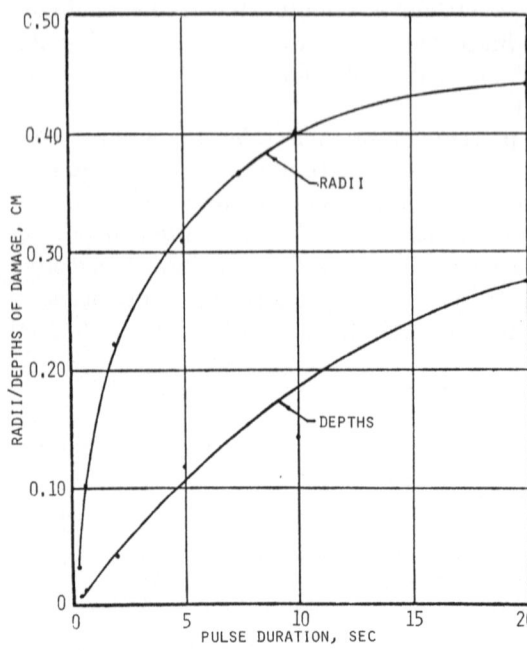

FIGURE 21

Radial and depth extent of burn damage produced by a CO_2 laser pulse to skin. Pulse power = 1.55 W, beam radius = 0.38 cm at $1/e^2$ point (see Ref. 78).

with depth into the skin. A set of implicit difference equations were developed to describe the temperature field and solved on a digital computer by the matrix inversion technique. The computed temperature–time history was used to calculate injury using a Henriques-type criterion. Threshold values for injury at the base of the epidermis ($\Omega = 1.0$) were used as the primary basis for evaluating the burn effect. Typical scald protocols considered were 70°C for 1 sec and 60.2°C for 5 sec.

Palla used this model to study the influence of a number of physiological and environmental parameters on the injury threshold. The initial tissue temperature profile was shown to be an important consideration in determining injury for scalds of short duration, with diminishing effect for increasing exposure times. For fluid exposure times greater than 5 sec, initial conditions were of practically no influence. Blood perfusion and metabolic heat generation were not considered to play important roles in regulating the degree of injury. However, the value assumed for the surface convective heat transfer coefficient was judged to be the most crucial factor in assessing the scald threat. The scald fluid temperature threshold for a significant burn injury was therefore a strong function of the surface heat transfer coefficient. Thus, establishing a criterion for the potential for scald injury is a complex task involving specification not only of fluid temperature and duration of exposure, but also requiring information concerning details of the flow conditions.

The works of Takata and Palla illustrate that the application of numerical modeling techniques implemented on the powerful digital computers that are now available enables a much broader range of burn scenarios to be simulated. The initial analytical formulations of Henriques,[58] Buettner,[61] and others have proved quite useful as a basis for subsequent numerical treatments. As the capability for handling both physical and physiological nonlinear effects increases, models that treat specific burn protocols will be feasible.

The Henriques damage integral has proved useful in predicting high-temperature thermal injury beyond the dermal and opthalmic applications previously described. For example, Link, Incropera, and Glover predicted the thermal response of tissue subjected to plasma scalpel heating with reasonable accuracy in comparison to *in vitro* and *in vivo* data.[81] Experimental and analytical results are shown in Fig. 22. The deviations of experiment and theory during cooling suggest there is a strong effect of model parameters on the relevence of the analytical results. Features of the surface heating not incorporated in the model include contraction and separation of tissue due to heat intensity, thus ablating or physically removing surface layers.

Cellular level hyperthermic injury kinetics were studied by Lloyd *et al.*[82] using a human erythrocyte model. Moussa *et al.*[83] also used the Henriques model to study the effects of hyperthermia on injury to HeLa cells. A stepped thermal history was used to approximate the burn process, consisting of an initial, linearly increasing temperature, which is then held constant for a period of time. This protocol was experimentally simulated on a controlled temperature microscope stage in which injury to the cells of interest could be observed directly and continuously.[84,85] Subsequently, a similar system

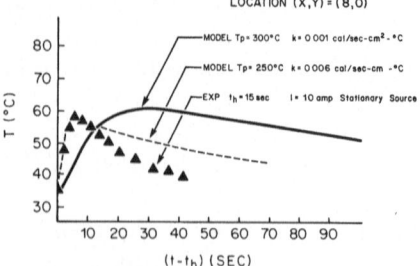

FIGURE 22

Comparison of predicted and measured temperature histories in liver for plasma scalpel heating. From Ref. 81 with permission.

was used to study the hemolysis kinetics of erythrocytes at elevated temperatures.[86] Good correlation was obtained between the kinetics model and the observed onset of cellular injury.

Recently, Diller and Hayes have used a finite element procedure to model the burn injury process in skin.[87] The finite element technique offers the advantages of handling problems involving complex geometries and nonlinear or discontinuous boundary or property conditions.[88,89] See Appendix 3 for a brief description of this method. A general code for solving transient, two-dimensional parabolic boundary value problems was applied.[90] The general form of the differential equation addressed is as follows:

$$\sum_{i,j=1}^{N} \frac{\partial}{\partial x_i}\left[k_{ij}(\mathbf{x}) \frac{\partial T}{\partial x_j}\right] + f_3(t)g_3(\mathbf{x})T + f_4(t)F_1 + F_1$$

$$+ f_5(t)F_2 f_1(t)g_1(\mathbf{x})\frac{\partial T}{\partial t} = f_2 z(t) \qquad (53)$$

where the $f(t)$'s are functions of time; $g(\mathbf{x})$'s are functions of the spatial coordinates; and F's are unspecified functions, which, in the present case, are interpreted as nonhomogeneous metabolic heat generation and blood perfusion. By proper selection of these functions, Eq. (47) may be stated directly in the format give by Eq. (53). The differential equation to be solved is developed in rectangular coordinates with constant property values.

The model simulates a physical process consisting of epidermal surface heating from a solid applicator of finite dimensions, for a specified temperature and time, followed by exposure to ambient air, characterized by a natural convection boundary condition; and finally, after a measured delay period, immersion into a cold water bath to simulate therapy (Fig. 23).[91] During the application of heat, the skin in the area surrounding the hot object is assumed to have a convective boundary heat flux. Due to the inherent flexibility of the finite element method in handling a grid mesh of general geometry, it was possible to generate a nonlinear network emphasizing a high concentration of elements in the region for which the steepest thermal gradients were expected. A sample grid is shown in Fig. 24. A composite system consisting of three parallel, unequal layers of epidermal, dermal, and subcutaneous fatty

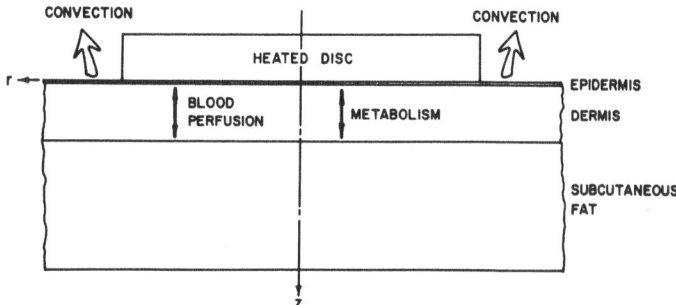

FIGURE 23

Burn created by contact of a finite-dimensioned heating source with the skin. From Ref. 87 with permission.

tissue is indicated. The initial temperature was assumed uniform at 34 °C, and blood perfusion was included only in the dermal layer.

Temperature contours are plotted for two different times during the preceding process, corresponding to the completion of the burn and following a subsequent 20 sec of surface cooling (Figs. 25, 26). The high gradient in the temperature profiles near the edge of the burn source is quite apparent. The generated times and temperature solutions were then used in a Henriques-type Eq. (36), to calculate values for the damage integrals. The resulting predicted injury contour regions are plotted in Fig. 27 for the thermal history described in Fig. 25.

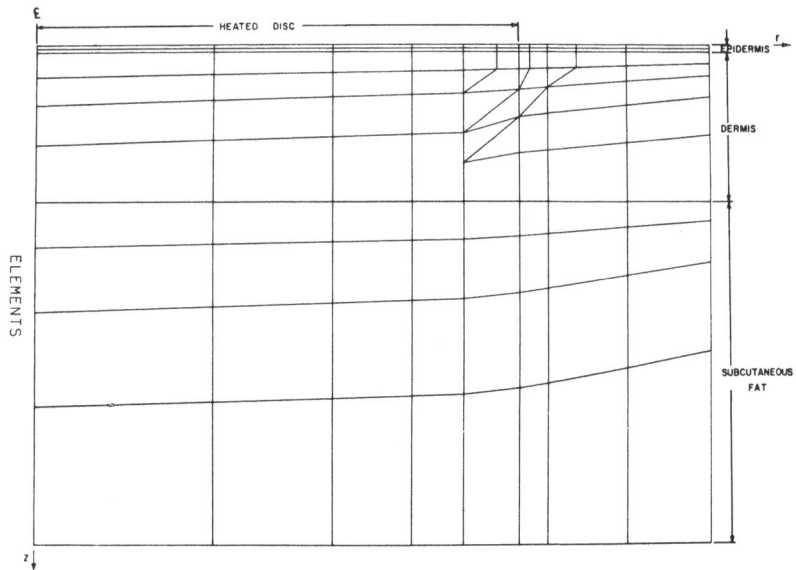

FIGURE 24

Nonlinear finite element grid to model the burn process depicted in Fig. 23. Accuracy is enhanced by increasing density of elements in the region of largest temperature gradients and by aligning element nodes with the predicted temperature profiles. From Ref. 87 with permission.

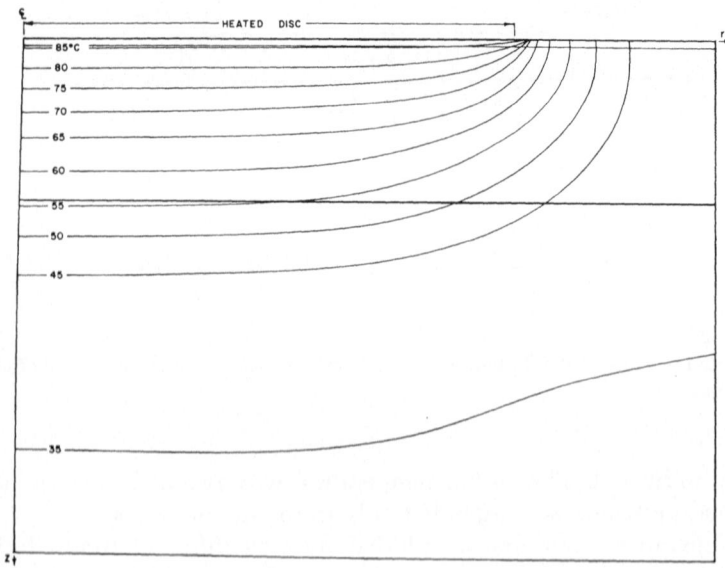

FIGURE 25

Temperature profiles predicted by finite element model at end of a burn at 85°C for 5 sec for the configuration shown in Fig. 23. Initial temperature is 34°C. For epidermis, $\alpha = 0.66 \times 10^{-7}\,\mathrm{m^2/sec}$, $w_b = 0$. For dermis, $\alpha = 1.3 \times 10^{-7}\,\mathrm{m^2/sec}$, $w_b = 0.02\,\mathrm{cm^3}$ blood/cm³ tissue sec. For subcutaneous fat, $\alpha = 0.81 \times 10^{-7}\,\mathrm{m^2/sec}$, $w_b = 0$. $h = 7\,\mathrm{W/m^2\,K}$. From Ref. 87 with permission.

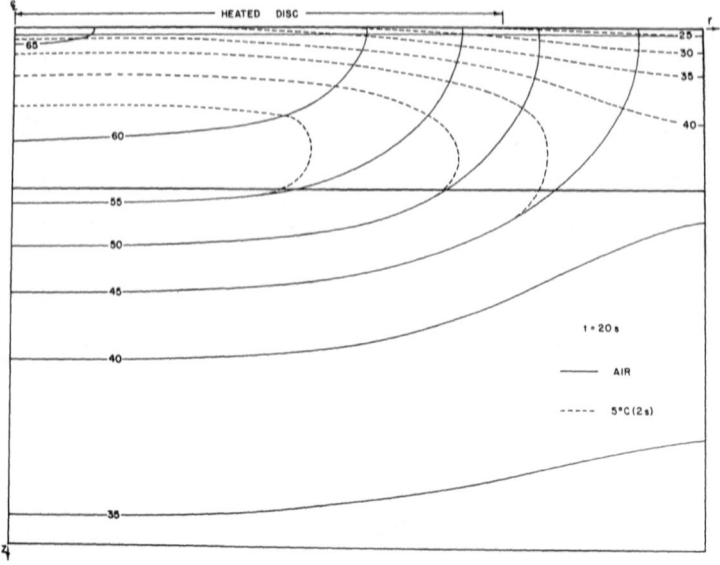

FIGURE 26

Temperature profiles at 20 sec predicted by finite element model for 85°C burn for 5 sec, followed by exposure to air for 1 sec with $h = 7\,\mathrm{W/m^2\,K}$ and exposure to cold water for 2 sec with $h = 700\,\mathrm{W/m^2\,K}$. From Ref. 87 with permission.

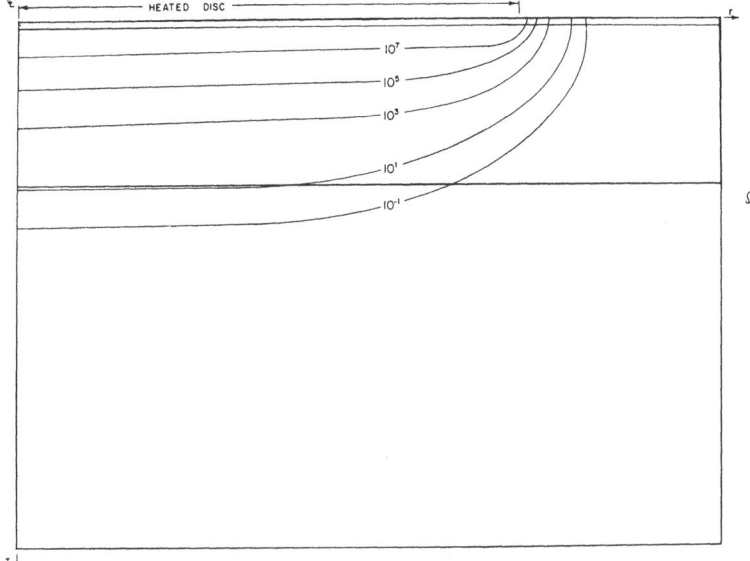

FIGURE 27
Contours of the damage integral calculated from the burn process depicted in Fig. 25. From Ref. 87
with permission.

5. COOLING THERAPY FOR BURN WOUNDS

Numerous reports have appeared in the literature describing the use of
cold application to the burn wound in the immediate postburn period to
reduce the extent of thermal injury; they have been reviewed by Wilson *et
al.*[92] Although cooling therapy for burns has long been recognized and
advocated,[93,94] neither the local nor systemic action and effects have been
completely identified.[95] Consequently, it has been a subject of continuing
investigation. Experimental studies have lead to the hypothesis that local
surface cooling subsequent to the primary insult acts to reduce (1) edema
formation, both locally and systemically[96,97]; (2) inflammatory reactions[98];
(3) local metabolism and the production rate of toxins[99]; and (4) pain.[100]
Cooling has been shown recently to specifically prevent release of histamine
from burned tissues, thereby inhibiting one of the primary inflammatory
mediators.[101]

It is generally agreed that cooling is most advantageous when applied
immediately following a burn injury and that this type of treatment loses its
effectiveness as the time between the insult and the treatment is prolonged.
The cooling temperatures that yield the most beneficial results have yet to be
clearly identified. The desired temperature for burn treatment on burns of
30% or less of the body's surface area is reported to vary from 0 to 30°C.[102,103]
The length of treatment is also widely disputed, with beneficial results reported for cooling delays ranging up to 3 hours postburn.[94,104] However, the
general belief is that this treatment loses its effectiveness after 1-hr postburn.

Recently, the efficacy of cooling procedures in modifying the physiological response to burn injury at both the microscopic and macroscopic levels has been quantified in a series of experiments by Ross, Thompson, and Diller.[105,106] Initially, studies were performed using the microvasculature of the hamster cheek pouch as a model tissue.[57] An *in vivo* hamster cheek pouch preparation was exposed on a computer-controlled microscope stage to a standardized thermal insult under direct microscopic observation and subsequently subjected to diverse conditions of postburn cooling. Following the standard heat insult, there was a delay of either 30 sec or 10 min before the cooling therapy was initiated. During this interval, the extent of microvascular stasis was measured. The criterion for evaluating the effects of postburn cooling was the recovery of flow in specific vessels after the burn. Specifically, the number of vessels temporarily occluded by the burn process that reopened following the cooling therapy and continued flowing throughout the remainder of the experiment was measured. The cooling regimen was defined according to the following independent parameters: (1) the minimum cooling temperature employed, (2) the duration of cooling, and (3) the delay in initiation of cooling following the burn. Tissue temperature in the cooling protocols was lowered by a constant flow of refrigerant fluid through the microscope stage until a desired minimum temperature between 3 and 25°C was reached; temperature was held there for either 5, 30, or 60 min, and then rewarmed to the physiological state. In the control burns that had no cooling, the microcirculation exhibited almost no signs of flow recovery.

For each duration of cooling, an optimum temperature was identified

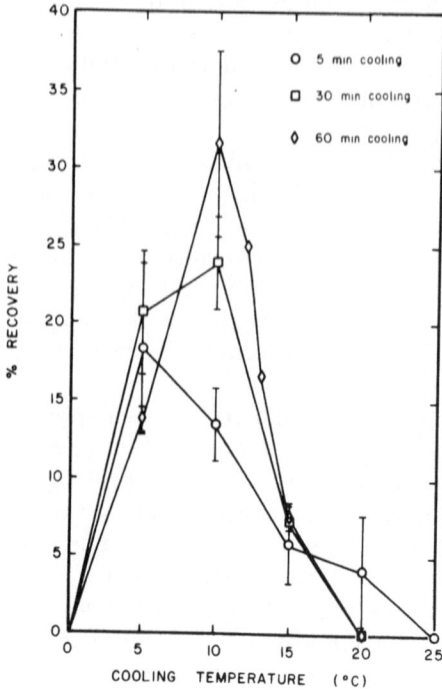

FIGURE 28

Recovery of microcirculatory blood flow for tissue cooling following a 10-min postburn delay (see Ref. 105).

that yielded a maximum recovery of flow to the occluded vessels. For 60 min cooling, the optimum temperature was 10°C, which produced a 32% recovery of the occluded vessels. As the duration of the cooling period was shortened, the optimum tissue temperature and the percentage of recovery decreased: this trend can be seen clearly in Fig. 28. When the initial delay was reduced from 10 min to 30 sec, the percentage of recovery increased significantly (Fig. 29). For temperatures above 8°C, the percentage of recovery is doubled by reducing the postburn cooling delay, and the total area under the recovery curve is increased by a factor of 2.5. These experiments indicate that with a proper postburn cooling protocol, the extent of injury to the local microcirculation due to thermal trauma can be significantly reduced. As the duration of cooling increases and the delay between the burn and initiation of cooling decreases, the therapy becomes more effective, and less severe depression of temperature is required to achieve a given benefit.

Cooling therapy was also evaluated on a macroscopic scale by using a modified dorsal scald burn technique on rats. The animals were prepared for an experimental trial by a depilatory procedure to ensure uniform skin thickness. Prior to a trial, the animals' backs were shaved, and then they were weighed to determine the depth of immersion into a hot water bath necessary to yield high temperature exposure over 20% of the body surface area.[107] The burn protocol was designed to produce a full-thickness skin burn and consisted of the prescribed immersion into a 90°C water bath for 15 sec. The subject was then removed, dried immediately, and after a predetermined delay period, immersed into a cold water bath at a set temperature and duration.

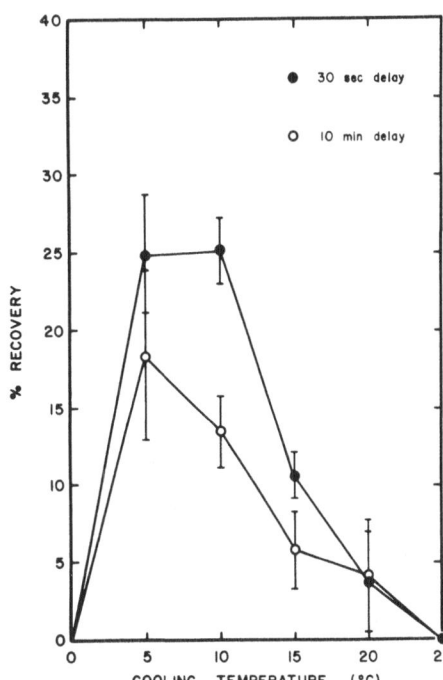

FIGURE 29
Recovery of microcirculatory blood flow for tissue cooling of 5-min duration (see Ref. 105).

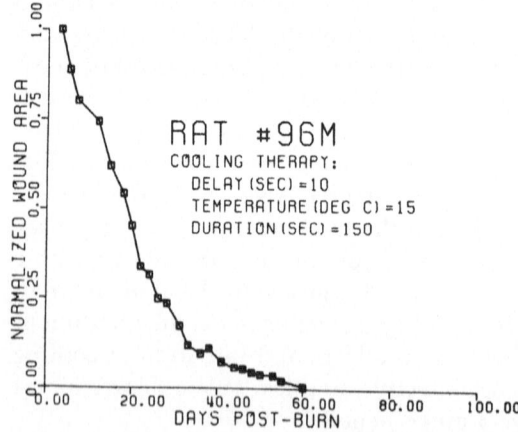

FIGURE 30

An unhealed burn wound area measured over a two-month postburn period by digital computer analysis of sequential photographs of the wound site. From Ref. 106 with permission.

Values of the cooling protocol delay ranged from 10 to 250 sec, for temperatures from 5 to 20°C, and cooling times from 30 to 750 sec.

The physiological response to the burn and therapy was quantified by measuring wound area. Serial photographs of the wound were taken over the duration of the healing process and analyzed by computer.[106] Compilation of the sequential, unhealed wound area data for each animal (Fig. 30) enabled the calculation of a value for the rate of healing. A single exponential decay function fit the healing data well; the correlation coefficient was in the range of 0.95–0.99. A rate constant for the healing process was then obtained from the exponential function and used to determine whether cooling affected the temporal characteristics of the healing process.

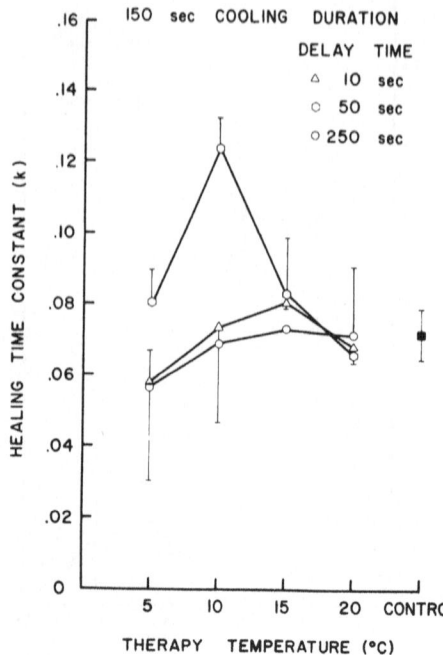

FIGURE 31

Measured time constant for wound healing for postburn cooling of a standard 90°C, 15-sec scald burn. From Ref. 106 with permission.

Results for delays of 10–250 sec and a cooling time of 150 sec are shown in Fig. 31. A substantial increase in the rate of wound healing is observed at a cooling temperature of 10°C. Overall data are compiled in Table 4 for all experiments. The results are not so clear-cut as in the microscopic study, but the trend suggests that a cooling temperature in the range 10–15°C is beneficial, particularly when used in conjunction with a therapy delay of about 1 min and a duration of about 2.5 min.

As previously noted, the biochemical basis for the efficacy of postburn cooling is yet to be demonstrated conclusively; indeed, it is likely that a number of factors may be involved. The results of the finite-element simulation[87] suggest that, in general, the tissue is returned to the physiological range of temperatures in a very short period in comparison to that for which thermal therapy might be considered. Thus, it is likely that cold inhibition of the release and action of chemical mediators for the inflammatory response are the major contributing factors to cold therapy.[101]

6. QUANTIFICATION OF THE MICROSCOPIC RESPONSE TO BURNS

The preceding analysis has focused on a macroscopic interpretation of burn injury. However, as noted in the earlier section on the physiological aspects of burn injury, the macroscopic response is dictated by microscopic

TABLE 4
Exponential Time Constants for Burn Wound Healing with Cooling Therapy[a]

Therapy delay (sec)	Therapy duration (sec)	Sex	Therapy temperature (°C)			
			5	10	15	20
10	30	M	0.0729	0.1007	0.0705	0.0630
		F	0.0917	0.0578	0.0597	0.0716
	150	M	0.0512	0.0970	0.0806	0.0662
		F	0.0650	0.0560	0.0826	0.0768
	750	M				0.0587
		F		0.0721	0.0995	0.0798
50	30	M	0.0604	0.1196	0.0496	0.0922
		F	0.0797	0.0784	0.0769	0.1214
	150	M	0.0737	0.1302	0.0743	0.0679
		F	0.0871	0.1182	0.0919	0.0645
	750	M		0.1348		
		F		0.0524	0.1133	0.0726
250	150	M	0.0381	0.0692	0.0429	0.0241
		F	0.0759	0.0214	0.0596	0.0719
Controls		M	0.0728			
		F	0.0754			

[a] From Ref. 106.

considerations. Modern techniques of engineering analysis can be applied to the evaluation of burn injury in the microvascular bed to quantify microscopic experimental data, obtain a better understanding of the mechanisms of burn injury, and afford a basis for correlation with macroscopic observations. In the following discussion, some modern engineering techniques, implemented to analyze and characterize the microscopic response to burns will be presented. In order to understand this analysis, a brief quantitative description of the inflammatory process in skin will be presented.

The development of inflammation in response to thermal injury follows a standard pattern involving predictably ordered events.[108] There is an immediate vasodilation concurrent with increased vessel wall permeability for macromolecules, lasting from $\frac{1}{2}$–1 hr. This phase is followed by a delayed and more prolonged response, starting at 1–8 hr and lasting 2–40 hr or longer. The delayed phase is characterized by migration of white cells to the wound site, with adhesion to the intima,[109,110] vascular stasis,[111] and a secondary increase in vascular permeability.[112] Figure 32 depicts a typical progression of the vessel wall permeability initiated by thermal injury. The inflammatory response is clearly biphasic. Extensive experimental evidence indicates that the immediate and delayed responses are precipitated by two distinct sequences of events that are uncoupled in their causative actions. The precipitating agents are thought to be endogenous chemical mediators released into the tissue at various stages of the injury process.[113,114]

The immediate response can be reasonably explained as a result of the action of a specific mediator, such as histamine[115] and/or 5-hydroxytryptamine (serotonin).[116] This explanation is offered, since antagonists of these mediators can be used to block the course of immediate inflammatory development.[117] The causes of subsequent increases in permeability during the delayed phase have not been so clearly described. There is a much greater variation in the time course of development of the delayed phase of inflammation in comparison with the immediate phase.[118] Experimental data suggest

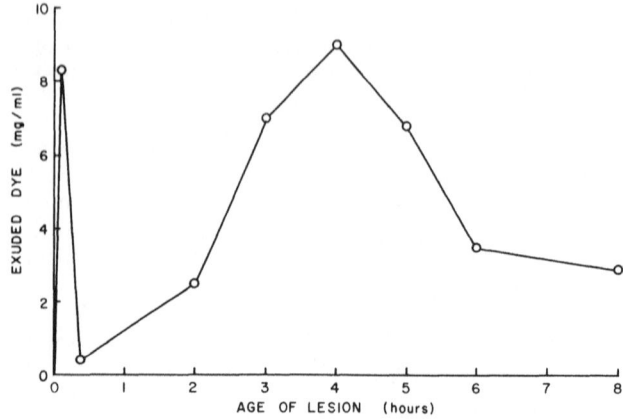

FIGURE 32

Variation in endothelial permeability with time following a burn injury. Replotted from Ref. 108 with permission.

that kinins may be responsible for the earlier portion of the delayed phase reaction,[119] while prostaglandins may cause the later inflammatory response.[120] The case for the action of these mediators is built on experimentally manipulating their ability to influence the development of inflammation by either depleting the tissue of the proposed mediator or administering known antagonists into the tissue or circulatory system.

The vessels involved in the immediate phase appear to be primarily the capillaries and/or venules. Vascular leakage from these vessels appears to be associated with a widening of the clefts between adjacent endothelial cells.[121,122] Electron micrographs do not show damage to the cells themselves.[123] A mechanism for spreading intercellular gaps and the associated chemical stimulants have not been defined specifically. The difficulty in resolving the nature of this response can be attributed in part to the involvement of numerous possible mediators.

On a somewhat larger scale, a quantitative microscopic analysis of the inflammatory response to burns was undertaken using a microvascular preparation in the hamster cheek pouch and gut mesentery.[124–126] Controlled burns were effected on a computer-regulated hot stage previously described.[57] Methods of mass transfer analysis were applied to quantify the leakage of plasma macromolecules into the interstitial space. Fluorescein isothiocyanate (FITC) conjugated to 3,000 MW Dextran was intoduced as a marker molecule into the circulation after completion of a burn via exposure and cannulation of the external jugular vein. Extravasation and interstitial diffusion of the dye was viewed with an epifluorescent illumination system and recorded on

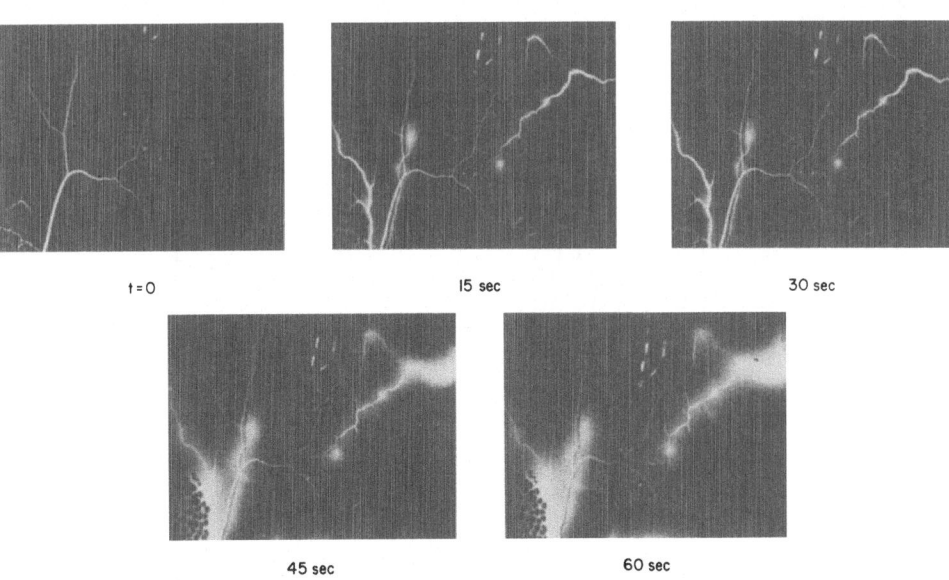

t = 0   15 sec   30 sec

45 sec   60 sec

FIGURE 33
Leakage of 3,000 MW Dextran–FITC in a control hamster gut mesentery tissue. Time is measured from the initial entry of dye into arterioles. Note that at the initial time, the dye has not yet appeared in venules. Subsequent extravasation sites are seen to be concentrated within the venular network (see Ref. 124).

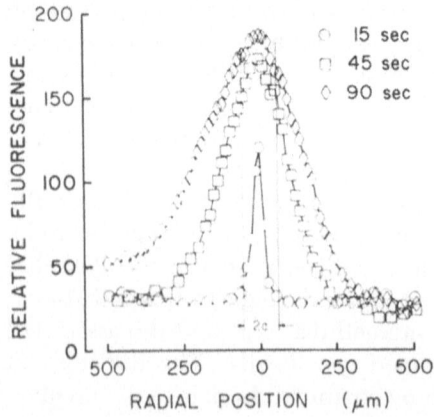

FIGURE 34
Digitized values of fluorescent intensities along a vertical line normal to the leaking venule seen in the top right corner of the microvascular bed in Fig. 33. The vessel diameter of 2a is denoted. Progressive extravasation and increase in interstitial dye concentration are apparent. From Ref. 124 with permission.

either film of video tape. The sequential process of filling the vessels followed by leakage and extravascular accumulation can be followed very clearly, as shown in Fig. 33. Although the response is visually very impressive in a qualitative sense, extraction of quantitative data from the picture is a challenging task. The dye concentration profile in the microvascular bed at specific postinjection times was obtained by videodigitizing the individual photomicrographs of the leakage process. The dye concentration gradients were evaluted in terms of a grayscale-based temporal and spatial distribution. Figure 34 illustrates profiles obtained by digital computer analysis of Figure 33. A dimensionless diffusion model was fit to the experimental data to estimate a value for the diffusion coefficient in the extravascular space.[126] The model was based on a one-dimensional, purely diffusional transport process, i.e., one

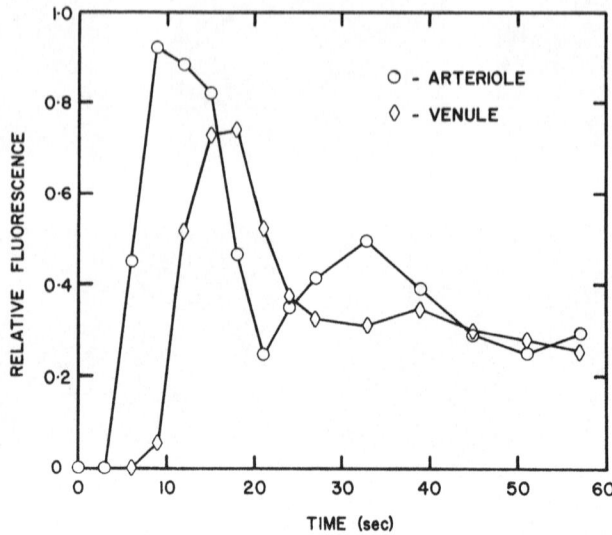

FIGURE 35
Measured intravascular fluorescence during mixing of 3,000 MW Dextran–FITC with the blood volume. From Ref. 124 with permission.

for which convective effects were assumed negligible. The coordinate system used was a function of the vascular geometry. In the cheek pouch preparation, which is considerably thicker than the vessel geometry, dye tracer concentration was most accurately described as a function of radial position in cylindrical coordinates, whereas in the very thin gut mesentery, the concentration was the function of a Cartesian coordinate, normal to the vessel wall. All transport coefficients were assumed constant, and the concentration immediately outside the vessel wall was assumed to undergo an initial step increase from zero to a finite and constant value c_0. This assumption represents a compromise in evaluating actual experimental data, as shown in Fig. 35, in order obtain an analytical solution for the problem.

In the case of the cheek pouch, for a vessel of radius a, the transient concentration is described by a conventional mass balance, analogous to the heat balance (Eq. 3), which relates storage and diffusion of mass in tissue.

$$\frac{\partial \bar{c}}{\partial \bar{t}} = \frac{\partial^2 \bar{c}}{\partial \bar{r}^2} + \frac{1}{\bar{r}} \frac{\partial \bar{c}}{\partial \bar{r}} \tag{54}$$

where

$$\bar{c} = \frac{c}{c_0}, \qquad \bar{r} = \frac{r}{a}, \qquad \bar{t} = \frac{D}{a^2} t$$

and D is the interstitial diffusion coefficient of the tagged macromolecule. The solution is[128]

$$\bar{c} = 1 + \frac{1}{(\bar{r})^{1/2}} \operatorname{erfc}\left[\frac{\bar{r} - 1}{2(\bar{t})^{1/2}}\right] + \frac{\bar{r} - 1}{4(\bar{r})^{3/2}} \bar{t}^{1/2} i \operatorname{erfc}\left[\frac{\bar{r} - 1}{2(\bar{t})^{1/2}}\right]$$

$$+ \frac{9 - 2\bar{r} - 7\bar{r}^2}{32(\bar{r})^{5/2}} \bar{t} i^2 \operatorname{erfc}\left[\frac{\bar{r} - 1}{2(\bar{t})^{1/2}}\right] + \cdots \tag{55}$$

In the gut mesentery, a one-dimensional tissue model can be applied, and the transient concentration of indicator is modeled by

$$\frac{\partial \bar{c}}{\partial \bar{t}} = \frac{\partial^2 \bar{c}}{\partial \bar{x}^2} \tag{56}$$

The solution is simply

$$\bar{c} = \operatorname{erfc}\left[\frac{\bar{x}}{2(\bar{t})^{1/2}}\right] \tag{57}$$

Equations (55) and (57) are plotted for comparison in Fig. 36 to illustrate the divergence of the contour profiles for these two coordinate systems with increasing values of t.

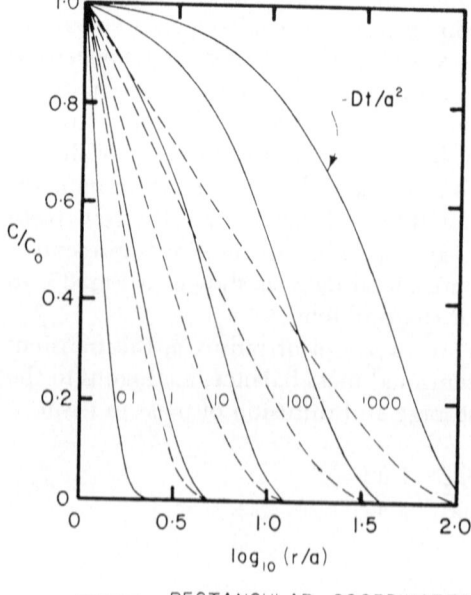

RECTANGULAR COORDINATES
CYLINDRICAL COORDINATES

FIGURE 36

Calculated interstitial concentration profiles for Cartesian and cylindrical coordinate systems, showing the effects of varying values of diffusivity D. From Ref. 126 with permission.

TABLE 5

Apparent Diffusivities of 3,000 MW Dextran–FITC for Control and Burned Hamster Gut Mesentery[a]

	Interstitial diffusivity of 3,000 MW Dextran–FITC $D\left(\dfrac{cm^2}{sec}\right) \times 10^6$		
Animal number	Control	Burn	n
1	2.49 ± 0.42		3
2	3.41 ± 0.59		5
3	3.37 ± 1.60		9
4	2.86 ± 0.41		4
5	2.31 ± 0.13		3
6	2.55 ± 0.86		3
7	2.40 ± 0.76		3
8	2.77 ± 0.25		3
9	3.05 ± 0.25		5
10		6 sec at 46°C 1.30 ± 0.40	3
11		6 sec at 48°C 0.98 ± 0.02	4
12		15 sec at 48°C 0.81 ± 0.48	4

[a] From Ref. 126 with permission.

Values for the diffusion coefficient D in the interstitial space were obtained by fitting the experimentally measured concentration profiles, such as in Figure 34, to the model, using a least squares technique. Approximately 20–30 data points were used for each curve; the average standard deviation of the data from the model was generally less than 10%. Thirty-eight control values for the interstitial diffusivity of 3,000 MW Dextran–FITC in the buffered saline-suffused gut mesentery were obtained for measurements in nine different animals at various locations in the vasculature and times post-injection. The control diffusivity was $D = 2.8$–$3.4 \times 10^{-6}\,\text{cm}^2/\text{sec}$, which is slightly lower than the value for free diffusion in water.[129,130] Limited data have been obtained for changes in diffusivity caused by localized burn insult (Table 5), but a consistent aspect of the data is the decrease in Dextran–FITC diffusivity, with the decrease being larger with greater severity of insult. The physiological basis for the burn-induced reduction of interstitial macromolecular diffusivity remains to be identified. It may be conjectured that the burn process causes a progressive denaturation of interstitial proteins resulting in an overall constrictive action. As the dimensions of the diffusive pathway are introduced, the resistance to macromolecular transport would be increased, causing the observed depression in D.

7. CONCLUSION

Although the physiological response to burns involves complicated and coupled reactions, even simple techniques of heat transfer analysis may contribute substantially to understanding the burn–wound process and how it may be treated. Quantitative analysis can be used to predict the extent and severity of injury due to prescribed environmental heating and cooling factors and to assess the physiological response to a given thermal insult. Recent advances in digital computing capacity and numerical analysis procedures have allowed the successful analysis of heat transfer in complex geometries. These methods hold the promise of providing more powerful and generally useful techniques of heat transfer analysis as applied to burn injury.

REFERENCES

1. Salisbury, R. E., and Pruitt, B. A., *Burns of the Upper Extremity* (Philadelphia: W. B. Saunders Co., 1975).
2. Monafo, W. W., *The Treatment of Burns: Principles and Practice* (St. Louis: Warren H. Green, 1971).
3. Artz, C. P., and Moncrief, J. A., *The Treatment of Burns*, 2d ed. (Philadelphia: W. B. Saunders Co., 1969).
4. Krizek, T. J., Robson, M. C., and Wray, R. C., Jr., Care of the burned patient, in *The Management of Trauma*, 2d ed. W. F. Ballinger, R. B. Rutherford, and G. D. Zuidema, eds. (Philadelphia: W. B. Saunders Co., 1973), pp. 650–718.
5. Pillsbury, D. M., *A Manual of Dermatology* (Philadelphia: W. B. Saunders Co., 1971).
6. Gray, H., *Anatomy of the Human Body*, 28th ed., C. M. Goss, ed. (Philadelphia: Lea and Febiger, 1969).

7. Zweifach, B. W., *Functional Behavior of the Microcirculation* (Springfield, IL: Charles C. Thomas, 1961).

8. Cope, O., and Moore, F. D., A study of capillary permeability in experimental burns and burn shock, using radioactive dyes in blood and lymph, *J. Clin. Invest.* 23, 241–275, 1944.

9. Moritz, A. R., Studies of thermal injury, III. The pathology and pathogenesis of cutaneous burns, an experimental study, *Am. J. Path.* 23, 915–934, 1947.

10. Sevitt, S., Local blood-flow changes in experimental burns, *J. Path. Bact.* 61, 427–442, 1949.

11. Wells, F. R., and Miles, A. A., Site of vascular responses to thermal injury, *Nature* 200, 1015–1016, 1963.

12. Spector, W. G., Walters, M. N.-I., and Willoughby, D. A., Venular and capillary permeability in thermal injury, *J. Path. Bact.* 90, 635–640, 1965.

13. Hardaway, R. M., Microcoagulation in shock, *Am. J. Surg.* 110, 289–301, 1965.

14. Order, S. E., *et al.*, Vascular destructive effects of thermal injury and its relationship to burn wound sepsis, *J. Trauma* 5, 62, 1965.

15. Robb, H. J., Dynamics of the microcirculation during a burn, *Arch. Surg.* (Chicago) 94, 776–780, 1967.

16. Branemark, P. I., *et al.*, Microvascular pathophysiology of burned tissue, *Ann. N. Y. Acad. Sci.* 150, 474–494, 1968.

17. Cotran, R. S., and Remensnyder, J. P., The structural basis of increased vascular permeability after graded thermal injury—light and electron microscopic studies, *Ann. N. Y. Acad. Sci.* 150, 495–509, 1968.

18. Schoen, R. E., Wells, C. H., and Kolmen, S. N., Viscometric and microcirculatory observations following flame injury, *J. Trauma* 11, 619–624, 1971.

19. Shea, S. M., Caulfield, J. B., and Burke, J. F., Microvascular ultrastructure in thermal injury: a reconsideration of the role of mediators, *Microvasc. Res.* 5, 87–96, 1973.

20. Moncrief, J. A., Burns, *N. Engl. J. Med.* 288, 444–454, 1973.

21. Ham, K. N., and Hurley, J. V., An electron-microscope study of the vascular response to mild thermal injury in the rat, *J. Path. Bact.* 95, 175–183, 1968.

22. Contran, R. X., and Majno, G., A light and electron microscopic analysis of vascular injury, *Ann. N. Y. Acad. Sci.* 116, 750–764, 1964.

23. Arturson, G., Pathophysiological aspects of the burn syndrome: with special reference to liver injury and alterations of capillary permeability, *Acta Chir. Scand. suppl.* 274, 1–35, 1961.

24. Courtice, F. C., and Sabine, M. S., The effect of different degrees of thermal injury on the transfer of proteins and lipoproteins from plasma to lymph in the leg of the hypercholesterolemic rabbit, *Aust. J. Exp. Biol. Med. Sci.* 44, 37–44, 1966.

25. Courtice, F. E., and Sabine, M. S., The effects of changes in local temperature on the transfer of proteins and lipoproteins from plasma to lymph in the normal and injured paw of the hypercholesterolemic rabbit, *Aust. J. Exp. Biol. Med. Sci.* 44, 23–36, 1966.

26. Arturson, G., Pathophysiology of acute plasma loss in burns, *Bibl. Haematologica* 23, 1130–1135, 1965.

27. Leape, L. L., Early burn wound changes, *J. Pediatr. Surg.* 3, 292–299, 1968.

28. Leape, L. L., Kinetics of burn edema formation in primates, *Ann. Surg.* 176, 223–226, 1972.

29. Leape, L. L., Initial changes in burns: tissue changes in burned and unburned skin of rhesus monkeys, *J. Trauma* 10, 488–492, 1972.

30. Cotran, R. S., The delayed and prolonged vascular leakage in inflammation, II. An electron-microscopic study of the vascular response after thermal injury, *Am. J. Pathol.* 46, 589–620, 1965.

31. McMantus, W. F., Eurenius, K., and Pruitt, B. A., Jr., Disseminated intravascular coagulation in burned patients, *J. Trauma* 13, 416–422, 1973.

32. Curreri, P. W., Eurenius, K., and Pruitt, B. A., Jr., A study of coagulation factors in the thermally injured patient, in *Transactions of the Third International Congress on Research in Burns*, Matter, P., Barclay, T. L., and Konickova, Z., eds. (Bern: Hans Huber, 1971), pp. 594–596.

33. Eurenius, K., and Rothenberg, J., Platelet aggregation after thermal injury, *J. Lab. Clin. Med.* 83, 344–363, 1974.

34. Fallon, R. H., and Moyer, C. A., Rates of insensible perspiration through normal, burned, tape-stripped, and epidermally denuded living human skin, *Ann. Surg.* **158**, 915–923, 1963.

35. Krizek, T. J., Topical therapy of burns—problems in wound healing, *J. Trauma* **8**, 276–290, 1968.

36. Blalock, A., Experimental shock, VII. The importance of the local loss of fluid in the production of the low blood pressure after burns, *Arch. Surg.* (Chicago) **22**, 31–41, 1931.

37. Baxter, C. R., Burns, in *Care of the Trauma Patient*, Shires, G. T., ed. (New York: McGraw-Hill 1966), pp. 197–222.

38. Moritz, A. R., Henriques, F. C., Dutra, F. R., and Weisiger, J. R., Studies of thermal injury, IV. An exploration of the causality producing attributes of conflagrations; local and systemic effects of general cutaneous exposure to excessive circumambient air and circumradiant heat of varying duration and intensity, *Arch. Pathol.* **43**, 466–488, 1947.

39. Epstein, B. X., Hardy, D. L., Harrison, H. N., Teplitz, G., Villareal, Y., and Mason, A. D., Hypoxemia in the burned patient. A clinical–pathological study, *Ann. Surg.* **158**, 924–932, 1963.

40. Pruitt, B. A., Jr., DiVincenti, F. C., and Mason, A. D., Jr., The occurrence and significance of pneumonia and other pulmonary complications in burned patients: comparison of conventional and topical treatment, *J. Trauma* **10**, 519–531, 1970.

41. Jelenko, C., III, and Ginsburg, J. M., Water-Holding lipid and water transmission through homeothermic and poikilothermic skins, *Proc. Soc. Exp. Biol. Med.* **1336**, 1059–1062, 1971.

42. Arturson, G., Evaporation and fluid replacement: research in burns, in *Transactions of the Third International Congress on Research in Burns*, Matter, P., Barclay, T. L., and Konickova, Z., eds. (Bern: Hans Huber Pub. 1971), pp. 520–531.

43. Nylen, B., and Wallenius, G., The protein loss via exudation from burns and granulating wound surfaces, *Acta Chir. Scand.* **122**, 97–100, 1961.

44. Soroff, H. S., Pearson, E., and Artz, C. P., An estimation of the nitrogen requirements for equilibrium in burned patients, *Surg. Gynecol. Obstet.* **112**, 159–172, 1961.

45. Blocker, T. G., Lewis, S. R., Kirby, E. J., Levin, W. C., Perry, J. A., and Blocker, V., The problem of protein disequilibrium following severe thermal trauma, in *Research in Burns*, Artz, C. P., ed. (Philadelphia: F. A. Davis, 1962), pp. 121–124.

46. Ehrlich, H. P., Tarver, H., and Hunt, T. K., Effects of vitamin A and glucocorticoids upon inflammation and collagen synthesis, *Ann. Surg.* **177**, 222–227, 1973.

47. Hellstrom, J. G., Vitamin E—a general review of the literature with assessment of its role in the healing of burns and wounds, *J. Med. Serv. Can.* **17**, 238–268, 1961.

48. Larson, D. L., Maxwell, R., Abston, S., and Dobrkovsky, M., Zinc deficiency in burned children, *Plast. Reconstr. Surg.* **46**, 13–21, 1970.

49. O'Neill, J. A., Jr., The influence of thermal burns on gastric acid secretion, *Surgery* **67**, 267–271, 1970.

50. Munster, A. M., Alterations of the host defense mechanisms in burns, *Surg. Clin. North Am.* **50**, 1217–1225, 1970.

51. Allgower, M., Cueni, L. B., and Stadtler, K., Burn toxin in mouse skin, *J. Trauma* **13**, 95–111, 1973.

52. Alexander, J. W., Serum and leukocyte lysosomal enzymes. Derangements following severe thermal injury, *Arch. Surg.* (Chicago) **95**, 482–491, 1967.

53. Stoll, A. M., Heat transfer in biotechnology, in *Advances in Heat Transfer*, vol. 4, Hartnett, J. P., and Irvine, T. F., Jr., eds. (New York: Academic 1969), pp. 65–141.

54. Moritz, A. R., and Henriques, F. C., Studies of thermal injury, II. The relative importance of time and surface temperature in the causation of cutaneous burns, *Am. J. Pathol.* **23**, 695–720, 1947.

55. Stoll, A. M., and Green, L. C., Relationship between pain and tissue damage due to thermal radiation, *J. Appl. Physiol.* **14**, 373–382, 1959.

56. Stoll, A. M., and Chianta, M. A., Heat transfer through fabrics as related to thermal injury, *Ann. N. Y. Acad. Sci.* **33**, 649–670, 1971.

57. Ross, D. C., and Diller, K. R., An experimental investigation of burn injury in living tissue, *Trans. ASME, J. Heat Transfer* **98**, 292–296, 1976.

58. Henriques, F. C., and Moritz, A. R., Studies of thermal injury, I. The conduction of heat to and through skin and the temperatures attained therein. A theoretical and an experimental investigation, *Am. J. Pathol.* **23**, 531–549, 1947.

59. Henriques, F. C., Studies of thermal injury, V. The predictability and the significance of thermally induced rate processes leading to irreversible epidermal injury, *Arch. Pathol.* **43**, 489–502, 1947.

60. Carslaw, H. S., *Introduction to the Mathematical Theory of the Conduction of Heat in Solids* (New York: Macmillan, 1921).

61. Buettner, K., Effects of extreme heat and cold on human skin, I. Analysis of temperature changes caused by different kinds of heat application, *J. Appl. Physiol.* **3**, 691–702, 1951.

62. Buettner, K., Effects of extreme heat and cold on human skin, II. Surface temperature, pain and heat conductivity in experiments with radiant heat, *J. Appl. Physiol.* **3**, 703–713, 1951.

63. Stoll, A. M., A computer solution for determination of thermal tissue damage integrals from experimental data, *Inst. Radio Engineers Trans. Med. Electronics* **7**, 355–358, 1960.

64. Smirnov, N., Table for estimating the goodness of fit of empirical distributions, *Ann. Math. Stat.* **19**, 279, 1948.

65. Stoll, A. M., and Chianta, M. A., Burn production and prevention in convective and radiant heat transfer, *Aerospace Med.* **39**, 1097–1100, 1968.

66. Stoll, A. M., and Chianta, M. A., A method and rating system for evaluation of thermal protection, *Aerospace Med.* **40**, 1232–1238, 1969.

67. Weaver, J. A., and Stoll, A. M., Mathematical model of skin exposed to thermal radiation, *Aerospace Med.* **40**, 24–30, 1969.

68. Stoll, A. M., and Chianta, M. A., Heat transfer through fabrics as related to thermal injury, *Ann. N. Y. Acad. Sci.* **33**, 649–670, 1971.

69. Shitzer, A., Studies of bioheat transfer in mammals, in *Topics in Transport Phenomena*, Gutfinger, C. C., ed. (New York: Halsted Press, 1975), pp. 211–341.

70. Weaver, J. A., Calculation of time–temperature histories and prediction of injury to skin exposed to thermal radiation, U.S. Naval Air Development Center Report No. NADC-MR-6623, 1967.

71. Mainster, M. A., White, T. J., Tips, J. H., and Wilson, P. W., Transient thermal behavior in biological systems, *Bull. Math. Biophys.* **32**, 303–314, 1970.

72. Carslaw, H. S., and Jaeger, J. C., *Conduction of Heat in Solids*, 2d ed. (London: Clarendon Press, 1959).

73. White, T. J., Mainster, M. F., Tips, J. H., and Wilson, P. W., Chorioretinal thermal behavior, *Bull. Math. Biophys.* **32**, 315–322, 1970.

74. Mainster, M. A., White, T. J., and Allen, R. G., Spectral dependence of retinal damage produced by intense light sources, *J. Opt. Soc. Am.* **60**, 848–855, 1970.

75. Priebe, L. A., and Welch, A. J., Asymptomatic rate process calculations of thermal injury to the retina following laser irradiation, *Trans. ASME, J. Biomech. Eng.* **100**, 49–54, 1978.

76. Priebe, L. A., and Welch, A. J., A dimensionless model for the calculation of temperature increase in biologic tissues exposed to nonionizing radiation, *IEEE Trans. Biomed. Eng.* **BME-26**, 244–250, 1979.

77. Welch, A. J., Wissler, E. H., and Priebe, L. A., Significance of blood flow in calculations of temperature in laser irradiated tissue, *IEEE Trans. Biomed. Eng.* **BME-27**, 164–166, 1980.

78. Takata, A. N., Zaneveld, L., and Richter, W., Laser-induced thermal damage of skin, SAM-TR-77-38, USAF School of Aerospace Medicine, 1977.

79. Takata, A. N., Thermal model of laser-induced skin damage: computer operator's manual, SAM-TR-77-37, USAF School of Aerospace Medicine, 1977.

80. Palla, R. L., A heat transfer analysis of a scald injury, U.S. National Bureau of Standards, Report No. NBSIR 81-2320, 1981.

81. Link, W. J., Incropera, F. P., and Glover, J. L., The thermal response of tissue subjected to plasma scalpel heating, ASME paper no. 73-WA/Bio-32, 1973.

82. Lloyd, J. J., Mueller, T. J., and Waugh, R. E., On *in vitro* thermal damage to erythrocytes, ASME paper no. 73-WA/Bio-33, 1973.

83. Moussa, N. A., McGrath, J. J., Cravalho, E. G., and Asimacopoulos, P. J., Kinetics of thermal injury in cells, *Trans. ASME, J. Biomech. Eng.* **99**, 155–159, 1977.

84. Diller, K. R., Quantitative low-temperature optical microscopy of biological systems, *J. Microsc.* **126**, 9–28, 1982.

85. McGrath, J. J., Cravalho, E. G., and Huggins, C. E., An experimental comparison of intracellular ice formation and freeze–thaw survival of HeLa S-3 cells, *Cryobiology* **12**, 540–550, 1975.

86. Moussa, N. A., Tell, E. N., and Cravalho, E. G., Time progression of hemolysis of erythrocyte populations exposed to supraphysiological temperatures. *Trans. ASME, J. Biomech. Eng.* **101**, 213–217, 1979.

87. Diller, K. R., and Hayes, L. J., A finite-element model of burn injury in blood-perfused skin, *Trans. ASME, J. Biomech. Eng.* **105**, 300–307, 1983.

88. Zienkiewicz, O. C., *The Finite-Element Method*, 3d ed. (New York: McGraw-Hill, 1977).

89. Becker, E. B., Carey, G. F., and Oden, J. T., *Finite Elements: An Introduction* (New York: Prentice Hall, 1981).

90. Hayes, L. J., A users' guide to PARAB: a two-dimensional linear time-dependent finite-element program, TICOM report 80-10, University of Texas at Austin, 1980.

91. Diller, K. R., Hayes, L. J., and Baxter, C. R., A mathematical model for the efficacy of cooling therapy for burns, *J. Burn Care Rehab.* **4**, 81–89, 1983.

92. Wilson, C. E., Sasse, C. W., Musselman, M. M., and McWhorter, C. A., Cold-water treatment of burns, *J. Trauma* **3**, 477–483, 1968.

93. Rose, A. Continuous water baths for burns, *JAMA* **47**, 1042, 1906.

94. Rose, H. W., Initial cold water treatment for burns, *NW Med.* **35**, 267–270, 1936.

95. Moncrief, J. A., Reply to "Cold-water immersion for burns," *N. Engl. J. Med.* **290**, 58–59, 1974.

96. Langohr, J. L., Rosenfeld, L., Owen, C. R., and Cope, O., Effect of therapeutic cold on the circulation of blood and lymph in thermal burns, *Arch. Surg. (Chicago)* **59**, 1031–1044, 1949.

97. King, T. C., and Price, P. B., Surface cooling following extensive burns, *JAMA* **183**, 151–152, 1963.

98. Wiedeman, M. P., and Brigham, M. P., The effects of cooling on the microvasculature after thermal injury, *Microvasc. Res.* **3**, 154–161, 1971.

99. Zifowitz, L., and Hardy, J. D., Influence of cold exposure on thermal burns in the rat, *J. Appl. Physiol.* **12**, 147–154, 1958.

100. Shulman, A. G., Ice water as primary treatment of burns, *JAMA* **173**, 1916–1919, 1960.

101. Boykin, J. V., Jr., Eriksson, E., Sholley, M. M., and Pittman, R. N., Histamine-mediated delayed permeability response after scald burn inhibited by cimetidine or cold-water treatment, *Science* **209**, 315–318, 1980.

102. Ofeigsson, O. J., Water cooling: first-aid treatment for scalds and burns, *Surgery* **57**, 391–400, 1965.

103. King, T. C., and Zimmerman, J. M., First-aid cooling of the fresh burn, *Surg. Gynec. Obstet.* **120**, 1271–1273, 1965.

104. Block, M., Cold water for burns and scalds, *Lancet* **1**, 695, 1968.

105. Ross, D. C., and Diller, K. R., Therapeutic effects of postburn cooling, *Trans. ASME, J. Biomech. Eng.* **100**, 149–152, 1978.

106. Thompson, K. F., and Diller, K. R., Use of computer image analysis to quantify contraction of wound size in experimental burns, *J. Burn Care Rehab.* **2**, 307–321, 1981.

107. Bailey, B. N., Lewis, S. R., and Blocker, T. G., Jr., Standardization of experimental burns in the laboratory rat, *Tex. Rep. Biol. Med.* **20**, 20–29, 1961.

108. Zwifach, B. W., Grant, L., and McCluskey, R. T., eds., *The Inflammatory Process*, 2d ed. (New York: Academic, 1974).

109. Allison, F., Smith, M. R., and Wood, W. B., Studies on the pathogenesis of acute inflammation, I: The inflammation reaction to thermal injury as observed in the rabbit ear chamber, *J. Exper. Med.* **102**, 655–668, 1955.

110. Schoefl, G. I., The migration of lymphocytes across the vascular endothelium in the living animal, *Am. J. Anat.* **57**, 385–438, 1972.

111. Zucker, M. G., Platelet agglutination and vasoconstriction as factors in spontaneous hemostasis in normal, thrombocytopenic, heparinized, and hypothrombinemic rats, *Am. J. Physiol.* **148**, 275–288, 1947.

112. Burke, J. F., and Miles, A. A., The sequence of vascular events in early infective inflammation, *J. Pathol. Bact.* **76**, 1–9, 1958.

113. Willoughby, D. A., Mediation of increased vascular permeability, in *The Inflammatory Process*, vol. 2, 2d ed. (New York: Academic, 1973), pp. 303–331.

114. Spector, W. G., and Willoughby, D. A., *The Pharmacology of Inflammation*, (London: English University Press, 1968).

115. Horakova, V., and Beaven, M. A., Time course of histamine release and edema formation in the rat paw after thermal injury, *Eur. J. Pharmacol.* **27**, 305–312, 1974.

116. West, G. B., Comparison of the release of histamine and 5-hydroxytryptamine from tissues of the rat, mouse, and hamster, *Int. Arch. Allergy Appl. Immunol.* **13**, 336–347, 1958.

117. Feldberg, W., and Talesnik, J., Reduction of tissue histamine by compound 48/80, *J. Physiol.* **120**, 550–568, 1958.

118. Wilhelm, D. L., Chemical mediators, in *The Inflammatory Process*, vol. 2, 2d ed. (New York: Academic Press, 1973), pp. 251–301.

119. Greaves, M., and Shuster, S., Responses of skin blood vessels to bradykinin, histamine, and 5-hydroxytryptamine, *J. Physiol.* **193**, 255–267, 1967.

120. Kaley, G., and Winger, R., Prostaglandin E_1: a potential mediator of the inflammatory response, *Ann. N. Y. Acad. Sci.* **180**, 338–350, 1971.

121. Majno, G., and Palade, G. W., Studies of inflammation, I. The effect of histamine and serotonin on vascular permeability: an electron-microscopic study, *J. Biophys. Biochem. Cytology* **11**, 571–605, 1961.

122. Wells, F. R., The site of vascular response to thermal injury in skeletal muscle, *Br. J. Exp. Pathol.* **52**, 292–306, 1971.

123. Majno, G., Shea, S. M., and Leventhal, M., Endothelial contraction induced by histamine-type mediators: an electron-microscopic study, *J. Cell Biol.* **42**, 647–672, 1969.

124. Green, D. M., and Diller, K. R., Measurement of burn-induced leakage of macromolecules in living tissue, *Trans. ASME, J. Biomech. Eng.* **100**, 152–158, 1978.

125. Evans, C. D., Diller, K. R., and Parsons, J. P., Burn-induced alterations of interstitial diffusion in mesentery tissue, in *1979 Advances in Bioengineering*, Wells, M. K., ed. (New York: ASME, 1979), pp. 161–164.

126. Evans, C. D., Diller, K. R., and Green, D. M., Interstitial macromolecular diffusivity in burned tissue, in *1979 Biomechanics Symposium*, Van Buskirk, W. C., (New York: ASME, 1979), pp. 93–96.

127. Diller, K. R., Evans, C. D., and Parsons, J. P., Quantification of the microcirculatory response to freezing, in *Biofluid Mechanics*, Schneck, D. J., ed. (New York: Plenum 1980), pp. 347–361.

128. Crank, J. *The Mathematics of Diffusion*, 2d ed. (London: Oxford University Press, 1975).

129. Garlick, D. G., and Renkin, E. M., Transport of large molecules from plasma to interstitial fluid and lymph in dogs, *Am. J. Physiol.* **219**, 1595–1605, 1970.

130. Nakamura, Y., and Wayland, H., Macromolecular transport in the cat mesentery, *Microvasc. Res.* **9**, 1–21, 1975.

Chapter 19

LASER IRRADIATION OF TISSUE

A. J. Welch

1. INTRODUCTION

The rapid development of laser technology has provided the biomedical community with a variety of intense sources of electromagnetic radiation in the visible and infrared spectrum. When laser radiation strikes tissue, it absorbs a portion of the incident energy. The absorbed energy elevates the temperature of the tissue, and if the temperature increase is sufficiently high, irreversible damage, such as enzyme inactivation or protein denaturation, occurs. Further increases in the amount of energy absorbed may burn or even vaporize the tissue. The high-power densities that can be realized with the focused laser beam provide a unique surgical tool for cutting or destroying tissue.

The thermal interaction of laser radiation with tissue is described in this chapter in terms of temperature fields and temperature-dependent rate processes. Mathematical representations of these processes are presented, which allow prediction of the occurrence and extent of damage due to laser irradiation. In addition to mathematical modeling, the uses of lasers in medicine and the hazards of laser irradiation are reviewed.

The basic interaction of laser radiation (or light) with tissue is illustrated in Fig. 1. When light strikes the tissue, a portion of the light is reflected; some of the light may be transmitted through the tissue; and the remainder of the light is absorbed by the tissue. The resulting temperature field in the tissue is a function of (1) the rate at which energy is deposited, (2) the duration of irradiation, (3) the volume of irradiated tissue, (4) transport losses by conduction and convection of heat, and (5) the heat capacity of the tissue. Despite the competing influences of these factors, it has been experimentally observed that the elevation of temperature is proportional to the light output power of the laser.[1] In the absence of vaporization or other nonlinear effects, the tissue temperature is proportional to laser power. However, the relationship of tissue temperature to other factors, such as the duration of irradiation, image size, and the absorptive properties of the tissue, is not obvious; nevertheless, these factors can be described mathematically.

A. J. Welch • Department of Electrical and Computer Engineering and Biomedical Engineering Program, University of Texas, Austin, Texas 78712.

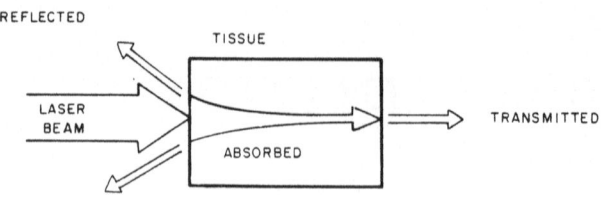

FIGURE 1
Interaction of laser light with tissue.

2. LASER PRINCIPLES

Atoms, molecules, and ions have internal vibrational resonances at certain frequencies. When radiative energy is absorbed near one of the resonant frequencies, there is a measurable emission at that frequency by the atoms, molecules, or ions. The overall response depends on population differences between the upper and lower quantum energy levels governing the particular transition. Under normal conditions, the net response is absorptive. If the population difference is inverted, then there is a net emission of energy. The stimulated emission represents an amplification of the applied signal.

The word *laser* is an acronym for *l*ight *a*mplification by *s*timulated *e*mission of *r*adiation. A lasing medium, such as a gas, liquid, or solid, is placed in an optical cavity and radiative energy is transferred into the medium to raise the energy level of atoms, molecules, or ions above the ground state. When a sufficient number of ions are energized to cause a population inversion, lasing begins.[2] As the ions decay from their excited state at energy E_X to their resting state at E_R, photons are emitted at a frequency f

$$f = (E_X - E_R)/h \tag{1}$$

where h is Planck's constant. Mirrors at the end of the optical cavity reflect the emitted photons and the entrained light beam through the medium, enhancing the stimulated radiative emission. The light in the laser cavity is amplified by repeated reflections from the mirrors. The simplest type of wave formed in the optical cavity is the elementary Gaussian beam. However, there is an infinite number of wave forms that will reproduce in the optical cavity. These wave forms are called higher order Gaussian beams, and they are usually described in terms of transverse electromagnetic modes (TEM_{mn}) along the plane perpendicular to the path of the beam. The subscripts indicate the shape of the intensity profile. A laser beam operating in the TEM_{00} mode has a Gaussian intensity profile, whereas the combination of TEM_{01} and TEM_{10} modes produces a volcano-shaped profile (center depressed). A small fraction of the emitted radiation, which is collimated by multiple reflections in the cavity, is transmitted through one of the partly silvered mirrors. If the excited or inverted population is obtained only on a transient basis, the laser output is intermittent or pulsed. If it is possible to maintain the population inversion on a steady-state basis, then the laser output may be continuous.

Laser action has been obtained at thousands of discrete wavelengths, ranging from the near ultraviolet ($\lambda = 300$ nm) to the far infrared ($\lambda = 800 \times 10^3$ nm).

The laser produces a unique type of electromagnetic radiation. The wavelength(s) of light emitted from the laser is (are) determined by the energy level(s) for population inversion of the lasing material. A single wavelength or line is emitted from He–Ne, ruby, Nd–YAG, and CO_2 lasers, whereas multiple lines are produced by the argon laser. The stimulated emissions are in phase with the applied signal within the laser cavity. This creates a narrow output beam; line broadening is extremely small, and typical ratios of line band-width to center frequency are of the order of 10^{-6} for most lasers. These unique properties of laser radiation are called spatial coherence and temporal coherence, respectively, and they are discussed in detail in reference 2. For most medical applications discussed in this chapter, the properties of the laser that make it the instrument of choice for creating precise lesions and removal of tissue are the following:

1. A stable output power that is easily varied by controlling laser pump power or by placing filters in the laser beam.
2. The ability to create high power densities by focusing the laser beam.
3. A small divergence angle of the laser beam, allowing precise control of the irradiated area.
4. The availability of a wide range of wavelengths from different lasers.

The wavelength, radiative power, and duration of the laser irradiation depend on the construction of the laser and the medium used for light amplification. Generally, lasers are classified as (1) solid state (ruby, Nd–glass, and Nd–yttrium–aluminum–garnet [YAG]), (2) gas (He–Ne), (3) molecular (CO_2), (4) ion (Ar, Kr), (5) diode (GaAs, GaInAs), (6) metal vapor (HeCd, HeSe), and (7) dye (rhodamine CG). A complete listing of commercial lasers is available in *Laser Focus*.[3] This magazine lists manufacturers of dermatological (Ar–Kr, Dye), dental (Ar), ophthalmological (Ar, Ar–Kr, CO_2), and surgical (Ar, ruby, Nd–YAG, Dye, CO_2) systems. Lasers most commonly used for medical applications are the ruby, CO_2, Nd–YAG, and Ar.

In the solid-state ruby laser, a xenon flash lamp is used to raise the energy level of Cr^{3+} ions above the ground state. When the population inversion is sufficiently pronounced, stimulated emission begins. The transition of ions emits photons at 694.3 nm. Another popular solid-state laser contains Nd^{3+} ions in a crystal of yttrium–aluminum–garnet. The laser is pumped with a continuous xenon lamp, and outputs of several hundred watts at 1,060 nm are possible.

The carbon dioxide laser is a molecular (gas) laser with high continuous output power. In this laser, at least 10% of the pumping power is converted to emitted radiation. The CO_2 molecules are excited to higher electronic and vibrational–rotational levels. The transition to lower energy levels produces a beam in the far infrared ($\lambda = 10,600$ nm). Ion gas lasers use dc or rf current to excite ions to lasing action. The He–Ne laser requires 5–10 W excitation power to produce a continuous red ($\lambda = 632.8$ nm) beam of 0.5–50 mW. The argon laser requires an excitation of several kilowatts to achieve up to 20 W

TABLE 1
***Typical Output Characteristics of Continuous Wave (CW) and Pulsed Lasers Used in
Ophthalmological and Surgical Systems[a]***

Laser	Wavelength (nm)	Beam diameter (mm)	Beam divergence (m rad)	CW output (W)	Pulsed output (W)	Pulse length (msec)	Repetition rate (pps)
Argon	330–448, 514.5	1–2	0.5–0.6	0.5–20	0.01–20 avg	0.1–500	1–250
Argon/krypton	450–670	1.3–2	<2	0.5–6			
Carbon dioxide	10,600	1.4–8	1.7	8–150	3.5 avg	0.1–250	1–300
Neodymium–YAG (Q-switched)	1,060	2–7	2–15	6–100	500 peak	150–300	1 K–50 K
Ruby (pulsed)	694	10–20	5–7		250–5000 peak	0.2–10	60
Ruby (Q-switched)	694	10–20	3–7		10^8 peak	5–50	30

[a] From Ref. 3.

total output over a number of discrete wavelengths. The primary output wavelengths of the argon laser are 488 and 514.5 nm. Typical output characteristics for lasers used for ophthalmological and surgical applications are presented in Table 1. Further information about the physical properties of surgical lasers has been provided by Fuller.[4]

3. LASER SAFETY

Safety standards for the use of all lasers are based on those irradiation conditions that may cause minimal damage to human tissue. Because of the eye's ability to focus light on a small spot and its inability to regenerate destroyed neural tissue, it is the organ of the body most vulnerable to laser irradiation. The coherent radiation of a laser appears as a point source to the eye and may be focused to a retinal image of approximately 10 μm. Even the light from a low-level He–Ne laser (5 mW) may be sufficient to burn the retina before the blink reflex (0.1 sec) shuts the eyelid. At lower power levels, the laser cannot damage the eye. This does not mean that low-level laser irradiation is safe. Gibbons and Allen[5] have produced damage with an exposure of 120 sec and corneal power of 0.54 mW. This power level would create less than a 1°C temperature rise.[6] Other evidence of nonthermal damage due to low-level, long-term, short wavelength irradiation has been published by Ham *et al.*,[7] and Lawwill *et al.*[8]

At the other extreme, high-power Q-switched lasers are believed to damage the retina by the creation of shock waves and mechanical displacement of material.[9-12] Thus, the laser is capable of damaging the eye by three processes: photochemical, thermal, or mechanical damage. In this chapter, we examine the process of thermal damage to the eye and other tissues. A complete review of the safety aspects associated with laser radiation has been published by Sliney and Wolbarsht.[13]

4. MEDICAL APPLICATIONS

Lasers have found widespread applications in general, neurological, gynecological, ophthalmological, otolaryngological, and plastic surgery. Applications for the removal of tissue generally use relatively high-power (20–100 W) continuous wave CO_2 lasers. The light power is absorbed at the surface of the tissue, and there is sufficient power density to either vaporize or "cut" a volume of tissue. At present, the CO_2 laser is limited to applications where the beam can be delivered by direct line of sight or reflection from mirrors.

Although the CO_2 laser is the instrument of choice for cutting or removing tissue, the argon and Nd–YAG lasers are required for applications involving either the transmission of the laser beam through an optical fiber or the ocular media, or the absorption of light energy below the surface of the tissue. Recent medical developments with all three laser systems will be described.

The first medical applications of the laser occurred in the early 1960s in the field of opthalmology. Early efforts involved the application of the ruby laser light for ocular photocoagulation. Compared to standard xenon arc coagulators, the ruby coagulators were smaller, simpler to use, produced less heating of the vitreous humor, and changed treatment from an operating room procedure to an outpatient treatment. The introduction of the continuous wave (CW) argon laser in the late 1960s provided the ophthalmologist with a beam that could be focused on a spot as small as 10 μm and with irradiation times that could be accurately controlled. Approximately 80% of the blue–green argon laser light was transmitted through the eye to the retina, where it was absorbed in the pigment epithelium and choroid. The coagulation of retinal material provides treatment for a number of macular diseases, "welding" of detached retina, and destruction of large portions of the paramacula for control of diabetic retinopathy.

More recent applications have included iridotomy and trabeculectomy for treatment of glaucoma. In these procedures, small holes are produced by vaporization of tissue, thereby reducing the pressure in the eye. The applications of the laser in ophthalmology are not cataloged in this chapter; they are expounded in such publications as *Investigative Ophthalmology, American Journal of Ophthalmology, Investigative Ophthalmology and Visual Science,* etc.

Kaplan and associates[14] began the first large-scale, multiple usage of the CW–CO_2 laser in 1972. A summary of the types of surgery where the CO_2 laser offers advantages over conventional modalities are:

1. Reduction in normally large blood losses in orthopedic and plastic surgery, including large excisions, mastectomies, mammoplasties, and lipectomies.[14–21]

2. Use in highly vascular areas, such as partial hepatectomies and partial nephrectomies.[19–23]

3. Extirpation of highly vascular tumors.[19,20,24–26]

4. Surgery on hemophiliacs and thrombocytopenics; the use of a laser

results in a striking reduction in blood loss and postoperative morbidity.[14,20]

5. Cancer surgery, which is accomplished with a minimal opening of blood vessels and lymphatics and extirpation of cancerous tissue, almost without touching the tissue.[19-21,23,25-27]

6. Excision of burns, synergistic gangrene, and decubitus ulcers and other highly infected tissue, with excellent healing.[19,20,28]

7. Operations on organs where simultaneous monitoring of signals, such as EKG and EEG, invalidates electrosurgery.[14]

8. Cavitational surgery where the CO_2 laser is combined with microscopic attachments as in the case of micro-neurosurgery.[29]

9. Incisions through retinal tissue, such as sclera, where there is no increase in intraoperative pressure and hemorrhage into the vitreous humor. Also, incisions through spinal meninges, where manipulation of the spinal cord with resultant damage to the nerve root is avoided.[30-31]

10. Microsurgery in otology requiring treatment of soft tissue and bone of the external ear canal, middle and inner ear.[32-34]

11. Treatment of cervical dysplasia by vaporization of cervical tissue and use of the CO_2 laser as a scalpel for conization of the cervix. Also vaporization of wart-like growths on the vulva, vagina, and/or cervix due to venereal infection. More recently, surgeons have used the laser to cauterize the lining of the uterus to stop prolonged and severe menstrual bleeding and for tubal reconstruction.[35-39]

Further advances in the use of CO_2 lasers are expected to result from several recent technical developments: a hollow fiber optic probe,[40] a hollow probe sealed at one end with an infrared transmitting window,[41] and the rapid-pulse CO_2 laser.[42] A fiber or probe that efficiently transmits 10,600 nm radiation would permit applications in such areas as the eye and gastrointestinal (GI) tract.[43] The high-intensity, rapid super pulse (RSP) laser introduces the possibility of non-linear interaction of laser light with tissue. It has been noted that this laser vaporizes tissue with a zone of coagulation or necrosis less than one-third that of the continuous wave (CW) CO_2 laser. The RSP laser is expected to find use where minimal tissue shrinkage is necessary and enhanced hemostasis is desired.[43]

The most successful applications of CW–CO_2 laser surgery involves combining the laser with a surgical microscope. Using the laser with a surgical microscope in neuro-, gynecologic, and otolaryngologic surgery offers a number of advantages: extreme precision and control over the extent of surgical lesions and the depth to which they are treated, excellent visualization of the surgical site, excellent hemostasis for small vessels, minimal scarring, faster healing rate than with conventional surgery, and less postoperative discomfort. A more detailed description of applications in otolaryngology follows.

4.1. The CO_2 Laser in Otolaryngeal Surgery

The first CO_2 laser irradiation of vocal cords, in a dog model, was accomplished by Jako in 1971.[44] Positive results for animal experiments led to a clinical program. Initially, benign lesions of the vocal cords were treated. Favorable results were obtained, and the laser procedure was extended to other lesions of the aerodigestive tract, e.g., laryngopharyngeal, respiratory, esophageal tracts.[45]

In these applications, a 10–40 W CW–CO_2 laser beam is focused with a 300–400-mm lens, coaxial with the optics of an operating microscope. The focal point of the invisible laser beam coincides with an aiming light that is directed through the microscope. Generally, the laser is focused on the tissue with an image of 0.5–2.0 mm. Due to the fluid content of the tissue, the 10,600 nm radiation is absorbed in a thin layer at the surface of the tissue. The intense heating in this layer destroys the tissue by vaporization and combustion. Since evaporation takes place at atmospheric pressure and soft tissue is 80–90% water, the temperature is thought to not exceed 100°C.

The view of the surgical field is not impaired by instruments or blood during CO_2 laser irradiation. The precision of tissue removal is limited by only the laser image size, which can be as small as 0.5 mm. A joystick–mirror assembly is used to direct the laser beam.

Tabulations of the depth and diameter of lesions of the vocal cords as a function of average power density in a canine model were published by Stern *et al.*[46] Lesion shape depended on the radiation mode of the laser. Lesions due to TEM_{00} (Gaussian irradiance profile) irradiation were "cup shaped," whereas TEM_{01} (bimodal irradiance profile) lesions possessed a "doughnut" appearance, in accordance with the relative intensity profiles of these modes. Stern *et al.* described the morphology of lesions in soft tissue.[46]

Mihashi *et al.*[47] carried out detailed studies of the irradiation of dogs' tongues with a standardized 20 W, 0.44-sec (8.8 J) dose from a CO_2 laser. This exposure produced a crater approximately 2 mm in diameter. Although image sizes in terms of $1/e^2$ diameter were not specified, the images were probably about the same size as the 2-mm craters that were produced. The crater was surrounded by a narrow band of charred material and a wider blanched zone.[47] Mihashi *et al.* documented the histology of the irradiated tissue and measured temperature profiles in the vicinity of the lesion with a chromel-alumel thermocouple that had a reasonably fast rise time (0.02 sec). Maximum temperature rise as a function of distance from the edge of the "crater" for a "standard" irradiation is illustrated in Fig. 2. At a distance of 0.2 mm, the maximum temperature rise was 33°C, and the relaxation time constant of the temperature was approximately 2 sec. Beyond 0.6 mm, the maximum temperature rise was less than 16°C. Mihashi *et al.* noted that vessels up to 0.5 mm in diameter were sealed by heating (cautery). Three factors probably account for the production of a distinct crater with relatively little damage to adjacent tissue: the limited temperature rise in the irradiated tissue, the relatively low lateral conduction of heat through the tissue, and the relatively bloodless tissue destruction.

Carruth *et al.*[48] state that damage to tissue adjacent to the irradiation image of a CO_2 laser is only a few cells thick. The lack of damage to normal tissue (and the avoidance of edema) permits excellent, pain-free healing, with minimal scar formation. Additional experiments have shown that no viable

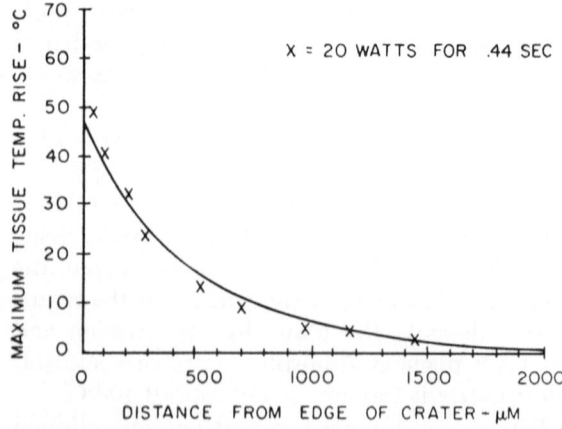

X = 20 WATTS FOR .44 SEC

FIGURE 2

Maximum tissue temperature for CO_2 irradiation plotted against the distance from the edge of the crater. Reprinted from Ref. 47 with permission.

cells are contained in the smoke and vapor produced by CO_2 laser irradiation in soft tissue. These promising results led to clinical applications; the laser has been subsequently used for treatment of malignancies of the larynx, nose, oral cavity, pharynx, nasopharynx, and trachea. An overview of additional applications of the CO_2 laser in otolaryngology can be obtained from references 49–52.

One hazard of laryngeal surgery by laser irradiation is ignition of the endotracheal tube that is used for intubation of the patient. Generally, endotracheal tubes are protected by wrapping them with a self-adhesive aluminum tape and shielding all flammable surfaces with wet gauze. However, a number of endotracheal tubes have been ignited by irradiation of unprotected portions of the tube.[53–55] These fires have led to a number of safety recommendations, alternative procedures for delivery of anesthesia[48,56–58], and modifications of endotracheal tubes.[59,60]

4.2. Argon and Nd–YAG Lasers in Surgery

Argon laser treatment has proven effective for skin lesions such as port wine stain[61–65] and for destruction of bladder tumors.[66,67] One of the most interesting applications of lasers in surgery from a thermal point of view is the use of Ar and Nd–YAG lasers in endoscopic surgery of the GI tract. Because these two lasers have different depths of tissue heating, there has been a continuing debate about the advantages and disadvantages of each laser.

Acute bleeding of the GI tract is one of the most common causes for hospital admission. The patients are usually elderly and often afflicted with age-related heart, circulatory, or respiratory diseases.[68] When medical management of acute GI bleeding is ineffective, surgical ligation of the bleeding vessels or excision of the bleeding tissue is necessary.[68,69] For all patients, the morbidity associated with the surgical procedure is high, and for some groups, the mortality rate approaches 65%. This has prompted the search for less invasive methods of managing GI tract bleeding.[69–73]

The fiberoptic endoscope allows visualization and direct treatment of bleeding sites in the GI tract. The endoscope is a smooth tube, 0.5–1.5 m long and 5–15 mm in diameter, which encloses two bundles of optically transparent fibers, one for illumination and a second, coherent bundle for viewing the reflected image. The endoscope also has one or more channels to accommodate passage of tools for biopsy, or specially adapted surgical tools, aspiration of fluids, injection of drugs, etc. The endoscope also contains a mechanism for manipulating the distal end so that virtually any region of the GI tract may be visualized. The endoscope may be passed through the natural openings in the body to allow for remote viewing and treatment of various tissues with a relatively small degree of discomfort or trauma. The endoscope has allowed several hemostatic methods to be tried, including direct application of tissue glues,[74-76] antifibrinolytic agents, or hemostatic clips,[77] electrocautery,[77] and laser cautery. Of these, only electrocautery[77-82] and laser cautery are regularly effective in stopping hemorrhage.

Electrocautery, guided and performed through an endoscope is a means of achieving hemostasis in which a high-frequency current is passed through a probe-type electrode to the bleeding site; the grounded patient completes the electrical circuit.[78] Electrocautery, however, poses several problems in managing GI tract bleeding, including ineffectiveness in cases of severe bleeding,[70] lack of adequate control over the depth of damage,[78] and the possibility of dislodging the coagulum on removing the probe.[80,83] Electrocautery is discussed in detail in Chapter 17.

Several endoscopic laser systems have been developed; two lasers, the Ar ($\lambda \approx 500$ nm) and the Nd–YAG ($\lambda = 1,060$ nm), have proved to be the most useful, since both laser wavelengths are efficiently transmitted through standard optical fibers.[70,84] The hemostatic effect of laser irradiation, like electrocautery, is due to the damage it causes by energy absorption in the tissue at a bleeding site. Unlike electrocautery, laser irradiation does not require intimate contact with the lesion. Experimental results show that morbidity and mortality rates are reduced with endoscopic laser therapy.[85] The tissue coagulation is localized, and the extent of tissue damage can be readily predicted.

In surgical applications, the desired goal is hemostasis with minimal damage to surrounding tissue. Since the laser cuts or treats tissue through the absorption of light energy in the tissue, some damage to normal surrounding tissue occurs. Minimizing this damage is a major concern of laser treatment, since excessive damage may cause delayed healing or even perforation of the treated tissue, e.g., stomach wall. The actual volume of tissue damaged by the laser depends primarily on the penetration depth of the radiation, the heat conductivity of the tissue, the exposure time, and the laser power.

There is nearly a five to one difference in the absorption coefficients of stomach tissue for the Ar and Nd–YAG laser wavelengths.[70,86] Thus, each laser offers certain advantages and disadvantages. Ninety percent of the Ar laser power is absorbed in the first 0.6 mm of tissue. Since the bulk of the energy absorption from this radiation is near the surface, power densities as low as 44 W/cm^2 have been effective in controlling slow venous bleeding.[69,85] Higher powers are required for hemostasis if there is pooled blood or severe bleeding.

Silverstein *et al.*[87] created standard lesions in the gastric mucosa and submucosa of anticoagulated (heparinized) dogs that produced bleeding rates up to 12 ml/min. An average of fourteen 5-sec bursts of Ar laser power were applied to the ulcers, which had a median bleeding rate of 3.59 ml/sec. Power densities in the range of 160–490 W/cm^2 were required to achieve a significant reduction in bleeding. Thirteen of 16 ulcers created and treated in this manner stopped bleeding. Treated ulcers had coagulum overlaying a layer of injured tissue in the ulcer base. Ulcers treated with the high-power laser had more extensive damage to the surrounding mucosal rim than damage produced by low-power Ar lasers. However the injury did not extend beyond the submucosa.[87] The problem associated with Ar irradiation of sites that were bleeding was the dissipation of the laser energy in the blood pool at the surface of the wound. Raising the laser power to increase the effectiveness of coagulation created an unacceptable depth of injury.[88] When the Ar laser was used to coagulate underlying bleeding vessels through the overlaying blood pool, a black char was formed, which insulated the bleeding site from the laser radiation.[70,88]

Fruhmorgen *et al.*[86] and Silverstein *et al.*[88] used a jet of CO_2 gas to remove the blood pool covering the ulcer. When the gas jet was applied in conjunction with the argon laser at a power density of 100 W/cm^2, bleeding was stopped permanently by the thermally induced coagulation in veins up to 2 mm and in arteries up to 1 mm in diameter. Silverstein *et al.*[88] noted that the CO_2 gas-jet-assisted Ar laser required one-fourth of the energy to stop bleeding from acute experimental gastric ulcers than was necessary with the Ar laser alone. A large number of bleeding lesions in the gastrointestinal tract were treated in this fashion by Fruhmorgen *et al.* using approximately 100 W/cm^2 applied for durations ranging from 2 to 20 sec in conjunction with a CO_2 gas jet.

In contrast to the Ar laser, the 1.06 μm radiation of the Nd–YAG laser penetrates blood and is absorbed over a greater depth of tissue. Kiefhaber *et al.*[70] demonstrated that bleeding lesions in the dog were more easily controlled with a Nd–YAG laser than with the Ar laser. In anticoagulated (heparinized) dogs, an average of seventeen 2-sec exposures of 100 W/cm^2 Ar radiation was required for coagulation in lesions produced by forceps biopsies, whereas only three 2-sec exposures of 300 W/cm^2 Nd–YAG radiation were required to stop the bleeding. When both lasers were used to irradiate 1.5-mm-diameter intact mesenteric veins, a deeper coagulation of the vessel wall and blood was noted with the Nd–YAG laser.[70] Kiefhaber *et al.*[70] used the Nd–YAG laser to treat gastrointestinal bleeding in approximately 200 patients as early as 1976.

The greater depth of penetration of the Nd–YAG laser produces greater depth of injury. Fruhmorgen[86] and others[87,89,90] have expressed concern that extensive necrosis and wall perforations are possible with the Nd–YAG laser. On the other hand, only the Nd–YAG laser can achieve hemostasis at sites of massive bleeding or bleeding within tissues.[70]

In establishing hemostasis, a minimal amount of tissue damage must occur. The key parameter in predicting the extent of damage produced by laser radiation is the temperature rise in the tissue. In the next section, we describe how temperature increases in the tissue may be calculated for various laser-operating conditions.

5. A THERMAL MODEL OF LASER IRRADIATION OF TISSUE

There are two parts to the modeling process for laser irradiation of tissue. First, the temperature field in the tissue caused by absorption of laser radiation is calculated. Then, heating damage is predicted with a rate process model; the scheme is depicted in Fig. 3. The accuracy of some models has been determined by comparison of predicted values with experimental measurements of temperature and extent of damage. The mathematical development of these models and examples of their use are given in the following sections.

In principle, the development of mathematical models for temperature rise in laser irradiated tissue is not difficult. Heat conduction is a well-understood phenomenon, and powerful analytical and numerical methods exist for solving transient-state conduction problems. However, the results are no better than the assumptions underlying the model. Therefore, it is essential that models be constructed with great care, so that important factors are not neglected.

For example, consider the structure of the eye. Laser light focused by the cornea and lens on the retina is absorbed by a thin layer called the pigment epithelium (PE), which is distal to the photoreceptor layer (see Fig. 4). Most of the light energy is absorbed and converted to heat by micron size melanin granules that are densely packed within the PE. The melanin granules in the PE are responsible for the black-appearing disk in the aperture of the iris. This layer absorbs light that has passed through the retina and prevents reflection and scattering of light throughout the eye. In the absence of the pigment granules, light is reflected from the blood vessels in the fundus, producing, for example, the red reflection seen in the eye of albino rabbits. Light that is transmitted through the PE enters the choroid. This layer is richly supplied by blood vessels that primarily absorb light in accordance with the absorption spectrum of hemoglobin. The light energy absorbed in the PE and choroid is converted to heat and conducted to surrounding tissue. If sufficient heat is absorbed and/or conducted to the retina, the photo receptors will be damaged.

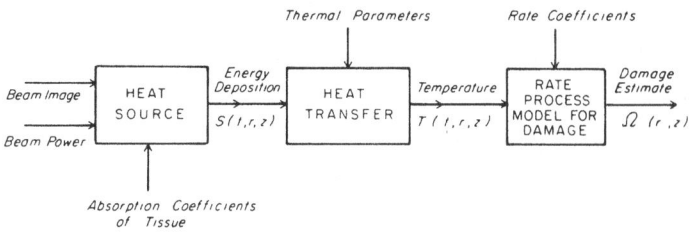

FIGURE 3
Model of thermal damage for laser-irradiated tissue.

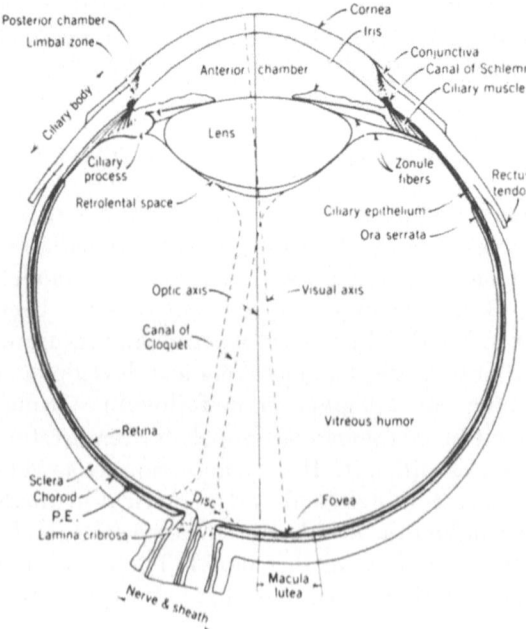

FIGURE 4
Structure of the eye. Approximately 2.5% of Ar or Nd–YAG laser light is reflected from the surface of the cornea. Another 19% of the energy is absorbed in the vitreous humor. Light reaching the retina is absorbed in the 12-μm thick layer of pigment epithelium and 168-μm thick layer of choroid. Only a small amount of energy reaches the sclera.

Although it is tempting to model each segment of the eye, starting with the heating of individual melanin granules, this is not necessary for the analysis of most irradiation conditions. Retinal laser images are sufficiently large (at least 10 μ in diam), so that they encompass a number of granules. The resulting temperature rise is created by the absorption of heat by the granules, with transfer by reradiation and conduction. However, the temperature rise of the tissue a few microns in front of or behind the granule layer is the same as would be produced by replacing the granules with a thin layer that possesses equivalent uniform absorption and emission and is subjected to the same conduction process. This representation of the PE and choroid as homogeneous layers of constant thickness and uniform absorption simplifies modeling the thermal response of the eye to laser radiation. However, computed temperatures based on this assumption may not be representative of the physical system. It is imperative that any thermal model be validated by experimental measurements before the model can be used to predict the thermal behavior of the system. Even so, unvalidated theoretical results can play an important role in predicting trends and provide a basis for formulating experimental procedures.

5.1. Equations and Boundary Conditions

Models of the transient- and steady-state temperature responses to laser irradiation of tissues are usually based on the heat conduction equation, which does not include heat convection (i.e., no heat loss by blood flow). If thermal conductivity k is constant in the region of interest,

$$\nabla^2 T + \frac{s}{k} = \frac{1}{\alpha} \frac{\partial T}{\partial t} \tag{2}$$

where k is the thermal conductivity (cal/cm sec °C), $\alpha = k/\rho c$ is the thermal diffusivity (cm^2/sec), c is the specific heat (cal/g °C), T is the temperature rise (°C), s is the heat source term (cal/cm^3 sec), and ρ is the density (g/cm^3).

Because of the circular symmetry of the laser beam, it is convenient to place the heat conduction equation in cylindrical coordinates

$$\frac{\partial^2 T}{\partial r^2} + \frac{1}{r} \frac{\partial T}{\partial r} + \frac{\partial^2 T}{\partial z^2} + \frac{s}{k} = \frac{1}{\alpha} \frac{\partial T}{\partial t} \tag{3}$$

Examples of such geometries for eye and skin are illustrated in Fig. 5. Models for the eye were developed by Mainster *et al.*[91–93] and Wissler[94]; both use a transparent layer to represent the retina and vitreous humor. The PE and choroid are modeled as one or more absorbing layers, and the sclera is depicted as either transparent or a layer of low absorption. The radius R and length L of the cylinder are usually chosen sufficiently large, so that heat is not conducted to the boundary during the time interval of interest. Thus, boundary and initial temperatures are set to constant values. When they are set to zero,

$$T(R, z, t) = 0 \tag{4a}$$

$$T(r, 0, t) = 0 \tag{4b}$$

$$T(r, L, t) = 0 \tag{4c}$$

$$T(r, z, 0) = 0 \tag{4d}$$

the temperature rise above normal tissue temperature is directly computed.

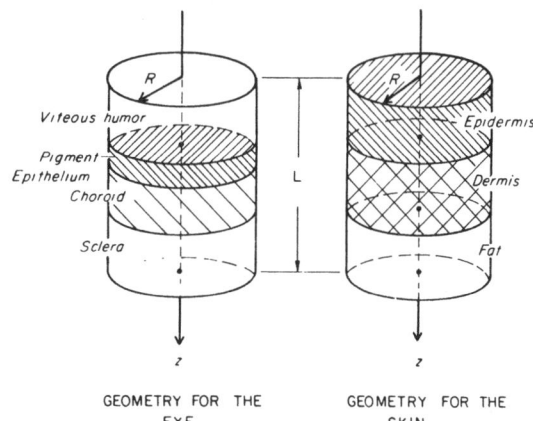

FIGURE 5
Cylindrical geometries for modeling eye and skin. The eye model has fluid–tissue interface at the front surface of the absorbing layer, whereas the skin has an air–tissue interface. The air-tissue interface is usually one of the boundaries of the skin model.

GEOMETRY FOR THE EYE

GEOMETRY FOR THE SKIN

For model geometries where heat is conducted beyond a boundary of the cylinder, the boundary conditions of Eq. (4) are replaced by the generalized Newtonian boundary condition

$$\frac{\partial T}{\partial n} + \frac{h}{k}(T - T_0) = 0 \tag{5}$$

where $h(\text{cal/cm}^2 \text{ sec } °C)$ is the coefficient of surface heat transfer, T is the temperature at the surface, T_0 is the temperature of the surrounding medium, and n is the dimension normal to the surface. The boundary condition Eq. (5) is necessary at air–tissue boundaries, such as shown in Fig. 5. Estimates of the coefficient of surface heat transfer for dry and wet surfaces have been reported by Takata et al.[95] to be 2×10^{-4} cal/cm^2 sec °C and 7×10^{-4} cal/cm^2 sec °C, respectively.

The temperatures in the tissue and at the boundaries are then computed from solution of the heat conduction equation including a source term, Eqs. (2) or (3), with the appropriate set of boundary conditions, Eqs. (4a)–(4c), and the initial condition, Eq. (4d). By redefining the tissue layers, the model for skin is suitable for modeling stomach, brain, and other configurations involving an air–tissue interface.

Perhaps the most important term in Eq. (2) is the source term s. As a first approximation, we can assume that light penetrates the tissue without scattering and that the local rate of absorption of radiant energy is proportional to the intensity (i.e., Beer's absorption law). This leads to an exponential decrease of both intensity and rate of absorption along the direction of propagation of a cylindrical beam as it passes through a homogeneous medium. For example, a beam of intensity $I_0(r)$ that is directed along the z-axis will have an intensity

$$I(r, z) = I_0(r) e^{-\beta z} \tag{6}$$

at a distance z from the surface of the absorbing layer. I_0 is the irradiance of the beam at the surface of the absorbing layer, and $\beta(1/\text{cm})$ is the absorption coefficient. The heat source in Eq. (2) or (3) then becomes

$$s(r, z) = \beta I_0(r) e^{-\beta z} \tag{7}$$

However, if the beam either converges or diverges, the intensity will vary accordingly. It should be noted that the absorption coefficient depends on the wavelength of the radiation, as well as the nature of the tissue, and, hence, light from different lasers will have different penetrating powers. The data of Welsch et al.[96] illustrate this phenomenon clearly for blood; shown in Fig. 6 are data used for the absorption of whole blood for wavelengths covering the range of 400–1,000 nm. The absorption coefficients for irradiation from an Ar laser ($\lambda = 488$ and 514 nm) and a Nd–YAG laser ($\lambda = 1,060$ nm) in oxygenated hemoglobin are 100 and 6.5 cm^{-1}, respectively. Assuming Beer's absorption law, the relative intensity as a function of distance from the surface

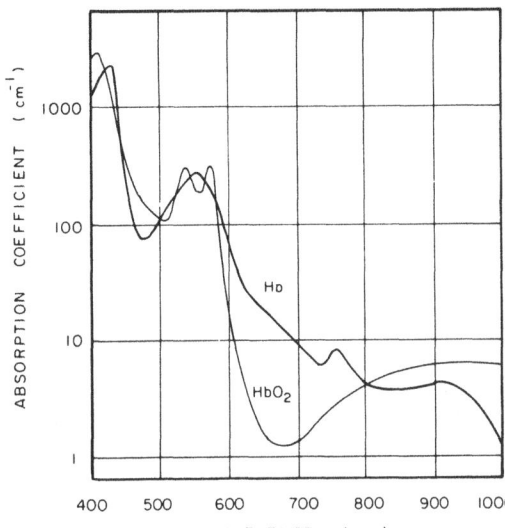

FIGURE 6

Absorption coefficient of oxygenated (HbO_2) and deoxygenated (Hb) hemoglobin solution as a function of wavelength. Reprinted from Ref. 96 with permission.

for Ar and Nd–YAG irradiation is presented in Fig. 7a. The intensity at the surface I_0 is reduced by 50% within 0.005 cm for the Ar radiation, whereas the half-intensity level occurs at approximately 0.11 cm for Nd–YAG radiation. The relative heating effect of the two lasers in hemoglobin in terms of their source $I_0 e^{-\beta z}$ is illustrated in Fig. 7b.

Radiative energy from an Ar laser would be deposited near the surface of a sample of hemoglobin, whereas Nd–YAG laser energy would be deposited more uniformly throughout the sample. In general, the blood content of the

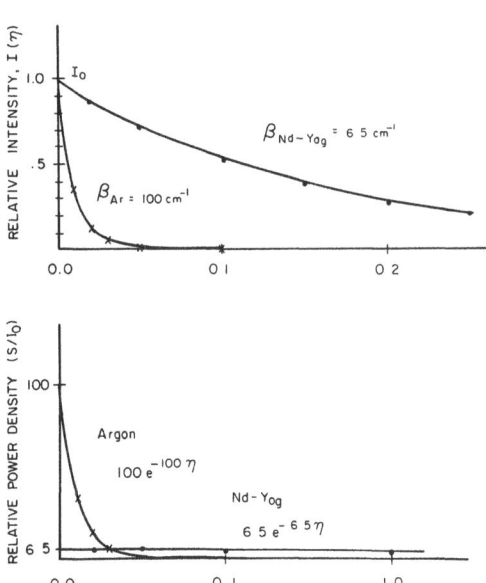

FIGURE 7

(a) Absorption of (Ar ×) and Nd–YAG (•) laser radiation as a function of depth in oxygenated hemoglobin. (b) Heat source created by absorption of Ar (×) and Nd–YAG (•) laser radiation in oxygenated hemoglobin. Reprinted from Ref. 96 with permission.

TABLE 2
Tissue Absorption Coefficients β *(1/cm)*

	Wavelength (nm)			
Tissue	500 (Ar)	1,100 (Nd–YAG)	110,000 (CO$_2$)	Reference
Water	0.00025	0.36	1,106	97
Skin				
Outer epidermis	55	231	911	97
Remaining skin	13	13		
Fat	13	10		
Damaged tissue	19	10		
Eye				
Pigment epithelium	1,545	169		99
	800	78		98
Choroid	169	107		99
	125	16		98
Stomach	28	6		70
Liver	50	12.5	200	70
Blood				
Oxygenated	105	9.9[a]		96
Deoxygenated	105	1.8[a]		96

[a] λ = 1,000 nm.

tissue will cause the argon radiation to be absorbed near the surface, while the Nd–YAG radiation will penetrate deeper into the tissue. The 15:1 ratio of the absorption coefficients for the wavelengths of these lasers in hemoglobin is not typical for most tissues. Absorption coefficients for a number of tissues are presented in Table 2.

5.2. Solutions for the Temperature Field

Because of the difficulty of computing the transient temperature response in a volume of tissue, some models compute only the steady-state solution to the heat conduction equation. These solutions are sufficient whenever the duration of the exposure is much longer than the rise time (time to reach 90% of the final value) of the transient response. For example, radiation of the eye, involving small images (less than 100 μ) and exposures of several seconds reach steady states within a few hundred milliseconds.[100] However, for short exposure durations or large images, a complete solution of the heat conduction equation is needed.

The finite difference technique developed by Mainster *et al.*[91–93] allows different thermal conductivities, specific heats, and absorption coefficients to be assigned to discrete layers of the cylinder. The model uses the Peaceman–Rachford alternating-direction method[101] to compute the temperature response along nonlinear radial and axial grid points for the cylindrical configuration in Fig. 5. The finite-difference program has an exponentially stretched grid to achieve an accurate, efficient computer algorithm. Source strengths for the model are calculated by assuming Beer's law for radiative absorption in one or more layers of the model. All media modeled by the cylinder are assumed to be thermally homogeneous and isotropic. The irradiance at the front surface of the tissue is computed as the total radiant energy striking the tissue minus the energy reflected from the surface. The cylinder is assumed to be sufficiently large, so that irradiation will not cause a significant increase in boundary temperature. The relative light intensity of the image is specified as a function of *r*, the radius of the image, as part of the input to the model. The model divides the image into a sequence of concentric rings and calculates the power entering the tissue in each ring. The grid is automatically adjusted in the radial direction with respect to the image radius.

The finite-difference technique permits solving the heat conduction equation for the interaction of radiation with biologic tissues for spatially varying thermal parameters and complex heat source distributions. The method described by Mainster *et al.*[91-93] has been expanded by Takata *et al.* into an eye model[99] and a skin model[95]. The skin model includes blister formation, vaporization of liquid, and heat transfer for either dry or moist skin at the air–skin interface.

Because of the computer's memory requirements and running time for these solutions, temperature responses have been computed for only a limited number of wavelengths, exposure durations, and image sizes. To reduce the number of solutions needed for a complete system solution, Priebe and Welch[102] modified the finite difference model in terms of dimensionless variables. With this technique, the thermal behavior of a simple laser irradiated system may be described with a set of solutions for a number of wavelengths, exposure durations, and image sizes.

5.3. A Dimensionless Model for the Temperature Field Calculations

By introducing dimensionless variables for a medium with absorption of radiation in one layer, the solution to the heat conduction equations becomes independent of conductivity, specific heat, and irradiance. Dimensionless radial and axial parameters are automatically scaled to compensate for the absorption parameter and image size variations. Dimensionless temperature θ is defined by the relationship

$$\theta = \frac{kT}{\alpha_r \sigma^2 I_0} \tag{8}$$

where I_0 (cal/cm^2 sec) is the irradiance at the center of the image on the surface of the absorbing layer, σ (cm) is the standard deviation of a Gaussian distribution or the radius of a uniform (disk) distribution, and α_r (1/cm) is the absorption coefficient at wavelength λ for Lambert–Beer absorption.

Dimensionless time τ is defined as

$$\tau = \frac{kt}{\rho c \sigma^2} \tag{9}$$

where t = real time. Dimensionless axial distance ξ and dimensionless radial distance η are defined by

$$\xi = \alpha_r z \tag{10}$$

$$\eta = \frac{r}{\sigma} \tag{11}$$

Substituting Eqs. (8)–(11) into Eq. (3) produces the dimensionless form of the heat conduction equation,

$$\frac{\partial^2 \theta}{\partial \eta^2} + \frac{1}{\eta} \frac{\partial \theta}{\partial \eta} + (\alpha_r \sigma)^2 \frac{\partial^2 \theta}{\partial \xi^2} + s(\eta, \xi) = \frac{\partial \theta}{\partial \tau} \tag{12}$$

The form of the source term for a uniform beam is

$$s(\eta, \xi) = e^{-\xi}, \qquad 0 < \eta \leqslant 1, \quad \xi > 0$$
$$= 0, \qquad \text{elsewhere} \tag{13}$$

For a Gaussian irradiance profile, truncated at a radius of $1/e^4$, the source term becomes

$$s(\eta, \xi) = e^{-(\eta^2/2 + \xi)}, \qquad 0 < \eta \leqslant 2\sqrt{2}, \quad \xi > 0$$
$$= 0, \qquad \text{elsewhere} \tag{14}$$

Coefficients for the dimensionless finite-difference equations and a solution for multilayered geometry have been published by Priebe and Welch.[102] From Eq. (12), the single parameter in the solution for dimensionless temperature is the product of the absorption coefficient α_r and the image radius σ. Although the range of absorption coefficients and image sizes is rather large, their product for most clinical applications is limited to the following range.[102]

$$0.4 < \alpha_r \sigma < 40 \tag{15}$$

Once dimensionless temperatures are computed, actual temperatures can be computed by rearranging Eq. (8). The irradiance term and image radius can be replaced by absorbed power P (cal/sec) assuming no reflection or transmission beyond the image plane

$$P = \pi \sigma^2 I_0, \qquad \text{(disk)} \tag{16a}$$

$$= 2\pi \sigma^2 I_0, \qquad \text{(Gaussian)} \tag{16b}$$

By substituting Eq. (16) into Eq. (8), actual temperature increase becomes a function of dimensionless temperature, the conductivity, the absorption para-

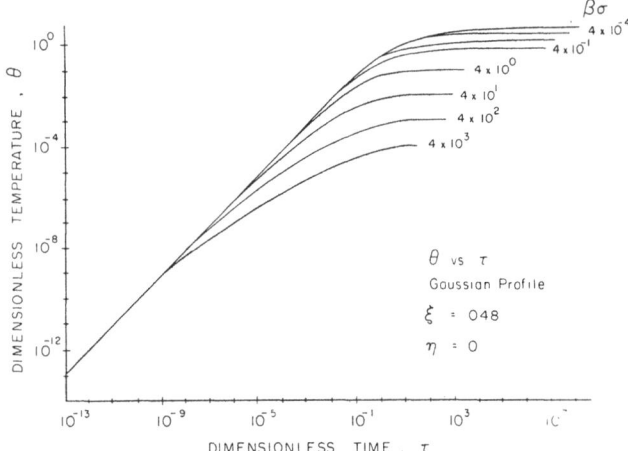

FIGURE 8

Dimensionless temperature vs. dimensionless time for various values of $\alpha_r \sigma (\beta \sigma)$ for a Gaussian irradiance profile. Reprinted from Ref. 102 with permission.

meter, and the laser power. The time t_i required to reach a given response depends on the dimensionless time, thermal conductivity, specific heat, and the image radius according to the expression obtained from Eq. (9)

$$t_i = \frac{\rho c \sigma^2 \tau}{k} \qquad (17)$$

The dimensionless temperature increase versus time for continuous laser irradiation with a Gaussian distribution is illustrated in Fig. 8 for eight values of $\alpha_r \sigma$. For small τ, conduction is insignificant, and temperature is equal to the product of the source term and time $\tau e^{-\xi}$. For a Gaussian image, the temperature is

$$
\begin{aligned}
\theta &\approx \tau e^{-(\eta^2/2+\xi)}, & 0 < \eta \leqslant 2\sqrt{2}, \quad \xi > 0 \\
&\approx 0, & \eta > 2\sqrt{2}, \quad \xi > 0
\end{aligned}
\qquad (18)
$$

These equations are useful for computing the peak temperatures for short duration exposures.

The axial and radial increase of temperature distributions for continuous irradiation are illustrated in Fig. 9 for a disk image and $\alpha_r \sigma = 40$. The axial temperature profile is plotted in the plane of the figure, while the radial profile is plotted in the isometric plane. The shortest time for which a profile is plotted is $\tau = 10^{-6}$. Heat is initially conducted in the axial direction, but as time progresses, the axial gradients become small ($\tau \approx 0.001$), and then energy is primarily stored. Conduction at later times is primarily in the radial direction. This pattern of axial conduction preceding radial conduction occurs for $\alpha_r \sigma > 1$. For $\alpha_r \sigma < 1$, the pattern is reversed, and the radial gradients are reduced before the axial gradients.

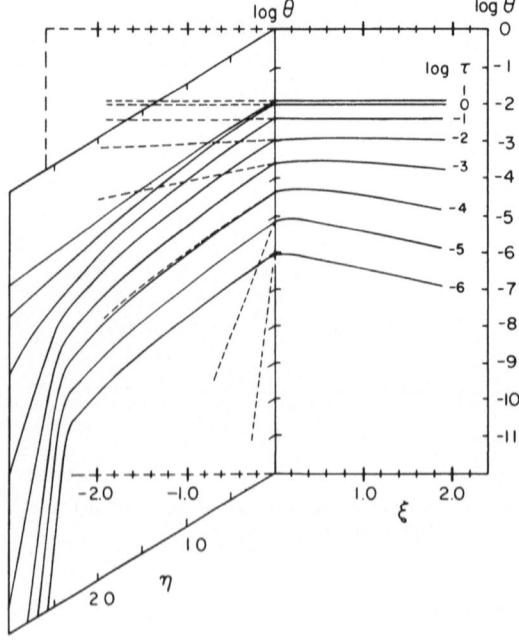

FIGURE 9

Isometric plot of log dimensionless temperature vs. dimensionless axial (ξ) and radial (η) distances for various values of dimensionless time τ. ($\alpha_r\sigma = 40$). The dotted profiles for $\xi < 0$ are axial temperatures in front of the absorbing layer ($\tau > 0$). Reprinted from Ref. 102 with permission.

To some degree, the temperature response of the tissue can be characterized by the steady-state and impulse-response asymptotic values. An example is given in Fig. 10 in terms of the ratio θ/τ vs. τ. Temperatures are computed at the center of the image and at a depth of $\xi = 0.048$ inside the absorbing layer. The dotted lines indicate where actual values differ from the impulse

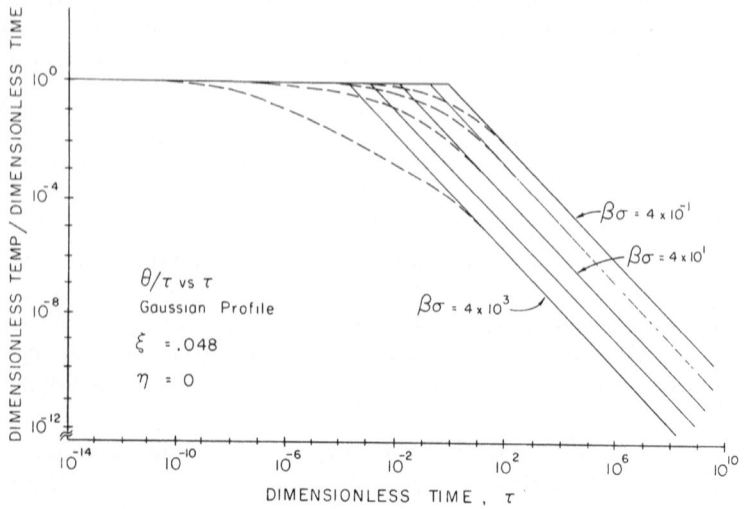

FIGURE 10

Impulse and steady-state asymptotes of θ/τ vs. τ for Gaussian irradiance profiles. Dotted lines indicate computed transient temperatures. Most surgical applications of lasers are limited to $0.4 < \alpha_r\sigma(\beta\sigma) < 40$. Reprinted from Ref. 102 with permission.

and steady-state asymptotes. The use of temperature asymptotes is discussed in more detail later in this chapter.

5.4. The Effects of Blood Flow

When tissue is modeled as an array of uniform layers, the absorption coefficient of a layer must include the absorption of energy by blood. Heat generated by the absorption of laser irradiation in blood is included in the source term s of Eq. (3), but the model assumes the blood is not moving. There is a convective cooling effect as the heated blood flows through the tissue layers. This heat loss, which may exert a strong influence on the temperature field, can be modeled by modifying the heat balance in Eq. (3).

When the heated region contains only capillary beds, the average convective heat loss can be expressed as (cf. Chaps. 6 and 7)

$$q_c = \rho_b c_b\, w_b (T_b - T) \qquad (19)$$

where T_b is the temperature of blood entering the region, T is the temperature of the heated blood leaving the region, and w_b is the perfusion rate. It is generally assumed that the temperature of blood entering a capillary bed is equal to the arterial temperature and the temperature of blood leaving is equal to the local tissue temperature (cf. Chap. 6). It will be shown that in many cases of laser heating, this term can be neglected without introducing appreciable error.

The second case that must be considered is one where a large blood vessel passes through the heated region. Since convection can be a much more effective heat transfer mechanism than conduction, the presence of such vessels may completely dominate the thermal response (cf. Chap. 7). If, during irradiation, the temperature of the lumen exceeds approximately 70°C, the vessels will constrict. At least two independent, secondary mechanisms may be involved in vessel closure. One is a complex process, which involves the destruction of erythrocytes with subsequent coagulation to form a thrombus[103,104]; the threshold temperature for this process is approximately 60°C. The second process involves partial melting and shrinkage of collagen fibres when the temperature exceeds approximately 76°C.[105]

A model has been proposed by Wissler and Gorisch[106] to predict the thermal response in the neighborhood of arteries and veins during laser irradiation. To reduce computer storage and computation time, they proposed the rectangular coordinate geometry shown in Fig. 11a. The model accounts for heating outside the vessel when the diameter of the laser beam is larger than the diameter of the vessel, as well as transport of heat by convection within the vessel.

Outside the vessel, the energy balance for the tissue has the usual form of the bioheat equation with a source s

$$\alpha \nabla^2 T + \frac{\rho_b c_b w_b}{\rho c}(T_b - T) + \frac{s}{\rho c} = \frac{\partial T}{\partial t} \qquad (20)$$

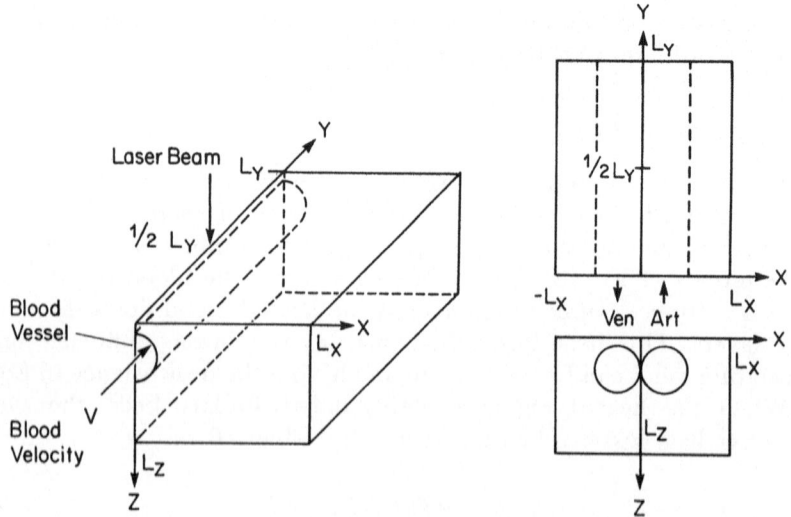

FIGURE 11

Model geometry for predicting temperatures in the neighborhood of laser-irradiated blood vessels. L_x, L_y, L_z are the rectangle dimensions. The beam impinges at $(0, \frac{1}{2}L_y, 0)$.

Within the vessel, heat transported by the blood is expressed as $\rho_b c_b v_b\,(\partial T_b/\partial y)$, where v_b is a local blood velocity. When a heat convection term for the blood stream is included in the energy balance, the approximation for q_c, Eq. (19), is omitted, yielding for the temperature field inside the vessel

$$\alpha_b \nabla^2 T_b + \frac{q_c + s}{\rho_b c_b} = \frac{\partial T_b}{\partial t} + v_b \frac{\partial T_b}{\partial y} \tag{21}$$

For simplicity, T_b in Eq. (21) denotes $T_b(x, y, z, t)$. At the interface between blood and the vessel wall, both temperature and thermal flux are continuous (cf. Chap. 7).

The axis of the laser beam is assumed to be perpendicular to the exposed surface, and the light is not scattered by the tissue nor blood. Assuming absorption according to Beer's law and allowing for different absorptivities in tissue and blood, the source term is

$$s(x, y, z) = I_0(x, y)\alpha_r(z) \exp\left[-\int_0^z \alpha_r(z')\,dz'\right] \tag{22}$$

where I_0 is the intensity of the incident laser beam.

The boundary condition at the upper surface ($z = 0$) is

$$\frac{\partial T}{\partial z} = \frac{h}{k}(T - T_0) \tag{23}$$

where T_0 is the ambient temperature.

Boundary surfaces not penetrated by the vessel are sufficiently far from the surface, so that temperature increases are small, but not zero. The rate of heat transfer from these surfaces is assumed to be proportional to the increase in temperature.[94] Hence, boundary conditions on the four surfaces defined by $x = L_x$, $y = 0$, $y = L_y$, and $z = L_z$ have the form

$$-k\frac{\partial T}{\partial n} = A_n[T(x, y, z, t) - T(x, y, z, 0)] \tag{24}$$

where $A_n = -1/L_n$ and L_n is the distance from the axis of the laser beam to the surface (see Fig. 11a). Conduction heat transfer in the axial direction in the blood stream may be neglected in comparison with axial heat convection. Thus, the boundary condition within the vessel is

$$\frac{\partial T}{\partial n} = 0 \tag{25}$$

The boundary condition at $x = 0$ is determined by the number and location of vessels. If there is only one vessel with its centerline in the plane $x = 0$, this becomes a plane of symmetry, and

$$T(-x, y, z, t) = T(x, y, z, t) \tag{26}$$

On the other hand, in the neighborhood of an irradiated artery–vein pair (Fig. 11b) with the laser beam centered at the point $(0, \frac{1}{2}L_y)$, rotation of the system through 180° about the axis of the laser beam leaves the temperature field unchanged. Hence,

$$T(-x, y, z, t) = T(x, L_y - y, z, t) \tag{27}$$

Since most experiments are conducted by allowing the system to come to equilibrium before irradiation begins, the initial temperature field is defined by the solutions of Eqs. (20)–(26) or (20)–(25), and (27) for $s = 0$. The problem is then simplified by computing only the change in temperature. If T_0 remains unchanged, then the temperature increase above the initial temperature field,

$$T' = T(x, y, z, t) - T(x, y, z, 0) \tag{28}$$

satisfies a set of equations that are completely homogeneous except for the term that accounts for heat generation due to radiative absorption.

Details of the finite-difference solution and sample results have been published by Wissler and Gorisch.[106] They analyzed the response of a small vessel having a diameter of 0.12 mm and a larger vessel having a diameter of 1.2 mm for both Ar and Nd–YAG irradiation. The absorption coefficients in blood were assumed to equal 93 and 12 cm^{-1} for Ar and Nd–YAG, respectively[96]; in the tissue, a value of 0.3 cm^{-1} was used for both lasers. The incident intensity profile was Gaussian, $I = I_0 \exp[-\frac{1}{2}(r/\sigma)^2]$. The beam

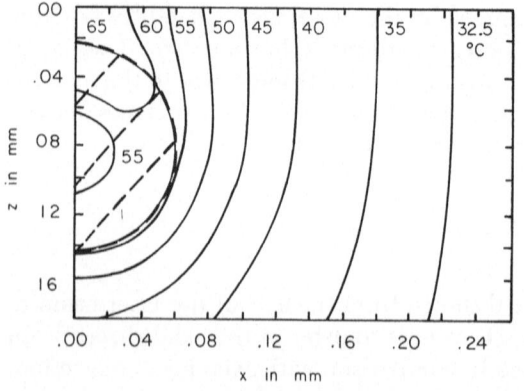

FIGURE 12

Isotherms produced after 80 msec of 0.25 W Ar laser irradiation of a single blood vessel that is 0.02 mm below the surface of the tissue. Reprinted from Ref. 106 with permission.

diameter, based on the $1/e^2$ point ($r = 2\sigma$), was equal to 4.72 times the vessel diameter. A mean blood flow velocity of 5.4 cm/sec was used for each vessel, and the velocity profile was assumed to be parabolic. Environmental temperature and initial tissue temperatures were set at 32.7°C to match the experimental system they were modeling.

Computed results displayed all of the expected characteristics. Figure 12 shows isotherms at the end of an 80-msec irradiation of the small vessel by a 0.25-W Ar laser. The maximum increase in temperature was displaced from the center of the laser beam because of the blood carrying the heat downstream. For this specific case, maximum temperature occurs on the upper surface of the vessel where the rate of heating is high and blood velocity is low. Irradiation of the same vessel with a 2.0-W Nd–YAG laser for 80 msec produces more uniform heating, as shown in Fig. 13, because of the lower absorption of the Nd–YAG laser irradiation.

The effect of perfusion on the temperature response of laser-irradiated tissue can be further illustrated by comparing solutions of the bioheat equation with and without perfusion.[107] As an example, assume a Gaussian radiation source is applied for time t_p. The corresponding radiative absorption term in cylindrical coordinates is

$$s = \alpha_r I_0 \exp\left[-\left(\frac{r^2}{2\sigma^2} + \alpha_r z\right)\right] g(t) \qquad (29)$$

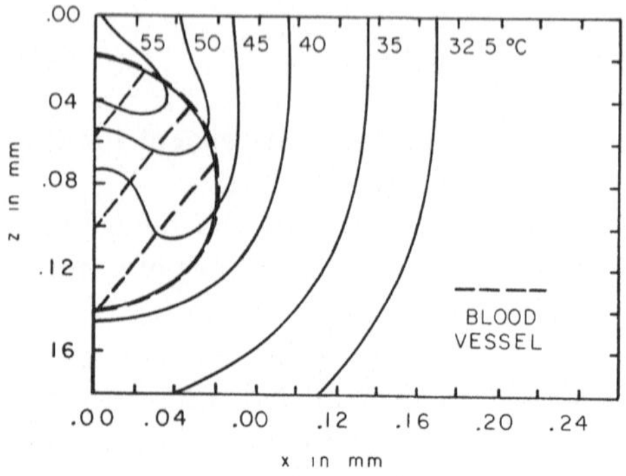

FIGURE 13

Isotherms produced after 80 msec of 2.0 W of Nd–YAG laser irradiation of a single blood vessel that is 0.02 mm below the surface of the tissue. Reprinted from Ref. 106 with permission.

where I_0 is the intensity at the center of the beam and $g(t) = 1$ for $0 \leq t \leq t_p$ and $g(t) = 0$ for $t > t_p$. To facilitate comparison of the solutions of the heat balance with and without perfusion, assume the solution with perfusion, e.g., that of the bioheat equation, Eq. (20), is $\phi(r, z, t)$, while the solution without perfusion, e.g., that of the heat conduction equation, Eq. (12), is $\theta(r, z, t)$. Consider the temperature increment $\phi(r, z, t)$ above that of a uniformly perfused field with $T_b = 0$, with homogeneous boundary conditions. Make the following standard transformation

$$\phi(r, z, t) = \psi(r, z, t) \exp(-w_b t) \tag{30}$$

Substitute for ϕ according to Eq. (30) into the bioheat equation, Eq. (20). Assume that

$$\frac{\rho_b c_b}{\rho c} \approx 1 \tag{31}$$

With the Gaussian irradiation profile, Eq. (29), Eq. (20) becomes

$$\frac{\partial^2 \psi}{\partial r^2} + \frac{1}{r} \frac{\partial \psi}{\partial r} + \frac{\partial^2 \psi}{\partial z^2} + \alpha_r I_0 \exp\left[w_b t - \frac{r^2}{2\sigma^2} - \alpha_r z \right] g(t) = \frac{1}{\alpha} \frac{\partial \psi}{\partial t} \tag{32}$$

where $g(t)$ is the time dependence of the intensity function. Equation (32) has the form of the heat conduction equation, except that the source term is modified by a factor $\exp(w_b t)$. For $t < t_p$ and short irradiations ($t_p \ll 1$) Eqs. (3) and (32) have nearly identical solutions. However, for $t > t_p$, w_b attenuates the perfusionless solution by the factor $\exp(-w_b t)$. Thus, if the temperature response to impulse heating without perfusion is $\theta_0(r, z, t)$, then the "impulse" response with perfusion $\phi_0(r, z, t)$ is

$$\phi_0(r, z, t) = \theta_0(r, z, t) \exp(-w_b t) \tag{33}$$

The effect of perfusion rate on the thermal response to impulse heating is illustrated in Fig. 14.

The solution for an exposure of duration t_p, including the effect of perfusion, can be obtained by the convolution of the impulse response $\phi_0(t)$ and intensity function $g(t)$. Thus,

$$\phi(r, z, t) = \int_0^t \phi_0(r, z, t - \tau) g(\tau) \, d\tau \tag{34}$$

Substituting from Eq. (33) into Eq. (34), we obtain

$$\phi(r, z, t) = \int_0^t \theta_0(r, z, t - \tau) \exp[-w_b(t - \tau)] g(\tau) \, d\tau \tag{35}$$

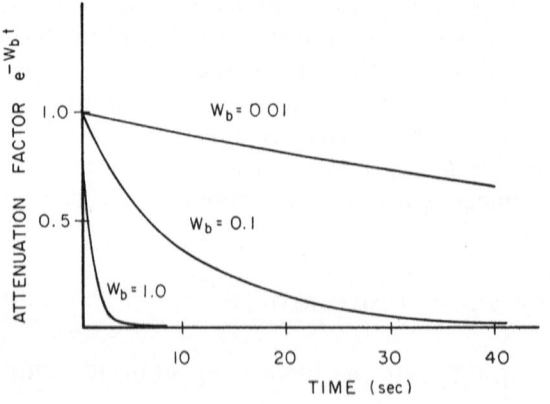

A bound on the difference between temperature fields defined by solutions to the energy balance with and without perfusion may be determined by using the inequality

$$\exp\left[-w_b(t - \tau)\right] \geq \exp\left(-w_b t\right), \quad 0 < t \leq \tau \tag{36}$$

Replacing $\exp\left[-w_b(t - \tau)\right]$ with $\exp\left(-w_b t\right)$ in Eq. (35) gives a lower bound for $\phi(r, z, t)$ of

$$\phi(r, z, t) \geq \int_0^t \theta_0(r, z, t - \tau) \exp\left(-w_b t\right) g(\tau) \, d\tau \tag{37}$$

After removing $\exp\left(-w_b t\right)$ from the integral,

$$\phi(r, z, t) \geq \exp\left(-w_b t\right) \int_0^t \theta_0(r, z, t - \tau) g(\tau) \, d\tau \tag{38}$$

The convolution integral in Eq. (38) is the temperature rise without blood flow $\theta(r, z, t)$. Thus,

$$\phi(r, z, t) \geq \exp\left(-w_b t\right) \theta(r, z, t) \tag{39}$$

The relative values of the three temperature functions, θ, ϕ, and $\theta \exp\left(-w_b t\right)$ for $t_p = 1$ and constant irradiation is illustrated in Fig. 15.

The error ε in calculating the temperature increase due to ignoring heat convection at uniform perfusion rate w_b is given by

$$\varepsilon = \theta(r, z, t) - \phi(r, z, t) \tag{40}$$

Substituting Eq. (39) into Eq. (40) provides an upper bound for the error

$$\varepsilon \leq \theta(r, z, t)[1 - \exp\left(-w_b t\right)] \tag{41}$$

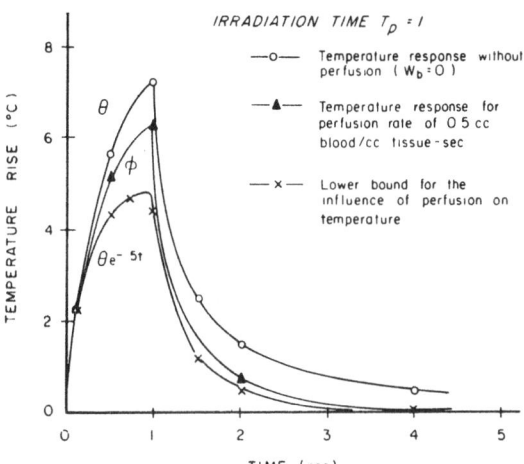

FIGURE 15
Temperature rise for an irradiation time of 1 sec. The top curve is the temperature response without perfusion, the middle curve is the response with perfusion, the bottom curve is the lower bound, given by $\theta \exp(-0.5t)$. Reprinted from Ref. 107 with permission.

For steady exposures, damage at a position (r, z) in the tissue is closely related to the temperature at the end of the irradiation period (i.e., at $t = t_p$). The relative temperature difference at the end of irradiation is also

$$\varepsilon \leq \theta(r, z, t)[1 - \exp(-w_b t_p)] \qquad (42)$$

which will be less than 0.1 if $w_b t_p < 0.1$. Thus, blood flow will affect the temperature rise by less than 10% if

$$w_b t_p < 0.1 \qquad (43)$$

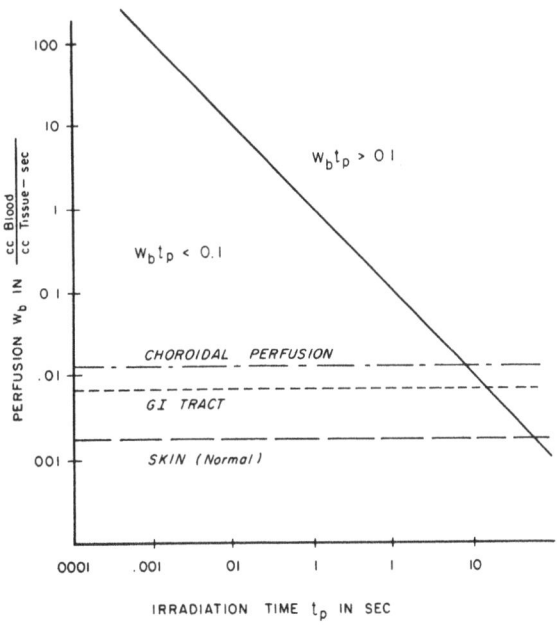

FIGURE 16
Relation of perfusion and irradiation time to damage threshold. Below the solid line ($w_b t_p = 0.1$), perfusion reduces the temperature rise with respect to the no-perfusion value at time t_p by less than 10%. Reprinted from Ref. 107 with permission.

The importance of convective heat loss for irradiation of retinal, stomach, and skin tissue is presented graphically in Fig. 16. Assuming perfusion rates $w_b = 0.012 \text{ cm}^3(\text{blood})/\text{cm}^3(\text{tissue})$ sec for choroid tissue, 0.0066 for stomach tissue, and 0.00165 for skin,[108] the exposure durations required for blood flow to influence the temperature rise significantly are 8, 16, and 50 sec, respectively. Thus, for exposure durations below these limits, neglecting blood flow should not significantly alter the temperature rise.

5.5. Validation of the Temperature Field Model

Modeling the irradiation of a living organism is a difficult task. Equally difficult are transient temperature measurements in thin layers of irradiated tissue. Acceptance of either the results of a mathematical model or experimental data requires some method of validation. The modeling and experimental procedures should not compete but should be used to complement one another. The model's results can identify important parameters and conditions and eliminate needless experiments. In turn, analysis of experimental data allows refinement of the model and modification of parameters. If sufficient confidence can be established in a model, it can be used to predict results for a wide range of irradiation configurations.

A complete validation of computed temperatures for a laser irradiated biological system are practically impossible. When computed and measured temperatures are compared, there are bound to be significant differences for some irradiation conditions and geometries. When differences occur, it is important to trace their origin. Errors in measured temperatures can be expected when laser images are small, compared with the size of the temperature sensor, or irradiation times are shorter than the response time of the sensor. Other differences may be due to errors in the model's representation of the biological system, the computer algorithm, selected values of thermal

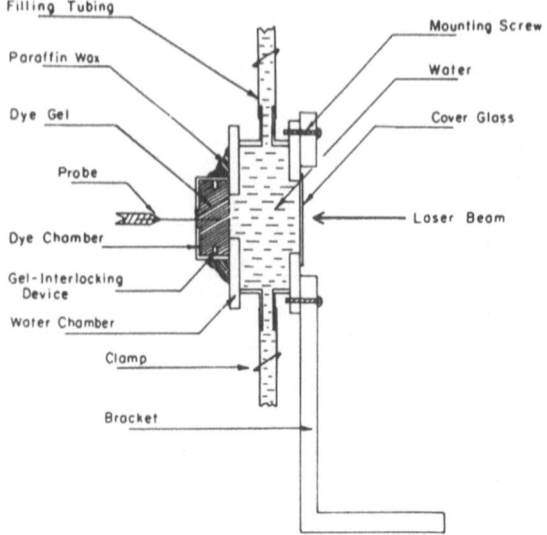

FIGURE 17

Two-chamber dye cell used to model eye. Dyed gel represents absorbing layers of the eye.

parameters, the calculation or measurement of image size and/or intensity, and the perturbation of the temperature field by the sensor.

The limitations of an experimental or theoretical model of a biological system can be established (to a degree) by results obtained in a physical model of the biological system. For example, temperature calculations and measurements for the laser-irradiated eye[1,97,109,110] were evaluated by studying laser irradiation in a two-chamber cell.[109,111] One cell was filled with water, and the second chamber contained a darkly pigmented gel (Fig. 17). Temperature and light intensity measurements were made in the gel with 15–30-μm thin film thermocouple probes.[112] These probes are described in detail in Chap. 27. Temperatures were measured for image radii at the gel–water interface from 50–750 μm for two wavelengths, 514.5 and 1,060 nm, and for two pulse durations, 30 msec and 10 sec. The experiment was used to test our ability to measure transient temperatures and evaluate our theoretical model.

The relative intensity profile, average absorption coefficient, power, probe size, and probe sensitivity were measured for each experiment. The intensity profile was measured at the water–dye interface using the direct absorption properties of the microthermocouple. The measured intensity profiles were approximated by a smooth profile (see Fig. 18). The smoothed profiles represented intensity at the surface of the absorbing layer in the theoretical model. Calculated and measured temperatures were obtained for a number of experiments, assuming k and c_p to be 0.00628 W/cm °C and 4.187 W sec/g °C, respectively. Typical experimental and model temperature responses at the center of the laser image are shown in Fig. 19. Radial temperature profiles for the same experiment are illustrated in Figs. 20 and 21. Radial temperatures after 0.01 and 0.35 sec for Ar irradiation are illustrated in

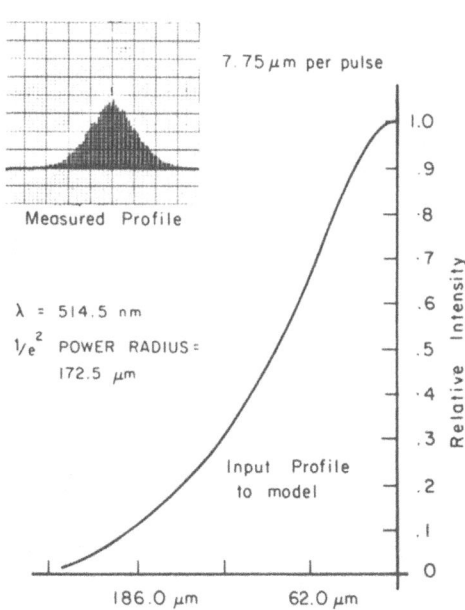

FIGURE 18

Power input profile. The measured intensity profile of the laser beam at the gel–water interface was obtained from direct absorption measurements from a thermocouple inserted in the gel. The laser beam was displaced in 7.75-μm steps to produce the profile. A smooth curve was drawn through the measured points to obtain the profile used in model computations.

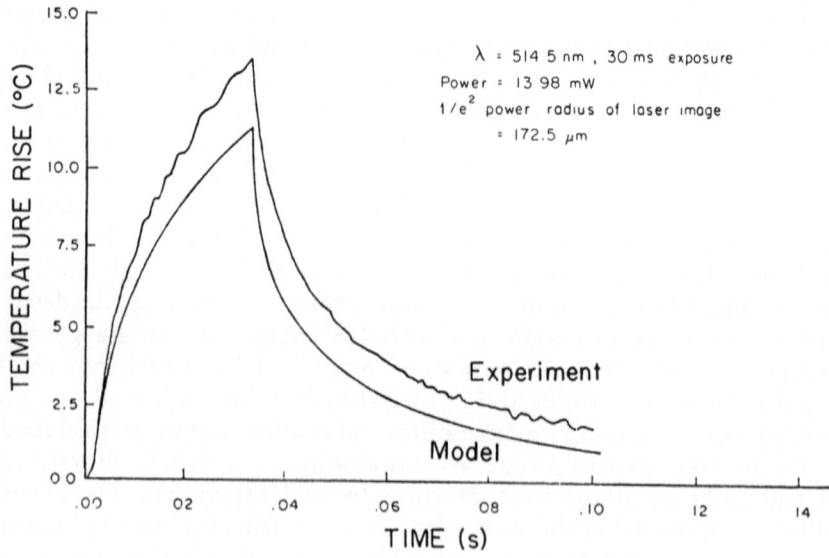

FIGURE 19
Experimental and model temperature response at the center of the laser image and a few μm deep in the dyed gel.

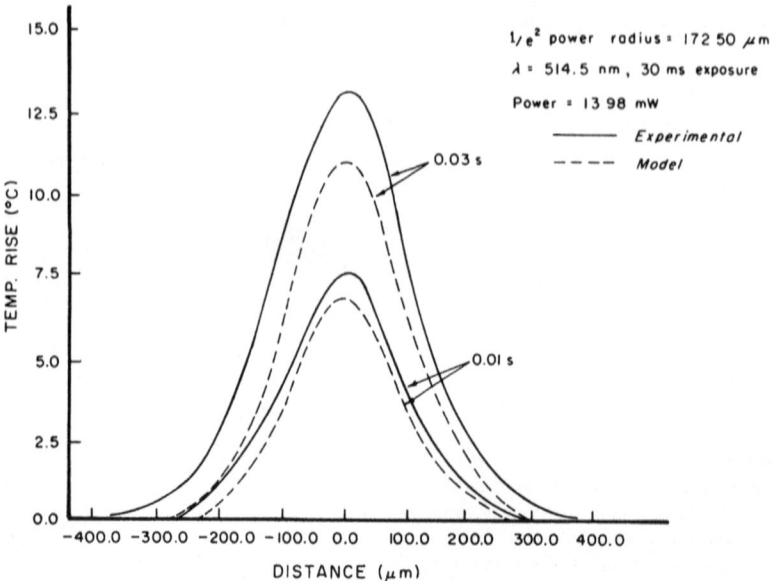

FIGURE 20
Experimental and model radial temperature profile a few μm deep in dyed gel during laser irradiation. Times indicated are elapsed times after irradiation began. Temperatures at $r = 0$ correspond to 0.01- and 0.03-sec temperatures in Fig. 19.

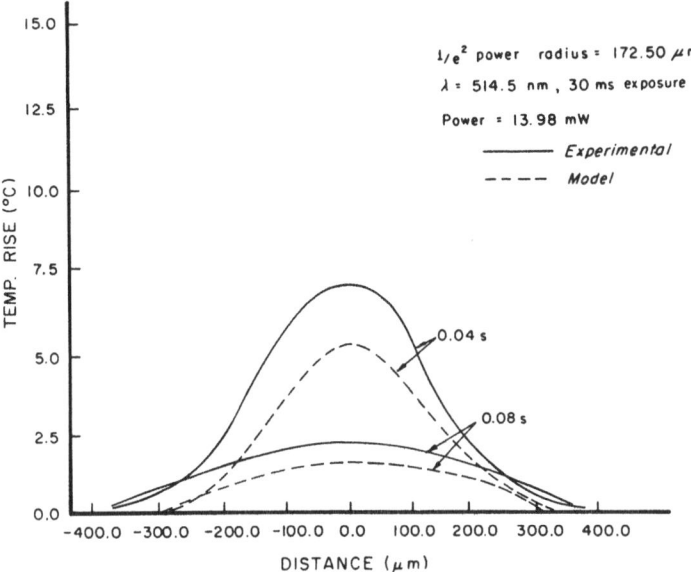

FIGURE 21

Experimental and model temperature profiles a few μm within the gel layer after laser irradiation. Times indicated are the elapsed times after irradiation began. Temperatures at $r = 0$ correspond to those at 0.04 and 0.08 sec in Fig. 19.

Fig. 20. Radial temperatures, 0.01 and 0.05 sec after irradiation had ended are shown in Fig. 21.

In order to compare the model and experimental temperatures from different experiments, temperatures were normalized to an irradiation power of 1 mW. Model and experimental values of the maximum temperature rise that occurred at the center of the image as a function of image radii are illustrated in Fig. 22 for 514.5 nm radiation. The averaged experimental and measured temperatures in the center of the beam and at the half-power radius at the axial "hot spot" are illustrated in Table 3. Measured and computed temperatures for the laser-irradiated dye cell were in general agreement. Although values for a single experiment sometimes differed by as much as 30%, average ratios of experimental temperature to model temperature at the axis were close to unity.

The relative spread of temperature profiles is also depicted in Table 3 by the ratio of temperature at the half-power radius to the temperature in the center of the beam. The average of the ratios for 0.03 sec argon radiation are 0.59 (SD = 0.08) for the model and 0.60 (0.1) for the experimental values. For 10 sec argon radiation, the average ratios are 0.73 (0.05) and 0.77 (0.1), respectively, for model and experiment. For 0.03 sec Nd–YAG radiations, the average ratios are 0.57 (0.08) for the model and 0.58 (0.1) for the experiment.

Close agreement was also obtained between experimental and model results for temperature changes in the axial direction. A few values where the error exceeded 50% illustrated the measurement artifact that occurred when

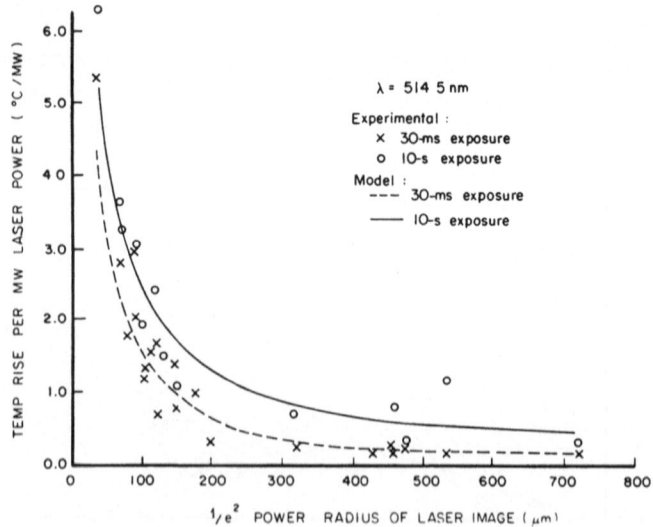

FIGURE 22
Peak normalized temperatures at the end of argon laser irradiation vs. image radius. Temperatures are normalized by laser power to allow experimental comparison in terms of °C/mW.

the probe was moved in the axial direction. This movement disrupted the gel, and repeated movement back and forth formed a hole in the gel that filled with fluid. Our results confirmed the basic mathematical model, the computer algorithm used in implementing the model, and the measurement of temperature in a laser-irradiated system.

5.6. A Damage Model

For the purpose of our discussion, tissue damage is defined as the denaturation or loss of function of biological molecules found in cells or extracellular fluid due to laser-induced heating. We assume that a critical temperature-related condition may be defined, above which such irreversible damage will occur.[113]

TABLE 3
Ratio of Experimental to Computed Temperature Profiles Due to Laser Irradiation

Irradiation condition		Temperature rise ratio	
Wavelength (nm)	Exposure duration (sec)	$r = 0$ z = axial hot spot	r = half-power radius z = axial hot spot
514.5	0.03	1.12 (0.22)[a]	1.01 (0.23)
514.5	10.0	0.98 (0.22)	1.05 (0.21)
1060	0.03	0.89 (0.17)	0.90 (0.17)

[a] Standard deviation.

Since the pioneering work of Henriques and Moritz,[114] thermal damage to tissue has been described as a temperature-dependent rate process, derived from first-order chemical reaction kinetics and the Arrhenius equation for the rate of a chemical reaction. The occurrence of damage is described by a damage function $\Omega(t)$. It depends on the reaction rate coefficient of the tissue and the temperature–time history of the tissue. That is, the damage rate at a point in tissue is expressed as

$$\frac{d\Omega(r, z, t)}{dt} = A \exp \left[\frac{E}{RT(r, z, t)} \right] \tag{44}$$

where A is a constant, E is the activation energy for the reaction (cal/mole), R is the universal gas constant 2(cal/mole K), and T is the absolute temperature (K). The damage function is obtained by integrating Eq. (44)

$$\Omega(r, z) = A \int_{t_i}^{t_f} \exp \left[\frac{-E}{RT(r, z, t)} \right] dt \tag{45}$$

Equation (45) is usually evaluated numerically from the onset of the laser-induced temperature rise (t_i) to the final recovery of the temperature transient (t_f). Based on experimental work with porcupine epidermis, Henriques and Moritz[114] selected rate coefficients such that complete cellular necrosis of the basal epidermal layer was indicated by a value of $\Omega = 1.0$. A value of $\Omega = 0.53$ was used as their criterion for irreversible threshold injury. Values for A and E were derived from experimental temperature measurements by prescribing a square temperature pulse in a period that corresponded to the heating time t_p.

$$\Omega = At_p \exp(-E/RT) \tag{46}$$

By plotting the experimental data in the form $\ln(t_p)$ vs. $1/T$ and assuming threshold damage occurs for $\Omega = 1$, Henriques and Moritz were able to compare their data with the modified Eq. (46)

$$\ln \Omega - \ln A = \ln(t_p) - E/RT \tag{47}$$

and obtain E/R and A from the slope and intercept of the curve, respectively. They obtained the coefficients

$$A = 3.1 \times 10^{98}, \quad (1/\text{sec}) \tag{48a}$$

$$E = 150{,}000, \quad (\text{cal/mole}) \tag{48b}$$

Since that time, the rate process model has been used by Stoll and Green[115] to model damage to skin due to thermal radiation. They determined a new set of empirical rate coefficients in order to better fit their data to the rate process model. Other applications of modeling have been suggested by Hu and Barnes,[116] Kach and Incropera,[117] Takata,[118] Vassiliadis,[119] and Priebe and Welch.[97,109,110] The rate of accumulation of damage $d\Omega/dt$ as a function of temperature for the original rate constants proposed by Henriques and Moritz is shown in Fig. 23. A step change of 14 °C above an initial tissue temperature of 51 °C would cause an accumulation of damage at a constant rate of 0.01 (1/sec). The time to produce threshold damage ($\Omega = 1$) would be approximately 100 sec. Likewise, if a step increase of 31 °C occurred, only 1 msec would be required for threshold damage, based on a damage rate of 1,000 (1/sec).

In general, the temperature response of tissue when it is irradiated is not a square pulse, i.e., a step increase when the laser is turned on and a step return to normal temperature when the laser is turned off. The shape of the temperature response depends on image size, exposure duration, and thermal properties of the tissue.[91,92,100,102] Typically, there is an exponential increase and decrease as illustrated in Fig. 24. The damage function can be determined for the temperature response in Fig. 24 by numerical integration based on Eq. (45). The procedure consists of dividing the time axis of the temperature response in Fig. 24 into intervals of length Δt. The rate of damage for each interval is approximated by reading the value of $d\Omega/dt$ that corresponds to the average temperature in the interval. These rates are plotted as bars of length Δt in Fig. 24. The area under each bar represents the fraction of damage occurring in that interval. The accumulation of damage with time $\Omega(k\Delta t)$,

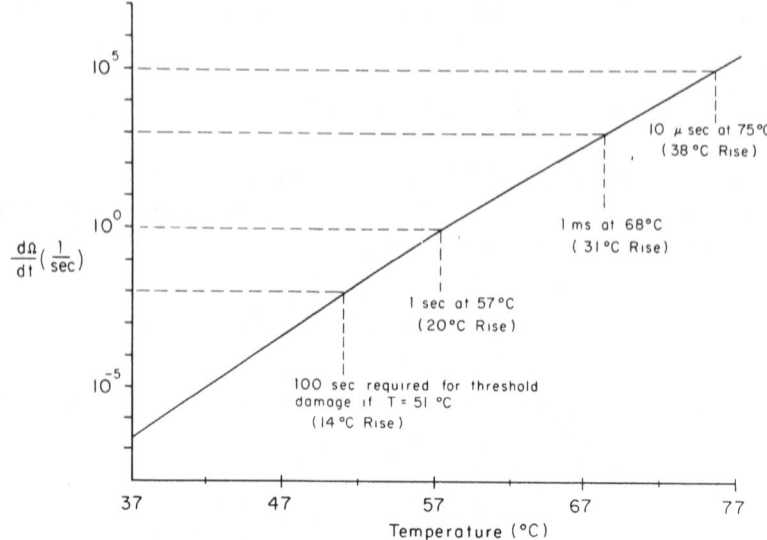

FIGURE 23
Rates of accumulation of damage in tissue based on the rate coefficients in Eq. (48). Times associated with dotted lines indicate the number of seconds to achieve the threshold damage for a step rise in tissue temperature.

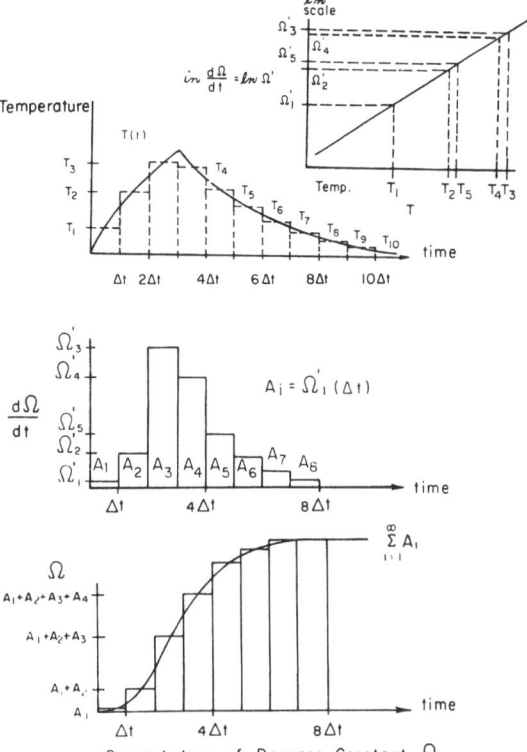

FIGURE 24
Computation of the damage constant Ω. This figure is not to scale. Because of the log scale of $d\Omega/dt$ vs. temperature, the damage rates A_3 and A_4 may be larger than the other rates.

which is approximated by $\Sigma_{i=1}^{k} A_i$ is illustrated by the bottom curve in Fig. 24. The damage function reaches its final value after the temperature response returns to normal. According to the theory, we would expect damage at every point in the tissue where $\Omega \geq 1$.

The sensitivity of the rate of damage $\dot{\Omega}$ to local tissue temperature may be expressed as the ratio of the incremental change in Ω due to an incremental change in $T(K)$. The sensitivity S is given by

$$S = \frac{d\dot{\Omega}/\dot{\Omega}}{dT/T} \tag{49}$$

Substituting Eq. (44) into Eq. (49) and differentiating with respect to T yields, for the coefficients of Eq. (48),

$$S = E/RT = 75{,}000/T \tag{50}$$

As temperature varies from normal (310 K) to boiling (373 K), S decreases from 242 to 201.

This number is so large that interpreting sensitivity as a percent change in $\dot{\Omega}$ for a percent change in temperature has little meaning. For example, a 1% increase in temperature from 320 to 323.2 K increases the rate of damage from 5.01×10^{-4} to 5.14×10^{-3}, which is over a tenfold increase in the rate of damage. Thus, it is difficult to predict the actual change in $\dot{\Omega}$ for variations in such parameters as image size, laser power, irradiation time, etc.

Perhaps a better understanding of the sensitivity of $\dot{\Omega}$ and Ω to temperature profiles resulting from laser irradiation can be illustrated by the following examples. First, let us consider irradiation conditions that produce a step response in tissue temperature. For a step temperature rise of 14 °C, the rate of damage is approximately 10^{-2} (1/sec); 100 sec are required to reach the threshold damage criterion of $\Omega = 1$. Increasing laser power increases the temperature rise and rate of damage and decreases the time to reach threshold damage. Assuming irradiation power is linearly related to induced temperature rise, a 43% increase in power increases the temperature rise from 14 to 20 °C, and the time to reach threshold damage is reduced to 1 sec. Thus, relatively small changes in laser power can produce dramatic changes in the damage function.

Large increases in temperature and $\dot{\Omega}$ occur when there is a decrease in image size. Steady-state temperature at the center of the image is inversely proportional to the square of the image radius for constant laser power. If T_0 is the temperature rise at the center of the image of radius r_0, the image radius to cause a temperature rise T_1 above baseline is given by the relationship

$$\frac{r_1}{r_0} \alpha \left(\frac{T_1}{T_0}\right)^{1/2} \tag{51}$$

If T_0 equals 14°C, then a 16% decrease in radius is required for a temperature rise of 20°C, and $\dot{\Omega}$ increases from 1.0 to 100.0.

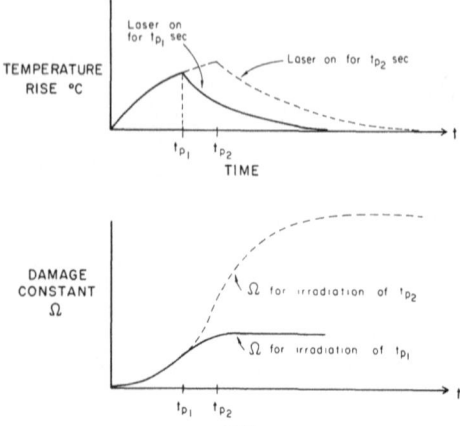

FIGURE 25
The effect of irradiation time on the transient temperature response and the damage constant.

The least sensitive parameter for constant temperature profiles is irradiation time t_p. If $\dot{\Omega}_1$ is the rate of damage for temperature step T_1, then $\Omega = \dot{\Omega}_1 t_p$. The sensitivity of Ω with respect to t_p is 1; i.e., a 10% change in t_p produces a 10% change in the final value of Ω.

As a second example, consider the transient temperature response in Fig. 25. Altering irradiation time from t_{p1} to t_{p2} while holding laser power constant has two effects. The maximum temperature rise and the length of time the tissue is at temperatures that cause significant accumulation of damage are both increased (decreased) if irradiation time is increased (decreased), as illustrated in the figure. Plots of the damage function for two irradiation times are also presented in Fig. 25. Note the dramatic changes in the final value of Ω accompanying an increase in irradiation time. In general, small changes in irradiation parameters of transient temperature responses may cause order of magnitude changes in the damage function. Remember, for transient temperature responses, changes in power, image size, and irradiation time alter both the peak temperature and the length of time the tissue is at a temperature where there is a significant rate of accumulation of damage.

So far, we have examined the damage function for step and transient (exponential) temperature responses. However, if the irradiation is short (for example, less than 0.1 msec), the peak temperature in the center of the beam is proportional to the absorbed energy. In the absence of nonlinear thermal effects, the temperature field for a 0.1 msec irradiation is the same as that for an irradiation of equal energy and image size, delivered in a shorter period of time. Temperature quickly reaches a peak value and then decays at a rate dependent on the thermal properties of the tissue and the size of the laser image (see Fig. 26). The time for accumulation of damage during the irradiation is insignificant. The magnitude of the damage function is determined in this case by the amount of energy delivered to the tissue and the temperature relaxation characteristics of the tissue.

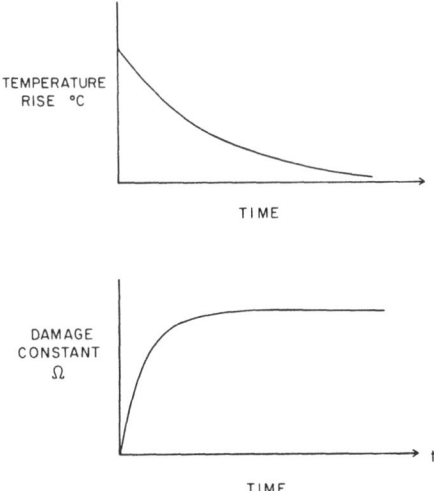

FIGURE 26
The impulse temperature response and the accumulation of damage.

6. MEASUREMENT AND PREDICTION OF THERMAL DAMAGE IN THE RETINA

In the early 1970s, the U.S. Air Force foresaw the rapid development of laser systems and the need for establishing a long-range hazard evaluation program. It was apparent that a program based solely on experimental threshold determinations was not practical for setting safety standards for each new laser system. Anticipating this difficulty, the Air Force sponsored a number of efforts to develop and validate models for the prediction of damage in laser-irradiated tissue.

The first significant modeling efforts resulted in the finite-difference solution reported by Mainster *et al.*[92] This model was expanded and modified by IITRI into a family of models that predicted temperature and damage for laser-irradiated fundus, cornea, and skin. The IITRI models included several absorbing surfaces, reflection, blood perfusion, and a number of other features.[95,99,120]

At the University of Texas, we compared model predictions with experimental measurements of temperature and damage in the eye of the monkey. Temperatures were measured with a unique thin film microthermocouple (cf. Chap. 27). Thin films of Cu and Ni were vapor-deposited on a quartz microcapillary that had a tip diameter of 10–30 μm. The thermocouple junction formed at the tip of the probe had a sensitivity of approximately 21 μV/°C and a rise time of 2 msec. System measurement error was estimated to be ±0.2°C.[112]

Adult monkeys (*Macaca mulatta*) weighing 2–4 kg were anesthesized and secured on an animal platform with ear bars. The backside of the eye was exposed using a surgical procedure that removed sections of the skull, brain tissue, fascia, muscle, bony orbit, periorbital fat, and the lacrimal gland.[110,121] The sclera was then cleared of all fascia, and the conjunctiva

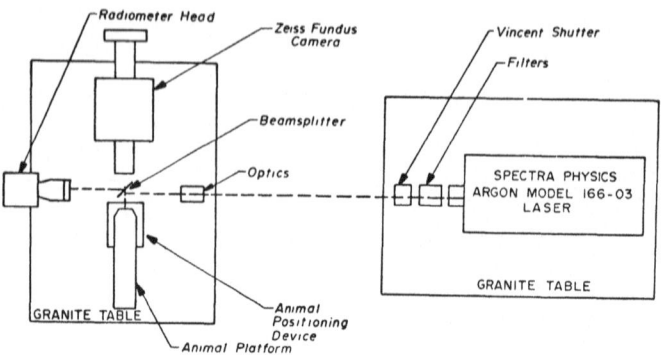

FIGURE 27

Equipment block diagram for experimental laser irradiation of the eye and measurement of temperature. The platform can be rotated with respect to the laser beam to change the relative positions of the thermocouple and the laser image at the retina.

was sutured to an eye holder. On completing the surgery, the platform was attached to a five-degree-of-freedom animal holder, and the eye was aligned with a fundus camera and laser, as shown in Fig. 27. A microthermocouple was inserted through either the macula or paramacula into the vitreous humor. When the probe tip was 100–200 μm anterior to the pigment epithelium (PE), it could be seen with the aid of the fundus camera. With the probe tip in the vitreous humor, the eye was irradiated with short (5–10 msec), low-power laser pulses. The animal was positioned at the center of the beam (position of maximum energy absorption), and the animal and probe were rotated through the laser beam in 25–100 μm radial steps, depending on the image size. At each step, we measured the direct-absorption temperature rise of the thermocouple. This profiled the shape of the intensity profile of the laser image at the retina.

After the intensity profile had been measured, the probe was retracted into the PE. Since the probe could not be visualized, it was again located in the position that gave maximum temperature rise for short (10–30 msec) irradiations. This location was used as a reference point for all subsequent radial and axial temperature measurements. The radial position of the probe corresponded to the center of the image, and the variation in the temperature transient with depth was a function of the true absorption property of the PE, since the direct-absorption component of temperature in the PE is, theoretically, insignificant with respect to the tissue temperature rise.

When the temperature measurements had been completed, the laser image was moved to macular tissue near the probe insertion site to determine threshold injury conditions. A marker lesion was placed on the fundus with an intense laser irradiation. This was followed by a series of irradiations of decreasing power, placed in several rows. The occurrence and sizes of ophthalmoscopically visible lesions were noted at 5 min and at 1- and 3-h exposures. Lesions were also identified, following dissection, by histologic methods.[109] The minimum power required to produce a discernible lesion by histologic method was designated as the threshold power.

A photograph of a typical lesion sequence placed in the paramacula of a monkey using an Ar laser is shown in Fig. 28. The $1/e^2$ radius of the laser image at the retina was approximately 60 μm; laser power was applied for 3 msec. Except for the marker lesion in the top right corner, power was decreased as the lesions were produced from lower left to upper right. The x's in the schematic of the photograph represent lesions visible in the histology preparation that were not seen with the ophthalmoscope. The radius of the threshold lesion in Fig. 28 was 75 μm and the threshold power was 78 mW. The threshold lesions were generally round and cratered, with whitened edges. In general, threshold powers determined by histological technique were 25% lower than those determined with the ophthalmoscope. In addition to the threshold data, the histology preparations graphically illustrated the effect of suprathreshold irradiations on the size of the lesion.

As illustrated in Fig. 29, thresholds measured in our laboratory have been reasonably consistent with data from other laboratories. Threshold energy densities at the retina as a function of retinal image diameters ($1/e^2$) are shown

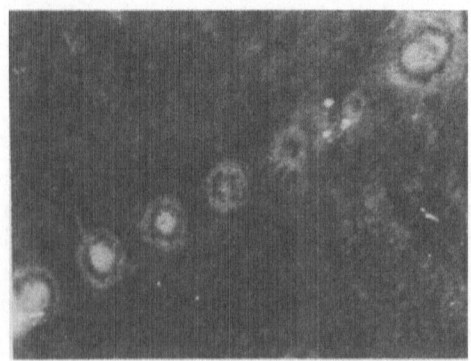

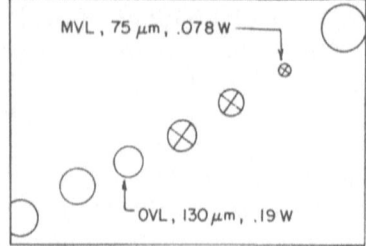

3 ms Exposures

MVL , 75 μm , .078 W

⊗

⊗

⊗

OVL , 130 μm , .19 W

FIGURE 28
(a) Microphotograph of a series of lesions placed in paramacula of monkey eye with argon irradiation for 30 msec. Laser power was reduced as lesions were produced from lower left to upper right. (b) Redrawn outlines of the lesions, OVL indicates lesions visible to the eye with the ophthalmoscope during the experiment; × indicates lesions observed by histological examination. The smallest lesion was 75 μm in diam., produced by 78 mW of laser power at the cornea.

for irradiation times from 30 nsec–1,000 sec. The figure reflects values reported by several groups.[6,7,109,122–125] Additional summaries of retinal threshold data for rabbit, monkey, and man for a wide range of irradiation conditions are available along with safety standards for the eye.[7,11,13]

Since threshold temperatures were not measured during damage threshold determinations, temperature fields for these irradiations were extrapolated from subthreshold temperature profiles as follows. Threshold temperatures were calculated from temperatures measured at subthreshold radiations by assuming a linear relationship between temperature and power. That is,

$$T_T = T_S \left(\frac{P_T}{P_S} \right)$$

where T_S is the temperature measured at subthreshold power P_S, and T_T is temperature measured at threshold corneal power P_T.

A summary of average retinal threshold temperatures from over 40 monkeys during four separate studies[100,109,125] is presented in Table 4. Some of the variation in the results was due to the location of the temperature measurement and criteria used to define damage. In the first set of experiments, temperatures were measured in the center of the image, and threshold power

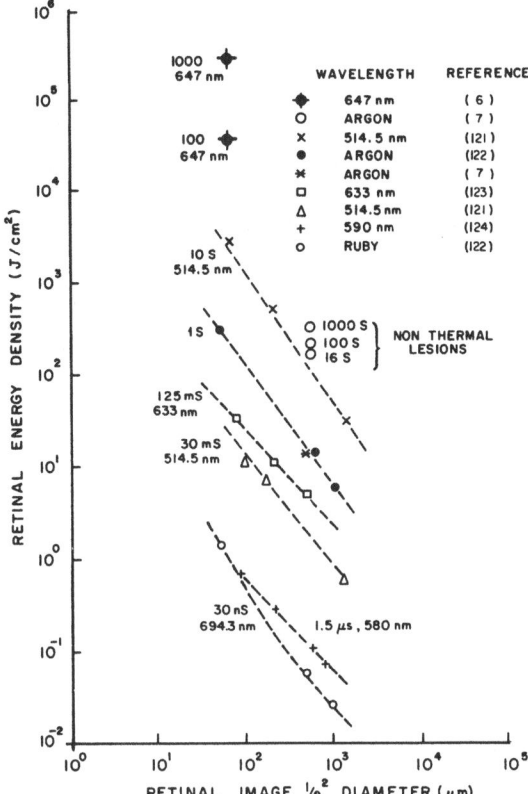

FIGURE 29
Retinal energy density damage
threshold as a function of image size.

was based on an ophthalmoscopic observation (OVL) 5 min after irradiation. The other studies used histologic preparations (MVL), which gave more sensitive assays. Threshold temperatures were determined at a radial distance equal to the lesion radius. All OVL thresholds and temperatures were extrapolated to MVL thresholds and lesion radius temperatures.

Rate coefficients A and E used in the damage function were determined for the eye using threshold data in Table 4. Threshold temperatures for MVL at the lesion radius were plotted on a $1/T$ vs. $\ln(t_p)$ scale in accordance with Eq. (47). The curve and associated values of A and E are shown in Fig. 30. For comparison, a curve based on the rate coefficients of Moritz and Henriques[114] for porcine skin and average threshold temperatures measured by Polhamus[6] for irradiations of 9, 100 and 1,000 sec with a CW krypton laser are included in the figure. For irradiation times below 100 msec, the temperature responses are not step functions. However, above 30 sec, there is no question about the "step" nature of the response. The two straight line segments for argon laser irradiation support arguments for the presence of at least two rate reactions: a thermally induced rate of damage and possibly a photochemical damage mechanism.[6] The threshold temperatures for long-term krypton laser ($\lambda = 647.1$ nm) irradiation suggest a thermal mechanism of injury.[6] Experimental values of the rate process coefficients A and E

TABLE 4

Average Temperature Measurement for Argon Laser Irradiation of the Monkey Eye

Exposure duration (sec)	Image half-power radius[a]	MVL threshold corneal power (mW)	Number of threshold measurements	OVL threshold corrected by factor	Normalized temperature rise at r = 0 and depth for maximum temperature (°C/mW)	Threshold temperature rise at r = lesion radius and depth for maximum temperature (°C)	Threshold temperature at r = 0 and corrected by factor	Lesion radius (μm)
0.002	M	310	1		0.14	32		50
0.003	S	73	2		1.0	21		38
0.003	M	230	1		0.33	31		65
0.01	S	57	3		1.2	37.0		33
0.01	M	97	4		0.56	34.0		37
0.03	S	23.0	5		1.6	21.0		34
0.03	M	34.0	8		1.3	41.0		35
0.03	L	142	1		0.34	46		25
0.1	M	70[b]	18	0.7	0.56	27.4	0.82	
0.1	L	73[b]	5	0.7	0.44	22.5	0.82	
1.0	M	26.6[b]	18	0.74	0.84	18.3	0.82	
1.0	L	28.1[b]	5	0.74	0.68	15.7	0.82	
10	S	12	7		1.7	14.0		39
10	M	7.4	7		1.7	15.0		29
10	L	16	1		1.1	21.0		104
30	M	22	1		0.95	14		100
60	M	22	4		0.94	15		120
100	M	16	2		0.67	8.5		145

[a] $S < 30$ μm, $M \geq 30$ μm and < 100 μm, $L \geq 100$ μm.
[b] MVL threshold computed from OVL threshold data.

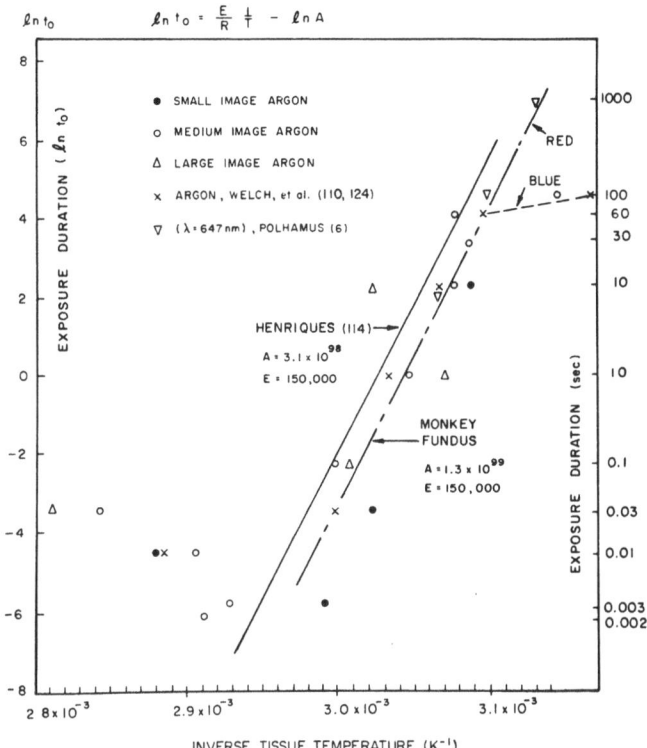

FIGURE 30

Threshold temperatures plotted in accordance with Eq. (47) to determine the rate coefficients A and E.

determined for retinal tissue have been used in Eq. (45) to predict damage for both experimentally measured temperature fields and temperatures calculated with Eq. (3). Results are described in Refs. 6, 100, 109.

Typical damage calculations are based on the maximum temperature rise of the tissue. This point occurs in the center of the laser image ($r = 0$) and near the front surface of the absorbing layer. The exact depth of the maximum in the temperature field is a function of the absorption characteristics of the tissue.

In practice, it is not possible to observe damage at a single point. Threshold damage detected in an experimental situation is a function of the observer's ability to detect some minimum volume of "altered" tissue. For example, we used histological preparations of the PE to evaluate damage to the eye produced by laser irradiation (Fig. 28). With the aid of a microscope, it was possible to measure the extent of damage in the radial direction. In Table 5, the size of the minimum visible lesion is presented for three laser image sizes and 0.03-sec exposures. Generally, we could not detect lesions smaller than 15 μm in radius. Typically, the observable damage radius for threshold irradiation increased with increasing image radius.

TABLE 5
Lesion Radius Measurements for 30 msec Exposures for
Three Image Sizes

$1/e^2$ image radius (μm)	Lesion radius (μm)
742, 94.5, 4[a]	57.5, 36.2, 4
97, 11.0, 12	30.5, 12.7, 12
41, 3.7, 9	20.2, 5.9, 7

[a] The three numbers represent the mean value, standard deviation, and number of measurements.

Thus, the experimental evidence associated with threshold damage ($\Omega = 1$) does not normally occur in the center of the laser image at the point of highest temperature, but at some radius $r = d$ that denotes the boundary of normal and damaged tissue. The irradiance or corneal power I_0 (W/cm^2) at the center of a Gaussian image required to produce damage at $r = d$ is obtained from the solutions to Eqs. (3) and (45).

Comparison of the calculated lesion radius and the damage radius measured from histological analysis of lesions in the monkey eye are presented in Fig. 31. The paramacula of the left eye and the macula of the right eye were irradiated for 10 sec with suprathreshold powers from an Ar laser. The points of zero image size indicate the theoretical threshold damage, located at the center of the image. A plot of the theoretical power per unit area at the retina to produce threshold lesions of various fixed radii for a 30 nsec irradiation are shown in Fig. 32 (data from Bergquist[126] and Ham et al.[127]

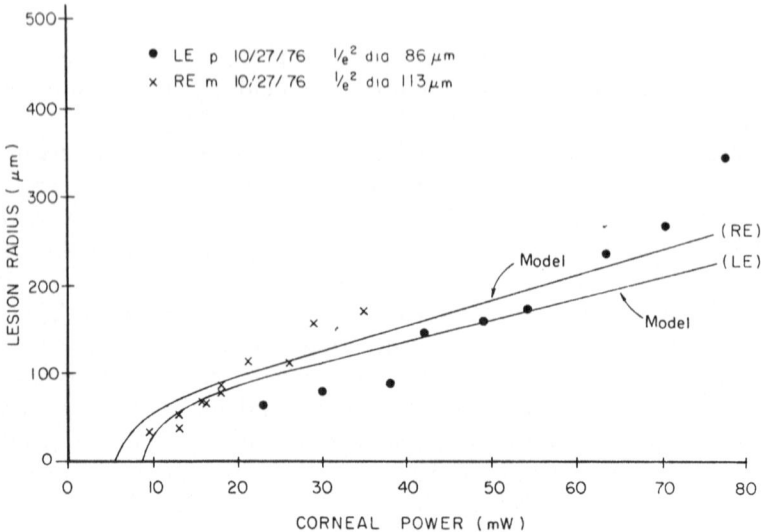

FIGURE 31
Lesion radius as a function of suprathreshold corneal power for a 10 sec argon irradiation. (*LE*—left eye, *RE*—right eye, *p*—paramacula, *m*—macula.)

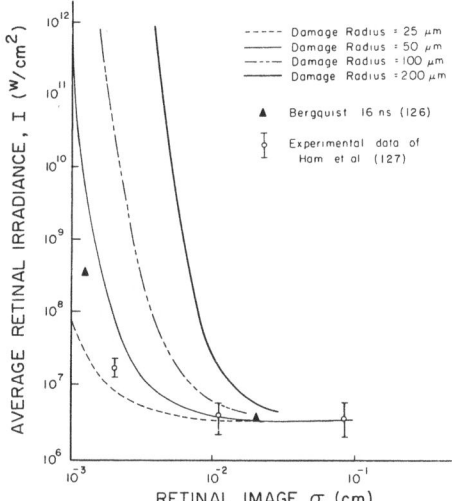

FIGURE 32
Predicted threshold power at the retina required to produce a lesion of specified diameter, expressed as the standard deviation of the retinal image, for a 3×10^{-8} sec exposure.

Approximately 25 times more power per unit area is required to produce a threshold lesion for an image with a standard deviation of 12.5 μm than for an image with a standard deviation of 250 μm.

For impulse and transient responses that produce threshold damage, the temperature responses at $T(r = 0, z, t)$ and $T(r = d, z, t)$ are sufficiently different that consideration should be given to the lesion radius for threshold damage. However, for irradiations that produce steady-state temperature responses, there is little difference in the temperature responses at the center of the image and at the lesion radius.

7. SUMMARY AND RECOMMENDATIONS FOR FUTURE WORK

The previous sections have described the modeling of temperature and damage of laser-irradiated tissue. Although the examples used Ar and Nd–YAG laser sources, the methods can be used to calculate temperature and damage fields due to CO_2 irradiation as well. Almost all of the radiative energy absorption occurs at the surface of the tissue; the source term in Eq. (3) is used only in the first layer of tissue in the model. When the source term is restricted to the surface, it is possible to define a Green's function for the temperature field produced by an impulse irradiation.[36] The temperature field is computed as the convolution of the Green's function with a unit step function representing the irradiation time of the CO_2 laser.

Although the models described in this chapter provide a method for calculating temperature fields and damage due to interaction of laser energy with tissue, most models do not account for the removal of tissue by vaporization nor include the effects of light scattering. A general model of these nonlinear processes is needed to improve our understanding of the surgical applications of lasers.

REFERENCES

1. Cain, C. P., and Welch, A. J., Measured and predicted laser-induced temperature rise in the rabbit fundus, *Invest. Ophthal.* 13, 60–70, 1974.
2. Siegman, A. E., *An Introduction to Lasers and Masers* (McGraw-Hill, New York, 1971).
3. 1981 *Laser Focus Buyer's Guide* 16th ed. (with *Fiberoptic Communications*) (Laser Focus Advanced Technology Publications, Newton, MA, Jan. 1981).
4. Fuller, T. A., The physics of surgical lasers, *Lasers Surg. Med.* 1, 5–14, 1980.
5. Gibbons, W. D., and Allen, R. G., Retinal damage from long-term exposure to laser radiation, *Invest. Ophthal.* 16, 521–529, 1977.
6. Polhamus, G. D., *In-vivo* measurement of long-term laser-induced retinal temperature rise, *IEEE Trans. Biomed. Eng.* BME-21, 617–622, 1980.
7. Ham, W. T., Jr., Mueller, H. A., and Sliney, D. H., Retinal sensitivity to damage from short wavelength light, *Nature* 260, 154–155, 1976.
8. Lawwill, T., Crockett, S., and Currier, G., Retinal damage secondary to chronic light exposure, *Doc. Ophthal.* 44, 379–402, 1977.
9. Marshall, J., Thermal and mechanical mechanisms in laser damage to the retina, *Invest. Ophthal.* 9, 97–115, 1970.
10. Cleary, S. F., and Hamrick, P. E., Laser-induced acoustic transients in the mammalian eye, *J. Acous. Soc. Am.* 46, 1037, 1969.
11. Sliney, D. H., and Freasier, B. C., Evaluation of optical radiation hazards, *Appl. Optics* 12, 1–24, 1973.
12. Ham, W. T., Mueller, H. A., Goldman, A. I., Newman, B. E., Holland, L. M., and Kuwabara, T., Ocular hazard from picosecond pulses of Nd:Yag laser radiation, *Science* 148, 362–363, 1974.
13. Sliney, D. H., and Wolbarsht, M. L., *Safety with Lasers and Other Optical Sources* (Plenum, New York, 1980).
14. Kaplan, I., Five years experience with the CO_2 laser, in *Laser Surgery*, vol. 2, I. Kaplan, ed. (Jerusalem Academic, Jerusalem, 1978) pp. 355–389.
15. Horch, H. H., McCord, R. C., Schaffer, E., and Rupracht, L., Aspects of laser osteotomy, *Proc. Lasers Med. Biol., GSF, Neuherberg, Bericht* BPT5, 40-1 to 40-12, 1977.
16. Morein, G., Bone growth alterations resulting from application of CO_2 laser beam to the epiphyseal growth plates, *Acta Orthop. Scand.* 49, 244–248, 1978.
17. Horch, H. H., Histological and long-term results following laser osteotomy, in *Laser Surgery*, vol. 2, I. Kaplan, ed. (Jerusalem Academic, 1978), pp. 319–325.
18. Farine, I., and Horoshowski, H., The use of the laser scapel in orthopaedic surgery, in *Laser Surgery*, vol. 2, I. Kaplan, ed. (Jerusalem Academic, 1978), pp. 351–354.
19. Glantz, G., and Korn, A., The use of the carbon dioxide laser in general surgery, in *Laser Surgery*, vol. 2, I. Kaplan, ed. (Jerusalem Academic, 1978), pp. 9–16.
20. Nimsakul, N., Nishimura, M., Tanino, R., Osada, M., and Hata, J., Our experiences with the Sharplan 791 CO_2 laser, in *Laser Surgery*, vol. 2, I. Kaplan, ed. (Jerusalem Academic, Jerusalem, 1978), pp. 59–75.
21. Kaplan, I., Sharon, U., and Ger, R., The carbon dioxide laser in clinical surgery, in *Laser Applications in Medicine and Biology*, vol. 2, M. L. Wolbarsht, ed. (Plenum, New York), pp. 295–308.
22. Hall, R. R., The carbon dioxide laser in nonendoscopic urological surgery, in *Laser Surgery*, vol. 2, I. Kaplan, ed. (Jerusalem Academic, 1978), pp. 197–202.
23. Barzilay, B., Perlberg, S., and Caine, M., Use of CO_2 laser beam for kidney surgery, in *Laser Surgery*, vol. 2, I. Kaplan, ed. (Jerusalem Academic, 1978), pp. 164–168.
24. Fidler, J. P., Slutzki, S., Shafir, R., znd Bornstein, L. A., Use of carbon dioxide laser for large excisions with minimal blood loss, *Plast. Reconstr. Surg.* 60, 250–255, 1977.
25. Kaplan, I., ed, *Laser Surgery*, vol. 2 (Jerusalem Academic, 1978).
26. Peled, I., Shohat, B., Gassner, S., and Kaplan, I., Excision of epithelial tumors: CO_2 versus conventional methods, *Cancer Lett.* 2, 41–46, 1976.
27. Aranoff, B. L., CO_2 lasers in surgical oncology, in *Laser Surgery*, vol. 2, I. Kaplan, ed. (Jerusalem Academic, 1978), pp. 133–158.
28. Stellar, S., Levine, N., Ger, R., and Levenson, S. M., Carbon dioxide laser for excision of burn eschars, *Lancet* 1, 945, 1971.

29. Friedman, E. W., The CO_2 laser in head and neck surgery, in *Laser Surgery*, vol. 2, I. Kaplan, ed. (Jerusalem Academic, 1978).

30. Karlin, D. B., Patel, C. K., Wood, O. R., and Rovere, J., CO_2 laser in vitreoretinal surgery, *Ophth.* (Rochester) 86(2), 290–298, 1979.

31. Peyman, G., and Sanders, D., Full-thickness eye wall resection, in *Advances in Uveal Surgery, Vitreous Surgery, and the Treatment of Endophthalmitis*, edited by Peyman and Sanders (Appleton-Century-Crofts, New York, 1975).

32. DiBartolomeo, J. R., and Ellis, M., The argon laser in otology, *Laryngoscope* 90, 1786–1796, 1980.

33. Perkins, C., Laser stapedotomy for otosclerosis, *Laryngoscope* 90, 228–241, 1980.

34. Escudero, L. H., Castro, A. O., Drumond, M., Porto, S. P., Bozinis, D. G., Penna, A. F., and Gallego-Lleusma, E., Argon laser in hyman tympanoplasty, *Arch. Otol.* 105, 252–253, 1979.

35. Bellina, J. H., and Polanyi, T. G., Management of vaginal adenosis and related cervicovaginal disorders in DES-exposed progeny by means of carbon dioxide laser surgery, *J. Repro. Med.* 16, 295–296, 1976.

36. Bellina, J. H., and Seto, Y. J., Pathological and physical investigations into CO_2 laser tissue with specific emphasis on cervical intraepithelial neoplasm, *Lasers Surg. Med.* 1, 47–69, 1980.

37. Burke, L., Covell, L., and Antonioli, D., Carbon dioxide laser therapy of cervical intraepithelial neoplasia: Factors determining success rate, *Lasers Surg. Med.* 1, 113–122, 1980.

38. Dorsey, J. H., Diggs, E. S., Microsurgical conization of the cervix by carbon dioxide laser, *Obstet. Gyn.* 54, 565–570, 1979.

39. Baggish, M. S., Carbon dioxide laser treatment for condylomata acuminata venereal infections, *Obstet. Gyn.* 55, 1980.

40. McCord, R. C., Medical applications of CO_2 laser fiber optics, *Proc. SPIE* 266, 1981.

41. Miller, J. B., and Smith, M. R., Transvitreal carbon dioxide photocautery—vitrectomy: A new instrument presentation, *Ophthalmology* 85, 1195–1200, 1978.

42. Beckman, H., and Fuller, T. A., Carbon dioxide laser scleral dissection and filtering procedure for glaucoma, *Am. J. Ophthal.* 88, 73–77, 1979.

43. Miller, J. B., Smith, M. R., and Boyer, D. S., Intraocular carbon dioxide laser photosurgery, *Lasers Surg. Med.* 1, 165–176, 1980.

44. Jako, G. J., Vaughan, C. W., Strong, M. S., and Polanyi, T. G., Surgical management of malignant tumors of the aerodigestive tract with carbon dioxide laser microsurgery, *Int. Adv. Surg. Oncol.* 1, 265–284, 1978.

45. Strong, M. S., The use of the CO_2 laser in otolaryngology: a progress report, *Trans. Am. Acad. Ophthal. and Otol.* 82, 595–602, 1976.

46. Stern, L. S., *et al.*, Qualitative and morphometric evaluation of vocal cord lesions produced by the carbon dioxide laser, *Laryngoscope* 90, 792–808, 1980.

47. Mihashi, S., *et al.*, Laser surgery in otolaryngology: interaction of CO_2 laser and soft tissue, *Ann. N. Y. Acad. Sci.* 267, 263–294, 1976.

48. Carruth, J. A. S., *et al.*, The carbon dioxide laser: Safety aspects, *J. Laryngol. Otol.* 94, 411–417, 1980.

49. Pratt, L. W., The CO_2 laser in otolaryngology, *J. Maine Med. Assoc.* 71, 39–45, 1980.

50. Putney, F. J., Carbon dioxide laser in otolaryngology, *Southern Med. J.* 72, 1385–1386, 1979.

51. LeJeune, F. E., Jr., Intralaryngeal surgery, *Laryngoscope* 84, 1815–1820, 1977.

52. Andrews, A. H., and Moss, H. W., Experiences with carbon dioxide laser in the larynx, *Ann. Otol.* 83, 462–470, 1974.

53. Snow, J. C., Norton, M. L., Saluja, T. S., and Estanislao, A. F., Fire hazard during CO_2 laser microsurgery on the larynx and trachea, *Anesth. Analg.* 55, 146–147, 1976.

54. Burgess, G. E., and LeJeune, F. E., Endotracheal tube ignition during laser surgery of the larynx, *Arch. Otol.* 105, 561–562, 1979.

55. Vourc'h, G., Tannieres, M., and Freche, G., Ignition of a tracheal tube during laryngeal surgery, *Anesthesia* 34, 685, 1979.

56. Vourc'h, G., Tannieres, M., and Freche, G., Anesthesia from microsurgery of the larynx using a carbon dioxide laser, *Anesthesia* 34, 53–57, 1979.

57. Norton, M. L., *et al.*, Endotracheal intubation and Venturi (JET) ventilation for laser microsurgery of the larynx, *Ann. Otol.* 85, 656–663, 1976.

58. Snow, J. C., Anesthesia for carbon dioxide laser microsurgery on the larynx and trachea, *Anesth. Analg.* 53, 507–512, 1974.

59. Patil, V., *et al.*, A modified endotracheal tube for laser microsurgery, *Anesthesiology* 51, 571, 1979.

60. Kalhan, S. *et al.*, A further modification of endotracheal tubes for laser microsurgery, *Anesthesiology* 53, 81, 1980.

61. Cosman, B., Clinical experience in the laser therapy of port wine stains, *Lasers Surg. Med.* 1, 133–152, 1980.

62. Apfelberg, D. B., Kosek, J., Maser, M. R., and Lash, H., Histology of port wine stains, *Br. J. Plastic Surg.* 32, 232–237, 1979.

63. Apfelberg, D. B., Progress report on extended clinical use of the argon laser for cutaneous lesions, *Lasers Surg. Med.* 1, 71–83, 1980.

64. Apfelberg, D. B., Maser, M. R., Lash, H., and Rivers, S. L., Extended clinical use of the argon laser for cutaneous lesions, *Arch. Dermatol.* 115, 719–721, 1979.

65. Ohshiro, T., Maruyama, Y., Nakajima, H., and Mima, M., Treatment of pigmentation of the lips and oral mucosa in Peutz–Jeghers syndrome using ruby and argon lasers, *Br. J. Plastic Surg.* 33, 346–349, 1980.

66. Staehler, G., and Hofstetter, A., Transurethral laser irradiation of urinary bladder tumors, *Eur. Urol.* 5, 64–69, 1979.

67. Staehler, G., Dosimetry for Nd:YAG laser applications in urology, *Lasers Surg. Med.* 1, 191–197, 1980.

68. Goodale, R. L., Okada, A., Gonzales, R., Borner, J. W., Edlich, R. F., and Wangensteen, O. H., Rapid endoscopic control of bleeding gastric erosions by laser radiation, *Arch. Surg.* 101, 211–214, 1970.

69. Yellin, A. E., Dwyer, R. M., Craig, J. R., Bass, M., and Cherlow, J., Endoscopic argon ion laser phototherapy of bleeding gastric lesions, *Arch. Surg.* 111, 750–755, 1976.

70. Kiefhaber, P., Nath, G., and Moritz, K., Endoscopical control of massive gastrointestinal hemorrhage by irradiation with a high-power Nd-Yag laser, *Prog. Surg.* 15, 140–155, 1977.

71. Meyer, H. J., Vonnahme, F. J., Haverkampf, K., and Huchzermeyer, H., Laser coagulation in the upper GI tract: A preliminary light and scanning electron-microscopic study, *Lasers Surg. Med.* 1, 103–112 1980.

72. Wirthlin, L. S., Van Urk, H., and Malt, R. A., Predictors of surgical mortality in patients with cirrhosis and nonvariceal gastroduodenal bleeding, *Surg. Gyn. Obst.* 139, 65–68, 1974.

73. Waitman, A. M., Spira, I., Chryssanthou, C. P., and Stenger, R. J., Fiberoptic-coupled argon laser in the control of experimentally produced gastric bleeding, *Gastrointest. Endosc.* 22, 78–81, 1975.

74. Dotter, C. T., Goldman, M. L., and Rosch, J., Instant selective arterial occlusion with isobutyl 2-cyanoacrylate, *Radiology* 114, 227–230, 1975.

75. Protell, R. L., Silverstein, F. E., Gulacsik, C., Martin, T. R., Dennis, M. B., Auth, D. C., and Rubin, C. E., Cyanoacrylate glue (flucrylate) fails to stop bleeding from experimental gastric ulcers, *Gastroenterology (abstr.)* 72, 1114, 1977.

76. Katon, R. M., Experimental control of gastrointestinal hemorrhage via the endoscope: A new era dawns, *Gastroenterology* 70, 272–277, 1976.

77. Sugawa, C., Shier, M., Lucas, C. E., and Walt, A. J., Electrocoagulation of bleeding in the upper part of the gastrointestinal tract: A preliminary experimental clinical report, *Arch. Surg.* 110, 975–979, 1975.

78. Blackwood, W. D., and Silvas, S. E., Electrocoagulation of hemorrhage gastritis, *Gastrointest. Endosc.* 18, 53–55, 1971.

79. Blackwood, W. D., and Silvas, S. E., Gastroscopic electrosurgery, *Gastroenterology* 61, 305–314, 1971.

80. Papp, J. P., Endoscopic electrocoagulation of upper gastrointestinal hemorrhage, *J. Am. Med. Assoc.* 236, 2076–2079, 1976.

81. Papp, J. P., Fox, J. M., and Wilks, H. S., Experimental electrocoagulation of dog gastric mucosa, *Gastr. Intest. Endosc.* 22, 27–28, 1975.

82. Volpicelli, N. A., McCarthy, J. D., Bartlett, J. D., and Badger, W. E., Endoscopic electrocoagulation: An alternative to operative therapy in bleeding peptic ulcer disease, *Arch. Surg.* 113, 483, 1978.

83. Laurence, B. H., Vallon, A. G., Cotton, P. B., Miro, J. R., Oses, J. C., LeBodic, L., Sudry, P., Fruhmorgen, P., and Bodem, F., Endoscopic laser photocoagulation for bleeding peptic ulcers, *Lancet* 124–125, 1980.

84. Brown, S. G., Salmon, P. R., Kelly, B. M., Calder, H., Pearson, H., Weaver, B. M. Q., and Read, A. E., Argon laser photocoagulation in the dog stomach, *Gut* **20**, 680–687, 1979.

85. Dwyer, R. M., Yellin, A. E., Craig, J., Cherlow, J., and Bass, M., Gastric hemostasis by laser phototherapy in man: A preliminary report, *J. Am. Med. Assoc.* **236**, 1383–1384, 1976.

86. Fruhmorgen, P., Bodem, F., Reidenbach, H. D., and Kaudk, B., Endoscopic laser coagulation of bleeding gastrointestinal lesions with report of the first therapeutic application in man, *Gastrointest. Endosc.* **23**, 73–75, 1976.

87. Silverstein, F. E., Protell, R. L., Piercey, J., Rubin, C. E., Auth, D. C., and Dennis, M., Endoscopic laser treatment, II. Comparison of the efficacy of high- and low-power photocoagulation in control of severely bleeding experimental ulcers in dogs, *Gastroenterology* **73**, 481–486, 1977.

88. Silverstein, F. E., Protell, R. L., Gulacsik, C., Auth, D. C., Deltenre, M., Dennis, M., Piercey, J., and Rubin, C., Endoscopic laser treatment, III. The development and testing of a gas-jet-assisted argon laser wave guide in control of bleeding experimental ulcers, *Gastroenterology* **74**, 232–239, 1978.

89. Staehler, G., Hoffstetter, A., Gorisch, W., Kieditsch, E., and Mussiggang, M., Endoscopy in experimental urology using an argon laser beam, *Endoscopy* **8**, 1–4, 1976.

90. Silverstein, F. E., Protell, R. L., Gilbert, D. A., Gulacsik, C., Auth, D. C., Dennis, M. E., and Rubin, C. E., Argon versus Neodymium–YAG laser photocoagulation of experimental canine ulcers, *Gastroenterology* **77**, 491–496, 1979.

91. Mainster, M. A., White, T. J., Tips, J. H., and Wilson, P. W., Refined temperature increases produced by intense light sources, *J. Opt. Soc. Am.* **60**, 264–270, 1970.

92. Mainster, M. A., White, T. J., Tips, J. H., and Wilson, P. W., Transient thermal behavior in biological systems, *Bull. Math. Biophys.* **32**, 303–314, 1970.

93. Mainster, M. A., White, T. J., and Allen, R. G., Spectral dependence of retinal damage produced by intense light sources, *J. Opt. Soc. Am.* **60**, 848–855, 1970.

94. Wissler, E. H., An analysis of chorioretinal thermal response to intense light exposure, *IEEE Trans. Biomed. Engr.* **BME-23**, 207–214, 1976.

95. Takata, A., Laser-induced thermal damage of skin, SAM-TR-77-38, USAF School of Aerospace Medicine (IIT Research Institute, Chicago, 1977).

96. Welsch, H., Birngruber, R., Boergen, K.-P., Gabel, V. P., and Hillenkamp, F., The influence of scattering on the wavelength-dependent light absorption in blood, *Proc. Lasers Med. Biol.* (*GSF Neuherberg*), vol. 6SF-Bericht BPT5 14-1 to 14-8, 1977.

97. Takata, A. N., Thermal model of laser-induced skin damage: computer program operator's manual, SAM-TR-77-37, USAF School of Aerospace Medicine (IIT Research Institute, Chicago, 1977).

98. White, T. J., Mainster, M. A., Wilson, P. W., and Tips, J. H., Chorioretinal temperature increases from solar observations, *Bull. Math. Biophys.* **33**, 1–17, 1971.

99. Takata, A. N., Goldfinch, L., Hinds, J. K., Kuan, L. P., Thomopoulis, N., and Weigandt, A., Thermal model of laser-induced eye damage, Report F-41609-74-C-0005, USAF School of Aerospace Medicine (IIT Research Institute, Chicago, 1974).

100. Welch, A. J., Cain, C. P., and Priebe, L. A., Temperature rise in fundus exposed to laser radiation, SAM-TR-75-32, USAF School of Aerospace Medicine (IIT Research Institute, Chicago, 1975).

101. Douglas, J., and Gunn, J., A general formulation of alternating direction methods, *Numer. Math.* **6**, 428–453, 1964.

102. Priebe, L. A., and Welch, A. J., A dimensionless model for the calculation of temperature increase in biologic tissues exposed to nonionizing radiation, *IEEE Trans. Biomed. Eng.* **BME-26**, 244–250, 1979.

103. Boergen, K. P., Birngruber, R., Gabel, V. P., and Hillenkamp, F., Experimental studies on controlled closure of small vessels by laser irradiation, *Proc. Lasers Med. Biol.* (*GSF Neuherberg* 1977).

104. Beibie, H. F., Frankhauser, F., Lotmar, W., and Roulier, A., Theoretical estimate of the temperature within irradiated retinal vessels, *Acta Ophthal.* **52**, 13–36, 1974.

105. Gorisch, W., and Boergen, K. P., Thermal shrinkage of collagen fibers during vessel occlusion, *Laser Surg.* 3rd International Congress for Laser Surgery, Graz, Austria, 1979.

106. Wissler, E. H., and Gorisch, W., A mathematical model for predicting thermal responses in the neighborhood of arteries and veins during laser irradiation, *Advances in Biomedical Engineering*, V. Mow, ed., (ASME, New York, 1980).

107. Welch, A. J., Wissler, E. H., and Priebe, L. A., Significance of blood flow in calculations of temperature in laser-irradiated tissue, *IEEE Trans. Biomed. Eng.* BME-27, 164–166, 1980.

108. Kelle, C. A., and Neil, E., *Samson Wright's Applied Physiology* (Oxford University Press, Oxford, Eng. 1971).

109. Welch, A. J., Priebe, L. A., Forster, L. D., Gilbert, R., Lee, C., and Drake, P., Experimental validation of thermal retinal models of damage from laser radiation, SAM-TR-79-9, USAF School of Aerospace Medicine, 1979 (IIT Research Institute, Chicago, 1979).

110. Priebe, L. A., Cain, C. P., and Welch, A. J., Temperature rises required for production of minimal lesions in the macula mulatta retina, *Am. J. Ophthal.* 79, 405–413, 1975.

111. Lee, C. F., Experimental validation of retinal temperature distribution model for laser irradiation (Masters thesis, University of Texas, Austin, 1977).

112. Cain, C. P., and Welch, A. J., Thin film temperature sensors for biological measurements, *IEEE Trans. Biomed. Eng* BME-21, 421–423, 1974.

113. Routh, J. I., *Introduction to Biochemistry*, (W. B. Saunders Col, Philadelphia, 1971).

114. Henriques, F. C., and Moritz, A. R., Studies of thermal injury, I. Conduction of heat to and through the skin, *Am. J. Path.* 23, 531–549, 1947.

115. Stoll, A. M., and Green, L. C., Relationship between pain and tissue damage due to thermal radiation, *J. Appl. Physiol.* 14, 373–382, 1959.

116. Hu, C. L., and Barnes, F. S., Thermal-chemical damage in biological material under laser irradiation, *IEEE Trans. Biomed. Eng.*, BME-17, 220, 1970.

117. Kach, E. A., and Incropera, F. P., Induction thermocoagulation: Thermal response and lesion size, *IEEE Trans. Biomed. Eng.* BME-21, 8, 1974.

118. Takata, A., Development of criterion for skin burns, *Aersospace Med.* 45, 634–637, 1974.

119. Vassiliadis, A., Ocular damage from laser radiation, *Laser Applications in Medicine and Biology*, vol. 1, M. L. Wolbarsht, ed. (Plenum, New York, 1971), pp. 125–162.

120. Mertz, A. R., Anderson, B. R., Bell, E. L., and Egbert, D. E., Retinal thermal model of laser induced eye damage: computer program operator's manual, SAM-TR-76-33, USAF School of Aerospace Medicine (IIT Research Institute, Chicago, 1976).

121. Welch, A. J., Cain, C. P., and Priebe, L. A., Temperature rise in fundus exposed to laser radiation, SAM-TR-75-32, USAF School of Aerospace Medicine (IIT Research Institute, Chicago, 1975).

122. Beatrice, E. S., and Frisch, G. D., Retinal laser damage thresholds as a function of image diameter, *Arch. Environ. Health* 27, 322–326, 1973.

123. Ham, W. T., Geeraets, W. J., Mueller, H. A., Williams, R. C., Clarke, A. M., and Cleary, S. F., Retinal burns threshold for the helium–neon laser in the rhesus monkey, *Arch. Ophthal.* 84, 797–808, 1970.

124. Schorner, J. Untersuchungen von Wechselwirkungs mechanismus an Biologischen Proben mit einem extrem schmälbandigen Farbstofflaser, GSF-Bericht AO 280, Gesellschaft fur Strahelen und Unwelfforschung GmbH, München, 1980.

125. Welch, A. J., Priebe, L. A., Polhamus, G. D., Mistry, G. D., and Drake, P., Limits of applicability of thermal models of thermal injury, final report for contract F41609-76-C-0005, USAF School of Aerospace Medicine, Brooks Air Force Base, Texas, 1976.

126. Bergquist, T., Kleman, B., and Tengroth, B., Laser irradiance levels for retinal lesions, *Acta Ophthal.* 43, 331–349, 1965.

127. Ham, W. T., Williams, R. C., Mueller, H. A., Guerry, D., Clarke, A. M., and Geeraets, W. J., Effects of laser radiation on mammalian eye, *Ann. N. Y. Acad. Sci.* 28, 4, 517, 1966.

Chapter 20

PRESERVATION OF BIOLOGICAL MATERIAL BY FREEZING AND THAWING

John J. McGrath

1. INTRODUCTION

Mankind is fascinated with the relationship between living systems and low temperature. This fascination has stimulated both scientific inquiry and literary license. As early as 1683, Robert Boyle found that fish and frogs could survive for short periods of time if a fraction of the body water remained unfrozen.[1] Looking to the possible future potential of cryopreservation, science fiction writers have envisioned the day when biological systems as complex as the human body will be preserved reversibly by freezing and storage at low temperatures. But what is the present state of affairs as we stand between early scientific observations on the one hand and the dreams of science fiction writers on the other?

To begin with, it is well known that decreased temperature generally results in the suppression of metabolic activity and, thus, in a reduction of the rate at which deterioration of an unnourished biological system would occur. However, to realize the long-term storage of a living system, freezing appears to be required, and temperatures in the range of −80 to −196°C are necessary for successful preservation for extended periods.

The freezing process is not as benign as one might assume. It generally induces extreme variations in chemical, thermal, and electrical properties that could be expected to alter intracellular organelles and delicate membrane systems. Given the extreme complexity of even the simplest biological cells, it is therefore remarkable that a reversible state of suspended animation by freezing is possible at all.

In fact, the chances of achieving reasonable recovery are very slim without employing the proper procedures for the addition and removal of chemical "antifreezes" and the careful control of the temperature history during freezing and thawing.

As a general rule, successful cryopreservation techniques are correlated inversely with the complexity of the system. Most of the existing, successful

---

John J. McGrath Ph.D. • Bioengineering Transport Processes Laboratory, Michigan State University, East Lansing, Michigan 48824.

cryopreservation techniques have been developed for individual cells or small tissue samples. Success with tissue is limited, and success with organs is essentially nonexistent: on the basis of this lack of success, whole-body freezing remains a very controversial area. A very small group of individuals is involved in body freezing, and no one has published reports of successful whole-body freezing with recovery.

At the present time, it is possible to cryopreserve many important smaller mammalian systems, including blood, embryos, spermatozoa, culture cells, hepatocytes, bone marrow cells, pancreatic tissue, cornea, skin, heart, and kidney. Other successfully frozen systems of interest include protozoa, parasites, insects, fish, microorganisms, plants, and algae.

Frozen blood is used in the clinic for transfusion and regulation of blood components. Embryos and spermatozoa are useful for animal-breeding purposes and studying basic genetics. Culture cells are routinely frozen in biological and medical laboratories as an economic alternative to continuous passage through culture. This procedure can also be used to prevent genetic drift and store rarely used but valuable strains. Hepatocytes and pancreatic tissue are used for hormone and diabetes research. Bone marrow cells are used in cancer therapy, and frozen–thawed skin is used in treating burn victims. Frozen parasites and micro-organisms are also frozen and thawed, and these systems are used in bioassays designed to screen for antiparasitic drugs and environmental mutagens and carcinogens.

The problems of freezing injury are intimately related to chilling and frost injury in plants. Thus, advances in cryobiology are expected to impact favorably on food production from the land (plants) and, in a related fashion, from the sea (fish, algae, etc.).

A preservation technique that is closely associated with cryopreservation but somewhat different is freeze drying (lyophilization). The fundamental importance of water is a characteristic shared by both processes. In the case of cryopreservation, water is "removed" as a result of solidification. Freeze drying involves the further step of sublimation from the solid to gaseous state after freezing. Lyophilization is an attractive process, because it would mean that biological systems could be stored at or near room temperature rather than at cryogenic temperatures. This is convenient and economical. However, it appears that no organism can be freeze-dried successfully if it can not be cryopreserved successfully.[2] Furthermore, it appears that freeze drying produces mutation in a number of biological systems, in contrast to cryopreservation, which is not mutagenic. A survey of this topic is given by Ashwood-Smith,[2] and further information is available in Refs. 3–5.

A complementary aspect of cryopreservation is the selective destruction of cells. If one knows the conditions required for preservation, one generally also knows those conditions that will destroy living material. This knowledge can be used to advantage in such clinical procedures as cryosurgery, where, for example, cryopreservation can be used to selectively kill tumors. Practical aspects of this technique are discussed in Chap. 21.

The author believes that thermodynamic and mathematical modeling of the response of biological cells to freezing has resulted in significant progress in cryobiology and will continue to do so in the future. Since the author is most familiar with this aspect of cryobiology, the topic of cryopreservation is presented here in a limited but somewhat unique fashion. Thus, this chapter emphasizes a review of the development and current state-of-the-art of thermodynamic modeling in the field of cryobiology.

Some details of the thermodynamic models are presented, but due to the limited scope of this chapter, detailed model development, mathematics, and computer techniques are omitted. The flavor of the chapter is therefore primarily qualitative, with references given for those interested in the details of such research. The thermodynamic models are presented after reviewing several of the basic phenomena involved in cryopreservation and after discussing the state-of-the-art in applied cryobiology.

The chapter begins with a discussion of basic aspects of low-temperature preservation. In this section, freezing, thawing, and simple models of water transport across semi-permeable membranes are presented in the context of cryopreservation. This section goes on to illustrate the general response of biomaterials to cryopreservation procedures and concludes with a discussion of many of the proposed mechanisms of freeze–thaw damage.

The next section, on applied cryobiology, outlines several practical preservation techniques and points out the advantages of cryopreservation applications. Limitations in current practice are given for a variety of biomaterials.

The development and scope of the application of thermodynamic modeling to problems in cryobiology are presented with reference to (1) the responses of biomaterials to freeze thawing and (2) the potential mechanisms of freezing injury. Critiques of existing models are offered in several instances.

For those readers interested in aspects of cryopreservation and cryobiology other than those given here, Ref. 6 contains an excellent review of journals, books, and societies related to all aspects of low-temperature biology and medicine.

2. BASIC ASPECTS OF LOW-TEMPERATURE PRESERVATION

2.1. Freezing

Biological tissues typically consist of large amounts of water and a complex mixture of solutes. Prior to considering models for such complex systems, several basic aspects of the liquid–solid phase change are considered for a very simple system model—the pure substance. Since water plays such an important role in biological systems, it is chosen as an example of the pure substance. Following an examination of this simple case, binary and ternary systems will be introduced to illustrate how additional information is included to more closely model a real mixture in a biological system.

2.2. The Pure Substance

In general, the state of a simple compressible pure substance can be defined by two independent properties of the system. However, when the two phases ($p = 2$), solid and liquid coexist in equilibrium, the Gibbs phase rule dictates that for the single component system ($n = 1$) consisting of water, there will be only one independent intensive variable f for the system

$$f = n + 2 - p \qquad (1)$$

Thus a pressure–temperature representation of the locus of two-phase equilibrium states for a pure substance that expands on solidification would be represented schematically as in Fig. 1. The Gibbs phase rule therefore implies a unique equilibrium freezing temperature for each pressure. Hence, as shown in Fig. 1, selecting a pressure of 1 atm defines the equilibrium freezing temperature, 273.15 K.

The large negative slope of the locus of solid–liquid equilibrium states indicates that extremely large increases in pressure are required to effect significant freezing point depressions. A freezing point depression may be gained by altering pressure but at the expense of an altered structure.[7] Crystalline structures are likely to cause more cellular damage, as will be discussed later. Some researchers have attempted to take advantage of pressure increases to depress the freezing point to such a low temperature that when solidification occurs, it will occur as an amorphous glass structure rather than a crystalline structure. The additional equipment required and potential safety hazards have tended to make this method an unattractive part of the cryo-preservation procedure, but some investigators believe that the potential benefits may outweigh the inconveniences.[8]

The typical temperature history observed experimentally during freezing pure water or a biological sample will be similar to that represented schematically in Fig. 2. In this situation, a bath temperature (shown as a dotted line) is being decreased at a constant rate. A similar sample temperature history would result from immersion into a low temperature "thermal sink" held at a constant temperature. It is assumed that there is good thermal contact between the bath and the sample (sample temperature shown as a solid line.)

During the first part of the freezing process a–b, sensible heat is removed from the sample, and the sample temperature tracks the bath temperature

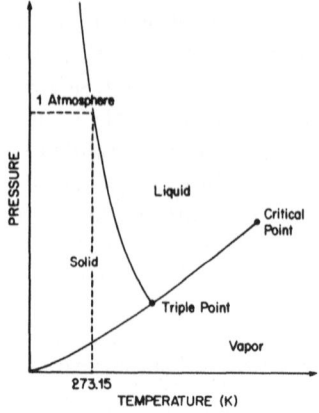

FIGURE 1

Pressure–temperature equilibrium states for water (shown schematically). A unique freezing temperature is associated with each pressure. The large negative slope of the solid–liquid loci of states indicates that large increases in system pressure are required to depress the freezing point significantly.

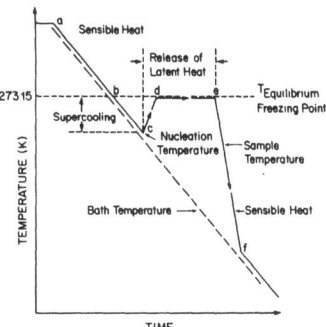

FIGURE 2

Typical thermal history of a sample undergoing freezing. Unless "seeded" with an external catalyst, most samples will supercool prior to nucleating. The large amount of latent heat released will return the sample temperature to close to the equilibrium freezing temperature. Large degrees of supercooling and rapidly decreasing temperatures following the release of latent heat have been linked with cell damage.

closely. Note, however, that solidification is not generally initiated at the equilibrium freezing temperature. The sample temperature is reduced (b–c) until the solid phase is nucleated at c. Between b and c, the sample is supercooled, but the release of the latent heat of fusion at nucleation sites tends to elevate the sample temperature (c–d) to its equilibrium freezing temperature (d). The latent heat is released to the bath (d–e), and then the sample temperature begins to decrease as sensible heat is extracted from it (e–f). It should be observed that the cooling rate during this latter phase may be substantially higher than the cooling rate of the bath unless the bath temperature is held constant (or increased) after nucleation is initiated.

States between b and c shown in Fig. 2 are metastable, since they are at temperatures below the equilibrium freezing temperature of the sample solution. The solidification process will be initiated in such metastable states by one of two modes of nucleation: homogeneous nucleation or heterogeneous nucleation. Both mechanisms require supercooling as a driving force. Supercooling increases, exponentially, the probability that a sufficient number of water molecules may be in a low enough free energy state to form a stable nucleus of the solid phase.[9]

As will be shown later in the chapter, cells are sensitive to the rate of freezing, and therefore rapid freezing after the complete release of latent heat may cause cell damage. For this reason, the magnitude of supercooling is normally controlled in practice by "seeding" the sample. The sample is supercooled to a small extent, at which point an ice crystal or cold wire is used as a nucleator in order to initiate solidification close to the equilibrium freezing temperature of the sample.

The experimental observation that supercooling is required for nucleation introduces the concept of nucleation activation energy. Homogeneous nucleation of pure water occurs in the absence of external nucleants and is a function of system size. For a water droplet of diameter 1 μm, this occurs at approximately $-40°C$.[10] Heterogeneous nucleation involves a lower activation energy (supercooling), since catalysts are present. Nucleation of ice within biological cells typically occurs in the temperature range -5 to $-15°C$,[3] which has led some researchers to propose that heterogeneous nucleation is responsible for intracellular ice formation. Mazur has reviewed the literature with respect to biological systems and concludes that cells contain no effective nucleants.[11] Morris and McGrath have recently obtained data that confirm this assertion and point to the importance of the presence of extracellular ice in nucleating intracellular ice.[12] Indeed, Mazur had suggested earlier that extracellular ice may be the nucleator for intracellular ice formation.[9]

Since the formation of intracellular ice is, in most cases, associated with cell freezing damage, it is important to understand the mechanism of its formation. Unfortunately, however, the mechanism of intracellular ice formation is not understood at the present time.

Another important aspect of the solidification process is the state of the newly formed solid phase. Classical theories of nucleation and solidification postulate that slow rates of cooling favor fewer nucleation sites and large rates of crystal growth,[9] producing many large crystals. Alternatively, if the cooling is done rapidly, large degrees of supercooling are realized, and many nucleation sites are formed; thus, crystal growth proceeds slowly, and many small crystals are observed as a result of fast freezing.

When very rapid freezing occurs, vitrification may be produced. This process is a glass transformation in which an amorphous ice is obtained, devoid of crystalline structure. This ultra rapid freezing technique is often used in electron microscopy to preserve morphological structure and avoid artifacts that might be introduced by crystal growth.

It is important to distinguish between crystallization and vitrification, since the consequences during freezing and thawing may be significantly different for the two cases. Recrystallization is defined as any change in the number, shape, orientation, or degree of perfection of crystals following completion of initial solidification. Much of the evidence relating intracellular ice formation to cellular injury has indicated that intracellular ice formation *per se* is not lethal. However, the recrystallization phenomena associated with high subzero storage temperatures or slow thawing rates apparently promote events that damage cells.[13] Thus, suppression of recrystallization would seem to be beneficial for preserving cells. However, when producing a true amorphous ice state in a "bulk sample," the size of cells may not be controllable.[7] Thus, recrystallization phenomena may always occur in frozen samples of biological cells.

2.3. The Binary Solution

The freezing process in biological fluids is much more complex than that in pure water. The intracellular and extracellular solutions of a biological cell consist of water and a complex assortment of solutes. As a first approximation, this solution is often modeled as a binary mixture of an electrolyte and water, such as $NaCl-H_2O$. This relatively simple model is more complex than the case of pure water, since the pressure alone is now insufficient thermodynamic information to specify the equilibrium freezing temperature of the solution.

The Gibbs phase rule dictates that both the pressure and the mole fraction of solvent (or solute) must be specified in this binary system in order to determine the equilibrium freezing temperature at which the solid and liquid states coexist. The loci of two-phase equilibrium states at atmospheric pressure are depicted in Fig. 3, using the NaCl–H₂O system as an example. In this case, the solid phase is assumed to be pure ice, and equilibrium solidification occurs over a range of temperatures rather than at the unique freezing temperature of the pure substance. The solute concentration will increase as pure ice is formed, and due to the constraints of the Gibbs phase rule, a temperature reduction below the equilibrium freezing temperature will occur (path *a* in Fig. 3). Alternatively, increasing the solute concentration before cooling will reduce the freezing temperature (path *b* in Fig. 3).

The quantitative thermodynamic description of a colligative property of solutions, such as freezing point depression, is given in terms of the Gibbs–Helmholtz relationship. This relates the water activity a_w to the latent heat of fusion L_f and the temperature of the solution.[14]

$$\frac{\partial \ln a_w}{\partial T} = \frac{L_f}{RT^2} \tag{2}$$

Unfortunately, the water activity is not directly related to conveniently determined measures of the concentration, such as mole fraction, molality, or molarity. Instead, it is related to these measures by the "activity coefficients" defined for each concentration scale.[15] The activity coefficients account for solution nonidealities and are typically not constant over the concentration range of interest. In addition, these coefficients can not usually be predicted, based on fundamental principles.

However, useful relationships between freezing point depression and solution concentration are available, based on various simplifying assumptions. One of the more common equations of this type relates the freezing point depression $(\theta = T_0 - T)$, where T_0 is the freezing temperature of pure solvent and the solution osmolality Ω

$$\theta = K_f \Omega \tag{3}$$

FIGURE 3
The loci of equilibrium states (phase diagram) for a binary solution. After ice has nucleated, the solute concentration in the unfrozen portion of the solution is directly related to temperature and increases as the temperature is reduced. Temperature and solute concentration are independent properties prior to the initiation of solidification.

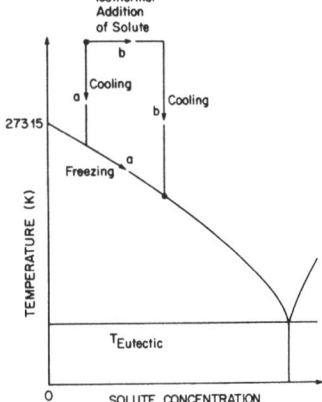

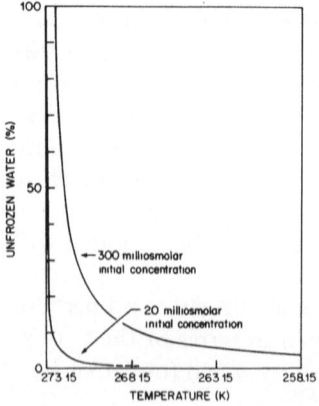

FIGURE 4

Percentage of unfrozen water in typical binary solution samples as a function of subzero temperature. Solutions with dilute initial solute concentrations typical of those observed in plants and mammalian cells will contain very little unfrozen water even at relatively high temperatures (263–256 K). The percentage of unfrozen water at a given temperature is directly proportional to the initial solute concentration of the sample.

For the case of an ideal dilute solution, K_f is a constant equal to 1.86°C/osmol for water.[15] The osmolality is defined as

$$\Omega = \sum_s \phi_s \nu_s m_s \tag{4}$$

where the summation is taken over all solutes present in solution,[15] and the osmotic coefficients ϕ_s refer to nonideal effects in the individual binary solutions of the specific solute s and the solvent. The solution molality is given as m_s, and the dissociation constant of the solute is given as ν_s.

For the ideal dilute case, all osmotic coefficients are equal to 1.0 by definition, so that Eqs. (3) and (4) predict that freezing point depression is only related to the type of solute to the extent that the solute dissociates (ν_s). Equation (3) can predict freezing point depression with reasonable accuracy, even for relatively high solution concentrations. By incorporating estimates of nonideality from room temperature data[15] into Eqs. (3) and (4), errors in predicted freezing point depression can be held to approximately 2–6% for solutions up to 2.0 M.[7]

These simplified equations have been used, for example, to estimate the unfrozen water fraction at any temperature during the freezing process. This information is equivalent to specifying the extent to which the solute in the solution has been concentrated. Figure 4 illustrates typical results of this type. A large fraction of the water in a biological sample would be frozen just below 0°C; this occurs because the typical biological sample is a dilute solution. The figure also shows that the unfrozen water fraction is inversely proportional to the initial osmolality of the sample. Thus, algae cells, frozen in a solution initially at 20 mosm will contain less than 1% unfrozen water at −5°C, whereas a sample of human erythrocytes frozen in plasma initially at 300 mosm will contain over 10% unfrozen water at the same temperature.

2.4. The Ternary Solution

Biological specimens usually can not be preserved by freezing unless a "cryoprotective" agent (CPA) is added to the specimen. Therefore, the

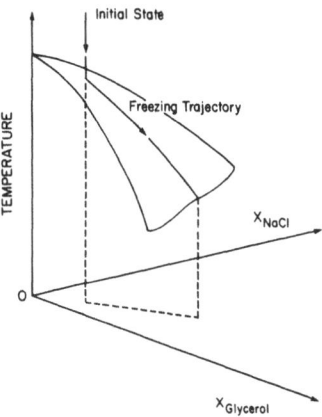

FIGURE 5
The loci of equilibrium states for a ternary solution. These states are represented as a surface, since the mole fractions of the two solutes present are independent properties. The freezing trajectory occurs along a line on the surface representing a constant ratio of the two solutes present. For a given initial ratio of solutes, the freezing process concentrates both solutes in the same proportion by the solidification of water.

simplest possible solution model that would take into account the effects of such a chemical, as well as the electrolyte and water of typical biological systems, would be a ternary system, such as glycerol–NaCl–H_2O. Since the number of components has increased from two to three, the Gibbs phase rule requires that in addition to the system pressure, the mole fractions of two of the components in solution must be specified in order to define the thermodynamic state and the equilibrium freezing temperature of the solution. The ratio of the mole fractions of electrolyte and CPA will remain constant even after solidification has been initiated, and the locus of equilibrium freezing states would be represented on a three-dimensional surface as shown schematically in Fig. 5 for the water-rich region.

Obviously, the situation becomes quite complex even at this relatively simple level. It is clear that biological specimens are much more complex than either the binary or the ternary solution models just presented. Nevertheless, models having the ability to predict the values of parameters that describe water transport during freezing have been developed.[16–22] These models are also useful in providing semiquantitative insight into processes occurring during freezing.

2.5. Melting

The melting process is similar to the solidification process in that a large fraction of the energy required for the sample phase change will be associated with the latent heat rather than the sensible heat. Freezing and thawing of samples is carried out in many cases by immersing the sample in a cold or hot medium. The sample is manipulated by altering the temperature or heat flux at the sample surface or in the medium in order to effect the required thermal history within the sample.[23]

This common technique, which is in effect boundary condition manipulation, may create more problems during thawing than during freezing. In thawing, the heat of melting must flow from the sample boundary through water to the phase boundary located near the still frozen center portion of the sample. However, as the system freezes, heat is extracted from the phase boundary through the already frozen phase (ice), which has a thermal conductivity approximately four times that of water and a thermal diffusivity of approximately an order of magnitude greater than that of water.

The heat transfer characteristics can be greatly enhanced by agitated immersion in a warm medium during thawing, thus minimizing problems associated with diffusion through a stagnant medium. This is a common technique for samples such as cultured cells in test tubes or small aliquots. However, it cannot be applied to larger, more complex systems, such as isolated rat hearts and kidneys, where internal thermal gradients may represent a serious problem for both freezing and thawing.

In the case of thawing, the boundary temperature can not be raised too high, since damage to the sample will result. This contrasts with the case for freezing, where very low temperatures may be applied at the system boundary to increase heat transfer rates if required. Microwave heating techniques have been applied as a means of attaining more uniform thermal conditions during thawing larger samples. This approach relies on control of internal heat generation rather than control of peripheral boundary conditions but suffers from various practical problems.[24]

Thawing rates also play an important role in the recovery of the sample, in that the thawing rate not only governs the rate of temperature rise in the sample, but it also controls the extracellular osmolality if extracellular thermodynamic equilibrium is assumed (see Figs. 3 and 5). Osmotic stress has been proposed as an important effect that could be responsible for freeze–thaw injury, as will be discussed later in the chapter.

The other important aspect of thawing has already been mentioned. Thawing rates can play a crucial role in determining the kinetics of recrystallization of any intracellular ice which may have formed within frozen samples.[13] Experimental evidence suggests that for cells which contain ice it is necessary to thaw rapidly in order to minimize the possibility of crystal growth and restructuring which may injure the cell during thawing.

3. OSMOSIS

3.1. Equilibrium Considerations

Living cells possess semipermeable membranes that control the passage of molecules between the cell and its environment. The selective nature of the cell membrane thus plays an all-important role in maintaining the proper chemical environment within the cell in the normal state and during the application of cryopreservation procedures. Water flux through the typical semipermeable cell membrane may be understood with reference to a simple osmotic system as described in Fig. 6a. The condition of osmotic equilibrium in this system, for which water passes through the membrane but solute does not, is

$$\mu_w^\alpha = \mu_w^\beta \tag{5}$$

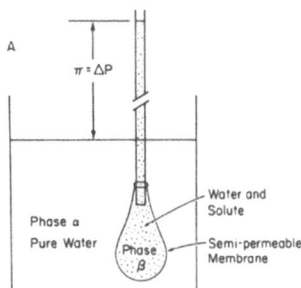

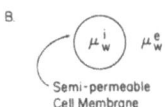

FIGURE 6
Osmosis. (a) When a membrane permeable to water but not to solute separates a pure water phase from a phase containing water and solute, the water will enter the latter phase, creating a hydrostatic pressure difference across the membrane. Water transport will cease when the chemical potential of water in the α phase is equal to that in the β phase. (b) The same general principles appear to be valid in living cells.

Equation (5) simply states that the chemical potential of water must be equal in the two phases on either side of the membrane in order for equilibrium to be satisfied. It is easily shown that water will enter phase β until the hydrostatic pressure in phase β exceeds that in phase α at any depth by the amount ΔP. The osmotic pressure of the β phase due to the presence of solute is defined as Π, which is numerically equal to ΔP. It should be borne in mind, however, that the osmotic pressure is best thought of as a measure of water activity or "concentration", since

$$\Pi = \frac{-RT \ln a_w}{\bar{V}_w} \tag{6}$$

where $\bar{V}_w$ is the partial molar volume of water.[14]

It is important to make this distinction in the general analysis of coupled flows of solvent and solutes across membranes when the difference in hydrostatic pressure ΔP and the difference in osmotic pressure Π represent independent driving forces for transmembrane fluxes.[25]

The laws of osmotic equilibrium were first developed by Van't Hoff. Soon after, the laws governing osmotic equilibrium in living cells were examined in light of Van't Hoff's work. The result is the Boyle–Van't Hoff law describing cell osmotic response.[26] This law states that the osmotic pressure of a solution varies inversely with the volume of solvent in which a quantity of the solute is dissolved

$$\Pi V_w = \text{const} \tag{7}$$

In this equation, V_w represents the volume of solvent (water).

Lucké and McCutcheon[26] point out that in the case of cells, if the intracellular solution can be considered ideal and if all of the solvent is osmotically active, then the total volume of the cell should vary inversely with the osmotic pressure of the extracellular solution. In addition, these authors point out that solute leakage can not occur if the law is to hold. They also state that it is important to correct the Boyle–Van't Hoff law for the presence of any osmotically inactive volume that may be present within the cell.

The modified Boyle–Van't Hoff law, corrected for osmotically inactive volume, may then be stated as

$$\Pi(V_{\text{cell}} - b) = \Pi^0(V_{\text{cell}}^0 - b) \tag{8}$$

where Π is the osmotic pressure (the extracellular and intracellular osmotic pressures are equal at equilibrium), V_{cell} is the total volume of the cell, and b is the "osmotically inactive" volume of the cell. The superscripted values correspond to a reference initial state. After normalizing with respect to the initial cell volume and performing some algebraic manipulation, the Boyle–Van't Hoff law is given as

$$\hat{V}_{\text{cell}} = \frac{\Pi^0}{\Pi}(1 - \hat{V}_b) + \hat{V}_b \tag{9}$$

where $\hat{V}_b = b/V_{\text{cell}}^0$ and $\hat{V}_{\text{cell}} = V_{\text{cell}}/V_{\text{cell}}^0$. A schematic graph of the relationship is given in Fig. 7.

The equilibrium osmotic characteristics of semipermeable cells are determined in practice by exposing cells to solutions of increasing osmotic pressure. The cells shrink, and the equilibrium cell volumes are determined; representative data are plotted as in Fig. 7. These experiments are performed in a range of osmotic pressure that does not damage the cell. However, the linear relationship between $(1/\Pi)$ and $\hat{V}_{\text{cell}}$ allows extrapolation to infinite osmotic pressure $(\Pi^{-1} \to 0)$ in order to estimate the normalized, osmotically inactive cell volume $\hat{V}_b$, as shown in Fig. 7.

The Boyle–Van't Hoff relationship is useful in several important respects. In the first place, equilibrium osmosis experiments yield values of $\hat{V}_b$ that are

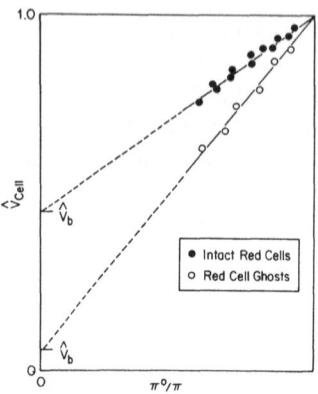

FIGURE 7
Schematic of Boyle–Van't Hoff plots for the equilibrium osmotic behavior of semi-permeable biological cells. Cells shrink in solutions of increasingly concentrated solutes in agreement with the behavior predicted by a simple thermodynamic model. The osmotically inactive cell volume V_b may be derived from such plots. Red cell ghosts containing little or no hemoglobin (which binds large amounts of water) have a much smaller inactive volume than intact red cells.

important for modeling the intracellular solution. When combined with dry weight measurements,[27] the total cell water volume V_w^T may be determined and compared to the cell's "osmotically free" water volume V_w^F, where

$$V_w^F = V_{cell}^0 - V_b \tag{10}$$

Note that the difference between the total cell water volume and the "osmotically free" water volume is the "osmotically bound" water volume V_w^B.

In the case of infinitely slow freezing, the extracellular solution and intracellular solution will be in equilibrium at all times, such that

$$\Pi^e = \Pi^i \tag{11}$$

Since the extracellular osmotic pressure Π^e would be known at any temperature from Eqs. (2) and (6), the intracellular osmotic pressure Π^i would also be known at this temperature. Hence, application of the Boyle–Van't Hoff equation would determine the cell volume at any subzero temperature during infinitely slow freezing. This equilibrium cell volume during freezing is usually included in graphical representations of cell volume during freezing (See Fig. 18). Departures from equilibrium values of cell volume for all finite freezing rates are measured with respect to this equilibrium curve. The Boyle–Van't Hoff relationship is also applied to link cell volume and intracellular osmotic pressure when the intracellular and extracellular solutions are not in equilibrium.[28,29] In this case, however, Π^i and Π^e are not equal.

Lucké and McCutcheon[26] indicate important restrictions on the application of the Boyle–Van't Hoff law to osmotic equilibrium in living cells. Dick[30] and Nobel[31] have emphasized the importance of a thermodynamic development of this law. Since the Boyle–Van't Hoff relationship is important in many applications in cryobiology, a development is given here that points out its limitations when applied to the description of "anomolous" osmotic behavior,[32] water transport models describing freezing[28,29], and proposed mechanisms of freezing injury.[33]

As shown in Fig. 6b and Eq. (5), the condition of osmotic equilibrium for a cell is the equality of water chemical potential across the cell membrane, here written as

$$\mu_w^i = \mu_w^e \tag{12}$$

The chemical potential of water in each phase may be expressed in the form[34]:

$$\mu_w = \mu_w^*(T) + \bar{V}_w P + RT \ln a_w \tag{13}$$

Assuming the cell and its environment are at the same temperature, the reference state chemical potentials μ_w^* will be identical inside and outside the cell. Thus,

$$\bar{V}_w^e P^e + RT \ln a_w^e = \bar{V}_w^i P^i + RT \ln a_w^i \tag{14}$$

Assuming that the solvent (water) is incompressible ($\bar{V}_w^e = \bar{V}_w^i = \bar{V}_w$) and using the defining equation for the osmotic pressure, Eq. (6),

$$P^i - P^e = \Pi^i - \Pi^e \tag{15}$$

If, as is the case for many cell types,[35] the hydrostatic pressure difference across the plasma membrane is approximately zero ($\Delta P = 0$), then the condition for osmotic equilibrium is

$$\Pi^i = \Pi^e \tag{16}$$

This is equivalent to stating that across the cell plasma membrane, there are no gradients in osmolality Ω, water chemical potential μ_w, or log water activity ($\ln a_w$), since all of these properties are proportional to each other.

For all equilibrium states, the intracellular and extracellular osmotic pressures will be equal: $\Pi^i = \Pi^e = \Pi$. The osmotic pressure of the intracellular solution is expressed as previously presented in Eq. (6)

$$\Pi = -\frac{RT}{\bar{V}_w} \ln a_w \tag{17}$$

or equivalently in Eq. (7)

$$\Pi = -\frac{RT\phi}{\bar{V}_w} \ln x_w \tag{18}$$

where ϕ is the average osmotic coefficient of the solution and x_w is the mole fraction of water in the solution.

Since the general solution will be composed of the solvent and a number of solutes, the mole fraction of water may be expressed as

$$x_w = 1 - \sum_{i=1}^{s} x_i \tag{19}$$

where the x_i are solute mole fractions summed over all species s in the solution.

Levin *et al.* have pointed out the importance of interpreting the mole fractions of water and solute on a "bound" water basis.[16] This interpretation will result in a distinction between total water volume and "osmotically free" water volume in the Boyle–Van't Hoff law. If the solution is assumed to be mole dilute, then

$$\ln\left(1 - \sum_{i=1}^{s} x_i\right) \simeq - \sum_{i=1}^{s} x_i \tag{20}$$

where

$$\sum_{i=1}^{s} x_i = \frac{\sum_{i=1}^{s} \nu_i n_i}{n_w + \sum_{i=1}^{s} \nu_i n_i} \tag{21}$$

Applying the dilute solution assumption again,

$$n_w \gg \sum_{i=1}^{s} \nu_i n_i \tag{22}$$

and

$$\sum_{i=1}^{s} x_i \simeq \frac{\sum_{i=1}^{s} \nu_i n_i}{n_w} \tag{23}$$

Letting

$$N_s = \sum_{i=1}^{s} \nu_i n_i \tag{24}$$

represent the total number of "osmolality distinct" species in solution, Eqs. (19)–(24) may be combined to reduce Eq. (18) to the approximate expression

$$\Pi \simeq \phi RT(N_s / V_w) \tag{25}$$

where

$$V_w = n_w \bar{V}_w \tag{26}$$

Equation (25) can be considered an equality in the limit of dilute solutions, so that

$$\Pi V_w = \phi RT N_s \tag{27}$$

If the "osmotically inactive" cell volume is subtracted from the total cell volume to yield the "osmotically active" or "osmotically free" water volume, then

$$V_w^F = V_w = V_{\text{cell}} - V_b \tag{28}$$

and Eq. (27) is written as

$$\Pi(V_{\text{cell}} - V_b) = \phi RT N_s \tag{29}$$

This equation is analogous to the Boyle–Van't Hoff law if the right-hand side of the equation is a constant and if the two assumptions used in its derivation are valid. Although cases may arise in which variations in the parameters ϕ, T, and N_s may occur while the right-hand side of Eq. (29) remains constant, it is more likely that this product invariance would be the result of constant values for each of the parameters. Therefore, the Boyle–Van't

Hoff law would be expected to be valid when

(a) No hydrostatic pressure gradients exist across the cell membrane.
(b) The intracellular solution is dilute on a mole basis.
(c) The solution osmotic coefficient is invariant with respect to changes in osmotic pressure.
(d) The temperature of the system remains constant during osmotic manipulation and is invariant across the membrane.
(e) The total number of osmotically active solute molecules in the intracellular solution N_s does not change during osmotic manipulation (usually taken to mean no solute leakage).

The Boyle–Van't Hoff law has been used successfully to describe cellular osmotic behavior, however, the law is usually applied over quite a limited range of osmotic pressures.[36,37,38] Attempts have been made to assign significance to nonlinearities in Boyle–Van't Hoff plots, such as those in Fig. 7[30,33] where nonlinearities are observed at extremes of osmotic pressure such that one or more of the assumptions just outlined may not be satisfied. Thus, this treatment suggests that care must be exercised in using the Boyle–Van't Hoff relationship in freezing models and in explaining potential mechanisms of freezing or osmotic injury.

3.2. Nonequilibrium Considerations

Freezing protocols are carried out rapidly in some cases, and as a consequence, the intracellular and extracellular osmotic pressures during the protocol are unequal. Extracellular solutions freeze first, which causes the extracellular osmotic pressure to be greater than that of the intracellular solution.

Due to the finite-transport impedance of the cell membrane, the water concentration inside the cell will always be greater than, or equal to, that outside the cell during freezing. If the cell membrane is only permeable to water, then

$$\frac{dV_w}{dt} = \frac{dV_w^F}{dt} = \frac{dV_{\text{cell}}}{dt} \tag{30}$$

and the volume flux of water J_w leaving the cell is

$$J_w = \frac{1}{A}\frac{dV_w}{dt} \tag{31}$$

J_w is usually related to the concentration driving force in the form[29,17]

$$J_w = P_w(\Pi^i - \Pi^e) \tag{32}$$

where Π^j are the osmotic pressures internal and external to the cell and P_w is the water permeability, a phenomenological coefficient.

If the cell membrane is also permeable to solute species, these will be transported across the membrane in addition to the solvent. In this situation, the analysis becomes considerably more involved, since in the general case the flow of the solvent is coupled with the flow of solute. The methods of irreversible thermodynamics have been applied successfully to handle some of these circumstances, and they are reviewed in more detail later in the chapter.

4. GENERAL RESPONSES OF BIOMATERIALS TO FREEZING AND THAWING

Most of the basic changes that occur in a simple solution during freezing and thawing have already been outlined. In addition, the osmotic considerations that lead to cell volume changes have been presented. With this information in mind, the general responses of biological cells to the thermal and chemical changes associated with freezing and thawing will now be examined. It is convenient to introduce several representations of the freezing process to illustrate the relevant responses. The simple case of freezing a binary solution is shown schematically in Fig. 8.

Figure 8a represents the temperature history of a controlled freeze–thaw process with constant freezing and thawing rates. Note that various combinations of fast or slow freezing and thawing rates are possible and that the subzero storage temperature could be varied. A "slow" freezing or thawing rate in practice might be 0.1–1.0°C/min,[11] while a "fast" freezing or thawing rate might be 100.0–1,000.0°C/min.[11] Obviously, part of the freezing procedure takes place above the freezing point of the sample solution. This part of the protocol may or may not be important in terms of cellular injury as will be seen shortly.

Figure 8b represents the freeze–thaw cycle on the binary phase diagram for a slow freeze and slow thaw protocol. Note that the loci of intracellular

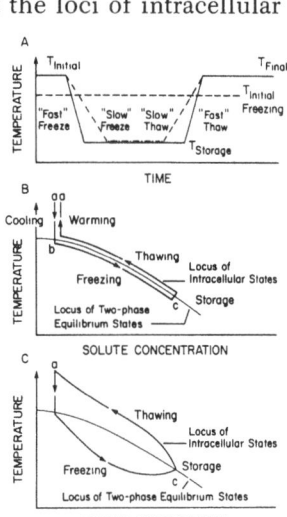

FIGURE 8
Temperature and concentration changes that occur during slow and rapid freezing and thawing. (a) Various combinations of freezing rates, thawing rates, and storage temperatures are possible. (b) During slow freezing and slow thawing, mass transfer of water across the cell membrane is fast enough, so that the intracellular concentration follows the extracellular concentration closely. (c) During rapid freezing and thawing, water efflux during freezing and influx during thawing are not large enough to maintain intracellular/extracellular equilibrium.

thermodynamic states remain close to the loci of extracellular states, indicating that the two solutions are esentially at equilibrium with respect to each other. The cryopreservation protocol is represented as a "reversible" cycle in Fig. 8b, and all phases of the procedure are labeled; this representation assumes that the extracellular solution always remains in thermodynamic equilibrium, a common but critical modeling assumption.[17] It should be pointed out that the only difference in a fast or slow freezing and thawing program in this representation would be the time taken to proceed from state *a* to state *c*, and from state *c* to state *a*, respectively. The loci of intracellular states during rapid freezing and thawing would be qualitatively the same but would deviate from the extracellular states due to nonequilibrium intracellular conditions. For cases involving fast freezing or thawing the nonequilibrium "lag" or hysteresis effect for the intracellular solution is shown schematically in Fig. 8c.

4.1. Chilling Injury

Some biological cells are extremely sensitive to temperature reduction, so that freezing the extracellular medium is not necessary to kill the organism. Temperature reduction *per se* (state *a* to *b* in Fig. 8b) is a sufficient insult to injure a number of organisms, such as the rapidly dividing cells of blue–green and green algae as well as tropical plant cells.[6] When this injury is independent of the cooling rate, it often does not manifest itself immediately. The damage is thought to result from membrane lipid phase changes, which affect cellular metabolism.[6] ‘

4.2. Thermal Shock or Cold Shock

Other organisms, such as bull sperm, are sensitive to temperature reduction to temperatures still above the freezing point, but the severity of damage is related to the rate of temperature reduction. This type of injury is also thought to be a result of membrane lipid phase changes, which cause changes in membrane permeability. In contrast to chilling injury, these changes are usually apparent immediately or soon after temperature reduction.[6]

4.3. Slow-Freezing Responses

The quantitative definition of a slow or rapid freezing process for a particular cell type will be a function of cell parameters, such as the surface area-to-volume ratio and the water permeability coefficient. In general, a slow freezing process is a mass transfer-dominated process in which sufficient transport occurs across the cell membrane, such that the intracellular and extracellular solutions are essentially in a condition of mutual equilibrium. The general response to slow freezing rates is severe dehydration, as depicted in Fig. 9. Figure 10a illustrates extensive dehydration during freezing and subsequent rehydration during thawing for a mouse ovum frozen on a cryomicroscope system (see section on cryomicroscopy). Figure 10b represents the dehydration due to freezing as observed with the electron microscope. Note the dense appearance of the nucleus and cytoplasm as well as the folding of the membrane.

FIGURE 9

Expected modes of attaining intracellular thermodynamic equilibrium at slow and rapid rates of freezing. For slow rates, mass transfer of water results in extensive cell dehydration. Rapid rates emphasize heat transfer resulting in rapid temperature reduction and little mass transfer. Little cell volume change occurs in this case, and the nucleation of large amounts of intracellular ice is expected.

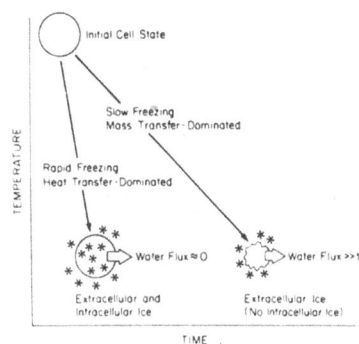

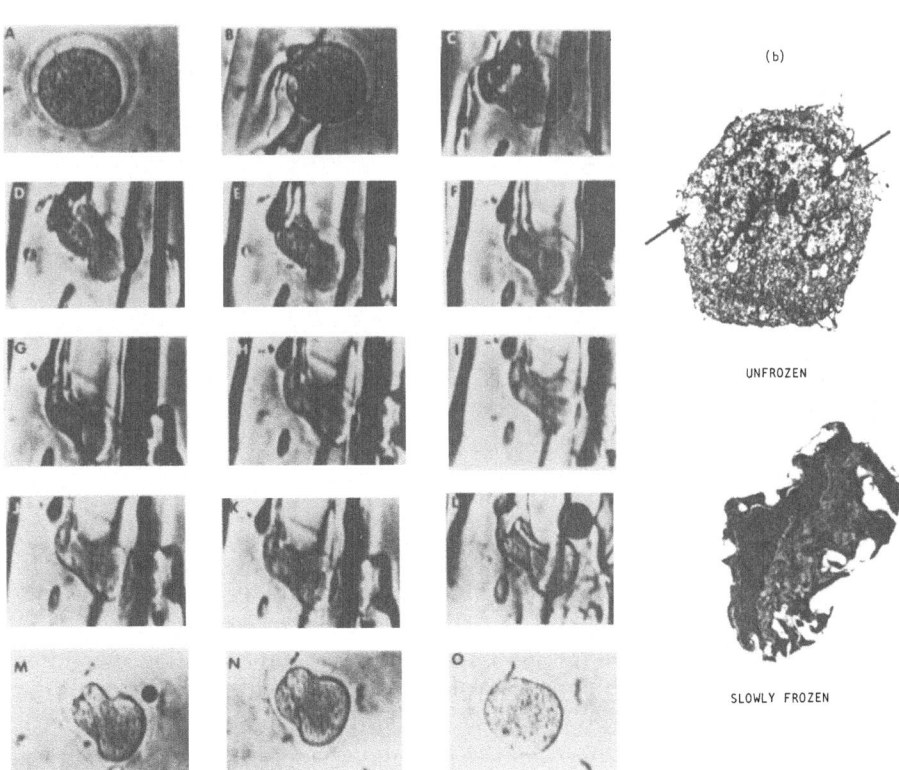

FIGURE 10

Examples of cellular dehydration resulting from slow freezing. (a) Mouse ovum as it dehydrates during slow freezing to −135°C, as observed with the aid of a cryomicroscope. The rehydration of the ovum on thawing is observed. (From Ref. 103, with permission.) (b) Electron micrographs of an unfrozen culture cell and a slowly frozen, dehydrated culture cell. (From Ref. 116, with permission.)

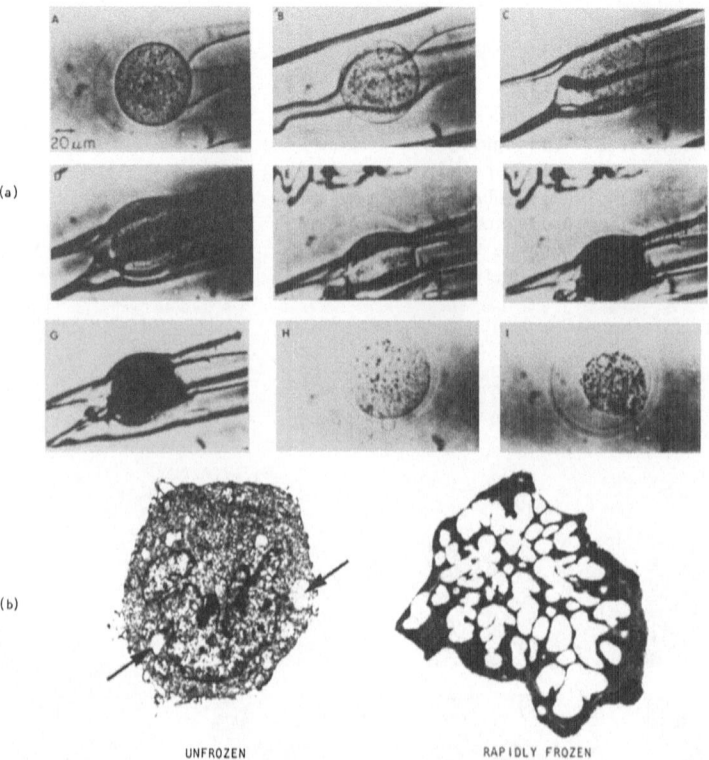

FIGURE 11

Examples of intracellular ice formation resulting from rapid freezing. (a) "Rapid" freezing for the ovum is 2.4°C/min, because the ovum has a low water permeability and a small surface area-to-volume ratio. Intracellular ice is present at −49°C (*F*) and −50°C (*G*). Note the degenerated appearance of the ovum at +20°C after thawing. (From Ref. 103, with permission.) (b) Electron micrographs of an unfrozen culture cell and a rapidly frozen cell with a large amount of intracellular ice. (From Ref. 116, with permission.)

4.4. Rapid–Freezing Responses

Rapid freezing will create high extracellular osmotic pressures quickly. Although large osmotic pressure differences across the cell membrane may result from fast cooling rates, the actual cell volume changes experienced may be relatively small, due to the low permeability to water and the short time elapsed during which the temperature drops. The result is that the intracellular water becomes supercooled and is likely to nucleate. This phenomenon is shown in Fig. 11a as it appears on the cryomicroscope. The darkening of the cell interior is interpreted as resulting from the formation of intracellular ice crystals that scatter light. Figure 11b shows a typical electron micrographic result when intracellular ice is present. Whether the nucleation is by a homogeneous or heterogeneous mechanism is still an open question; much evidence points to the important role played by extracellular ice in the nucleation of intracellular ice.[13]

Rapid freezing of cells produces a result that qualitatively mimics the behavior of freezing in a bulk solution. As would be expected from the nucleation and crystal growth theory mentioned earlier, cells frozen moderately rapidly are likely to have larger intracellular crystals than those cells frozen very rapidly.

4.5. Storage Temperature

The basic premise of low-temperature preservation is to place the sample in a state of suspended animation by lowering the temperature of the system below the point where any significant biochemical activity occurs, and specifically below the temperature that might allow irreversible damage to occur. The lower storage temperatures would also be desirable if there were a chance that intracellular ice were present, since the potentially injurious recrystallization occurs more rapidly at higher temperatures.[9] Figure 12 illustrates the presence of a very fine ice structure present in rapidly frozen cells that changes structure at elevated temperatures. These electron micrographs of freeze-cleaved cells reveal the growth of large ice crystals from very small crystals. As shown, the occurrence of larger crystals can be correlated with decreased

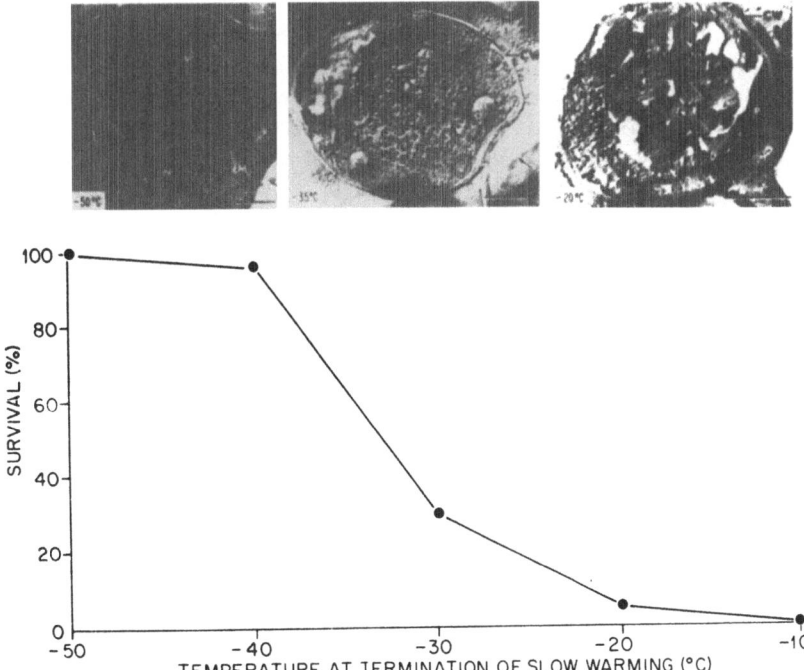

FIGURE 12

Recrystallization of intracellular ice is inversely related to cell survival. While the presence of intracellular ice per se is not lethal, circumstances allowing the growth of very fine crystals is generally associated with cell destruction. Such circumstances include storage at high subzero temperatures or slow warming. (From Ref. 13, with permission.)

cell survival. Storage temperatures will generally range between −80°C (dry ice) and −196°C (liquid nitrogen).[39] Human red blood cells have been successfully preserved for as long as 12 years with no observable biochemical or functional deterioration.[40] It is likely that storing biomaterials at low temperatures, e.g., −196°C, would enable storage periods of decades[9] or even centuries.[41]

There is no apparent lower limit to the storage temperature, since a variety of biological systems have been successfully frozen almost to absolute zero temperature (~0.2 K) with successful recovery.[42] These lower temperatures are more expensive to produce and maintain. Since they do not appear to be required in most cases, these temperatures are more of an academic than a practical interest.

4.6. Thawing Response

The thawing response of biological cells is more complex than the freezing response. Some appreciation of this complexity may be acquired by considering the schematic representation of the response to thawing in Fig. 13. Some cells fail altogether to response osmotically after having been frozen (path *a*),[43] which is an extreme case of those cells with an abnormally "low" osmotic response (path *b*). Other cells display a normal osmotic response but over only a limited range and they lyse (path *c*). Still other cell types appear to have an abnormally "high" osmotic response, as in path *d*.[44] Figure 10a illustrates a case where the cell rehydrates in a "normal" fashion. Figure 14 is an example of two pairs of epithelial cells that were frozen and thawed on the cryomicroscope. The pair of cells on the left have rehydrated in a normal fashion, whereas the pair on the right have abnormally low degrees of rehydration (path *a* or *b* in Fig. 13).

The response of a frozen sample to the thawing rate is generally a function of the previous history of the sample.[45] While there appear to be conflicting suggestions for the best thawing procedures, there seem to be several instances where slow thawing should be imposed for slowly frozen samples and rapid thawing applied to rapidly frozen samples in order to achieve the best recoveries possible.[45]

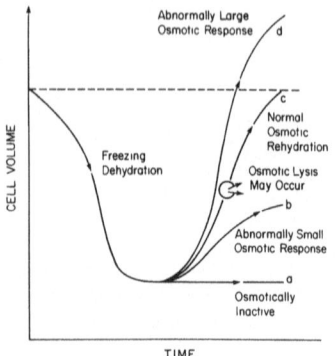

FIGURE 13
Experimentally observed osmotic responses of biological cells during thawing. Some cells rehydrate in a "normal" fashion in the sense that they return to their isotonic volumes. Other cells remain dehydrated and appear to be osmotically inactive or only partially rehydrate. A fraction of the cells begin to rehydrate normally but lyse prior to attaining isotonic volume. In other cases, cells behave as if they were loaded with solutes during dehydration, since they return to abnormally large volumes.

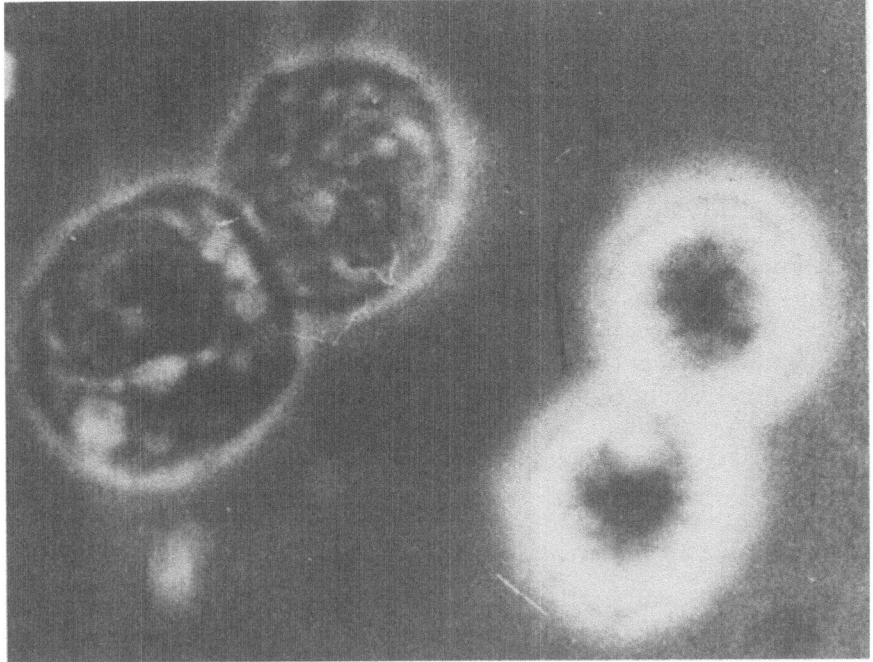

FIGURE 14
An example of the types of rehydration behavior observed in a frozen–thawed sample of HeLa S-3 cells. The cell pair on the left rehydrated completely, whereas the cell pair on the right remains contracted. The pronounced "halo" is an artifact common to the phase contrast technique used to obtain this cryomicroscope photograph.

4.7. Cryoprotective Agents

In most cases, little or no survival of frozen–thawed biological samples can be obtained unless a chemical additive is mixed with the cell sample in order to protect it during the attempted cryopreservation procedure. The modern era of cryobiology began in 1949 with the accidental discovery that glycerol protected spermatozoa during freezing,[46] since that time, scores of compounds have been tested as potential cryoprotective "antifreezes" (CPA). The most common agents used in practice are divided into those that penetrate cells and those that do not penetrate cells. Popular penetrating compounds include glycerol, dimethylsulfoxide, methanol, ethylene glycol, and dimethylacetamide. Frequently used nonpenetrating chemicals include poly-vinylpyrollidone (PVP), hydroxyethyl starch (HES), polyethylene glycols (PEG), dextrans, and albumin.

The general effects of these compounds are classified as follows. These chemicals will tend to enhance cell recovery, primarily at the slower cooling rates, and often will expand the range of cooling rates that will produce acceptably high recovery.[45] This broadened optimal cooling rate behavior is important for tissues or organs when simultaneous satisfaction of optimal freezing rates for a number of different cell types is required. An important consideration when using these compounds is the fact that while increased concentrations of CPA may lead to higher frozen–thawed cell survival, the CPA may itself be toxic to the cell.[45] There is evidence that suggests that cryoprotective agents may reduce the recovery of cells at fast freezing rates.[13]

4.8. Removal of Cryoprotective Additives

All practical cryopreservation procedures at the present time require the presence of one or more chemical antifreezes during the preservation procedure. Generally speaking, most, if not all, of the CPA must be removed prior to transfusion or transplantation. Postthaw processing of this type must be carried out carefully, since there are numerous published examples that indicate the possibility of significant osmotic damage being induced at this stage of the cryopreservation protocol.[41,45,47]

4.9. Assessing Cryopreservation End Points

In the final analysis, there must be some criteria applied to the cryopreservation protocol to determine whether or not it was successful. In the ideal case, a procedure would be developed to process the sample to a low-temperature state of suspended animation. In this state, metabolic processes are suppressed to such an extent that irreversible damage, e.g., cell or organic membrane disruption, lysosomal enzyme release, etc., does not take place. At the appropriate time, the process would be reversed and the sample retrieved exactly as it had been in its initial state. In thermodynamic terms, the goal for the viability of the sample is complete reversibility of processing, returning the sample to its initial state.

Practically speaking, cryobiologists will normally test for "reversibility" in a very limited sense: if a cell retains a fluorescent dye or excludes a vital dye, the cell is assumed to have been recovered. Other end points, such as chemotaxis, enzyme activity, motility or cell cleavage may also represent end points signifying "recovery" or "survival."

Because the typical cell system is so complex, the safest criteria for recovery is a direct test for the end point of interest.[48] The point is that survival or recovery is relative to the indices of cell viability being probed, and these indirect probes can often lead to misleading results.

4.10. The General Recovery Response

To summarize the preceding discussion, the recovery of frozen–thawed biomaterials is affected by a number of obvious factors, such as freezing rate, storage time, storage temperature, thawing rate, and chemical composition of the suspending medium as well as biochemical factors that can be indicated by indirect probes of cell viability. Generally, these factors are interactive and yield a bell-shaped curve, such as those shown in Fig. 15. It is important

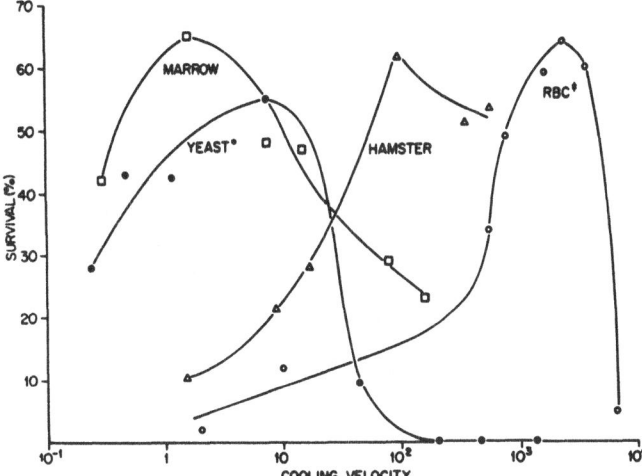

FIGURE 15

Representative "survival signatures" for several cell types. Plots of cell recovery as a function of cooling rate typically reveal an optimum cooling rate. The figure illustrates that optimum recovery may vary from one cell to another. Due to differences in cell characteristics, the optimum cooling rate for one cell type may differ by several orders of magnitude when compared to another cell type. (From Ref. 11, with permission.)

to realize that there exists an "optimal" procedure for each cell type and large discrepancies in the optimal procedures for different cell types may make it difficult to freeze even relatively small tissues and organs, such as isolated rat hearts or rabbit kidneys. These organs contain a number of cell types that may possess quite different optimal freezing and thawing rates. These "small" samples may in fact be very large in a diffusion sense, so that significant chemical and thermal gradients may be produced in the sample. Therefore, satisfactory "optimal" thermal and chemical trajectories may be difficult or impossible to achieve at all spatial locations throughout the cryopreservation process. Hence, the size and complexity of organs makes this area of cryobiology an especially challenging one.[49]

5. MECHANISMS OF FREEZE–THAW DAMAGE

Some aspects of freezing damage mechanisms have been mentioned in previous sections. In the present section, the subject is considered in more detail. Many freeze–thaw damage mechanisms have been proposed, including the following:

5.1. Injury at Slow Freezing Rates

Due to the complexity of biological systems, there are many potential sites and possible mechanisms of cell-freezing injury. Some of the hypotheses proposed are presented at this time to indicate the range of possibilities.

(a) Lovelock developed a theory of freezing damage, proposing that freezing produces an increased concentration of electrolytes that could break down membrane lipoprotein complexes[50] and cause dissolution of the cell membrane.[51]

(b) Levitt has proposed that increased concentrations of solutes induced by freezing eventually reach a point where sulfhydryl (SH) and disulfide (SS) groups on protein molecules may interact irreversibly, leading to lethal consequences.[52]

(c) Heber and Santarius postulate irreversible disruption of photophosphorylation and thus adenosine triphosphate (ATP) synthesis in mitochondria as the cause of freezing damage.[53] They point out the occurrence of organelle membrane damage but are unclear about the role of the phosphorylation defect with respect to membrane damage.

(d) A so-called "structured water" hypothesis has been suggested.[54] Death due to freezing in this view is thought to occur primarily as a result of removing "bound" water from vital cellular structures, such as membranes and proteins. This hypothesis stresses the importance of lattice-structured water for cell integrity.

(e) Meryman has proposed a "minimum volume" hypothesis for freezing injury.[55] This theory suggests that when a critical minimum cell volume is reached, an irreversible and lethal increase in membrane permeability occurs for normally impermeable solutes. It should be noted that this hypothesis is not linked specifically to electrolytes nor to a universal critical solute concentration.

(f) Mazur and his colleagues have proposed the now classic "two-factor" hypothesis of freezing damage.[56] This hypothesis is illustrated schematically in Fig. 16 and is based on the experimentally observed fact that cryopreserved cells require an intermediate freezing rate for optimal recovery. Assuming that two independent major damage modes are operative, e.g., intracellular

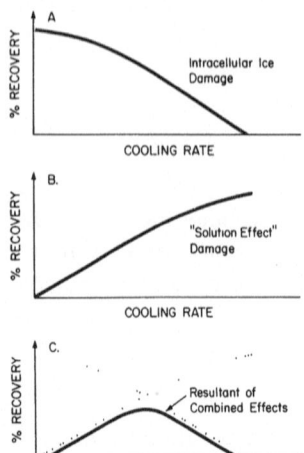

FIGURE 16

The two-factor hypothesis of freezing injury. (a) If freezing injury were the result of intracellular ice formation alone, then faster cooling rates would result in less recovery. (b) if long exposures to concentrated solutions caused freezing injury, slow freezing rates would result in less recovery. (c) A combination of these two effects would yield an optimum recovery at intermediate rates (cf. Fig. 15).

ice formation and "solute damage," would account for the observed data. Fast freezing results in large amounts of supercooled intracellular water and an increased likelihood of intracellular ice. On the other hand, freezing can not be performed too slowly. Slow freezing rates imply extended exposure time of cells to concentrated solutions, which results in less recovery for the slower freezing rates. Although this hypothesis is in qualitative agreement with experimental results, it does not address the basic mechanisms by which intracellular ice or the so-called "solute effects" cause cell damage.

Recent research has attempted to identify the mechanisms by which solute effect damage occurs. In a system as complex as a biological cell, it has proven difficult to distinguish primary damaging events from pathological secondary effects. While cytoplasmic alterations may represent primary sites of cellular damage in some cases, much of the current evidence points to the plasma and organelle membranes of the cell as the primary sites of alteration. In fact, it appears that biophysical rather than biochemical alterations may be responsible for cellular freeze–thaw injury.[57]

Williams and his colleagues have published a number of papers that propose that increased osmotic pressures, which cause cell shrinkage, induce increased surface pressure within the plane of the membrane.[33,58,59] The compression of this membrane may result in a number of alterations, including membrane phase separations, membrane collapse, and reversible or irreversible loss of membrane components.

Weist and Steponkus have emphasized the importance of incremental cell surface area changes in relation to cellular injury.[37] Using the isolated spinach protoplast as a model system, these researchers have shown that damage due to osmotic shrinkage is not related to an absolute minimum cell volume. Instead, there appear to be alterations induced in the membrane during shrinkage that would limit the surface area expansion potential of the cell during subsequent thawing. These data are consistent with the hypothesis that the severe dehydration that occurs during freezing may result in the physical loss of membrane material. Thus, during thawing, there may be insufficient membrane material to provide the required surface area for the cell to resume its original volume.

Morris has recently published results of studies using liposomes as simple models to study freezing injury.[57] He has obtained data indicating increased membrane pressure during osmotic dehydration and has postulated a model for injury, consistent with those of Williams and Steponkus.

Recently, McGrath has reviewed developments in thermodynamic and kinetic modeling of membrane damage resulting from simulated slow freezing.[60] This approach relies on experimental evidence such as that previously given regarding the mechanisms of damage and is therefore consistent with it. Major advantages of this thermodynamic modeling approach are the application of a systematic methodology to interpret existing data; a theoretical basis for identifying important missing data; and the potential for quantitative computer simulations of membrane-freezing damage.

5.2. Injury at Rapid Freezing Rates

There is only limited knowledge about the mechanism or mechanisms responsible for cellular injury at characteristically fast freezing rates, and the picture remains unclear. However, much of the existing evidence suggests that recrystallization of intracellular ice, rather than the presence of ice *per se* is involved. There is evidence that ice may be present within cells and these cells may survive if the thawing rate is rapid enough.[13] A number of researchers have published results that are consistent with the theory that damage is caused by the interaction of large crystals with the cell membrane during recrystallization.[13] A recent ultrastructural study has demonstrated that membrane lesions occur close to intracellular ice crystals.[61]

Farrant has suggested that the total amount of intracellular ice may be important and also that damage to rapidly frozen cells may occur during warming, due to osmotic imbalances rather than a direct mechanical effect of the intracellular ice.[45]

6. APPLIED CRYOBIOLOGY

Some types of biomaterials have been frozen and thawed on a routine basis, with excellent recovery. However, for many important materials, it has not yet been possible to carry out cryopreservation successfully. In this section, a brief review of the state-of-the-art is undertaken in order to give an appreciation of currently feasible techniques and applications. The advantages of various cryopreservation techniques are also outlined. It should be kept in mind that, more often than not, the procedures were developed on an empirical basis and in response to a clinical or research need.

6.1. Blood

The need for frozen blood and blood components results from the clinical imbalance between the supply and demand for blood: the nonfrozen shelf life of blood cells and components is 72 h–4 weeks, depending on the type of blood component. The demand for blood components is especially evident. Red cells are needed in the clinic for oxygen transport; platelets are required for preventing bleeding disorders; leucocytes are needed for immunological integrity; and plasma protein is required for various special functions.

6.1.1. Erythrocytes (Red Blood Cells)

Blood was one of the first biomaterials to be successfully cryopreserved.[62] Mollison and colleagues were the first to transfuse cryopreserved red blood cells into a patient in 1951.[63]

There are two major cryopreservation procedures applied to red cells in current blood bank practice. The high-glycerol/slow-cooling method uses glycerol concentrations in the range of 40–80% and freezing rates on the order of 1°C/min or less.[40] Recoveries obtained with this type of technique are generally in the range of 85–95%.[64]

The second major method is the so-called "low-glycerol/rapid-cooling technique".[40] This protocol uses glycerol concentrations of approximately 10–20% and cooling rates of approximately 100°C/min.

A promising third technique, of current research interest, is a rapid-freezing technique using nonpermeating extracellular additives, such as HES (hydroxyethyl starch), PVP (polyvinyl pyrrolidone), and various sugars. A major advantage of this technique would be its simplicity, since it may be possible to directly transfuse cells protected in this manner. The glycerol techniques previously mentioned involve time consuming and expensive washing procedures to remove the cryoprotectant.

6.1.2. Lymphocytes

Cryopreservation of this type of cell is now routine, and available techniques can produce essentially 100% recovery.[48] Current techniques are quite similar to that of the original method for lymphocytes. These cells are usually protected with 10% DMSO (along with 10–20% serum) and frozen at slow rates (~1°C/min). In some procedures, the freezing rate is increased once the sample has reached temperatures of −30 to −40°C. In either case, the cells are stored in liquid nitrogen (−196°C).

These cells are also successfully frozen using what is termed the "two-step" freezing procedure. Such protocols freeze the cells to some intermediate subzero temperature (−30 to −40°C). The cells are held at the intermediate temperature in order to allow cell dehydration (thermodynamic equilibrium) prior to rapid freezing to liquid nitrogen temperature.

6.1.3. Platelets

Freeze–thaw cycle recoveries of 50–80% can be obtained for platelets frozen slowly (−1°C/min) in 5% DMSO to −80°C prior to rapid cooling to −150°C or −196°C.[40] Although these recovery figures are respectable, several leading cryobiologists, representing the American Red Cross, have pointed out that none of the existing procedures are suitable for routine use in a blood bank.[40,65] Problems associated with existing techniques include the fact that they are technically demanding, expensive, and/or inconvenient.

6.1.4. Granulocytes

Of the formed blood elements, granulocytes have been the most resistant to successful freeze preservation. The best recoveries to date suggest that only 10–30% of a granulocyte population will survive the cryopreservation procedures available for this cell type. These procedures involve very low freezing rates (~0.3°C/min) in ~5% DMSO. Such low recoveries are not considered adequate for clinical transfusions.

6.2. Gametes and Embryos

The importance of being able to store gametes and embryos at a low temperature was recognized at an early stage. Polge points out that experimental observations in 1866 indicated that human spermatozoa could survive freezing.[66] A benchmark date in the history of cryobiology occurred in 1949 when Polge, Smith, and Parkes discovered (accidentally) that glycerol protected the spermatozoa of several species during freezing.[46] Three years later Smith demonstrated that it was possible to recover fertilized one-cell rabbit eggs that could be successfully divided after freezing to −196°C and subsequent thawing.[67]

Successful cryopreservation of spermatozoa was immediately applied in breeding farm animals, particularly cattle, the spermatozoa of over 50 different species have been frozen successfully.[66] Very high recoveries (~98–99%) can be achieved with these techniques. The fertility of frozen-thawed cattle spermatozoa is good, since the conception rate from artificial insemination using frozen–thawed sperm is comparable with that using fresh semen. The methods used for spermatozoa are similar to those used for blood. Slow freezing (at 5–7°C/min) in glycerol (4–9%) yields high recovery, as does rapid freezing (~100°C/min) in the presence of nonpermeating solutes.[66]

Embryo freezing has the same practical advantages as spermatozoa freezing with respect to mammalian genetics and breeding. These techniques make it possible to use recipient foster mothers to accelerate the production of superior genetic stock. It is also possible with these techniques to study the effects of xenogenously fertilized embryos, in order to distinguish between genetic and environmental factors of importance in reproductive biology. Rarely used but important laboratory animals might be stored as embryos in the frozen state, to be thawed and used only when needed, thereby reducing the expense of continuously maintaining generation after generation in the laboratory. Genetic drift might be controlled by using these techniques.

Although Audrey Smith demonstrated that embryo cryopreservation was possible in 1952, this conclusion was based on the postthaw division of a one-cell rabbit egg *in vitro*. Twenty years later, two independent reports were published indicating that approximately 80% recovery could be expected from all stages of preimplantation mouse ova and embryos frozen to −196°C. What was more, dramatic proof of the viability of the embryos was demonstrated by the birth of mouse pups from frozen–thawed embryos that had been surgically implanted in the womb of a foster mother mouse.[68]

The basic procedure used for embryos is to equilibrate them with the cryoprotectant (generally 1.5 M DMSO), seed the suspending medium at −5°C, cool at 0.2–0.8°C/min to −80°C, and then directly transfer to −196°C. The original basic technique for freezing embryos has been retained but modified somewhat from situation to situation. It is presently possible to cryopreserve embryos derived from a number of species incuding cattle, sheep, goats, rabbits, and rats.[66]

6.3. Cultured Cells

It is standard practice in biomedical and biological research laboratories around the world to maintain cultures of active, reproducing cells for experimental purposes. It is often the case that frozen storage of an infrequently used but important cell line represents an economical alternative to continuous passage through culture. This procedure is also valuable in providing a safeguard against contamination or loss of strains of actively reproducing cell cultures. Tissue culture banks of mammalian cells as well as algae and protozoa are now feasible in many cases. Genetically unstable strains that may be high yielding or disease resistant can be stored for future breeding or crop development. Frozen preservation also makes it possible to store quantities of cell types that may be impossible to test at the time of acquisition. This list is not exhaustive, but serves to illustrate some of the advantages of cryopreserved cell cultures.

Cultured mammalian cells are typically frozen in standard tissue culture media augmented with 10–15% serum and 5–10% dimethyl sulfoxide or glycerol, at rates beteen 0.3 and 10°C/min. The samples are generally frozen to −50−−60°C then plunged directly into liquid nitrogen. Thawing is done rapidly (100–200°C/min) by agitation in a water bath maintained at 37°C. Postthaw dilution of permeating cryoprotective agents may be continuous, step wise, or abrupt, depending on the biological system.[45] Generally, abrupt dilutions are tolerated more readily at 20–37°C, as compared to 0°C.[45] In any case, dilution procedures should be considered part of the overall cryopreservation process, which can produce as much or more damage than the freezing and thawing processes.

6.4. Hepatocytes

These liver cells are receiving increased attention, since cryopreserved hepatocytes represent a convenient means for the *in vitro* study of liver metabolism and the action of hormones.[49] Survival rates of 80–85% are possible using techniques similar to those given for mammalian cells in general.[49]

6.5. Bone Marrow Stem Cells

One of the major uses of bone marrow stem cells is in the management of malignant diseases. Since rejection of transplanted grafts would normally occur for allogenic transplants due to immunological incompatibility, the present use of such transplants is limited to autologous cases.[49] Indications from several species, including mice, dogs, monkeys, and man, suggest that for well-controlled conditions, survival rates of 65–100% of bone marrow cells can be expected with the use of glycerol, dimethylsulfoxide, or PVP.[49] The cryopreservation procedures are similar to those described for other mammalian cells. Recovery is evaluated by testing for the repopulating capacity (haemopoietic activity) following a lethal irradiation of the recipient.

This type of therapy has been applied in a limited way to human patients with acute leukemia. Remissions of 2–9 months have been achieved after transplantation of cryopreserved bone marrow cells following supralethal chemoradiotherapy.[69] Thus the cryobiology of this system does not appear to be a limiting factor in the therapy.

It has been claimed that there are few problems associated with the successful cryopreservation of mammalian cells from tissue culture. However, this claim masks the fact that it is seldom necessary to have the majority of the frozen cells survive in order to make the process worthwhile. Furthermore, it is not uncommon for the changes induced by freezing to be underestimated, since it is normal practice to perform limited end point assays.

In summary, it is now possible to cryopreserve successfully many important cell types. However, since these successful procedures have been developed empirically, there is no sound fundamental basis for the design of successful cryopreservation protocols for cell types of interest that have not been frozen. Generally, such cases are handled by selecting a "similar" cell type and using as a starting point the protocol that works for the similar cell. Empirical perturbations about this initial procedure are attempted.

At the present time, the design of cryopreservation protocols can not be based on an adequate understanding of the fundamental problems involved in freezing damage. Added to this limitation is the fact that as the biological system becomes more complex, more conditions must be satisfied in order to achieve recovery. Thus, the next several sections dealing with tissues and organs are rather brief. These areas are certainly important and active research programs continue to study these problems but success has been quite limited to date.

6.6. Cornea and Skin

Both cornea and skin have been successfully cryopreserved for subsequent grafting applications.

In the case of corneal tissue, it has been shown that the success of a penetrating keratoplasty procedure depends on the viability of the endothelium of the graft.[70] Capella *et al.*[70] reported in 1965 that 100% of the endothelial cells in human corneal tissue appeared histochemically normal after 8 months of preservation in the frozen state. The recommended cryopreservation procedure for this system is to use 7.5% dimethylsulfoxide and 10% sucrose (with serum), freeze slowly (~5°C/min), and thaw rapidly (~100°C/min).[70] Storage at −160°C or −196°C is required, since storage at −79°C for 2 months resulted in the loss of approximately one-third of the normal oxidative enzyme content. Clinical experience with penetrating keratoplasties using cryopreserved corneal tissue demonstrated that there was "no significant difference between fresh corneal tissue and the living preserved corneal tissue frozen in liquid nitrogen."[71] The average storage time was 2 months, but some tissue was stored for as long as 14 months. In comparison, donor tissue stored at 4°C must be used within 48 hr.

Skin was first frozen successfully in 1939, and research in this area up to 1966 has been summarized by Perry.[72] As of that time, it was known that a relatively wide range of slow freezing rates between 0.4 and 8.0°C/min would yield good skin recovery. On the other hand, the thawing rate appeared to be a more critical parameter. Moderately rapid warming rates of 50–70°C/min were necessary for successful recovery. Glycerol is the most commonly used cryoprotective chemical, and it is normally used at concentrations of 15%. Dimethylsulfoxide is also used, and storage is often at −196°C, although −79°C appears to be acceptable.[72] These general procedures have been confirmed in more recent studies (1980) where the usefulness of skin allografts or xenografts in the treatment of burn victims is emphasized.[73]

6.7. Pancreatic Tissue

Some cases of human diabetes could potentially be treated by transplantation of pancreatic islets.[49] Mazur and colleagues have been successful in cryopreserving intact fetal rat pancreases, which will subsequently synthesize protein and yield viable allografts.[74] Other researchers have worked with isolated pancreatic islet cells and have demonstrated insulin secretion *in vitro* and the abolition of symptoms of hypoglycemia from diabetic rats injected with the islet cells. Dimethylsulfoxide was the principal cryoprotectant used in both cases. This procedure for cryopreserving isolated islet cells was quite similar to that used for other mammalian cells, with slow cooling (at ∼1°C/min) followed by rapid warming (∼150°C/min). The larger intact pancreatic tissue requires somewhat slower freezing (∼0.2°C/min) and significantly slower thawing (∼4–15°C/min). This is presumably a reflection of constraints imposed by the longer diffusion time required to equilibrate the larger tissue system as compared to individual cells.

6.8. Heart and Kidney

With the exception of the rare case of a very small organ, such as the rat pancreas just described, early attempts as well as more recent attempts to freeze isolated organs have not been successful. This is perhaps not surprising, since two important characteristics of these tissues would be expected to create problems.[49] These organs are comprised of a number of cell types, each of which has its own diverse requirements for optimal cryopreservation. This situation is likely to create problems when the requirements for all cell types present must be satisfied simultaneously. In addition, the size of these large organs suggests that significant thermal and concentration gradients will exist within them for many protocols. These gradients may cause damage by osmotic stress as well as make controlling thermal and chemical histories extremely difficult.

6.9. Other Applications

The preceding examples have focused on mammalian tissues. However, there are other species and other aspects of applied cryobiology that are of significance. They include the preservation of protozoa and parasites, insects, fish, microorganisms and plants, including algae. In many cases, these applications are considered for health-related reasons. Pharmaceutical companies use cryopreservation to store parasites for screening potential antiparasitic drugs. Microsomal metabolic organelles have been cryopreserved for use as a part of bioassays used to detect environmental mutagens and carcinogens.

Other applications include cryopreservation of fish gametes and biomaterials for controlling the stocking of fish. Related studies of plant cold hardiness and freezing stress are important in terms of the world's food supply and development of food-freezing procedures (cf. Ref. 75).

6.10. Preservation Techniques

A summary of the preservation techniques just discussed is presented in Table 1.

A more extensive treatment of standard cryopreservation procedures may be found in Ref. 39, which is an excellent source and includes extensive bibliographies at the end of each chapter as well as a list of relevant journals and books.

7. THERMODYNAMIC MODELS AND CRYOBIOLOGY

Although the study of low-temperature biology remains to a large extent a descriptive and empirical activity motivated by immediate practical problems, a significant improvement in the current understanding of problems in cryobiology has been obtained as a result of mathematical modeling and the application of the principles of thermodynamics. This section reviews the development and current status of these types of thermodynamic models. Ultimately, such models could be used to design optimal cryopreservation protocols based on fundamental principles.

7.1. Development of the First Model

Mazur made a major contribution to the field of cryobiology in 1964 when he published the first thermodynamic model describing cell water transport during freezing.[17] Part of the significance of this paper was that it provided a framework for interpreting the experimental freezing response of a variety of cells. Because the framework was in the form of a thermodynamic model, Mazur's work made it possible to quantitatively predict the response of a cell while being frozen. Computer simulations of cell volume response during freezing could then be compared with experimental data. In addition, parameter studies using the computer model could quantify the importance of certain freezing parameters.

The basic features of Mazur's and subsequent thermodynamic models are illustrated in Fig. 17. The three essential elements of such models are: 1)

TABLE 1
A Summary of Cryopreservation Procedures

Biological system	Cryoprotective additive	Cooling rate (°C/min)	Thawing rate (°C/min)	Recovery (%)	References
Cells					
Erythrocytes	40–80% glycerol	~1	~25	85–95	7, 40, 64
	10–20% glycerol	~100	~100	90–95	40
Lymphocytes	10% DMSO[a] + 10–20% serum	1–10	2.5–100	~100	48
Granulocytes	~5% DMSO	~0.3	~40	10–30	48, 114
Platelets	5% DMSO	1–30	500–1000	50–80	40, 115
Spermatozoa	4–9% glycerol	~5	1.0–60	~99	7, 66
	nonpermeating solutes	~100	1.0–60	~99	7, 66
Embryos	1.5M DMSO	0.2–0.8	4–25	~80	66
Cultured mammalian cells (see text for references to specific systems)	5–10% DMSO or glycerol with 10–15% serum	0.3–100	~100	Variable	See text
Tissues					
Cornea	7.5% DMSO and 10% sucrose	~5	~100	~100	70, 71
Skin	15% glycerol or DMSO	0.4–8.0	50–70	~100	72, 73
Organs					
Fetal rat pancreas	2.0 M DMSO	0.3	4–15	80–100	74
Heart	Little success has been achieved with these systems				
Kidney					49

[a] DMSO = dimethylsulfoxide.

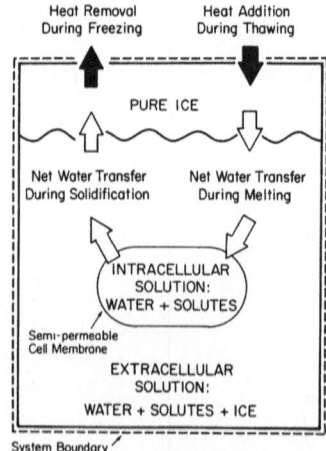

Heat Removal
During Freezing

Heat Addition
During Thawing

PURE ICE

Net Water Transfer
During Solidification

Net Water Transfer
During Melting

INTRACELLULAR
SOLUTION:
WATER + SOLUTES

Semi-permeable
Cell Membrane

EXTRACELLULAR
SOLUTION:
WATER + SOLUTES + ICE

System Boundary

FIGURE 17

A simple system definition for a typical thermodynamic model describing the response of cells to freezing and thawing. The behavior of the intracellular and extracellular solutions are modeled and a rate equation is included that relates water transport across the cell membrane to the thermodynamic properties of the intracellular and extracellular solutions.

the intracellular solution, 2) the extracellular solution, and 3) a rate equation that models the transport characteristics of the cell membrane and links the intracellular solution with the extracellular solution.

The first thermodynamic model published by Mazur was based on a number of simplifying assumptions; they have been retained in subsequent modeling studies performed by a number of researchers. However, some of the most important parameter values, which were poorly defined or unknown at the time Mazur published his model, remain in a similar state.

Table 2 lists nine assumptions in approximately the order of their influence on the computer simulation of water transport during freezing. The development of a typical model follows and is discussed in light of these assumptions.

7.1.1. Intracellular Solution

Mazur modeled the intracellular solution as an ideal solution in the sense of Raoult's law. Accordingly, the partial pressure of water vapor in the cell is directly related to the mole fraction of water in solution

$$p_w = x_w p_w^0 \tag{33}$$

where p_w^0 is the partial pressure of pure water at the temperature of the solution. This statement is equivalent to stating that the water activity and mole fraction are equal

$$a_w^i = x_w^i \gamma_w^i = x_w^i \tag{34}$$

where the activity coefficient γ_w is identically equal to 1.0 for the ideal solution. In addition, the intracellular system is taken to be in a state of internal equilibrium, such that there are no spatial gradients of temperature, pressure, or chemical composition. Alternatively, the intracellular and

TABLE 2
A Summary and Evaluation of Thermodynamic Modeling Assumptions

Modeling assumption	Effect on predictions	Uncertainty associated with the assumption	References
1. Temperature dependence of membrane water permeability	Major	Large	13, 18, 19, 78, 79
2. Extracellular solution remains in a state of internal equilibrium	Major	Moderate to large	88–91
3. Concentration (or chemical) dependence of membrane water permeability	Moderate to major	Large	19, 80, 81
4. Intracellular solution remains in a state of internal equilibrium	Small to moderate	Small	20, 82
5. Constant membrane surface area during freezing and thawing	Small to moderate	Moderate	13, 21, 60, 84, 85
6. Transport during freezing is due to water and not solute (i.e., no solute leakage or transport of cryoprotective agent)	Small in some cases major in others	Large	22, 43, 44, 86
7. Cell and environment are in thermal equilibrium	Small to moderate	Small	13
8. Ideal intracellular solution behavior	Small	Small	18, 21, 87
9. Latent heat of fusion of water is constant	Small	Small	17, 62

extracellular solutions may be described in terms of water chemical potential μ_w,[22] osmotic pressure Π,[29] or osmolality Ω,[76] for use in freezing models.

7.1.2. Extracellular Solution

The extracellular solution is modeled as an ideal solution, which is assumed to be in a state of internal equilibrium. The temperature of extracellular solution is assumed uniform and equal to that within the cell. The pressure of the extracellular solution is assumed to be equal to that within the intracellular solution and spatially uniform. All of the solute is considered to be rejected by the ice, and the concentration of the solute in the extracellular solution is taken to be uniform in space. Since the ice and liquidus in the extracellular solution are assumed to be in thermodynamic equilibrium, a Gibbs–Helmholtz relationship can be applied to relate the partial pressure change of the extracellular water p_w^e to the change in temperature of the system

$$\frac{\partial(\ln p_w^e)}{\partial T} = \frac{L}{RT^2} \tag{35}$$

where L is the latent heat of fusion. This latent heat is usually approximated as that of pure water at the system temperature. Alternatively, this model of the extracellular solution is cast directly into the form[14]

$$\frac{\partial(\ln a_w^e)}{\partial T} = \frac{L}{RT^2} \tag{36}$$

which is equivalent to Eq. (35).

7.1.3. The Water Transport Equation

Mazur adopted a phenomenological expression for the water transport through the cell membrane, which had been described earlier by Davson and Danielli.[77] The form of this equation is

$$\frac{dV_{cell}}{dt} = KA(\Pi^i - \Pi^e) \tag{37}$$

where it is assumed that the membrane is perfectly semipermeable (i.e., only water transport occurs). In this expression, K is the membrane water permeability, A is the membrane surface area, and the osmotic pressures inside and outside the cell are given as Π^i and Π^e, respectively.

For constant cooling rates B, where

$$B = dT/dt \tag{38}$$

the time differential in Eq. (37) may be expressed in terms of the cooling rate and the temperature differential dT

$$\frac{dV_{cell}}{dT} = (KA/B)(\Pi^i - \Pi^e) \tag{39}$$

The osmotic pressures are directly related to the equivalent partial pressures of water vapor inside and outside the cell. Equations (34), (36), and (39) then form a set of equations that may be integrated numerically with respect to temperature, given the initial condition of the system, the cooling rate, and various cell parameters (area, osmotically active cell volume, and water permeability)

$$V_{cell} = V_{cell}^0 - \int_{T_i}^{T_f} \frac{KA}{B}(\Pi^i - \Pi^e)\, dT \tag{40}$$

Mazur presented the results of his computer simulations in the form of Fig. 18, which represents the predicted "osmotically free" water content of the cell, normalized with respect to the initial osmotically free water volume of the cell, as a function of subzero temperature. For the case shown in Fig. 18, the family of curves demonstrates the effect of cooling rate on water

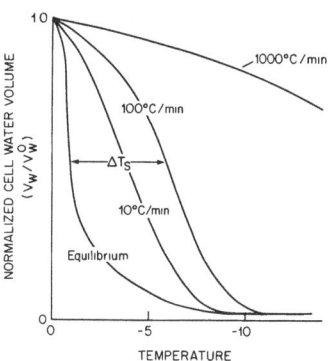

FIGURE 18

Typical computer predictions of cell water volume as a function of temperature during freezing. Cells are expected to shrink as the temperature decreases, and the extracellular solution becomes concentrated. An equilibrium volume can be described at a given temperature, corresponding to infinitesimal cooling rates. Finite cooling rates cause departures from equilibrium, and ΔT_s, the instantaneous super-cooling of the intracellular solution, is a measure of the departure from equilibrium.

transport during freezing. The curve labeled equilibrium represents the case when freezing occurs at an infinitely slow rate. These predictions clearly show that faster cooling rates result in less cell dehydration. At higher subzero temperatures, a finite rate will cause the volume of water retained within the cell to be greater than that associated with the cell in equilibrium. This departure from equilibrium leaves metastable water within the cell. The greater the departure from equilibrium, the more likely the formation of intracellular ice.

Mazur defines the degree of supercooling with reference to the horizontal distance ΔT_s between the intracellular water volume at any temperature and the temperature corresponding to the same volume if the intracellular water were at equilibrium (see Fig. 18). Knowing that most cells form intracellular ice between -10 and $-15°C$,[9] Mazur assumes a critical degree of supercooling. With this assumption, he is able to predict (semiquantitatively) the rates of freezing that will result in intracellular ice formation for cells with such widely different physical parameters as yeast cells and red blood cells.[11]

Of all the assumptions in this initial model, those that exerted the most significant effect on the predicted cell response were cell size (surface area to volume ratio), cooling rate, and the membrane water permeability. The sensitivity of the model to changes in these parameters is readily examined by computer simulations and represented in terms of parameter studies similar to the results given in Fig. 18.

Mazur was aware that membrane water permeability was a strong function of temperature. He used an exponential function of the form

$$K = K_g \exp\left[b(T - T_g)\right] \qquad (41)$$

to relate the water permeability K at any temperature T to K_g, a reference value at the reference temperature T_g. The thermodynamic model revealed that water permeability plays a dominant role in the predicted water transport. The model also pointed out the need for refinements in the expression for the temperature dependence of the membrane water permeability. Data, curve-fitted in the temperature range 0 to $+30°C$, were used to define Eq. (41). This relationship was then used for freezing predictions by extrapolating to subzero temperatures. This assumption and others given in Table 2 are now discussed in more detail.

7.1.4. A Critique of the Modeling Assumption

Uncertainty regarding the numerical value of the membrane water permeability at subzero temperatures remains a major problem in cryobiological modeling at the present time. There is some experimental evidence that suggests that extrapolation of data obtained at higher temperatures into the subzero temperature range may be valid, at least for human erythrocytes down to temperatures of $-10°C$.[78]

However, Levin has developed a membrane model[79] that suggests that such extrapolation must be undertaken with care, since at least two processes may limit the rate of water permeation through the membrane. As each process may be dominant within different temperature regimes, typical Arrhenius correlations of membrane water permeability may be significantly nonlinear, leading to problems if a single linear plot is extrapolated. It is also the case that phase separation within the membrane may occur, which could greatly affect water permeability. Little is known about this possibility, particularly on a quantitative basis.

It appears that in the same sense that temperature affects membrane water permeability, the extracellular and intracellular chemical composition may affect the state of the membrane and therefore its water permeability. This is a controversial topic, as pointed out by Mazur,[13] but several investigators, including Silvares[18] and Diller[19] have shown that accounting for this effect is important in the successful computer simulation of freezing. An improved understanding of this effect is apparently necessary. Some researchers have claimed that water permeability depends on the osmolality of the suspending medium,[80] while others have claimed that water permeability is independent of it.[81] In either case, the problem is similar to that of the temperature dependence of the membrane water permeability, in that data are available only over a small range of osmolalities compared to those experienced during freezing. Thus, rather large extrapolations are again required.

Pushkar et al.[20] and Levin et al.[82] have examined theoretically the consequences of assuming uniform solute concentration within a cell during freezing. They both concluded that Mazur's initial assumption of a uniform concentration is likely to be a very good one, with the possible exception of cells with very high water permeabilities, such as human red cells. In these cases, the Peclet number may be significant, and solute polarization effects should be included. The Peclet number Pe here is defined by Levin[83] as

$$Pe = \frac{\bar{V}_w J_w l_c}{D^v} \tag{42}$$

where $\bar{V}_w$ is the partial molar volume of water, J_w is the mole flux of water, l_c is the cell half-thickness, and D^v is the volume diffusivity (see Ref. 83 for details).

It is not clear whether membrane area remains invariant during freezing. Mansoori[21] has questioned Mazur's assumption of constant cell surface area during freezing. Furthermore, there is evidence to suggest that only small surface area changes can be tolerated before cell lysis occurs,[84] at least for hypotonic osmotic exposures. While Mazur claims that this will exert only a small influence on the cellular freezing response, this aspect of the problem deserves further attention. As recently reviewed by McGrath, there is growing evidence that cell membrane material is lost during freezing.[60] Boroske *et al.* have postulated the formation of submicroscopic "blebs" on egg-lecithin vesicles during osmotic shrinkage.[85] It is unclear how these various membrane modifications may influence the effective area available for water transport.

The assumption that water transport is the dominant transport process during freezing appears to be correct. Diller has developed a model that predicts the freezing response of human red cells immersed in a ternary solution including the permeable cryoprotectant, glycerol.[22] The results of Diller's computer simulations show that glycerol transport through the cell membrane is negligible when compared to water transport.

In addition to considering the relative magnitudes of the transport of water and other permeable solute species, there is the related problem of the transport of normally impermeable solute species. These species may enter the cell should the membrane barrier function be compromised during freezing. There is certainly evidence that exposure to hypertonic solutions can induce solute leakage in human red cells.[86] Indeed, this leakage begins to occur at osmolalities that are much below the maximum concentrations encountered during freezing. Diller has obtained data with a cryomicroscope that document abnormally large cell volumes during thawing.[44] As noted by Diller, this behavior would be consistent with a net solute leakage into the cell at some point during low-temperature storage or thawing. Since there are other cases where thawed cells exhibit "anomalous" osmotic behavior,[43] it would seem appropriate to reexamine the assumption of no solute leakage, anticipating that leakage of "impermeable" solutes may lead to larger effects on water transport than those that were initially assumed.

Both Silvares[18] and Levin[87] have challenged the assumption of ideal behavior in the intracellular and extracellular solutions. Their conclusions, based on analytical considerations, are similar to those of Mansoori.[21] It is therefore concluded that the assumption that the subtleties of nonideal solution behavior may be ignored in the freezing models is valid.

Typically, the heat fluxes present in samples of individual cells in a suspending medium during freezing and thawing are sufficiently small, such that local temperature gradients are negligible in the vicinity of the cell. Hence, assuming the cell to be at thermal equilibrium with its surrounding medium seems to give a good approximation. As noted earlier, this assumption is likely to be in error for larger systems, such as tissues and organs.

Mazur[17] and Silvares[18] have checked the assumption of constant latent heat of fusion; the errors introduced by this approximation are insignificant in their view.

The assumption that extracellular ice and solution remain in equilibrium, such that changes in concentration, temperature, and pressure differences within the extracellular solution are ignored, has received considerable attention recently. O'Callaghan et al.,[88] Levin,[89] and Korber et al.[90] have shown, theoretically and experimentally, that significant concentration gradients may arise in the liquid phase during planar and dendritic solidification. What is more, Brower et al.[91] propose that cells suspended in solution may be encapsulated by ice or moved by the ice interface as it grows into the sample. The concentration histories experienced by the cells will differ in these two cases, and the spatially nonuniform concentrations are thought to have significantly different effects on the freezing response of the cells.

Successful cryopreservation requires understanding the processes occurring during thawing as well as those occurring during freezing. The fact that the present discussion has emphasized models for the freezing part of cryopreservation procedures reflects the fact that little has been published concerning thawing models. One computer simulation of thawing is available.[44] An impediment to work in this area is the uncertainty in the accuracy of the freezing portion of the simulations, which is important in determining initial conditions for the subsequent thawing process.

7.2. Other Models

Several other models are now described briefly, which go beyond those of Mazur and his school.

7.2.1. The Pseudobinary Intracellular Solution

Levin has developed a detailed intracellular solution model for the human erythrocyte.[16] Based on rather complete knowledge of the intracellular composition for the red cell, Levin defines an ideal pseudobinary aqueous solution. Considering the relative simplicity of the pseudobinary model compared to the complexity of the real intracellular solution, it is remarkable that predictions of the osmotic behavior of red cells with this model agree well with experimental data obtained by other investigators.[27] An attractive feature of this model is that the modeling assumptions directly address the question of "bound water." Hence, the agreement between experimental data and this model sheds some light on the possible mechanisms involved in the "osmotically inactive" water volume determined empirically from Boyle–Van't Hoff experiments (see Fig. 7). Unfortunately, extension of such a pseudobinary model to other cells and tissues seems unlikely at present, since detailed intracellular composition and other required information are unavailable for most biological systems. Thus, in later work with other tissues, Levin makes use of the empirical Boyle–Van't Hoff data for intracellular solution models.[29]

7.2.2. Cryoprotective Additives

The modeling work published by Lynch and Diller[22] which includes the effects of ternary solutions, is important for improving our understanding of the response of cells frozen in the presence of cryoprotective additives. Cryoprotective agents are thought to protect cells on a colligative basis, i.e., the unfrozen CPA within the cell dilutes the electrolytes in the intracellular solution, even when water is leaving the cell to freeze in the extracellular volume. The model developed by Diller clearly shows that the presence of glycerol significantly lowers the intracellular electrolyte concentration during freezing.

Computer simulations performed with this model also permit examination of the effect of the cryoprotective additive on the probability of forming intracellular ice. For at least some cell types,[22] it is experimentally observed that the addition of CPA lowers the survival percentage at higher than optimal freezing rates. This observation is taken to mean that the likelihood of intracellular ice formation is enhanced by the added glycerol. The computer model suggests that this may be caused by the depression of the freezing temperature of the solution by the added glycerol. Since membrane permeability for water is exponentially related to temperature, it may be decreased significantly, even prior to ice formation. Increased supercooling may be expected with this lower membrane water permeability because of the increased impedance of the water transport, and this in turn increases the probability of intracellular ice formation.

7.2.3. Intracellular Ice Formation

The exact mechanism by which the formation of ice causes cell damage is unknown, nor is it known how intracellular ice nucleation occurs. In an attempt to explore some of the theoretical aspects of this problem, Mazur has derived a thermodynamic explanation, based on the possibility that extracellular ice penetrates membrane pores after some degree of supercooling. The supercooling is thought to provide the driving force for the "breakdown" of the membrane as a barrier.[9] Mazur was able to show that the experimentally observed intracellular nucleation temperatures of -10 to $-15°C$ are in close agreement with the theoretical model where extracellular ice grows through membrane pores to nucleate intracellular ice.

Toscano *et al.* have published a nucleation model that considers both homogeneous and heterogeneous mechanisms of intracellular nucleation.[92] These authors suggest that intracellular proteins, such as hemoglobin, may represent sites for heterogeneous catalysis, although they do not rule out the possibility of ice propagation through the cell membrane.

Morris and McGrath have obtained experimental data that demonstrate the important effect of extracellular ice on the intracellular ice nucleation temperature.[12] Freezing the algae cell Spirogyra resulted in an intracellular nucleation temperature of $-7.7°C$ when extracellular ice was present throughout the freezing process. This nucleation temperature did not change ($P < 0.05$) with variation of the freezing rate. If, on the other hand, the cells were carefully cooled, such that both the cells and the extracellular solution were supercooled, then intracellular ice did not form until immediately after extracellular ice appeared at a mean temperature of $-12.4°C$. This was significantly different ($P < 0.01$) from the first case, where intracellular nucleation occurred at $-7.7°C$. No current model explains this experimental observation, and the results cast doubt on the existence, at least for some cell types, of a critical amount of supercooling that triggers nucleation on a heterogeneous intracellular catalyst.

7.2.4. Thermodynamics of Membrane Systems

Before discussing systems larger than individual cells, which would represent tissues or organs, it should be mentioned that some of the recent thermodynamic modeling efforts in cryobiology have focused on the cell membrane.[60] This research contrasts with the modeling work presented so far, in that the cell membrane itself is considered the system, as compared to the intracellular and extracellular solutions, which were the systems of interest in the previous cases.

The motivation for this research is to develop a thermodynamic model capable of describing state changes within the membrane, resulting from changes of state in the environment of the membrane. McGrath has recently reviewed this work.[60] He discussed the potential value of this approach in terms of understanding membrane alterations resulting from thermal, chemical, mechanical, and electrical extracellular perturbations, which might occur during freezing.

7.2.5. Cell Clusters

It might be expected that clusters of cells, and larger scale cellular organizations (tissues and organs), would respond differently to freezing than individual cells. In particular, the larger size of such clusters limits heat and mass transfer rates through them. Data obtained by McGrath *et al.* indicate that cells in the center of cell clusters may be less susceptible to solute damage, and more susceptible to the formation of internal ice, than individual cells forming the clusters.[93]

Levin has also considered the effect of cell clusters on freezing response. He modified his water transport model to consider one-dimensional water transport through cell clusters.[94] This model predicts that cells in the center of a cluster lose less water than those at the periphery of the cluster. Solute is therefore less concentrated in these cells, and the water in them may be more supercooled. Thus, "interior" cells would be less susceptible to "solute effect" freezing damage and more susceptible to "intracellular ice" freezing damage.

7.2.6. Osmotic Stress and Perfusion

Cryobiologists have emphasized the problems of introducing and removing cryoprotective additives (CPAs) from cells, tissues, and organs. They have focused on the osmotic manipulations associated with cryoprotective additives at the cellular level. These manipulations are assumed to be bounded by two limits. Osmotically induced volume changes may lead to "solute effect" damage at high concentrations, or such changes may burst cells at low concentrations. Levin has published several models dealing with the cellular response to CPA addition and removal,[47,95] and he has used these concepts to define optimal techniques for adding and removing CPA from cells.[95]

Rubinsky and Cravalho[96] have analyzed the dynamics of perfusing a tissue, using a simple unit model, and defined useful guidelines describing the equilibration times expected in various situations. The results are reported in terms of nondimensional parameters, thus illustrating which non-dimensional parameters are important and providing a means for analyzing a variety of cases. Levin has also developed a three-compartment model to conduct computer studies of organ perfusion, which includes coupled transport of solvent and cryoprotective additive.[47]

8. CRYOMICROSCOPY

One of the experimental methods for studies in cryopreservation makes use of a computer-controlled, low-temperature, light microscope. The so-called cryomicroscope consists of a computer-controlled heat transfer system, which subjects the sample to a desired freeze–thaw history while allowing real time observation of the process with a light microscope. The microscope may be linked to a video camera and digital image processing and storage system. Diller has recently written an excellent review on the subject, which outlines the historical development as well as the current capabilities of such devices.[97]

The cryomicroscope has been applied to study many facets of cryopreservation including, but not limited to, the following general types of studies:

(a) Determining cell volume changes during freezing[29,44,98–100] and thawing.[44]
(b) Determining freezing rates required for intracellular ice formation.[12,93,101–103]
(c) Determining nucleation temperatures for intracellular ice formation.[12,93,101–104]
(d) Correlating intracellular ice formation and cell recovery.[102]
(e) Detecting recrystallization phenomena.[93,105]
(f) Detecting bubble formation due to gas supersaturation.[12,106]
(g) Quantifying nonequilibirium solute concentrations in the unfrozen liquid solution during solidification.[90]
(h) Examining the encapsulation and/or movement of cells due to interaction with the ice/solution interface.[91]

As pointed out recently by Callow and McGrath,[100] a quantitative correlation between theoretical predictions of cell volume changes during freezing and experimental results obtained with the cryomicroscope has not been established yet. There are two reasons for this: Firstly, it appears that no one has conducted computer simulations of freezing using a complete set of model parameters that were determined in the same laboratory as the one performing the cryomicroscopy. Secondly, there are experimental uncertainties in both temperature and volume determinations that may limit quantitative comparisons severely.[100] These limitations have not been dealt with rigorously and have, in fact, been ignored in many cases.

The scope of this chapter precludes a more extensive discussion of the cryomicroscope as an experimental tool. This instrument has proven to be very valuable in both semiquantitative and quantitative modes as a means of developing or verifying models that describe the freeze–thaw response of biological systems. The interested reader is referred to the Callow and McGrath[100] for more detail.

There have been a number of recent improvements in the analysis, design, and use of cryomicroscope systems which make them even more powerful research tools. Diller has led the way in implementing digital image analysis techniques on cryomicroscope systems.[44,107] This advancement is of significant value in the automation of quantitative microscopy tasks such as determining cell volume changes which occur during cryopreservation processing as well as the mapping of solute concentration fields. Another recent improvement in cryomicroscopy involves the use of microprocessor and microcomputer technology. In the past, cryomicroscope temperature controllers were relatively limited with respect to programming various types of cryopreservation protocols. Evans and Diller[108] and Cosman *et al.*[109] have described microprocessor-based controllers which are considerably more flexible. McGrath has interfaced a laboratory microcomputer to a cryomicroscope system for the purpose of thermal control and general data acquisition.[12] In addition to more programming flexibility, an improved understanding of the behavior of such systems is developing. Cosman *et al.* recently developed the first model describing the dynamic response of the cryomicroscope heat transfer stage and its associated control system.[109] This analysis will be valuable to those who wish to build such a system or improve the performance of an existing system. In a similar vein, the range of freezing/thawing rates attainable and the magnitude of potential temperature measurement errors are important considerations in cryomicroscope design. Recent thermal analyses of cryomicroscope heat transfer stages by Schiewe and Körber[110] and McGrath and Khompis[111] have led to an improved understanding of the behavior of such systems. Körber *et al.*[90] have improved the quantitative capabilities of their cryomicroscope system by interfacing it with a photomultiplier system. Laser-excited fluorescence and photobleaching experiments will be possible on the author's cryomicroscope system in the near future.

Probably the three most important limitations of current cryomicroscope systems are: (a) the optical resolution of the light microscope; (b) the "two-dimensional" nature of the microscope image; and (c) the accuracy and resolution of cryomicroscope temperature measurements. Theoretically, the optical resolution of the light microscope is approximately 0.33μm (at 500 nm).[12] However, optical techniques such as phase contrast microscopy which are useful in enhancing the contrast between the cell and its environment often produce a "halo" artifact around cell boundaries and organelle boundaries. Thus the optical resolution at potential sites (membranes) of freeze/thaw injury is much worse than 0.33 μm. The Nomarski differential interference contrast technique offers some important advantages in this respect since it has been shown that this type of system is capable of resolving separation distances on the order of 440 Å[112] to 0.1 μm.[113] Steponkus has made good use of this increased resolving power by adapting Nomarski optics to his cryomicroscopic studies.[43,106] Thus, this technique can be expected to yield more accurate determinations of cell boundary locations and cell volumes that are derived from such determinations. In addition to increased resolving power, Nomarski optics offer another potential advantage to those cryomicroscopists involved in determining cell volume changes during freezing. Specifically, such optics have a very shallow depth of field.[112] There is therefore the possibility to recreate a "three-dimensional" image of the cell by "optical sectioning" with the light microscope. This could be quite useful in terms of improving the information available in the form of the two-dimensional projected area available from bright-field and phase-contrast optical systems which have much larger depths of field. Finally, it should be understood that determining cryomicroscopic temperatures accurate to a fraction of a degree is no trivial matter. Such tolerance will be necessary to make good quantitative comparisons between cryomicroscopic volume determinations and computer simulations of cell volume changes during freezing.[100] Temperature measurement inaccuracies may arise from a number of sources including calibration inaccuracies, limitations of electronic linearization or compensation circuits, and electronic drift. Perhaps even more importantly, all current cryomicroscope designs make use of only one thermocouple to effect thermal control and to monitor the sample temperature. For a number of reasons it is often necessary that the thermocouple be displaced horizontally or vertically from the sample. Since thermal gradients can be significant in such systems, only those samples very close to the thermocouples can be used. It is therefore proper to critically assess the validity of any claims that cryomicroscope temperature measurements are more accurate than 0.5 to 1.0°C. While this accuracy may be realized in some cases, documentation for the claims should be provided. Recent heat transfer analyses of such systems are one way to examine this problem,[110,111] and the use of multiple thermocouples in the sample could lead to more confidence in temperature measurements.

While there are some limitations to the use of cryomicroscopes which are important to recognize, the technique has proven to be tremendously useful to date, and there is every reason to believe that it will continue to play an important role in the progress of cryobiological science.

9. SUMMARY

Cryopreservation represents an important means of storing living bio-material in a state of suspended animation. An improved understanding of the effects of low temperatures on biological systems would have an important impact on a large number of medical problems, animal-breeding practices, food production and preservation techniques. The complexity of even the simplest biological systems, coupled with the complexities of the freezing–thawing phenomena that are a part of cryopreservation, make it difficult to separate and evaluate the many parameters involved.

Empirical approaches, motivated by immediate clinical needs, have led to some remarkable and impressive success stories, such as blood and embryo freezing. Even though such impressive achievements have been realized, appropriate cryopreservation protocols do not exist for recovering a great many biological tissues of interest from frozen storage.

This situation is due in part to the fact that relatively little effort has been devoted to analyzing the cryopreservation process with methods that could lead to quantitative predictions. Quantitative methods can be developed from computer simulations and verified with careful microscope and microprobe studies. Experimentally validated computer simulations might then be used to uncouple parametric relationships and thus perform studies to determine the relative importance of the various parameters in a given model. In this way, it would be possible to develop accurate models for the design and optimization of cryopreservation procedures for an arbitrarily chosen bio-material of interest. These predictions would be based on fundamental principles incorporated into the model.

In this chapter, heat and mass transfer, as well as thermodynamic considerations relevant to current problems in cryopreservation, have been discussed. The state of the art with respect to thermodynamic modeling has been reviewed in detail. These thermodynamic models, and the computer simulations of cryopreservation processes using such models, are extremely useful in the author's view. There is reason to believe that such research will continue to play a vital role in improving the understanding of the many phenomena involved in cryopreservation.

Basic models describing cell freezing and thawing focus at present on predictions of water transport and the likelihood of intracellular ice formation. Nearly 20 years of modeling research have identified those physiochemical parameters that are likely to play the most important roles in determining cell water transport during freezing. Environmental thermal and chemical changes that occur during cell freezing are expected to alter significantly the membrane's permeability to solutes and solvent in the general case. More research is warranted in this area, both in terms of experimentally defining the magnitudes of such effects and, if possible, modeling them.

The assumption of constant cell membrane area during freezing and thawing deserves a closer look, in light of recent results documenting membrane alterations, including the loss of membrane material. Similarly, it would appear that membrane breakdown and solute leakage may occur during freezing or storage, and this possibility should also be examined carefully.

Comparatively little work has been published that describes the thawing response of cells, either experimentally, with a cryomicroscope, or theoretically, using computer simulations. The fact that much more research of this type has been devoted to freezing rather than thawing may be due in part to the uncertainty in various parameters, such as those described in Table 2 that affect the observed cell volume changes that occur during freezing. Thus, the initial conditions for thawing still remain uncertain.

The mechanisms by which the nucleation of intracellular ice occurs remain unclear. Several thermodynamic models describing intracellular nucleation have been developed that are consistent with available experimental data. No models exist that describe the mechanical interaction between ice and cell membranes.

The cell membrane appears to be the site of freezing injury for solute effect damage at slow-freezing rates. Thus, a kinetic/thermodynamic model has been developed to describe the expected changes of membrane state, resulting from extracellular temperature and chemical changes, such as those that might occur during freezing. More research is required in this area to further develop this model and check its validity.

Several aspects of the effects of the presence of cryoprotective agents have been modeled. In particular, theoretical studies have been performed that predict cell volume changes, electrolyte concentration build-up, and the degree of supercooling to be expected for red cells frozen in the presence of glycerol. These studies are useful for understanding why cells may be less susceptible to solute effect damage and more susceptible to intracellular ice damage when a cryoprotective compound is present during freezing. The osmotic changes associated with the addition and removal of the cryoprotective chemical to and from the sample have been considered theoretically. Future research combining such theoretical studies with experimental studies could be quite valuable.

Finally, a start has been made in linking together these freezing studies to consider what differences would be expected between single-cell behavior and the behavior of a larger biological sample, such as a cell cluster, which may represent a tissue or organ. So far, membrane water transport and the effect of perfusion with cryoprotective agents have been the only aspects of the problem studied theoretically on this scale.

Thermodynamic modeling has several advantages that will make it useful for future progress in this field. The methodologies involved require clear statements of assumptions and therefore careful thought about the fundamental biological processes involved. Modeling provides a conceptual framework for identifying missing information. The fact that the models are quantitative means that computer simulations may distinguish the important and negligible parameters in a given model. Finally, the quantitative aspect of this approach offers the promise that, one day, cryopreservation procedures for specific tissues may be designed from scratch and optimized, based on fundamental principles.

ACKNOWLEDGMENTS. The author would like to express his thanks to Dr. E. G. Cravalho for his critical review of this manuscript. He has been a pioneer in the field of cryobiology and is responsible in many ways for whatever contributions the author may have made to the field in this manuscript or otherwise.

This work was supported in part by a grant from the Whitaker Foundation (ORD 26018) and in part by a grant from the National Institutes of Health (GM 31202).

REFERENCES

1. Boyle, R., *New Experiments and Observations Touching Cold* (R. Davis, London, 1683).
2. Ashwood-Smith, M. J., Preservation of micro-organisms by freezing, freeze drying, and dessication, in *Low-Temperature Preservation in Medicine and Biology*, edited by M. J. Ashwood-Smith and J. Farrant (University Park Press, Baltimore, 1980), pp. 219–252.
3. Meryman, H. T., Freeze drying, in *Cryobiology*, edited by H. T. Meryman (Academic, New York, 1966), pp. 610–664.
4. Fry, R. M., Feezing and drying of bacteria, in *Cryobiology*, edited by H. T. Meryman (Academic, New York, 1966), pp. 665–1966.
5. Greiff, D., and Rightsel, W., Freezing and freeze drying of viruses, in *Cryobiology*, edited by H. T. Meryman (Academic, New York, 1966), pp. 698–728.
6. Morris, G. J., Plant cells, in *Low-Temperature Preservation in Medicine and Biology*, edited by M. J. Ashwood-Smith and J. Farrant (University Park Press, Baltimore, 1980), pp. 253–283.
7. Fennema, O. R., Powrie, W. D., and Marth, E. H., *Low-Temperature Preservation of Foods and Living Matter* (Dekker, New York, 1973).
8. Fahey, G. M. and Hirsch, A., Prospects for organ preservation by vitrification, in *Organ Preservation: Basic and Applied Aspects*, edited by D. E. Pegg, I. A. Jacobson, and N. A. Halasy (M.T.P. Press, Lancaster, England, 1982), pp. 399–404.
9. Mazur, P., Physical and chemical basis of injury in single-celled micro-organisms subjected to freezing and thawing, in *Cryobiology*, edited by H. T. Meryman (Academic, New York, 1966), pp. 213–315.
10. Thomas, D. G., and Staveley, L. A. M., Supercooling of drops of some molecular liquids, *J. Chem. Soc.* **1952**, 4569–4577, 1952.
11. Mazur, P., Cryobiology: the freezing of biological systems, *Science* **168**, 939–949, 1970.
12. Morris, G. J., and McGrath, J. J., Intracellular ice nucleation and gas bubble formation in spirogyra, *Cryo-Letters* **2**, 341–352, 1981.
13. Mazur, P., The role of intracellular freezing in the death of cells cooled at supra-optimal rates, *Cryobiology* **14**, 251–271, 1977.
14. Denbigh, K., *The Principles of Chemical Equilibrium*, 3d ed. (Cambridge University Press, Cambridge, England, 1971).
15. Robinson, R. A., and Stokes, R. H., *Electrolyte Solutions*, 2d ed. rev. (Butterworths, London, 1959).
16. Levin, R. L., Cravalho, E. G., and Huggins, C. E., Effect of hydration on the water content of human erythrocytes, *Biophys. J.* **16**, 1411–1426, 1976.
17. Mazur, P., Kinetics of water loss from cells at subzero temperatures and the likelihood of intracellular freezing, *J. Gen. Physiol.* **47**, 347–369, 1963.
18. Silvares, O. M., Cravalho, E. G., Toscano, W. M., and Huggins, C. E., The thermodynamics of water transport for biological cells during freezing, *Trans. ASME, J. Heat Transfer* **98**, 582–588, 1975.
19. Diller, K. R., A simple computer model for cell freezing, *Proc. 1977 Summer Computer Simulation Conf.* (Simulation Councils, La Jolla, CA, 1977), pp. 455–459.
20. Pushkar, N. S., Itkin, Y. A., Bronstein, V. L., Gordiyenko, E. A., and Kozmin, Y. V., On the problem of dehydration and intracellular crystallization during freezing of cell suspensions, *Cryobiology* **13**, 147–152, 1976.
21. Mansoori, G. A., Kinetics of water loss from cells at subzero centigrade temperatures, *Cryobiology* **12**, 34–45, 1975.
22. Lynch, M. E., and Diller, K. R., Analysis of the kinetics of cell freezing with cryophylactic additives, Trans. ASME, 81-WA/HT-53, 1981.
23. Meryman, H. T., Review of biological freezing, in *Cryobiology*, edited by H. T. Meryman (Academic, New York, 1966), pp. 1–114.
24. Macklis, J. D., Kelterer, F. D., and Cravalho, E. G., Temperature dependence of the microwave properties of aqueous solutions of ethylene glycol between +15°C and −70°C, *Cryobiology* **16**, 272–286, 1979.

25. Kedem, O., and Katchalsky, A., Thermodynamic analysis of the permeability of biological membranes to nonelectrolytes, *Biochim. Biophys. Acta* **27**, 229–246, 1958.

26. Lucke', B. and McCutcheon, M., The living cell as an osmotic system and its permeability to water, *Physiol. Rev.* **12**, 68–139, 1932.

27. Savitz, D., Sidel, V. W., and Solomon, A. K., Osmotic properties of human red cells, *J. Gen. Physiol.* **48**, 79–94, 1964.

28. Terwilliger, T. C., and Solomon, A. K., Osmotic water permeability of human red cells, *J. Gen. Physiol.* **77**, 549–570, 1981.

29. Levin, R. L., Water permeability of yeast cells at subzero temperatures, *J. Mem. Biol.* **46**, 91–124, 1979.

30. Dick, D. A. T., Osmotic properties of living cells, in *International Review of Cytology*, vol. 3, G. H. Bourne and J. F. Danielli, eds. (Academic, New York, 1959), pp. 388–433.

31. Nobel, P. S., The Boyle–Van't Hoff relation, *J. Theor. Biol.* **23**, 375–379, 1969.

32. Meryman, H. T., Osmotic stress as a mechanism of freezing injury, *Cryobiology* **8**, 489–400, 1971.

33. Williams, R. J., Frost dessication: an osmotic model, in *Analysis and Improvement of Plant and Cold Hardiness*, edited by C. R. Olien and M. N. Smith (CRC Press, Boca Raton, LA, 1981), pp. 89–115.

34. Castellan, G. W., *Physical Chemistry*, 2d ed. (Addison-Wesley, Reading, PA, 1971).

35. Rand, R. P., and Burton, A. C., Mechanical properties of the red cell membrane, I. Membrane stiffness and intracellular pressure, *Biophys. J.* **4**, 115–135, 1964.

36. Leibo, S. P., Water permeability and its activation energy of fertilized and unfertilized mouse ova, *J. Mem. Biol.* **53**, 179–188, 1980.

37. Weist, S. C., and Steponkus, P. L., Freeze thaw injury to isolated spinach protoplasts and its simulation at above-freezing temperatures, *Plant Physiol.* **62**, 699–705, 1978.

38. Hempling, H. G., Permeability of the Ehrlich ascites tumor cell to water, *J. Gen. Physiol.* **44**, 365–379, 1960.

39. Farrant, J. and Ashwood-Smith, M. J., Practical aspects, in *Low-Temperature Preservation in Medicine and Biology*, edited by M. J. Ashwood-Smith and J. Farrant (University Park Press, Baltimore, 1980), pp. 285–310.

40. Rowe, A. W., Lenny, L. L., and Mannoni, P., Cryopreservation of red cells and platelets, in *Low-Temperature Preservation in Medicine and Biology*, edited by M. J. Ashwood-Smith and J. Farrant (University Park Press, Baltimore, 1980), pp. 85–120.

41. Whittingham, D. G., Principles of embryo preservation, in *Low Temperature Preservation in Medicine and Biology*, edited by M. J. Ashwood-Smith and J. Farrant (University Park Press, Baltimore, 1980), pp. 65–83.

42. Cravalho, E. G., Huggins, C. E., Diller, K. R., and Watson, W. W., Blood freezing to nearly absolute zero temperature: −272.29°C, *J. Biomech. Eng.* **103**, 24–26, 1981.

43. Steponkus, P. L., Evans, R. Y., and Singh, J., Cryomicroscopy of isolated rye mesophyll cells, *Cryo-Letters* **3**, 101–114, 1982.

44. Knox, J. M., Schwartz, G. S., and Diller, K. R., Volumetric changes in cells during freezing and thawing, *J. Biomech. Eng.* **102**, 91–97, 1980.

45. Farrant, J., General observations on cell preservation, in *Low-Temperature Preservation in Medicine and Biology*, edited by M. J. Ashwood-Smith and J. Farrant (University Park Press, Baltimore, 1980), pp. 1–18.

46. Polge, C., Smith, A. U., and Parkes, A. S., Revival of spermatozoa after vitrification and dehydration at low temperatures, *Nature* **164**, 666, 1949.

47. Levin, R. L., Osmotic effects of introducing and removing cryoprotectants: perfused tissues and organs, in *1981 Advances in Bioengineering Trans. ASME*, pp. 131–134, 1981.

48. Knight, S. C., Preservation of leukocytes, in *Low Temperature Preservation in Medicine and Biology*, edited by M. J. Ashwood-Smith and J. Farrant (University Park Press, Baltimore, 1980), pp. 121–138.

49. Ashwood-Smith, M. J., Low temperature preservation of cells, tissues and organs, in *Low Temperature Preservation in Medicine and Biology*, edited by M. J. Ashwood-Smith and J. Farrant (University Park Press, Baltimore, 1980), pp. 19–44.

50. Lovelock, J. E., The denaturation of lipid–protein complexes as a cause of damage by freezing, *Proc. Roy. Soc. London Ser. B* **147**, 427–433, 1957.

51. Lovelock, J. E., Physical instability of human red blood cells, *Biochem. J.* **60**, 692–696, 1955.

52. Levitt, J., A sulfhydryl disulfide hypothesis of frost injury and resistance in plants, *J. Theor. Biol.* **3**, 355–391, 1962.

53. Heber, U. W., and Santarius, K. A., Loss of adenosine triphosphate synthesis caused by freezing and its relationship to frost hardness problems, *Plant Physiol.* **39**, 712–719, 1964.

54. Doebler, G. F., and Rinfret, A. P., The influence of protective compounds and cooling and warming conditions on hemolysis of erythrocytes by freezing and thawing, *Biochim. Biophys. Acta.* **58**, 449–458, 1962.

55. Meryman, H. T., Modified model for the mechanism of freezing injury in erythrocytes, *Nature* **218**, 333–336, 1968.

56. Mazur, P., Leibo, S. P., and Chu, E. H. Y., A two-factor hypothesis of freezing injury evidence from Chinese hamster tissue culture cells, *Exp. Cell. Res.* **71**, 345–355, 1972.

57. Morris, G. J., Liposomes as a model system for investigating freezing injury, in *Effects of Low Temperatures on Biological Membranes*, edited by G. J. Morris and A. Clarke (Academic, London, 1981), pp. 241–262.

58. Williams, R. J., and Hope, H. J., The relationship between cell injury and osmotic volume reduction, III. Freezing injury and frost resistance in winter wheat, *Cryobiology* **18**, 133–145, 1981.

59. Williams, R. J., Willemot, C., and Hope, H. J., The relationship between cell injury and osmotic volume reduction, IV. The behavior of hardy wheat membrane lipids in monolayer, *Cryobiology* **18**, 146–154, 1981.

60. McGrath, J. J., Thermodynamic modelling of membrane damage, in *Effects of Low Temperatures on Biological Membranes*, edited by G. J. Morris and A. Clarke (Academic, London, 1981), pp. 335–377.

61. Fujikawa, S., Freeze-fracture and etching studies on membrane damage on human erythrocytes caused by formation of intracellular ice, *Cryobiology* **17**, 351–362, 1980.

62. Smith, A. U., Prevention of haemolysis during freezing and thawing of red blood cells, *Lancet* **259**, 910–911, 1950.

63. Mollison, P. L., and Sloviter, H. A., Successful transfusion of previously frozen human red cells, *Lancet* **261**, 862–864, 1951.

64. Huggins, C. E., A general system for the preservation of blood by freezing, in *Long-Term Preservation of Red Blood Cells*, NASNRC Publ. 160–180, 1965.

65. Meryman, H. T., and Burton, J. L., Cryopreservation of platelets, in *The Blood Platelet in Transfusion Therapy* (Alan R. Liss, New York, 1978), pp. 153–165.

66. Polge, C., Freezing of spermatozoa, in *Low Temperature Preservation in Medicine and Biology*, edited by M. J. Ashwood-Smith and J. Farrant (University Park Press, Baltimore, 1980), pp. 45–64.

67. Smith, A. U., *In-vitro* experiments with rabbit eggs, in *Mammalian- Germ Cells* (Ciba Foundation Symposium, London, 1953), pp. 217–225.

68. Whittingham, D. C., Leibo, S. P., and Mazur, P., Survival of mouse embryos frozen to $-196°C$ and $-269°C$, *Science* **178**, 411–414, 1972.

69. Schaefer, U. W., Bone marrow stem cells, in *Low Temperature Preservation in Medicine and Biology*, edited by M. J. Ashwood-Smith and J. Farrant (University Park Press, Baltimore, 1980), pp. 139–154.

70. Capella, J. A., Kaufman, H. E., and Robbins, J. E., Preservation of viable corneal tissue, *Cryobiology* **2**, 116–121, 1965.

71. Kaufman, H. E., and Capella, J. A., Preserved corneal tissue for transplantation, *J. Cryosurg.* **1**, 125–129, 1968.

72. Perry, V. P., A review of skin preservation, *Cryobiology* **3**, 109–130, 1966.

73. May, S. R., and DeClement, F. A., Skin banking methodology: an evaluation of package format, cooling and warming rates, and storage efficiency, *Cryobiology* **17**, 33–45, 1980.

74. Mazur, P., Kemp, J. A., and Miller, R. H., Survival of fetal rat pancreases frozen -78 to $-196°$, *Proc. Nat. Acad. Sci. USA* **73**, 4105–4109, 1976.

75. Christiansen, M. N., and St. John, J. B., The nature of chilling injury and its resistance in plants, in *Analysis and Improvement of Plant Cold Hardiness*, edited by C. R. Olien and M. N. Smith (CRC Press, Boca Raton, LA, 1981), pp. 1–16.

76. Sidel, V. W., and Solomon, A. K., Entrance of water into human red cells under an osmotic pressure gradient, *J. Gen. Physiol.* **41**, 243–257, 1957.
77. Davson, H., and Danielli, J. F., *The Permeability of Natural Membranes*, 2d ed. (Cambridge University Press, Cambridge, England, 1952).
78. Papanek, T. H., The water permeabiity of the human erythrocyte in the temperature range +25°C to −10°C (Ph.D. dissertation, M.I.T., Cambridge, MA, 1978).
79. Levin, R. L., Cravalho, E. G., and Huggins, C. E., A membrane model describing the effect of temperature on the water conductivity of erythrocyte membranes at subzero temperatures, *Cryobiology* **13**, 415–429, 1976.
80. Rich, G. T., Sha'afi, R. I., Romualdez, A., and Solomon, A. K., Effect of osmolality on the hydraulic permeability coefficient of red cells, *J. Gen. Physiol.* **52**, 941–954, 1968.
81. Farmer, R. E., and Macey, R. I., Perturbation of red cell volume: Rectification of osmotic flow, *Biochim. Biophys. Acta* **196**, 53–65, 1970.
82. Levin, R. L., Cravalho, E. G., and Huggins, C. E., The concentration polarization effect in a multicomponent electrolyte solution—the human erythrocyte, *J. Theor. Biol.* **71**, 225–254, 1978.
83. Levin, R. L., Cravalho, E. G., and Huggins, C. E., The concentration polarization effect in frozen erythrocytes, *J. Biomech. Eng.* **99**, 65–73, 1977.
84. Rand, R. P., Mechanical properties of the red cell membrane, II. Visco-elastic breakdown of the membrane, *Biophys. J.* **4**, 303–316, 1964.
85. Boroske, E., Elwenspoek, M., and Helfrich, W., Osmotic shrinkage of giant egg-lecithin vesicles, *Biophys. J.* **34**, 95–109, 1981.
86. Farrant, J., and Woolgar, A. E., Human red cells under hypertonic conditions: A model system for investigating freezing damage, 1. Sodium chloride, *Cryobiology* **9**, 9–15, 1972.
87. Levin, R. L., Cravalho, E. G., and Huggins, C. E., Effect of solution nonideality on erythrocyte volume regulation, *Biochim. Biophys. Acta* **465**, 179–190, 1977.
88. O'Callaghan, M. G., An analysis of the heat and mass transport during the freezing of biomaterials (Ph.D. dissertation, M.I.T., Cambridge, MA, 1978).
89. Levin, R. L., The effect of solute polarization on the freezing and thawing of aqueous solutions, *Trans. ASME*, 80-WA/HT-21, 1980.
90. Korber, C., Schiewe, M. W., and Wollhover, C., Solute polarization during planar freezing of aqueous salt solutions, *Int. J. Heat Mass Transfer* **26**, 1241–1253, 1983.
91. Brower, W. E., Freund, M. J., Baudino, M. D., and Ringwald, C., An hypothesis for survival of spermatozoa via encapsulation during plane front freezing, *Cryobiology* **18**, 277–291, 1981.
92. Toscano, W. M., Cravalho, E. G., Silvares, O. M., and Huggins, C. E., The thermodynamics of intracellular ice nucleation in the freezing of erythrocytes, *Trans. ASME J. Heat Transfer* **97**, 326–332, 1975.
93. McGrath, J. J., Cravalho, E. G., and Huggins, C. E., An experimental comparison of intracellular ice formation and freeze–thaw survival of HeLa S-3 cells, *Cryobiology* **12**, 540–550, 1975.
94. Levin, R. L., Cravalho, E. G., and Huggins, C. E., Water transport in a cluster of closely packed erythrocytes at subzero temperatures, *Cryobiology* **14**, 549–558, 1977.
95. Levin, R. L., and Miller, T. E., An optimum method for the introduction or removal of permeable cryoprotectants: isolated cells, *Cryobiology* **18**, 32–48, 1981.
96. Rubinsky, B., and Cravalho, E. G., An analytical model for the prediction of the local concentration of cryophylactic agents in perfused organs, *Cryobiology* **16**, 362–371, 1979.
97. Diller, K. R., Quantitative low temperature optical microscopy of biological systems, *J. Microscopy* **126**, 9–28, 1982.
98. Leibo, S. P., Fundamental cryobiology of mouse ova and embryos, in *The Freezing of Mammalian Embryos*, edited by K. Elliot and J. Whelan, Ciba Foundation Symposium no. 52 (Elsevier, Amsterdam, 1977), pp. 69–92.
99. Watson, W. W., Volumetric changes in human erythrocytes during freezing at constant cooling velocities (S. M. Thesis, M.I.T., Cambridge, MA, 1974).

100. Callow, R. A., and McGrath, J. J., Mass transfer response of cell-sized semipermeable vesicles during freezing: A comparison of computer simulations and cryomicroscopic data, in *Advances in Bioengineering*, L. Thibault, ed. (ASME, New York, 1982), pp. 104–107.

101. Callow, R. A., and McGrath, J. J., Unilamellar liposomes as a model system to study freezing damage to cells, in *Proceedings of the Tenth Annual Northeast Bioengineering conference*, edited by E. W. Hansen, IEEE 82CH1747-5, pp. 269–272, 1982.

102. Diller, K. R., Cravalho, E. G., and Huggins, C. E., An experimental study of freezing in erythrocytes, *Med. Biol. Eng.* 14, 321–326, 1976.

103. Leibo, S. P., McGrath, J. J., and Cravalho, E. G., Microscopic observation of intracellular ice formation in unfertilized mouse ova as a function of cooling rate, *Cryobiology* 15, 257–271, 1978.

104. Rall, W. F., Mazur, P., and McGrath, J. J., Depression of the ice nucleation temperature of rapidly cooled mouse embryos by glycerol and dimethyl sulfoxide, *Biophys. J.* 41, 1–12, 1983.

105. Rall, W. F., Reid, D. S., Farrant, J., Innocuous freezing during warming, *Nature* 286, 511–514, 1980.

106. Steponkus, P. L., and Dowgert, M. F., Gas bubble formation during intracellular ice formation, *Cryo-Letters* 2, 42–47, 1981.

107. Schwartz, G. J., and Diller, K. R., Design and fabrication of a simple, versatile cryomicroscope stage, *Cryobiology* 19, 529–538, 1982.

108. Evans, C. D., and Diller, K. R., A microprocessor-programmable, controlled-temperature microscope stage for microvascular studies, *Microvasc. Res.* 24, 314–325, 1982.

109. Cosman, M. D., Cravalho, E. G., and Huggins, C. E., A cryomicroscope data acquisition system: design and performance, *Cryobiology*, submitted.

110. Schiewe, M. W., and Korber, C., Thermally defined cryomicroscopy and some applications on human leucocytes, *J. Microsc.* 126, 29–44, 1982.

111. McGrath, J. J., and Khompis, V., A numerical heat transfer analysis of a cryomicroscope conduction stage, *Trans. ASME*, 81-WA/HT-56, 1981.

112. Lang, W., Nomarski differential interference—contrast microscopy, Carl Zeiss reprint, S41-210.2-5-e, 1975.

113. Allen, R. D., David, G. B., and Nomarski, G., The Zeiss–Nomarski differential interference equipment for transmitted-light microscopy, *Z. Wiss. Mikrosk.* 69, 193–221, 1969.

114. Crowley, J. P., Rene, A., and Valeri, C. R., The recovery, structure, and function of human blood leukocytes after freeze-preservation, *Cryobiology* 11, 395–409, 1974.

115. Dayian, G., and Rowe, A. W., Cryopreservation of human platelets for transfusion, A glycerol-glucose, moderate rate cooling procedure, *Cryobiology* 13, 1–8, 1976.

116. Farrant, J., Walter, C. A., Lee, H., and McGann, L. E., Use of two-step cooling procedures to examine factors influencing cell survival following freezing and thawing, *Cryobiology* 14, 273–286, 1977.

Chapter 21

THERMAL ANALYSIS FOR CRYOSURGERY

George J. Trezek

1. INTRODUCTION

Although analytical methods for predicting the growth of cryogenic lesions are mostly products of the last decade, various forms of cryosurgical procedures have been in practice for over a century. Reviews of the subject of cryogenic surgery are given by Cooper,[1] and Van Leden and Cahan,[2] which chronicle the pioneering efforts in establishing this modality and the number of areas in medicine where it has been applied. A brief summary of these developments is provided to serve the needs of the unfamiliar reader and to offer a perspective for the analytical methods that follow.

2. BACKGROUND

Depending on one's viewpoint, the birth of cryosurgery could be attributed to either Openchowski or White. In 1883, the effects of using a low-temperature system for freezing portions of the cerebral cortex of dogs was reported by Openchowski.[3] A short time later, in 1899, Dr. A. Campbell White,[4] reported to be the first clinical cryosurgeon, used liquid air and a cotton-tipped applicator to treat various dermatological disorders. The procedure, referred to as the swab method, is still in use. In the beginning of the nineteenth century, Openchowski began the treatment of gynecologic diseases with cryosurgery when he circulated cold water through the vagina of a patient with a gynecologic malignancy. Developments in the application of low-temperature treatment were enhanced after the commercialization of liquid air and the cryogenic storage concepts developed by Sir James Dewar.

By 1930, the first monograph on cryosurgery had appeared.[5] During the period of the early 1940s, most of the medical research in the field virtually ceased. It was not until the 1960s, and largely due to new bioengineering and technological innovations fostered by the efforts of Dr. I. S. Cooper and the Linde Co., that a renaissance occurred in the use of cryogenics in medicine.

George J. Trezek • Department of Mechanical Engineering, University of California, Berkeley, California 947200

In addition to its use in dermatology and gynecology, applications of cryosurgery have been found in urology, orthopedics, otology, neurosurgery, opthalmology, the management of cancer, surgery of the head and neck, plastic surgery, and other specialities. Although cryosurgery appears to be a useful and effective treatment modality, there are differing viewpoints as to the degree of its applicability. As summarized by Cooper,[1] the use of extreme cold as a surgical tool is relatively safe, with little or no general body reaction to localized freezing. The ability to create a reversible lesion with a cryosurgical technique offers a unique opportunity for evaluating the effectiveness of surgery. This is done in functional tissue, such as the central or peripheral nervous system. The consequence of a lesion is assessed by initially applying a "test" lesion at a lower degree of cold, thereby reversibly blocking a nervous disorder. If the test lesion is effective, the region can then be cooled to cryogenic temperatures, thereby destroying the tissue. Disadvantages of the cryosurgical method result from the fact that it is sometimes difficult to tell if the desired tissue has been frozen and irreversibly damaged. Furthermore, in some applications, it is difficult to control and adjust the fixed probe configuration to the formation of an irregular-shaped lesion. Thus, normal as well as diseased tissue may have to be destroyed to an undesirable degree.

As the number of applications and frequency of use increased, it became apparent that analytical dosimetry methods that cryosurgeons could use to predict the rate of growth and size of cryological lesions for various situations would be of great value. The merit of the analysis lies in its ability to eliminate largely the need for basing treatment solely on clinical empiricism that is generally not transferable to diverse situations. The analytical approach and its use in cryosurgery follows.

3. BIOHEAT TRANSFER MODELS

Modeling a phase change or freezing process in conventional materials has received considerable attention in the classical heat transfer literature. Although these developed analytical techniques can be used, a modification of the governing energy equation is necessary in order to include the effects of tissue perfusion and metabolic heat generation. The relationship describing heat conduction in tissue, often called the bioheat equation, is simply Fourier's law with two additional terms and appears in Chap. 7; namely,†

$$\rho c \frac{\partial T}{\partial t} = \nabla(k \nabla T) + q_m + q_b \tag{1}$$

In Eq. (1), q_m accounts for metabolic heat generation, which is a function of the tissue oxygen consumption rate, and q_b describes the energy transfer between blood and tissue according to

$$q_b = \rho_b c_b w_b (T_b - T) \tag{2}$$

† See Nomenclature.

The expression for q_b is analogous to that of a perfect heat exchanger in that it assumes that the blood acquires the tissue temperature as it leaves a unit volume of tissue. The validity of this assumption has been recognized in various bioheat transfer applications and is discussed in detail in Chap. 7. In general, q_m and w_b, the blood perfusion rate, are temperature dependent. In order to obtain closed-form solutions, these terms are treated as constants, as shown in Chap. 12. Temperature dependence of q_m, with a typical Q_{10} between 2 and 3, can be included in a numerical solution. Similarly, as shown in Chap. 12, nonuniform perfusion can easily be treated by numerical methods.

For ease of solution, Eq. (1) can be rewritten as

$$\rho c \frac{\partial T}{\partial t} = \nabla(k\nabla T) + \rho_b c_b w_b (T_0 - T) \tag{3}$$

where

$$T_0 = T_b + \frac{q_m}{\rho_b c_b w_b} \tag{4}$$

As an example, in the normothermic human brain, T_0 is approximately 0.5°C higher than the systemic arterial blood temperature T_b. The rate of lesion growth and its ultimate size are controlled by a number of parameters. They include the probe geometry, size and surface temperature, the thermal conductivity, density, and heat capacity of both the frozen and unfrozen tissue, the tissue blood perfusion rate and metabolic heat generation rate, the blood temperature, the initial tissue temperature, and the phase change temperature.

4. MAXIMUM LESION SIZE

Due to the physical nature of the freezing problem, the steady-state solution of the bioheat equation (Eq. 3) in frozen and unfrozen regions subject to the appropriate boundary conditions will yield the steady-state temperature fields in these regions as well as the location of the ice front developed for a particular cryoprobe configuration. These analytically determined temperature fields will allow the cryosurgeon to acquire a feeling for the extent of the temperature field that will develop in the tissue as a result of applying a given size probe at a given temperature. More importantly, these results will allow the surgeon to select, beforehand, the correct probe surface temperature needed to create a preselected lesion size. For ease of use, a set of nomograms to determine lesion size have been prepared. They may be used to determine the steady-state position of the ice front as a function of probe size, probe temperature, tissue blood perfusion rate, and metabolic heat generation rate and the thermal conductivities of the frozen and unfrozen phases.

Steady-state solutions are developed for planar, cylindrical, and spherical probe geometries, which reflect the two basic types of lesions, namely, those formed by an external application of the probe and those formed by inserting the probe deep into the tissue. After appropriate nondimensionalization, the governing equations for each region take the following forms:

Frozen phase

$$\frac{d(R^n d\theta_f/dR)}{dR} = 0 \tag{5}$$

Unfrozen phase

$$\frac{1}{R^n} \frac{d(R^n d\theta/dR)}{dR} - \beta\theta = 0 \tag{6}$$

where $R = r/\mathcal{R}$, and the exponent n has a value of 0, 1, or 2 for planar, cylindrical, or spherical coordinates, respectively. Thus, for a planar probe (Cartesian coordinates), Eq. (5) is simply $d^2\theta/dX^2 = 0$, that is, R is replaced by X. The appropriate nondimensional groups are listed in the Nomenclature. The boundary conditions are: (1) At the probe surface, $R = 1$, $\theta_f = 1$; (2) at the frozen–unfrozen-tissue interface, $R = r^*$, $\theta_f = \theta_{pc} = \theta$, and $k_f \, d\theta_f/dR = k \, d\theta/dR$; (3) at great distances from the probe, $R \to \infty$, $\theta \to 0$. In the Cartesian system, r, $\mathcal{R}$, R, and r^* are replaced by x, L, X, and x^*, respectively.

When these boundary conditions are applied to the general solutions of Eqs. (5) and (6), the following sets of solutions are obtained for the three-coordinate systems:

Planar probe

$$\theta_f = \left[\frac{\theta_{pc} - 1}{x^* - 1}\right]X + \left[\frac{x^* - \theta_{pc}}{x^* - 1}\right] \tag{7}$$

$$\theta = \theta_{pc} \exp\left[-\sqrt{\beta}(X - x^*)\right] \tag{8}$$

$$x^* = 1 - \frac{\phi}{\sqrt{\beta}} \tag{9}$$

Cylindrical probe

$$\theta_f = [\theta_{pc} - 1](\ln R/\ln r^*) + 1 \tag{10}$$

$$\theta = \theta_{pc} \frac{K_0(\sqrt{\beta}R)}{K_0(\sqrt{\beta}r^*)} \tag{11}$$

The solution of the following transcendental equation yields the ice-front location:

$$\phi = \sqrt{\beta}r^* \frac{K_1(\sqrt{\beta}r^*)}{K_0(\sqrt{\beta}r^*)} \ln r^* \tag{12}$$

Spherical probe

$$\theta_f = 1 + (\theta_{pc} - 1)\left[\frac{r^*}{1 - r^*}\right]\left[\frac{1 - R}{R}\right] \tag{13}$$

$$\theta = \frac{\theta_{pc}r^*}{R} \exp\left[-\sqrt{\beta}(R - r^*)\right] \tag{14}$$

$$r^* = \frac{(\sqrt{\beta} - 1)}{2\sqrt{\beta}} + \left[\frac{(1 - \sqrt{\beta})^2}{4\beta} + \frac{(1 + \phi)}{\sqrt{\beta}}\right]^{1/2} \tag{15}$$

5. RATE OF LESION GROWTH

The development of the transient temperature field emanating from the cryoprobe yields the rate of lesion growth. Approximate analytical solutions of Eq. (3) for cylindrical and spherical probes embedded in tissue are presented. Since general transient analytical solutions do not exist, the approximate solution is an attempt to generalize the problem of lesion growth, so that an extensive numerical computer solution will not have to be generated for each individual situation of probe geometry and surface temperature. The approximate solutions are generated by neglecting the heat capacity effects of both the frozen and unfrozen phases, a technique adopted from the solution of classical phase change problems. The nondimensional governing equations and boundary conditions are

Frozen phase

$$\frac{1}{R^2}\frac{\partial[R^2(\partial\theta_f/\partial R)]}{\partial R} = \frac{k\rho_f c_f(T_0 - T_{pc})}{k_f L}\frac{\partial\theta_f}{\partial\tau} \tag{16}$$

Unfrozen phase

$$\frac{1}{R^2}\frac{\partial[R^2(\partial\theta/\partial R)]}{\partial R} - \beta\theta = \frac{\rho c(T_0 - T_{pc})}{L}\frac{\partial\theta}{\partial\tau} \tag{17}$$

Boundary conditions

$$\theta_f = 1, \quad R = 1 \tag{18}$$

$$\theta_f = \theta_{pc} = \theta, \quad R = r^* \tag{19}$$

$$\frac{k_f}{k}\frac{\partial\theta_f}{\partial R} = \frac{\partial\theta}{\partial R} - \theta_{pc}\frac{dr^*}{d\tau}, \quad R = r^* \tag{20}$$

$$\theta \to 0, \quad R \to \infty \tag{21}$$

The fact that the ice-front location changes with time is expressed through Eq. (20).

General solutions to Eqs. (16) and (17) that satisfy boundary conditions (Eqs. 18–21) are not known. However, approximate solutions can be obtained if the heat capacity effects in both the frozen and unfrozen phases are neglected; that is, if $k\rho_f c_f(T_0 - T_{pc})/k_f L$ and $\rho c(T_0 - T_{pc})/L$ are small, then Eqs. (16) and (17) take the simplified steady-state forms, given by Eqs. (5) and (6). As before, the corresponding temperature fields in the frozen and unfrozen zones are given by Eqs. (10) and (11) and by Eqs. (13) and (14) for the cylindrical and spherical probes, respectively.

The fundamental difference here stems from the fact that the temperature profiles θ_f and θ are transient in nature by virtue of the time dependence of r^*. In order to solve for r^* in terms of τ, a boundary condition (Eq. 20) is applied. For the case of cylindrical geometry, a first-order, nonlinear, ordinary differential equation relating r^* to ϕ results when the expressions for θ_f and θ (Eqs. 10 and 11) are substituted into Eq. (20) and differentiated, i.e.,

$$\frac{dr^*}{dr} + \frac{\sqrt{\beta}\,K_1(\sqrt{\beta}\,r^*)}{K_0(\sqrt{\beta}\,r^*)} - \frac{\phi}{r^* \ln r^*} = 0 \tag{22}$$

This equation can be separated and integrated to give

$$\tau = \int_1^{r^*} \frac{K_0(\sqrt{\beta}\,\mu)\mu \ln \mu}{\phi K_0(\sqrt{\beta}\,\mu) - \sqrt{\beta}\,K_1(\sqrt{\beta}\,\mu)\mu \ln \mu}\, d\mu \tag{23}$$

noting that

$$r^* = 1, \qquad \tau = 0 \tag{24}$$

and r^* may not exceed the steady-state value that is given by the solution to the transcendental equation

$$\sqrt{\beta}\,r^* \ln r^* K_1(\sqrt{\beta}\,r^*) - \phi K_0(\sqrt{\beta}\,r^*) = 0 \tag{25}$$

Since Eq. 23 cannot be integrated exactly, it must be evaluated numerically. The results of the numerical integration are presented graphically in the following section.

Although a similar situation exists for the spherical case, the resulting expression for τ, unlike the cylindrical case, can be obtained in a closed form. Thus, when expressions for θ_f and θ, Eqs. (13) and (14), are substituted into Eq. (20) and differentiated, the following differential equation is obtained, and upon solution yields an expression for τ, namely,

$$\frac{dr^*}{d\tau} + \left[\frac{1 + \sqrt{\beta}\,r^*}{r^*}\right] + \frac{\theta}{r^*(1 - r^*)} = 0 \tag{26}$$

Equation (26) is readily separated into

$$d\tau = \frac{r^*(1 - r^*)}{\sqrt{\beta} r^{*2} + (1 - \sqrt{\beta})r^* - (1 + \phi)} \, dr^* \tag{27}$$

Equation (27) may now be integrated, noting the two conditions

$$r^* = 1, \qquad \tau = 0 \tag{28}$$

$$r^* \to \frac{\sqrt{\beta} - 1}{2\sqrt{\beta}} + \left[\frac{(1 - \sqrt{\beta})^2}{4\beta} + \frac{1 + \phi}{\sqrt{\beta}} \right], \qquad \tau \to \infty \tag{29}$$

There results

$$\tau = \int_1^{r^*} \frac{\mu(1 - \mu) \, d\mu}{\sqrt{\beta} \mu^2 + (1 - \sqrt{\beta})\mu - (1 + \phi)} \tag{30}$$

where μ is a dummy variable of integration. This integral can be evaluated exactly to give

$$\tau = \frac{1}{2a^2} \ln \left[\frac{ar^{*2} + (1 - a)r^* + c}{1 + c} \right] + \left[\frac{1 - a(1 + 2c)}{2a^2 b} \right]$$

$$\times \ln \left\{ \left[\frac{a(2r^* - 1) + b + 1}{a(2r^* - 1) - b + 1} \right] \left[\frac{a - b + 1}{a + b + 1} \right] \right\} - \left(\frac{r^* - 1}{a} \right) \tag{31}$$

where

$$a = \sqrt{\beta}, \qquad b = [(1 - a)^2 - 4ac]^{1/2}, \quad \text{and} \quad c = -(1 + \phi)$$

6. STEADY-STATE RESULTS AND APPLICATIONS

A comparison of the nondimensional temperature profiles for the planar, cylindrical, and spherical systems is shown in Fig. 1 for a value of β equal to 0.1. The quantity $\beta = \rho_b c_b w_b r_0^2 / k$ results from nondimensionalizing the governing bioheat transfer equation. It is important to note that β is geometry dependent and hence cannot be construed as a property of the system. Rather, it is a relative comparison of convective effects due to blood flow and tissue conductive effects. Thus, it is analogous to the classical Biot number. The β value used in Fig. 1 is typical for cryosurgical applications.

The location of the ice–tissue interface for each of the three coordinate systems is shown in Fig. 2. This type of representation can be used as a convenient surgical nomogram. For example, assuming typical values of $T_0 = 38°C$, $T_{pc} = 0°C$, $T_s = -75°C$, $k_f/k = 3$, and $\beta \approx 0.1$ for a probe radius or half-thickness of 2 mm, the following ice-front locations are found (1) for

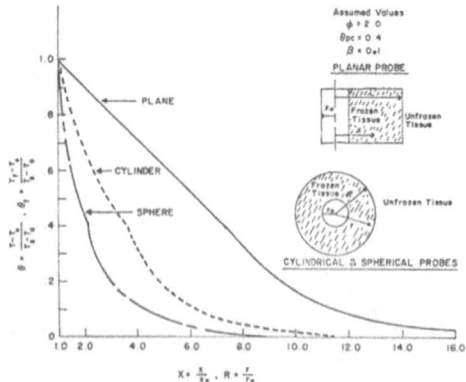

FIGURE 1

Frozen phase and tissue temperature profiles for planar, cylindrical, and spherical cryoprobes for a nondimensional probe surface temperature ϕ of 2, nondimensional phase-change temperature θ_{pc} of 0.4, and blood flow parameter β of 0.1. From Ref. 9 with permission.

the planar probe $L = 39.6$ mm, (2) for the cylindrical probe $R = 15.2$ mm, and (3) for the spherical probe $R = 7.4$ mm. The assumed thermal conductivity ratio of 3 for frozen and unfrozen tissue is consistent with existing biological thermal conductivity data.[6-8]

Additional temperature profiles for cylindrical and spherical cryoprobes as well as ice-front locations over a wider and more finely incremented range of β are given in Ref. 9.

Two examples will be presented as a means of demonstrating how the analysis can be used to estimate the setting of probe temperature and to determine the lesion size. These examples are intended primarily to demonstrate the usefulness of the analysis without being concerned with mathematical details.

Example 1: A standard Linde cryoprobe, with a radius of 2.38 mm, is to be used to form a 5-mm radius lesion in brain tissue that is perfused with blood at a systematic arterial temperature of 37°C. Determine the proper value of the probe surface temperature (T_s) to form such a lesion.

Solution: The active portion of the Linde cryoprobe is hemispherical in shape. The lesion formed in the region directly beneath the hemisphere will be essentially the same as the lesion formed in the region below a sphere of the same radius, so that the results obtained for the spherical solution may be used.

The appropriate nondimensional ice-front radius corresponding to a 5-mm radius lesion around a 2.38-mm radius spherical probe is

$$r^* = \frac{R}{r_0} = \frac{5\text{ mm}}{2.38\text{ mm}} = 2.10 \tag{32}$$

The next step is to obtain a value for β. For human brain tissue, values of $\rho_b w_b$ range from 0.0067 to 0.016 g/cm^3 sec,[10,11] and a typical value of the thermal conductivity of whole-brain tissue is 0.0013 cal/cm °C sec.[6] If we assume that the normal value of blood flow is 0.0083 g/ml sec and a linear decrease in this flow rate occurs with decreasing temperature, then the average blood flow rate in the temperature range 0°–37°C is 0.0041 g/ml sec. β is then calculated as follows:

$$\beta = \frac{\rho_b c_b w_b r_0^2}{k} = \frac{(0.0041 \text{ g/cm}^3 \text{ sec})(1 \text{ cal/g } °C)(0.238 \text{ cm})^2}{0.0013 \text{ cal/cm } °C \text{ sec}} \tag{33}$$

The value of ϕ corresponding to $r^* = 2.10$ and $\beta = 0.179$ is approximated from Fig. 2 and found to be about 2. Greater accuracy can be obtained by using a more finely graduated representation for β in Fig. 2.[9] For the case of a spherical probe, a more exact value of ϕ can be readily obtained by rearranging Eq. (15) to form

$$\phi = \sqrt{\beta}\, r^{*2} + (1 - \sqrt{\beta})r^* - 1 \tag{34}$$

Using this expression

$$\phi = \sqrt{0.179}(2.1)^2 + [1 - \sqrt{0.179}](2.1) - 1$$

$$= 2.08 \tag{35}$$

The value of T_s corresponding to ϕ of 2.08 can now be obtained from the expression

$$\phi = -\frac{k_f}{k}\left(\frac{T_{pc} - T_s}{T_{pc} - T_0}\right) \tag{36}$$

where $T_0 = T_b + q_m/\rho_b c_b w_b)$. The value of q_m is given by the brain's O_2 consumption rate and is of the order of 0.006 cal/cm^3 sec.[12] Therefore, the term $q_m/\rho_b c_b w_b$, or the contribution of metabolism, adds approximately

$$\frac{q_m}{\rho_b c_b w_b} = 0.7°C \tag{37}$$

This is added to the initial value of the systemic arterial blood temperature T_b of 37°C; thus T_0 is 37.7°C. This is consistent with the results reported by Melzack,[12] which indicate that in the steady state, the brain temperature is

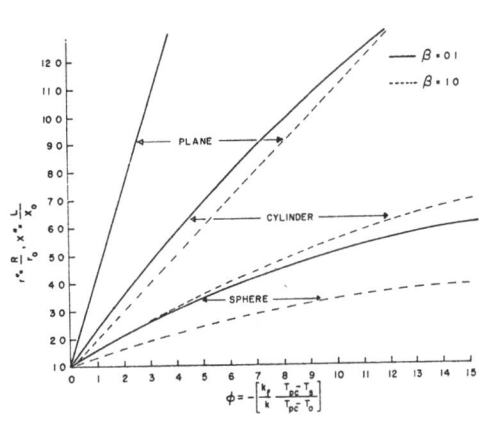

FIGURE 2
Lesion size nomograms for planar, cylindrical, and spherical cryoprobes for blood flow parameter β of 0.1 and 1.0. From Ref. 9 with permission.

on the order of 0.5°C higher than the systemic arterial blood temperature. If the value for water is used for the phase change temperature ($T_{pc} = 0°C$) and the ratio of the thermal conductivities is taken as that for ice and water ($k_f/k \approx 3$), then the preceding expression for ϕ may be rearranged to solve for a value for T_s

$$\phi = -\left(\frac{k_f}{k}\right)\left(\frac{T_{pc} - T_s}{T_{pc} - T_0}\right) \tag{38}$$

or

$$T_s = \frac{k}{k_f}\phi(T_{pc} - T_0) - T_{pc} \tag{39}$$

$$= \tfrac{1}{3}(2.08)(0 - 37.7) - 0 \tag{40}$$

$$= -26.1°C \tag{41}$$

Therefore, if the 2.38-mm radius Linde cryoprobe is set at a surface temperature of −26.1°C and steady-state conditions are allowed to develop, a lesion with radius 5 mm will be formed in the region below the probe.

Example 2: The inverse of the situation considered in Ex. 1 can also be treated; that is, what size lesion will be created if a probe is set at a fixed surface temperature and a steady-state field is allowed to develop in the tissue.? For example, suppose the 2.38-mm hemispherical probe has a surface temperature $T_s = -175°C$. What is the radius of the lesion that will be formed?

Solution: Calculate the ϕ that corresponds to $T_s = -175°C$

$$\phi = -\frac{k_f}{k}\left(\frac{T_{pc} - T_s}{T_{pc} - T_0}\right)$$

$$= (-3)\left[\frac{0 - (-175)}{0 - 37.7}\right]$$

$$= 13.92 \tag{42a}$$

Figure 2 is next used to find r^*. Recall that $\beta = 0.179$. From Fig. 2, $r^* = 5.2$. Equation (15) can be used to calculate the exact value of r^* if desired ($r^* = 5.28$). The lesion radius is then calculated as

$$R = r_0 r^* = 5.28(2.38 \text{ mm}) = 12.56 \text{ mm} \tag{42b}$$

7. EVALUATING THE RATE OF LESION GROWTH

An indication of how the lesion growth rate is affected by the probe surface temperature and blood flow can be seen by evaluating Eqs. (23) and

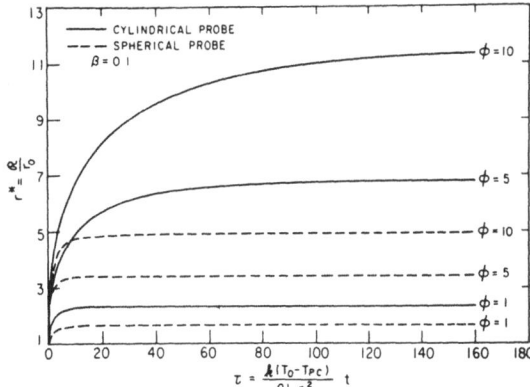

FIGURE 3
Rate of lesion growth for cylindrical and spherical cryprobes for a blood flow parameter β of 0.1. From Ref. 13 with permission.

(31), for cylindrical and spherical probe geometries respectively, for various values of ϕ and β. The results of this parameter study are shown in Figs. 3 and 4. Note that for a fixed value of the blood flow parameter, the time required to reach steady state is increased as the nondimensional probe surface temperature is increased. On the other hand, for a fixed probe surface temperature, the time required to reach steady state is decreased as the value of the blood flow parameter is increased. Additional evaluations are given in Ref. 13.

Numerical methods were also used to generate transient temperature profiles for a few selected spherical and hemispherical cryoprobe configurations. Unfortunately, when numerical methods are used to solve problems such as those involving a phase change, generality is lost, in that solutions are generated for specific cases. Further, due to the excessive amount of computer time involved in this type of calculation, only a few cases indicative of actual practice were evaluated. Specifically, transient temperature fields were generated for (1) spherical cryoprobes having a radius r_0 of 2.38 mm and 1.68 mm and surface temperatures T_s of $-25°C$, $-50°C$, $-75°C$, and $-100°C$; and (2) a vacuum-insulated stem, hemispherical tip cryoprobe having a radius r_0 of 2.38 mm (dimensions of a standard Linde cryoprobe) and a

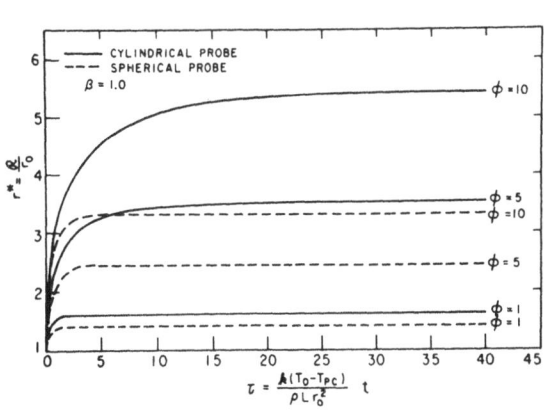

FIGURE 4
Rate of lesion growth for cylindrical and spherical cryoprobes for a blood flow parameter β of 1.0. From Ref. 13 with permission.

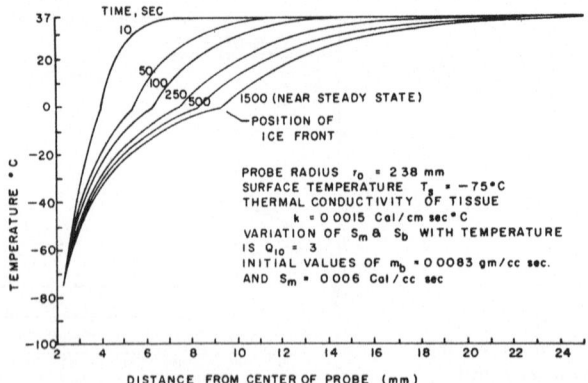

FIGURE 5

Typical transient temperature profile for a spherical probe having a surface temperature of −75°C in tissue.

surface temperature T_s of −75°C and −100°C. The details of the nodal network and computational methodology that were used can be found in reference 14. These numerical solutions, although restricted to the evaluation of special cases, are in a certain sense more general than the analytical solutions because heat capacity effects as well as the temperature dependence of blood flow and metabolic heat generation are included.

A typical example of a transient temperature field for a spherical probe showing the details of lesion formation is shown in Fig. 5. The discontinuity in the temperature profile at 0°C, the phase change temperature, indicates the lesion size at a particular time after the initiation of cooling. The rate of lesion growth for these probe dimensions is shown in Fig. 6. This representation, similar to that shown in Figs. 3 and 4 and somewhat analogous to the steady-state nomograms, can be used to predict the length of time required to create a particular size lesion for a given size probe r_0 and tip surface temperature T_s. The following example indicates how this information can be used.

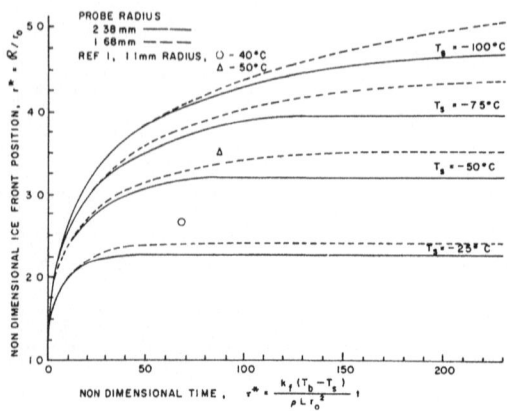

FIGURE 6

Position of ice front with time for spherical probes.

Example 3: How long must a 2.38-mm radius hemispherical Linde cryoprobe be maintained at a tip temperature of −26.1°C to form a 5-mm radius lesion in brain tissue (continuation of Ex. 1)?

Solution: Refer to Fig. 6. For practical purposes, the −25°C band may be used. Noting that the solid line represents the rate of growth of the lesion around a 2.38-mm probe, it is found that the steady-state value of τ^* is approximately 40. To solve for the real time this represents, we rearrange the expression for τ^* in the following form:

$$t = \frac{\rho L r_0^2}{k_f(T_b - T_s)} \tau^* \tag{43}$$

Treating the frozen phase as having icelike quantities, the quantity $(\rho L_f / k_f)$ = - 16,000°C sec/cm². Therefore,

$$t = \frac{(16,000°C \ sec/cm^2)(0.238 \ cm^2)(40)}{37°C + 50°C}$$

$$= 417 \ sec \tag{44}$$

Hence, if the 2.38-mm Linde cryoprobe is maintained at a temperature of −26.1°C for approximately 0.8 min in brain tissue, a lesion of radius 5 mm will be formed in the region below the cryoprobe.

The position of the ice front for the hemispherical stem probe for values of T_s of −75°C and −100°C is shown in Fig. 7. These curves are the result of solving the two-dimensional transient problem with a $Q_{10} = 3$ temperature dependency of both w_b and q_m and a temperature-dependent heat capacity in the ice phase. The solution treats the stem as being insulated; thus, the isotherms are perpendicular to the stem, so that the ice front merely travels up the stem with time. Since a two-dimensional solution represents a considerable increase in computation time, hence cost, over that required to generate a one-dimensional solution, a comparison was made with the one-dimensional

FIGURE 7

Comparison of the transient temperature profiles for hemispherical and spherical probes at surface temperatures of −75 and −100°C in tissue (variation of q_m and w_b is $Q_{10} = 3$).

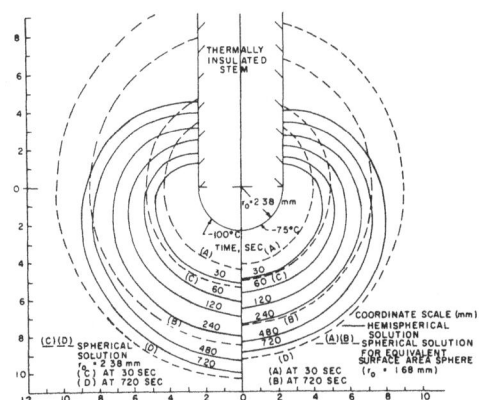

transient spherical solution to determine the feasibility of using it to predict the ice front in the hemispherical region. These results (shown as the dashed line) are in close agreement in the hemispherical region at early times (30 sec) and deviate by only 0.5 mm at 720 sec. Thus, using the simpler one-dimensional spherical solution yields acceptable results in the hemispherical region. A comparison was also made with the 1.68-mm radius probe one-dimensional solution, noting that the total surface area of the 1.68-mm probe is equivalent to the 2.38-mm radius hemispherical surface area. These results do not compare favorably. In fact, using this equivalent area concept, the ice front lags by approximately a factor of 3 in time or is about 2 mm smaller than it should be in the hemispherical region.

8. COMPARISON OF LESION GROWTH COMPUTATIONAL METHODS

In order to illustrate the applicability of various computational methods, a comparison between the approximate analytical solution for the rate of ice growth around a sphere (Eq. 31) and a numerical solution that includes heat capacity effects for values of $\beta = 0.336$ and $\phi = 7.0$ is shown in Fig. 8. Notice that the approximate solution slightly overestimates the rate of growth of the ice region. For example, at a value of nondimensional time equal to 1.0, the approximate solution gives the ice-front location as $r^* = 2.85$, while the numerical solution yields $r^* = 2.65$, a difference of approximately 7.5%. For a value of τ equal to 3.0, the difference is only about 3%.

If we examine the error from the other point of view, that is, how long it takes to reach a given percentage of the total growth, the error between the approximate and numerical solution is larger. For example, the approximate analytical solution indicates that 80% of the total growth will occur at $\tau = 1.15$, a difference of 35%. This difference is reduced to 20% if one examines the time required to reach 95% of the total growth. The approximate solution yields $\tau = 2.90$, while the numerical solution yields 3.60.

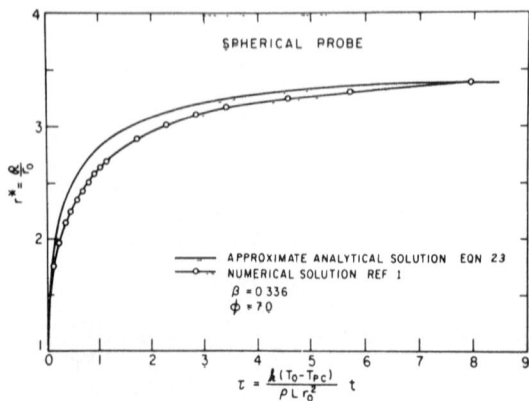

FIGURE 8

Comparison between the approximate analytical and exact numerical prediction techniques for the rate of lesion growth for a spherical cryoprobe.

It is suggested that the approximate solution be used in the following manner to calculate lesion growth rate. First, using Eq. (29) or (25), calculate the steady-state location of the ice front (total lesion size) for the particular probe. Next, calculate a certain percentage of the steady-state value. Finally, use Eq. (31) or (23) or, if possible, the plots of r^* vs. τ (Figs. 3 or 4) to calculate the amount of time necessary to reach the given percentage of the steady-state ice-front location. This procedure can best be illustrated by a numerical example.

Example 4: A Linde hemispherical cryoprobe with a radius of 1.75 mm is to be used to form a lesion in brain tissue that is perfused with blood at a systemic arterial temperature of 37°C. If the probe is maintained at a constant surface temperature of −125°C, how large will the final lesion be, and how long will it take to reach 95% of the final value?

Solution: The active portion of the Linde cryoprobe is hemispherical in shape. However, the lesion formed in the region directly beneath the hemisphere will be essentially the same as the lesion formed below a sphere of the same radius. As such, the solution for the sphere may be used.

The first step in the procedure is to calculate a steady-state value for r^*. To do this, values of β and ϕ are needed. For human brain tissue, values of w_b, the blood flowrate, range from 0.0067 to 0.0110 ml(blood)/ml(tissue) sec, and the value of the thermal conductivity of whole brain tissue is 0.0013 cal/cm °C sec. The specific heat of blood c_b is close to that of water, 1 cal/g °C. If we assume that the normal value of blood flow is 0.0083 sec^{-1} and that a linear decrease in this flow rate occurs with decreasing temperature, then the average blood flow rate on the temperature range 0°–37°C is 0.0041 sec^{-1}. β is then calculated as follows:

$$\beta = \rho_b c_b w_b r_0^2 / k$$

$$= \frac{(0.0041 \text{ g/ml sec})(1 \text{ cal/g °C})(0.175 \text{ cm})^2}{0.0013 \text{ cal/cm °C sec}}$$

$$= 0.096 \tag{45}$$

The value of ϕ corresponding to a probe surface temperature of −125°C is obtained from the expression

$$\phi = -\frac{k_f}{k}\left(\frac{T_{pc} - T_s}{T_{pc} - T_0}\right) \tag{46a}$$

where $T_0 = T_b + (q_m/\rho_b c_b w_b)$. The value of q_m is given by the brain's O_2 consumption rate and is on the order of 0.006 cal/ml sec. Therefore, the term $q_m/\rho_b c_b w_b$, or the contribution of metabolism, adds approximately

$$\frac{q_m}{\rho_b c_b w_b} = \frac{0.006 \text{ cal/ml sec}}{(0.0083 \text{ g/ml sec})(1 \text{ cal/g °C})}$$

$$= 0.7°C \tag{46b}$$

to the value of the systemic arterial blood temperature T_b of 37°C; thus, $T_0 = 37.7°C$. If the value for water is used for the phase change temperature $T_{pc} = 0°C$ and if the ratio of the thermal conductivities of the frozen and unfrozen tissue is taken as being the same as that for ice and water ($k_f/k \approx 3$), then the value of ϕ corresponding to a probe surface temperature of $-125°C$ is, following Eq. 46a,

$$\phi = (-3)\left[\frac{0°C - (-125°C)}{0°C - 37.7°C}\right]$$

$$= 10 \tag{47}$$

The steady-state location of the ice front can now be calculated from Eq. (29).

$$r^* = \frac{\sqrt{\beta} - 1}{2\sqrt{\beta}} + \left[\frac{(1 - \sqrt{\beta})^2}{4\beta} + \frac{1 + \phi}{\sqrt{\beta}}\right]^{1/2}$$

$$= \frac{\sqrt{0.096} - 1}{2\sqrt{0.096}} + \left[\frac{(1 - \sqrt{0.096})^2}{4(0.096)} + \frac{1 + 10}{\sqrt{0.096}}\right]^{1/2} \tag{48a}$$

$$= 4.94$$

This value could have also been obtained by using the steady-state lesion size nomograms that were previously developed. The actual lesion radius corresponding to $r^* = 4.94$ for a probe with a radius of 0.175 cm is

$$R = r_0 r^* = 4.94(0.175 \text{ cm}) = 0.864 \text{ cm} \tag{48b}$$

To calculate the time required to reach 95% of the steady-state value, we first calculate the nondimensional radius corresponding to 95% growth

$$r^*|_{95\%} = 4.94(0.95) = 4.69 \tag{48c}$$

Equation (31) is next used to calculate the nondimensional time required to reach 95% of steady-state conditions. For the problem being considered, the constants a, b, and c in Eq. (31) have the following values:

$$a = 0.31, \qquad b = 3.75, \qquad c = -11 \tag{49}$$

Substituting the numerical values in Eq. (31) yields

$$\tau = \frac{1}{2(0.31)^2} \ln\left[\frac{(0.31)(4.69)^2 + (0.69)(4.69) - 11}{-10}\right]$$

$$+ \left[\frac{1 + (0.31)(21)}{2(0.31)^2(3.75)}\right] \cdot \ln\left\{\left[\frac{(0.31)(8.38) + 3.75 + 1}{(0.31)(8.38) - 3.75 + 1}\right]\left(\frac{0.31 - 3.75 + 1}{0.31 + 3.75 + 1}\right)\right\}$$

$$- \left(\frac{3.69}{0.31}\right)$$

$$\tau = 8.86 \tag{50}$$

This value of τ can also be obtained using the results shown in Fig. 1 for the value of $\beta \simeq 0.1$ and $\phi = 10$. To calculate the real time corresponding to $\tau = 8.86$, we rearrange the expression for τ into the following form:

$$t = \left[\frac{\rho L r_0^2}{k(T_0 - T_{pc})} \right] \tau \qquad (51)$$

The latent heat of fusion of tissue is assumed to be the same as water, namely, $L_f = 79.7$ cal/g. Therefore,

$$t = \left[\frac{(1 \text{ g/ml})(79.7 \text{ cal/g})(0.175 \text{ cm})^2}{(0.0013 \text{ cal/cm °C sec})(37.7°C - 0°C)} \right] 8.686$$

$$= 441 \text{ sec} \simeq 7.4 \text{ min} \qquad (52)$$

Thus, if a 1.75-mm radius hemispherical cryoprobe with a surface temperature of $-125°C$ is embedded in brain tissue at an initial temperature of $37°C$, 95% of the total lesion size of 0.864 cm will be formed in about $7\frac{1}{2}$ min.

Another aspect of cryosurgery involves the surface application of cylindrical cryoprobes. If it is assumed that the cylindrical probe is pressed with sufficient force against the tissue, such that half of the cylindrical surface is in contact with the tissue, then the transient cylindrical solution can be used to estimate the ice-front location or the depth of penetration. In order to illustrate how results of the approximate analytical solution can be used to predict the cylindrical probe ice-front location, Ex. 5 is considered.

Example 5: A 1.1-cm diam cylindrical cryoprobe is to be used to create a surface lesion in tissue initially at $37°C$. If the probe surface temperature is $-125°C$, how far will the ice front have penetrated after an elapsed time of 10 min?

Solution: Using the same value of the thermal physical properties cited in Ex. 4, it is found that $\beta = 1.0$, $\phi = 10$, and $\tau = 1.2$ for $t = 10$ min. Since Fig. 4 corresponds to $\beta = 1.0$, a value of r^* corresponding to $\tau = 1.2$ and $\phi = 10$ is 3.4. This yields a penetration depth of

$$R = r^* r_0$$

$$= 3.4(0.55 \text{ cm})$$

$$R = 1.87 \text{ cm} \qquad (53)$$

It should be noted that the actual frozen region excludes the volume occupied by the probe. Thus, the thickness of the frozen band is 1.87 cm $-$ 0.55 cm or 1.32 cm.

9. CRYOSURGICAL ATLAS

Cooper and Trezek[9] found that it was possible to display the results of the analytical and numerical computations in a manner that would allow the

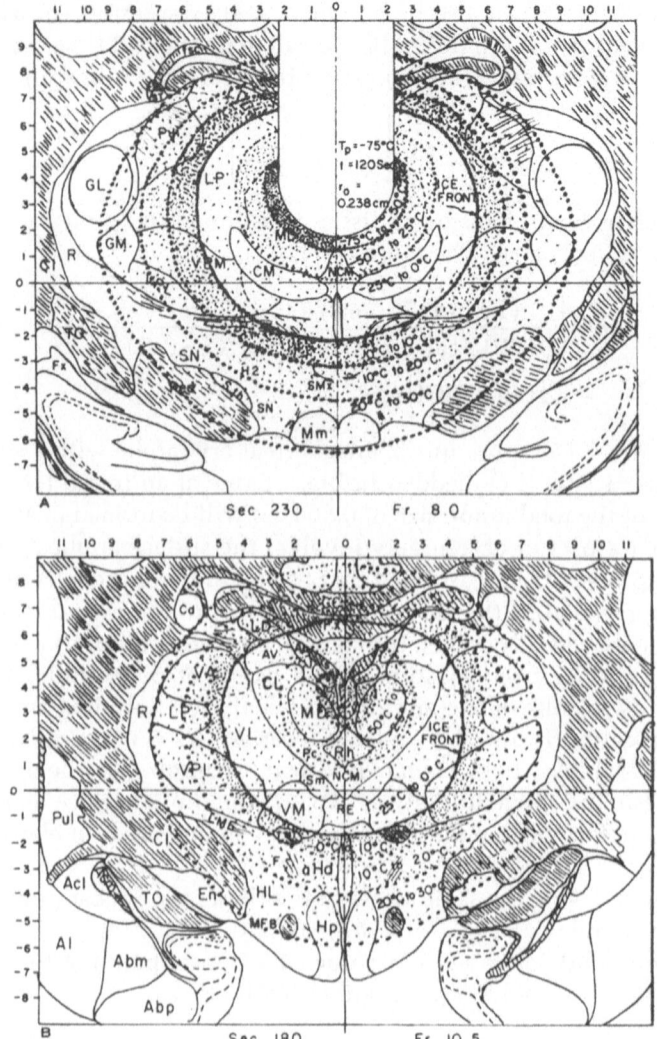

FIGURE 9
Cryosurgical atlas (base plane taken at the interaural line and the line of the inferior orbital ridge). (a) Probe located 8 mm anterior to base plane. (b) Tissue section at 10.5 mm anterior to base plane.

cryosurgeon to obtain a quantitative feel for the rate of growth and lesion size without performing prior calculations. From the suggestion of Cooper and Gionino[15] that analytical predictions could lead to a so-called "cooling atlas," Cooper and Trezek developed the following arrangement that would take a form similar to a stereotaxic atlas, that is, a cryosurgical atlas.

The hemispherical transient temperature fields are superimposed over a regular stereotaxic atlas. Various sections from a cat brain atlas[16] have been used for illustrative purposes and are shown in Figs. 9a–9f. The reference system shown in the sections is formed by (1) the interaural line that connects

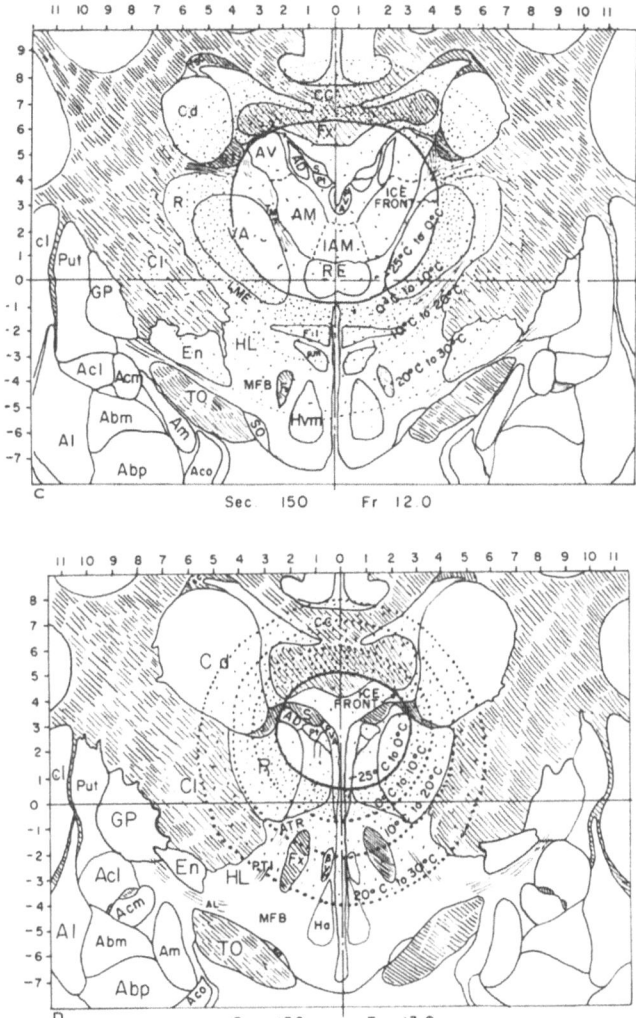

FIGURE 9 *(cont.)*
(c) Tissue section at 12.0 mm anterior to base plane. (d) Tissue section at 13.0 mm anterior to base plane. (Continued on p. 258.)

the center of each external auditory meatus; (2) the center line of the inferior orbital ridge; and (3) the third dimension, which is in the direction perpendicular to the plane containing the interaural line and the line of the inferior orbital ridge. This dimension is indicated in the various sections as Fr. The origin of the coordinate system is arbitrarily taken 1 cm above the point where the interaural line intersects the plane of the inferior orbital ridge. Figure 9a shows the 2.38-mm hemispherical probe located in a region 8 mm anterior to the plane containing the interaural line and the line of the inferior orbital ridge. The probe has a surface temperature of −75°C. The temperature of various regions of tissue (at a time of 120 sec) is depicted by the dot-shading

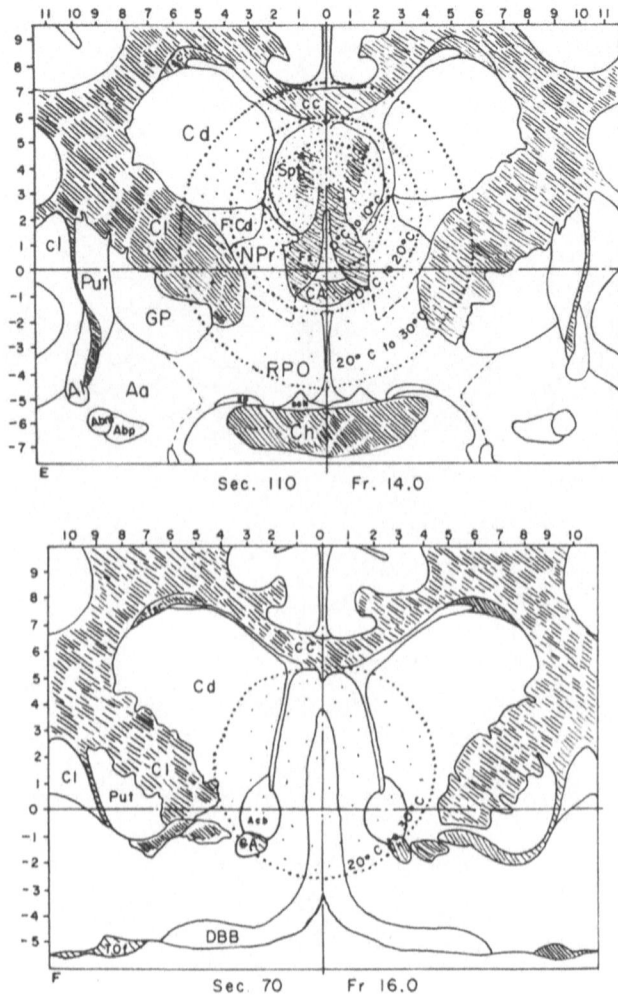

FIGURE 9 (cont.)
(e) Tissue section at 14.0 mm anterior to base plane. (f) Tissue section at 16.0 mm anterior to base plane.

technique. The intensity of the shading indicates the particular temperature range and is also labeled in the figures. In the frozen tissue, temperature bands of −75–−50°C, −50–−25°C, and −25–0°C are shown. In the unfrozen tissue, temperature bands of 0–10°C, 10–20°C, and 20–30°C are depicted. The unshaded regions are at temperatures in excess of 30°C. The third dimension in the field can be seen by looking at successive pages in the atlas. Figures 9b–9f show the temperature fields at locations 10.5, 12.0, 13.0, 14.0, and 16.0 mm anterior to the base plane, respectively, as a result of embedding the probe in the position shown in Fig. 9a. By using this technique, a surgeon can acquire a feeling for the extent of the temperature field and determine which structures will be destroyed or affected in addition to those that are intended for destruction.

ACKNOWLEDGMENT. Dr. Tom Cooper deserves special credit. He performed the vast majority of the analysis, including many conceptual contributions during his graduate studies and later while a temporary faculty member at the University of California (Berkeley) and a regular faculty member at the U.S. Naval Post Graduate School. His Ph.D. thesis[14] and publications have become classics in the field and are a tribute to his involvement and contribution.

REFERENCES

1. Cooper, I. S., Cryogenic surgery, in *Engineering in the Practice of Medicine*, Segal, B. L. and Kilpatrick, D. G., eds. (Williams and Wilkins Co., Baltimore, 1967), pp. 122–140.
2. Van Leden, H., and Cahan, W. G., eds.*Cryogenics in Surgery* (Medical Examination Publishing Co., Flushing, 1971).
3. Openchowski, S., Sur l'action localisée du froid appliqué à la surface de la region cortirale du cerveau, *C. R. Soc. Biol.* 5, 38–43, 1883.
4. White, A. C., Liquid air in medicine and surgery, *Med. Rec.* 56, 109, 1889.
5. Lortat-Jacobs, L., and Solente, G., *La cryotherapie*, (Maisson et Cie, Paris, 1930).
6. Cooper, T. E., and Trezek, G. J., A probe technique for determining the thermal conductivity of tissue, *ASME Trans., J. Heat Transfer, Ser. C* 94, 133–140, 1972.
7. Cooper, T. E., and Trezek, G. J., Correlation of thermal properties of some human tissue with water content, *Aerosp. Med.* 42, 24–27, 1971.
8. Poppendiek, H. F., Randall, R., Breeden, J. A., Chambers, J. E., and Murphy, J. R., Thermal conductivity measurements and predictions for biological fluids and tissue, *Cryobiology* 3, 318–327, 1966.
9. Cooper, T. E., and Trezek, G. J., Analytical prediction of the temperature field emanating from a cryogenic surgical cannula, *Cryobiology* 7, 79–93, 1970.
10. Ganong, W. F., *Review of Medical Physiology* (Lange Medical Publications, Los Altos, CA 1967), p. 484.
11. Rosomoff, H. L., and Holaday, D. A., Cerebral blood flow and cerebral oxygen consumption during hypothermia, *Am. J. Physiol.* 179, 85–88, 1954.
12. Melzack, R., and Casey, K. L., Localized temperature changes evoked in the brain by somatic stimulation, *Exp. Neurol.* 17, 276–292, 1967.
13. Cooper, T. E., and Trezek, G. J., Rate of lesion growth around spherical and cylindrical cryoprobes, *Cryobiology* 7, 183–190, 1971.
14. Cooper, T. E., Bioheat Transfer Studies, (Ph.D. dissertation, University of California, Berkeley, 1970).
15. Cooper, I. S., and Gionino, G., Temperature gradients during cooling and freezing in the human brain, *Cryobiology* 1, 341–344, 1965.
16. Jasper, H. H., and Ajmone-Marsan, C., *A Stereotaxic Atlas of the Diencephalon of the Cat*, (The National Research Council of Canada, Ottawa, Canada).

Chapter 22

ANALYSIS OF HEAT EXCHANGE DURING COOLING AND REWARMING IN CARDIOPULMONARY BYPASS PROCEDURES

Robert M. Curtis and George J. Trezek

1. INTRODUCTION

For over two decades, whole-body hypothermia has been widely used to reduce metabolic demand and protect vital organs during cardiopulmonary bypass (CPB) for open heart surgery. Some form of hypothermia is currently being used in over 85% of the 275,000 CPB procedures performed annually in the world today. This procedure is among the most widely used and least quantified, from a modern engineering standpoint, of all modern medical uses of hypothermia.

In early practice, hypothermia was induced by the application of ice packs to the skin surfaces[1]. This inefficient practice requires long cooling times and may cause deleterious alterations of organ perfusion[2]. Modern CPB is accomplished using a variety of disposable components (oxygenator, blood reservoir, filter, heat exchanger) connected to a permanent heart lung (HL) console that typically contains blood roller pumps and a temperature-regulated heat exchange water supply (Fig. 1).

At the heart of the CPB system lies the blood oxygenator, the primary purpose of which is to add oxygen to, and remove carbon dioxide from, venous blood drained from the vena cavae of the patient while on CPB. The majority of the CPB procedures performed today are done with disposable bubble oxygenators, in which oxygen is bubbled directly into the venous blood via a gas exchange column. The gas bubbles are removed in a defoaming section, and the arterialized blood is collected in a reservoir from which it is returned to the patient.

State-of-the-art bubble oxygenators contain integral heat exchangers, which allow temperature-regulated water from the HL console to be circulated

Robert M. Curtis • Shiley Inc., Irvine, California 92714. *George J. Trezek* • Department of Mechanical Engineering, University of California, Berkeley, California 94720.

HEAT EXCHANGE CIRCUIT

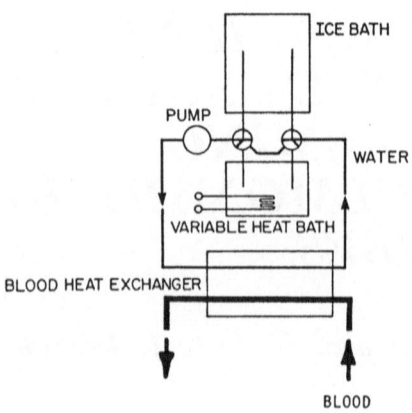

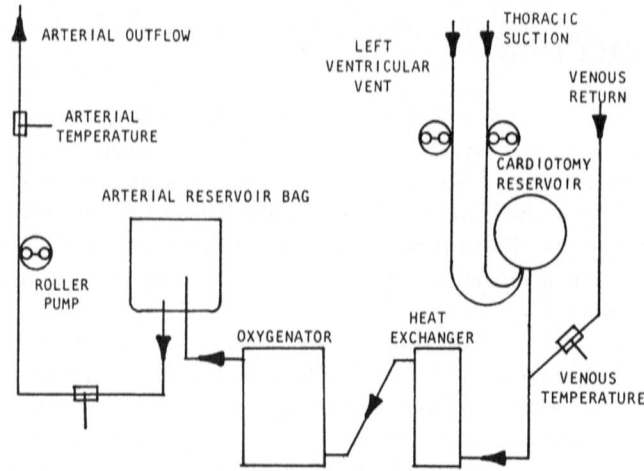

CARDIOPULMONARY BYPASS CIRCUIT

FIGURE 1

(a) Configuration of a blood-heat exchanger, providing cooling water for inducing hypothermia (ice bath), and a regulated heater assembly for rewarming. (b) Configuration of a typical cardiopulmonary bypass circuit for cardiac surgery. The blood heat exchanger in Fig. 1a interfaces with this system as indicated.

through the devices. Several models of bubble oxygenators are currently available that contain the heat exchanger within the gas exchange column.

Hypothermia is induced at the beginning of the CPB by the circulation of cold (5–10°C) water through the oxygenator–heat exchanger immediately following the initiation of bypass. Typically, rectal and/or esophageal temperatures are monitored to assess the patient's temperature throughout the procedure. Also, some models of bubble oxygenators have in-line temperature probes, which allow estimation of the temperatures of blood entering and leaving the oxygenator.

On attaining the desired rectal temperature, normally 20–30°C depending on the surgeon in charge and the surgical procedure being performed, the cooling is stopped and the patient is maintained at hypothermic conditions. Gas exchange is still maintained by the blood oxygenator, while the heart defect is corrected. Near the end of the procedure, rewarming is begun by circulating warm water through the heat exchanger. A maximum water temperature of 42°C is (nearly universally) used during rewarming to prevent overheating the blood and patient. Also, as a rule of thumb, a gradient between the patient's rectal temperature and the water inlet temperature no greater than 8–10°C is usually specified in order to reduce thermal shock during rewarming. The basis for these choices is unclear, but they are certainly widely accepted values.

On completing the heart surgery and when the patient's core temperature reaches 35–37°C, the bypass is terminated. Heart function and other indices of physiological condition are closely monitored during removal of the cannulating tubes, chest closure, and other procedures, and throughout the operation and postoperative periods. There is often a slight hypothermic rebound during the terminating phases of the surgery in which core temperature dips below normal. Normal temperature is usually regained within the first postoperative hours[3].

The total duration of CPB or "pump time" in present procedures ranges from 10–180 min, with duration a function of the type of procedure being performed, the number of complications encountered, and the technical skill of the surgeon. A typical coronary artery bypass operation will take approximately 30–60 min of pump time to complete. Of this, 5–7 min are spent cooling, and 15–20 min are spent in rewarming. Rewarming times are variable from patient to patient and from one open heart center to another because of the wide variety of blood flow rates, water flow rates, anaesthesia regimens, perfusion techniques, and degrees of hemodilution in use today. These parameters, coupled with the limited temperature gradient available to accomplish rewarming, as was previously noted, make the prediction of patient cooling or rewarming times from classical heat exchanger analyses nearly impossible without further analysis of the heat exchanger itself and the coupled effect of the exchanger and the patient.

2. HEAT EXCHANGER ANALYSIS

Several investigators have reported on the *in-vitro* performance of blood oxygenator heat exchangers[4,5]. These analyses have unfortunately been limited to "black-box" presentations of the classical heat exchanger effectiveness as a function of important operating variables. The variables are usually blood flow and heat exchanger water flow. For several state-of-the-art bubble oxygenators, which contain heat exchangers within the gas exchange column, gas flow and the degree of hemodilution have been considered as well.[6] The more detailed analysis of the heat exchange in these devices has been hampered

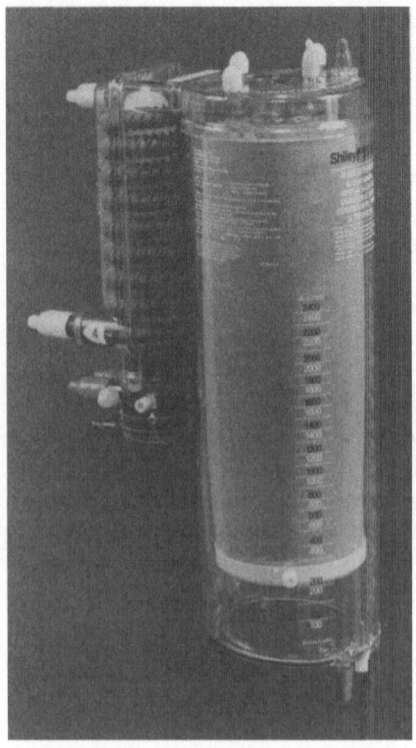

FIGURE 2
A typical blood oxygenator of the bubble type,
with attached heat exchanger.

by the complicated flow geometries and two-phase (blood–gas) flow in the heat exchange section.

In this chapter, a deeper analysis of heat exchanger performance is carried out for one particular design of oxygenator. The relative importance of the convective and conductive heat exchange resistances is analyzed. An empirical correlation between the Nusselt, Reynolds, and Prandtl numbers is developed to enable general conclusions to be drawn from this somewhat complex design.

Figure 2 shows the disposable blood oxygenator with integral heat exchanger that will be considered here, the Shiley model S-100A (Shiley Inc., Irvine, CA). The oxygenator consists of two major subassemblies, a bubble–heat exchange column and a defoaming column. Figure 3 shows the helically coiled, helically fluted heat exchanger coil and its solid center core. The heat exchanger coil is an aluminum tube that has been helically twisted to form hollow fins along the length of the coil. The twisted finned tube·is then coiled into a helix and coated by·hard anodization. Water is circulated through the tube, and a mixture of gas bubbles and blood flows over the outside of the coil in the spaces defined by the center core, heat exchange coil, and outer housing.

An *in vitro* test circuit was used in this study (Fig. 4). The circuit consists of two closed loops, one to maintain a blood inlet temperature of approximately 30°C and the other, a heated system to maintain a water inlet temperature of approximately 42°C, typical of values found clinically at the

FIGURE 3
Spiral volute blood channel for a blood heat
exchanger.

beginning of rewarming. Blood and water inlet and outlet temperatures and
heat exchanger coil surface temperatures were measured as a function of blood
flow, heat exchange water flow, gas flow, and blood hematocrit. Values listed
in Table 1 for these variables were used to span a range of conditions normally
found in adult clinical CPB procedures.

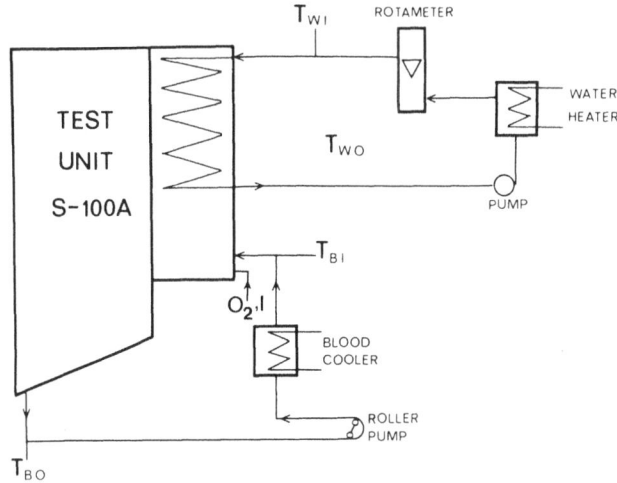

FIGURE 4
Schematic of heat exchanger test circuit. Water flow is generated by a centrifugal pump. Pulsatile
blood flow is generated by a roller pump, set for total occlusion of the pump tubing. Flow is
calibrated by timed collection. Blood inlet temperature is maintained by a secondary heat
exchanger (not shown). Thermistors for temperature measurement are calibrated to ±0.05°C.

TABLE 1
Heat Exchanger Design and Operating Conditions

D_s = Effective tubing diam.	0.015 m
D_{OH} = Housing i.d.	0.077 m
D_{IH} = Center core o.d.	0.041 m
D_{maj} = Tubing flute diam.	0.075 m
D_{min} = Tubing body diam.	0.0126 m
Blood flow rate (W_b)	2, 3, 4, 5, 6 l/min
Water flow rate (W_w)	11.6 and 14.9 l/min
Gas:blood flow ratio ($\dot{V}_g/W_b$)	0 and 1.0

Tests were performed using both bovine blood (hematocrit adjusted to 31%) and water as the blood-side fluid. To determine the effects of gas flow, tests were repeated with zero oxygen flow and an oxygen: blood flow ratio of 1:1. The zero oxygen flow situation, while not feasible in a clinical bypass, was used to establish an absolute lower limit for the effect of gas flow on the heat exchange process. Other details of the experimental procedure can be found in Ref. 6.

The steps involved in clinical heat exchanger analysis will be described in some detail in order to illustrate for the student and life scientist the logic of this type of analysis. Overall heat exchanger performance may be evaluated in either terms of the actual rate of heat transferred q or more conveniently in terms of the heat exchanger effectiveness ε. The actual rate at which heat is delivered to or from the blood passing through the exchanger at each data point can be calculated from the expression†

$$q = \rho_b c_b W_b (T_{bo} - T_{bi}) \tag{1}$$

Heat exchanger effectiveness ε is defined as the ratio of actual heat transferred to the maximum possible heat transferred. The effectiveness is a useful measure for comparing heat exchanger data, since it compensates for variations in inlet blood–water temperature gradients. Theoretically, the maximum heat transfer would be obtained if the fluid with the minimum heat capacity were to undergo a temperature change equal to the maximum temperature difference in the exchanger, i.e., the temperature difference between the entering hot and cold fluids.

$$q_{max} = (\rho_b c_b W_b)_{min} (T_{wi} - T_{bi}) \tag{2}$$

Since the blood flow is always smaller than the heat exchanger water flow, blood is usually the fluid with the minimum heat capacity in blood–water heat exchangers. The effectiveness is given by the ratio of Eq. 1 to Eq. 2

$$\varepsilon = \frac{q_{actual}}{q_{max}} = \frac{\rho_b c_b W_b (T_{bo} - T_{bi})}{\rho_b c_b W_b (T_{wi} - T_{bi})} = \frac{T_{bo} - T_{bi}}{T_{wi} - T_{bi}} \tag{3}$$

To enable a more detailed analysis of the heat exchange process, it must be realized that overall mechanisms for transfer are both conductive and

† See Nomenclature.

convective in nature. Figure 5 shows a schematic of the temperature profile at any point within an idealized heat exchange device. Depicted is the transfer of heat from water into blood across a two-layer conductor, which approximates the experimental situation.

Under steady-state conditions, the convective heat transfer from the water to the wall, the conductive heat transfer across each of the conductors, and the convective transfer from the wall to the blood are all equal.

The convective heat transfer may be expressed according to Newton's law of cooling as

$$q = h_w A_s (T_w - T_1), \quad \text{(water-side)} \tag{4}$$

$$q = h_b A_s (T_3 - T_b), \quad \text{(blood-side)} \tag{5}$$

where h is the convective heat transfer coefficient for the blood (h_b) and water (h_w), and A_s is the surface area of the conductor. By Fourier's law of heat conduction, the heat flow through the wall is

$$q = \frac{k_1 A_s}{t_1} (T_1 - T_2), \quad \text{(conductor 1—aluminum)} \tag{6}$$

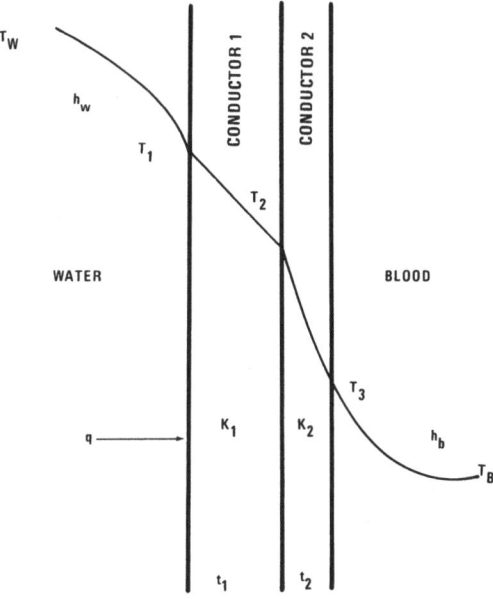

FIGURE 5
Idealized view of a two-conductor heat exchanger with layer thicknesses t_1 and t_2 with superimposed temperature profile. The thermal resistance of the system is analogous to the indicated electrical resistance network. The h and k are heat transfer coefficients and thermal conductivities, respectively. Heat flow is shown positive from left to right.

$$q = \frac{k_2 A_s}{t_2}(T_2 - T_3), \qquad \text{(conductor 2—anodized coating)} \qquad (7)$$

where k_1 and k_2 are the thermal conductivities of the conductors and t_1 and t_2 are the conductor thicknesses. At steady state, the heat flow in all sections is equal. Thus, equating Eqs. (5–8)

$$q = h_w A_s (T_w - T_1) = \frac{k_a A_s}{t_a}(T_1 - T_2) = \frac{k_c A_s}{t_c}(T_2 - T_3) = h_b A_s (T_3 - T_b) \quad (8)$$

Rearranging and canceling in Eq. 8

$$q = \frac{1}{\dfrac{1}{h_w A_s} + \dfrac{t_a}{k_a A_s} + \dfrac{t_c}{k_c A_s} + \dfrac{1}{h_b A_s}}(T_w - T_b) \qquad (9)$$

Overall heat transfer by combined convection and conduction may also be expressed in terms of an overall heat transfer coefficient U, defined by

$$q = U A_s (T_w - T_b)$$

$$U = \frac{q}{A_s(T_w - T_b)} \qquad (10)$$

Substituting Eq. 11 into the preceding expression for U,

$$U = \frac{1}{\dfrac{1}{h_w} + \dfrac{t_a}{k_a} + \dfrac{t_c}{k_c} + \dfrac{1}{h_b}} = \frac{1}{R_n} \qquad (11)$$

Thus, the overall heat transfer coefficient is equal to the reciprocal of the sum of the resistances to heat flow, R_n.

In practical application, the temperature gradient between water and blood is not constant along the flow paths, and thus heat flow varies along the length of the exchanger. Consequently, the incremental rate of heat transfer must be integrated over the surface area along the length of the heat exchanger in order to accurately determine the actual rate of heat transfer. This results in the use of the well-known log mean temperature difference (ΔT_{lm}) for the entire exchanger.[7] Using this concept and expression, overall heat transfer may be given by

$$q = U A_s \Delta T_{lm} \qquad (12)$$

Thus, knowing the blood, water, inlet, and outlet temperatures, and the heat flow calculated from the temperature rise of the blood (Eq. 11), the overall heat transfer coefficient may be calculated from Eq. 12.

To enable an estimation of the contribution of the various resistances, the overall blood convective heat transfer coefficient is defined by

$$q = \bar{h}_b A_s (\bar{T}_s - \bar{T}_b) \tag{13}$$

$$\bar{h}_b = \frac{q}{A_s(\bar{T}_s - \bar{T}_b)} \tag{14}$$

where, again, the overall heat transfer rate q is related to the temperature gradient between the wall and the bulk fluid flowing past the wall and the surface area A_s available for heat exchange. Since the heat transfer coefficient h varies from point to point along the conductor surface, the convective coefficient h_b is an average value based on the mean surface and fluid temperatures. In subsequent calculations, the mean surface temperature T_s is the mean of three surface temperatures measured at each data point. The mean blood temperature T_b is the average of the inlet and outlet blood temperatures. From the data, the mean blood-heat transfer coefficient is calculated from Eq. 14. From Eq. 11, the overall heat transfer coefficient is calculated. Using Eq. 11 and values for the conductive parameters h_w is calculated.

To enable a clearer discrimination between fluid physical parameter variations and the effects of gas flow, a classic correlation of the Nusselt (Nu), Reynolds (Re), and Prandtl (Pr) numbers was performed (cf. App. 1).

The Nusselt number is defined as

$$\bar{\text{Nu}}_b = \frac{\bar{h}_b D_s}{k} \tag{15}$$

where $\bar{h}_b$ is the average convective heat transfer coefficient, D_s is the diameter at the wall where T_s is measured, and k is the thermal conductivity of the fluid flowing past the exchanger surface. The Nusselt number is a convenient measure of the convective heat transfer coefficient, because once its value is known, the convective coefficient can be directly determined. The Reynolds number is defined as

$$\text{Re} = \frac{\rho v D_s}{\mu} \tag{16}$$

where ρ is the fluid density, v is a reference fluid velocity based on a minimum area available for flow ($v = Q/A_{\min}$), D_s is the diameter measured at the heat exchanger surface, and μ is the fluid viscosity. The Prandtl number relates the flow and thermal fields and is defined as

$$\text{Pr} = \frac{c_p \mu}{k} \tag{17}$$

where c_p, μ, and k are the specific heat, viscosity, and thermal conductivity of the fluid, respectively.

The relationship between the Nusselt number, Prandtl number, and Reynolds number has been experimentally determined to be

$$\bar{\text{Nu}} = a\text{Re}^m\text{Pr}^n \tag{18}$$

where a, m, and n are constants. The constants a, m, and n are evaluated by experimentally determining the values of the Nusselt number for different flow rates and different fluids. A log–log plot of the Nusselt number vs. the Reynolds number is then made for one fluid to estimate the dependence of heat transfer on the Reynolds number, i.e., to find an approximate value of m. This is done for one fluid, so that the influence of the Prandtl number will be relatively constant. Then $\log(\text{Nu}/\text{Re}^m)$ is plotted vs. $\log(\text{Pr})$ in order to determine the dependence on the Prandtl number. After a value for n has been determined, the Nusselt number divided by the Prandtl number raised to the nth power (Nu/Pr^n) is plotted as a function of the Reynolds number to ascertain the final values of the exponent m and the constant a.

Special consideration must be given to the situation where both liquid and gas are flowing. Heat transfer with gas flow is correlated to fluid flow using average or effective fluid properties of the two blood-side fluids and neglecting specific parameters related to the gas phase. The effective value of a property p can be expressed in terms of a weighted fraction of the properties of each constituent

$$p = \rho_t \sum_j \frac{x_j p_j}{\rho_j} \tag{19}$$

where p_j, ρ_j, and x_j are the fluid property, density, and mass fraction of the jth constituent, respectively; and ρ_t is the combined density of the j fluids and may be calculated from the expression

$$\rho_t = \frac{1}{\Sigma_j(x_j/\rho_j)} \tag{20}$$

Equation (20) may also be used to calculate values for the effective thermal conductivity, viscosity, and specific heat for blood and oxygen, and water and oxygen. These effective fluid properties can then be used to calculate the effective Nusselt and Prandtl numbers

$$\bar{\text{Nu}} = \frac{\bar{h}_b D_s}{\bar{k}} \tag{21}$$

$$\bar{\text{Pr}} = \frac{\bar{c}_p \bar{\mu}}{\bar{k}} \tag{22}$$

The Reynolds number with gas flow is calculated using a reference velocity based on the total mass flow rate of both the liquid and the gas and the effective viscosity

$$\overline{Re} = \frac{\rho_t W_t D_s}{\bar{\mu}_t A_{\min}} \tag{23}$$

Figure 6 shows the heat exchanger effectiveness (ε) as a function of the blood-side mass flow rate (ρw_b). This graph demonstrates the effects of gas, water-side and blood-side flow rates on heat transfer efficiency. Note that while the effectiveness decreases, the actual rate of heat transfer increases for increasing mass flow. Water-side flow rate does not significantly affect the heat exchanger's performance. However, blood-side liquid and gas flow rates do have a profound effect on the effectiveness. Heat exchanger effectiveness is greater with water than blood on the blood side, and from 10 to 30% greater when oxygen is bubbled in the liquid.

Figures 7a and 7b show the dependence of the blood-side ($r_b = 1/h_b$) and water-side ($r_w = 1/h_w$) heat transfer resistances on the blood-side mass flow rate. The r_b decrease with increasing flow in all cases and are much more flow dependent than are the r_w. Water flow rate has little effect on the r_b, but its effect increases with increasing r_b. Tests with gas flowing in the blood-side fluid result in an 8–21% decrease in r_b for water and a 25–30% decrease for blood as the blood-side fluid. The r_w is nearly independent of blood flow. Water-side flow rate or gas flow have little effect on the water-side resistance.

The convective resistances are at least two orders of magnitude higher than the conductive resistances, as demonstrated by comparing the aluminum and anodized coating values ($k_a/t_a = 1.7 \times 10^{-6}$, $k_c/t_c = 1.1 \times 10^{-6}$ m^{2}K/W, respectively) with the values for $1/h_b$ (Fig. 7).

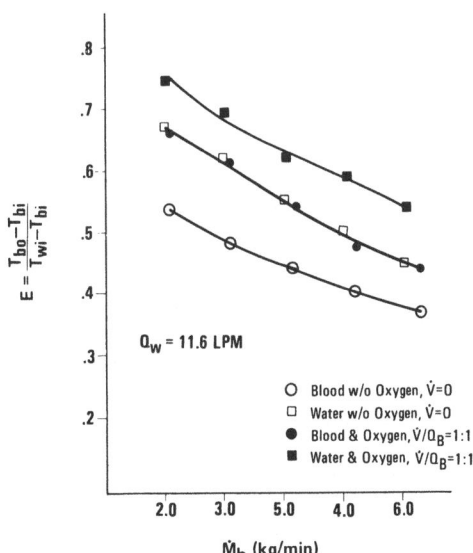

FIGURE 6

Dependence of heat transfer effectiveness on blood-side mass flow rate. Symbols represent data taken without oxygen bubbles ($\dot{V} = 0$) and with superimposed bubbles at a gas–blood flow ratio = 1.0. Water flow rate is constant.

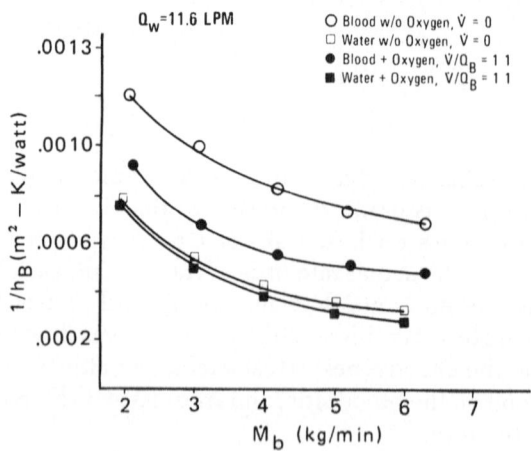

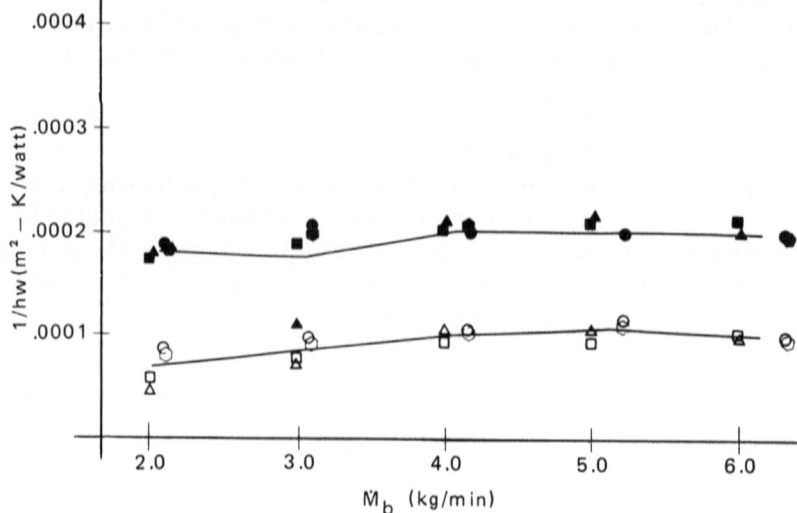

FIGURE 7

(a) Water-side convective resistance $(1/h_w)$ vs. blood-side flow rate. (b) Blood–side convective resistance $(1/h_b)$ versus blood-side flow rate. Symbols list defines test conditions for both single-phase (no oxygen bubble flow) and two-phase flow.

The results of the dimensionless analysis are shown in Fig. 8. The $\log(\mathrm{Nu}/\mathrm{Pr}^{0.38})$ is plotted as a function of the $\log \mathrm{Re}$. The Prandtl number exponential value was found to be 0.38, somewhat higher than the classical value for tube flow. Two distinct correlations exist

$$\bar{\mathrm{Nu}} = 0.217\, \bar{\mathrm{Re}}^{0.67}\, \bar{\mathrm{Pr}}^{0.38} \tag{24}$$

for both water and blood with gas flowing, and

$$\bar{\mathrm{Nu}} = 0.072\, \mathrm{Re}^{0.74}\, \mathrm{Pr}^{0.38} \tag{25}$$

for liquids without gas flow.

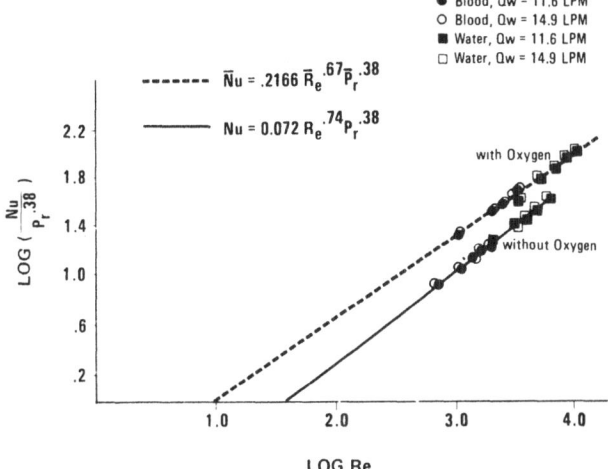

FIGURE 8

Plot of dimensionless heat transfer $(Nu/Pr^{0.38})$ vs. dimensionless blood flow (Re) for two-phase flow at gas–blood flow rate = 1.0.

These results show that the convective resistances, particularly the blood-side resistance, are at least an order of magnitude greater than the conductive resistances and dominate the heat transfer process in the heat exchanger coil. In this case, the conductive resistances may be neglected. However, this should not be generalized to other coatings, which may be thicker or have different thermal conductivities, and each coating resistance should be considered separately.

The convective resistances are most dependent on blood-side flow rates and gas flow. Increasing the blood-side mass flow and using gas decreases the convective resistances and increases heat exchange with the patient. The degree of turbulent flow and fluid mixing, as evidenced by the effect of gas flow, has a profound effect on the heat transfer process. Blood is seen to offer an overall larger resistance to heat flow than water. This would seem to indicate that the degree of hemodilution is a determining factor in the heat exchange process. Data accumulated here encompass a reasonable limiting range for the effects of hemodilution. Study with additional variables, including the measurement of fluid surface tensions, is necessary to further elucidate the dependence of heat exchange on the degree of hemodilution in heat exchangers of this type.

Of more general applicability are the two dimensionless correlations. The exponential dependence of the Nusselt number on the Prandtl number indicates that there is a nonnegligible effect of fluid thermal properties on heat exchange, perhaps due to the relatively complicated flow geometry. The clear-cut separation between the gas and no-gas data indicates that the heat transfer is indeed augmented to a greater degree than one would expect from the simple difference in mass flow due to the presence of the gas. It is believed that the scrubbing of the thermal boundary layers by the gas–blood mixture is the dominant factor in this augmentation. Further experiments with different types of spiral tube heat exchangers are necessary to determine the range

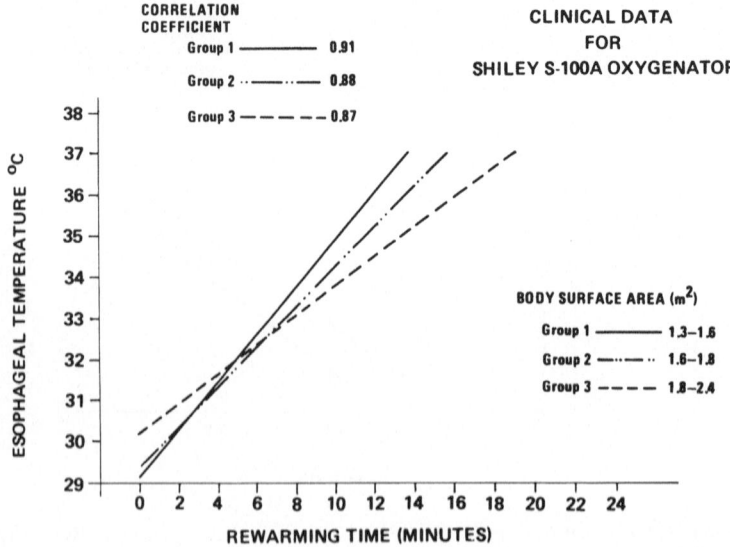

FIGURE 9

Dependence of clinical rewarming on patient surface area. Constant blood flow rate index = 2.4 l/m² min.

and applicability of this correlation. However, these results are a useful basis for further study.

Further inspection of Fig. 6 points out the asymptotic decrease of the blood-side resistance as blood flow increases. It would appear that one would have to further decrease both the blood-side and water-side resistances to heat flow in order to realize an additional increase in heat exchange performance.

To further decrease blood-side resistance, a compromise between increased blood damage as a result of greater turbulence and a decrease in clinical rewarming times would have to be considered. Also, considering that the heat exchanger performance in a particular design does not increase substantially with water-side flow rate, the most straightforward way to further increase performance would be to increase heat exchanger surface area. The effect of such an increase on the decrease in clinical rewarming time may not be determined easily from this data.

As seen in Fig. 9, the clinical rewarming times measured with this device are reasonably consistent. The dependence of the rewarming time on patient size (surface area) indicates that an increase in heat exchange performance may have beneficial effects on large patients but detrimental effects on smaller patients, with the possibility of "thermal hemolysis" arising if the device is made too efficient.

3. WHOLE-BODY HEAT EXCHANGER MODELS

To further study the dynamics of whole-body cooling and rewarming during CPB procedures, whole-body heat transfer models were constructed.

The behavior of single and multiple pool whole-body models interfaced with the blood oxygenator–heat exchanger are analyzed. An evaluation of the parameters controlling the dynamics of body cooling and rewarming was facilitated through the application of overall body-heat transfer models in conjunction with the previously described heat exchanger–oxygenator. The behavior of single- and multiple-pool models interfaced with the blood heat exchanger–oxygenator are analyzed. Emphasis is placed on illustrating the effects that various body parameters, such as metabolic rate, blood flow, body geometry (size and weight), and heat exchange surface area have on heat exchanger performance. In addition, oxygen consumption and carbon dioxide production are evaluated.

3.1. Single-Pool Model

As the name implies, the single pool considers that the entire body behaves as a lumped system or single homogeneous compartment. It is important to realize that the purpose of the model is to predict cooling and rewarming the body where the predominant mode of energy exchange occurs between the blood and water in the heat exchanger–oxygenator. Consequently, the adaptation of overall body heat transfer models to analysis of heat exchange during cardiopulmonary bypass procedures differs distinctly from most other applications. In other heat transfer considerations in this book, the major potential for energy exchange occurs between the body surface and the surroundings, such as in a space suit application, where a garment is placed on the body through which heat is exchanged with an external source or sink.

The single-pool body heat exchanger arrangement is shown schematically in Fig. 10a. The rate of change of energy in the lumped system of total mass M is given by the following energy balance:

$$\bar{c}M\frac{dT}{dt} = \dot{Q}_{\text{gen}} + \dot{Q}_{\text{env}} + \dot{Q}_{\text{blood}} \qquad (26)$$

The energy transfer terms in Eq. (26) are described as follows: the perfect heat exchange approximation is used to describe the energy transfer between the blood and tissue, namely,

$$\dot{Q}_{\text{blood}} = \rho_b c_b W_b (T_{b0} - T) \qquad (27)$$

The usual modes of energy exchange between the body and surroundings are convection, radiation, and evaporation. In this model, these three effects are represented by one overall heat transfer coefficient h, so that $\dot{Q}_{\text{env}}$ is given by

$$\dot{Q}_{\text{env}} = hS(T_{\text{amb}} - T) \qquad (28)$$

where body surface area can be obtained from body height and weight

$$S = 7.18 \times 10^{-3} H_{\text{body}}^{0.725} M_{\text{body}}^{0.425} \qquad (29)$$

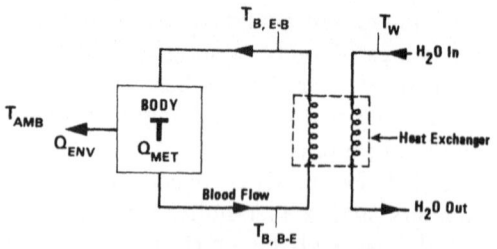

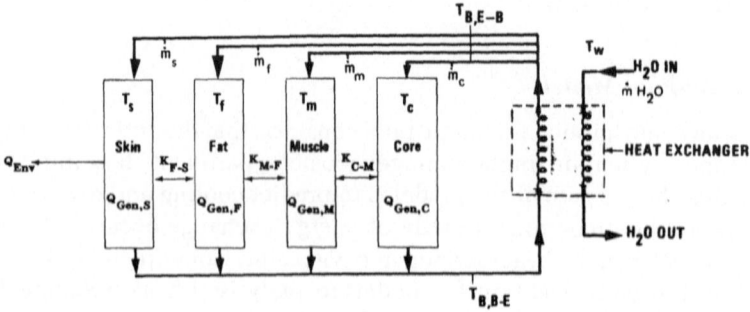

FIGURE 10
Schematics of lumped parameter patient's blood-heat exchanger models, incorporating metabolic heat exchange, regional blood flow $(\rho_i W_i = \dot{m}_i)$, and intermass heat conductivity K_{i-j}. (a) Single-pool patient, (b) multiple-pool patient.

In Eq. (27), the blood flow rate W_b is given by the product of the cardiac index (CI) and the body surface area (S), i.e.,

$$W_b = (CI)(S) \tag{30}$$

When the time rate of change of temperature (dT/dt) is approximated by

$$\frac{dT}{dt} = \frac{T' - T}{\Delta t} \tag{31}$$

Equation (26) can be rearranged to yield T', i.e.,

$$T' = \frac{\Delta t}{M}[\dot{Q}_{\text{gen}} + \rho_b c_b W_b(T_{bo} - T) + hS(T_{\text{amb}} - T)] + T \tag{32}$$

Typically, metabolic heat generation $\dot{Q}_{\text{gen}}$ is temperature dependent. In this case, the Q_{10} ranging between 2 to 3 is used as the functional representation of $\dot{Q}_{\text{gen}}$.

Heat exchanger dynamics enter the overall model through the effectiveness given by the ratio of the change in blood temperature across the heat exchanger T_b to the difference between the blood T_{bi} and water T_{wi} entering the heat exchanger, i.e.,

$$\varepsilon = \frac{T_{bi} - T_{b0}}{T_{bi} - T_{wi}} \tag{33}$$

Since values of effectiveness are available from the previous experimental data, Eq. (33) can be solved for T_{b0}, so that Eqs. (32) and (33) form the governing equations of the single-pool model.

This model was coded for digital computer evaluations over a range of the following parameters: body height and weight, cardiac index, performance factor, and heat exchanger operation. The latter parameter refers to heat exchanger operation under either various water inlet temperatures or constant temperature gradients between the inlet water and blood temperatures. Examples of the behavior of this model for a typical 70-kg, 180-cm man undergoing cooling and rewarming are shown in Figs. 11 and 12, respectively. As expected, cooling with constant water temperature exhibits eventual asymptotic behavior, while cooling under the constant temperature gradient condition is linear. Furthermore, higher cardiac index or blood flow rates yield faster responses in each case. During rewarming, a similar effect is observed for the cardiac index and heat exchanger operation either under

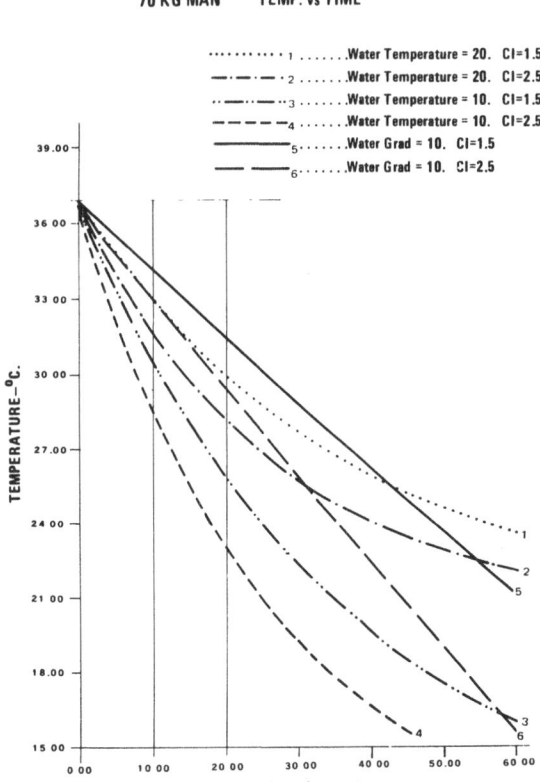

FIGURE 11

Cooling profile for single-pool patient model for various water temperatures and blood flow rates (CI = cardiac index, l/min m^2 body surface area). 70-kg patient.

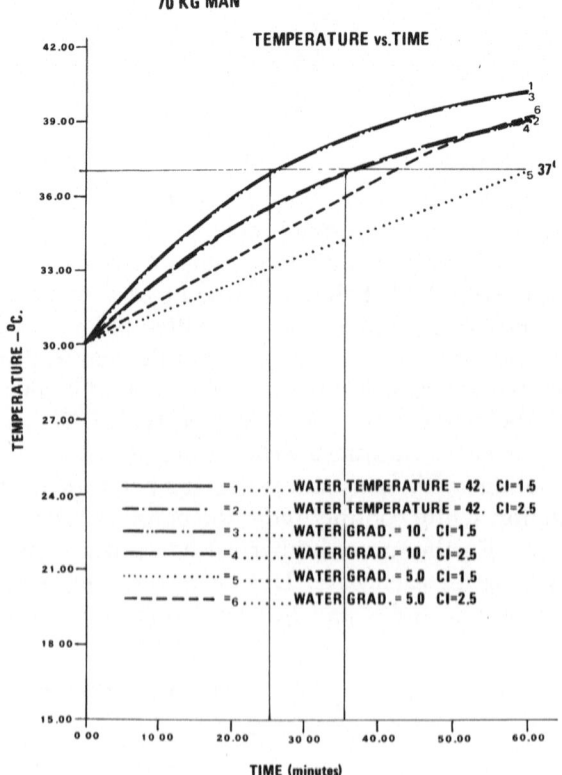

FIGURE 12
Warming profiles for the model patient and conditions in Fig. 11.

constant temperature or gradient. In the example illustrated in Figure 12, operation under a gradient of 10°C is linear for only a relatively short period, and then the rewarming response exhibits an asymptotic behavior, similar to that of constant temperature operation. Essentially, for a heat exchanger water inlet temperature of 42°C or a blood–water gradient of 10°C, the response is controlled by the cardiac index. It is interesting to note that operation under constant water temperature conditions exposes the blood to elevated temperatures more rapidly than would occur under constant gradient conditions on the order of 5°C, but only at the sacrifice of increased rewarming time. However, as long as the water temperature does not exceed 42°C, the potential of thermal hemolysis is much less than that caused by increased bypass time.

Although the single-pool model is an expedient means of qualitatively evaluating the effect of various parameters, the response times predicted by the model, particularly for rewarming, are considerably higher than those clinically observed. It is important to emphasize that energy exchange between circulating blood and tissue is the dominant mechanism for altering the body temperature. Order of magnitude changes in metabolic heat generation and environmental energy exchange have an insignificant influence on the thermal response times.

3.2. Multiple-Pool Model

In order to improve the predictive capability of the body model as well as attempt to characterize the effects of cooling and rewarming on major components of the body, a four-compartment model was developed. This model, patterned after the work at NASA, consists of four pools or compartments, i.e., core, muscle, fat, and skin. This model was developed for surface-induced heat exchange and was modified, for this application, to account for the external blood-heat exchanger. The pools and their connection with the heat exchanger are shown schematically in Fig. 10b. Flexibility is ensured by the use of a model formulation, in that heat generation and blood flow terms as well as the thermal parameters can be specified for individual compartments and for communication between compartments. The model, comprised of energy balances for each compartment and also the blood-heat exchanger, is as follows:

Core component:

$$c_c m_c \frac{dT_c}{dt} = K_{c-m}(T_m - T_c) + \dot{Q}_{\text{gen},c} + \rho_b c_b W_{b,c}(T_{b0} - T_c) \qquad (34)$$

Muscle component:

$$c_m m_m \frac{dT_m}{dt} = K_{m-f}(T_f - T_m) + K_{c-m}(T_c - T_m) + \dot{Q}_{\text{gen},m} + \rho_b c_b W_{b,m}(T_{b0} - T_m) \qquad (35)$$

Fat component:

$$c_f m_f \frac{dT_f}{dt} = K_{f-s}(T_s - T_f) + K_{m-f}(T_m - T_f) + \dot{Q}_{\text{gen},f} + \rho_b c_b W_{b,f}(T_{b0} - T_f) \quad (36)$$

Skin component:

$$c_s m_s \frac{dT_s}{dt} = K_{f-s}(T_f - T_s) + \dot{Q}_{\text{env}} + \dot{Q}_{\text{gen},s} + \rho_b c_b W_{b,s}(T_{b0} - T_s) \qquad (37)$$

Blood pool:

$$c_b m_b \frac{dT_{bi}}{dt} = \rho_b c_b W_{b,c}(T_{b0} - T_c) + \rho_b c_b W_{b,m}(T_{b0} - T_m)$$

$$+ \rho_b c_b W_{b,f}(T_{b0} - T_f) + \rho_b c_b W_{b,s}(T_{b0} - T_s) \qquad (38)$$

As in the single-pool model, the rate of change in temperature dT/dt in Eqs. (34)–(37) are approximated by $(T' - T)/\Delta t$, so that values of T_c, T_m, T_f, and T_s are computed at the new time increment. In addition, the working

TABLE 2
Heat Exchange and Multiple-Pool Model Constants

	Core	Muscle	Fat	Skin
Heat capacity (K cal/liter °C)	0.90	0.90	0.90	0.90
Blood flow ($\rho_b W_b$) (kg/min)	0.76	0.14	0.04	0.06
Heat generation ($\dot{Q}_{gen}$) (K cal/min)	0.80	0.14	0.05	0.01
Weight (M) (kg)	0.38	0.42	0.16	0.04
Conductivity constant (K cal/min °C)	165	1,200	4,200	

relationship for Eq. (38) takes the form

$$T_{bi} = (W_{b,c} T_c + W_{b,m} T_m + W_{b,f} T_f + W_{b,s} T_s)/W_b \tag{39}$$

and the heat exchanger and blood flow are coupled through the performance factor expression as in Eq. (33). Metabolic heat generation $\dot{Q}_{gen}$ can be expressed in terms of Q_{10}, namely,

$$\dot{Q}_{gen} = \dot{Q}_{gen,i} Q_{10}^{(T_c - 37)/10} \tag{40}$$

where the initial heat generation ($\dot{Q}_{gen,i}$) is given by the basal metabolic rate (BM)

$$\dot{Q}_{gen,i} = (\text{BM})(S) \tag{41}$$

and values of Q_{10} on the order of two to three are typical. Oxygen consumption $\dot{V}_{O_2}$ is proportional to $\dot{Q}_{gen}$, and carbon dioxide production $\dot{V}_{CO_2}$ is the product of $\dot{V}_{O_2}$ and the respiratory quotient. No provision is made for autoregulation of blood flow by the subject in this model.

The multiple-pool model equations are readily adaptable to digital computer evaluation. Values for the heat conductivity, constants and component masses, heat generation, blood flows, and thermal properties are summarized in Table 2.

3.3. Results of Model Studies

The model was used to evaluate the effect of various parameters, such as body weight, cardiac index, heat exchanger performance factor, and the mode of operation on the rewarming time of the core and other compartments. An example of the rewarming pattern of the various compartments is shown in Fig. 13. Of particular interest is the rewarming of the core, which for this model is approximately 16 min for a 70-kg person with a cardiac index of 2.5 being rewarmed from 28 to 37°C when the heat exchanger is operating with a constant water inlet temperature of 42°C. The effect of body weight on the

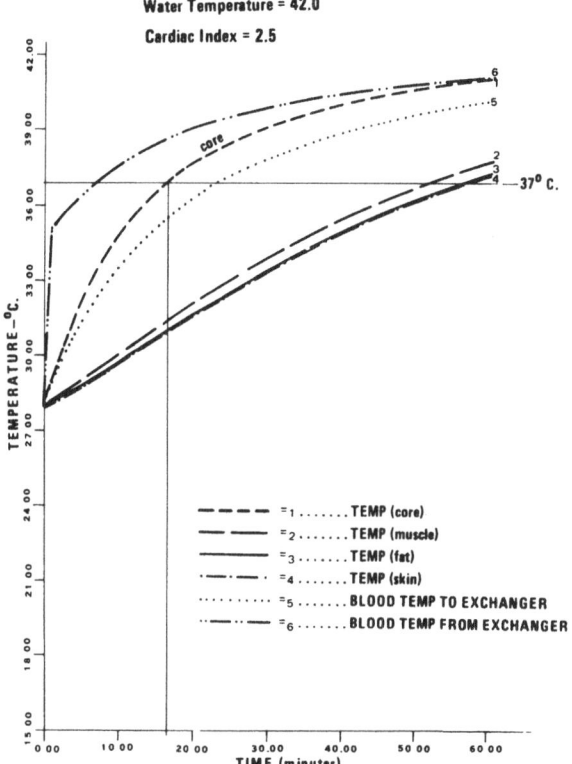

FIGURE 13
Warming profiles for the compartments of the multiple-pool patient model, 70-kg patient, with CI = 2.5 l/m^2 min and T_w = 42°C.

TABLE 3
Effect of Body Weight on Rewarming Time[a]

Body weight (kg)	Cardiac index (l/m^2 min)	Rewarming time (min)
50	1.5	15.5
50	2.5	13.5
70	1.5	19.5
70	2.5	16.0
90	1.5	25.0
90	2.5	20.0

[a] Heat exchanger performance factor was determined from Figure 6 for the appropriate blood flow rate. Rewarming time is the time required to rewarm the core from 28° to 37°C under constant water temperature conditions of 42°C.

TABLE 4
Effect of Performance Factor on Rewarming Time[a]

P.F.	Constant water temperature						Constant gradient								
	T_w (°C)	t^a (min)	T_c (°C)	T_m (°C)	T_f (°C)	T_s (°C)	T (°C)	t^a (min)	T_c (°C)	T_m (°C)	T_f (°C)	T_s (°C)	t^b (min)	T_c^c (°C)	T_c^d (°C)
0.30	42	48	37.0	33.4	32.5	32.1									
0.30	47	26	37.0	32.2	31.4	31.2									
0.30	57	14	37.0	31.4	30.8	30.7									
0.50	42	24	37.0	31.9	31.4	31.3	10	25	37.0	31.8	31.4	31.3	9	32.8	33.2
0.50							12	24	37.0	31.9	31.4	31.3	3	30.6	30.6
0.50							14	21	37.0	31.7	31.2	31.1	0[e]	29.0	29.0
0.50							15	19	37.1	31.6	31.2	31.1	0[f]	29.0	29.0
0.55	42	22	37.1	31.8	31.4	31.3	10	23	37.1	31.8	31.4	31.3	8	32.8	33.2
0.595	42	20	37.1	31.7	31.3	31.2	10	21	37.1	31.7	31.3	31.2	8	33.0	33.5
0.65	42	18	37.1	31.6	31.1	31.0	10	19	37.1	31.6	31.2	31.1	7	32.9	33.3
0.70	42	16	37.0	31.4	31.0	30.9	10	17	37.0	31.5	31.0	30.9	6	32.6	33.0
0.80	42	14	37.0	31.5	30.9	30.8									

[a] Time required for core to reach 37°C, rewarming from 29°C; CI = 1.5; body weight = 70 kg.

[b] Time for T_w to reach 42°C.

[c] Core temperature at time T_w reached 42°C under constant gradient conditions.

[d] Core temperature at the particular time under constant water temperature = 42°C conditions.

[e] T_w = 43°C.

[f] T_w = 44°C.

rewarming time is summarized in Table 3. As expected, the rewarming time increases with body weight and decreases with increased cardiac index. Increasing the body weight from 50 to 90 kg (80% increase) increases the rewarming time from 6.5 (48% increase) to 9.5 (61% increase) min. depending on the blood flow rate. Further, the effect of increased cardiac index on the reduction in rewarming time is more pronounced at higher body weight.

The effect of the heat exchanger performance factor and mode of operation on the rewarming time is summarized in Table 4. The two methods of introducing water into the exchanger, i.e., at either constant temperature or under constant gradient conditions, are compared. As expected, the rewarming time decreases with increased performance factor. However, the decrease in time is relatively small compared to the rather considerable and perhaps unachievable improvement in heat exchanger design that would be required to significantly affect the performance factor. Operation under either constant temperature or constant gradient conditions does not significantly influence the rewarming time. Various adjustments can be made to affect rewarming time. For example, at a performance factor of 0.3, constant water temperature on the order of 50°C would be required to produce rewarming times in the neighborhood of 20 min. Increasing the temperature gradient will also reduce the rewarming time. In fact, a 4-degree increase in temperature gradient reduces the rewarming time by about 4 min, which is effectively the same as increasing the performance factor from 0.5 to 0.595.

Temperatures of each compartment are also compared in Table 4 for the various operating conditions. At core temperatures of 37°C, the other components are nearly 6°C cooler. A comparison is also given in Table 4 for core

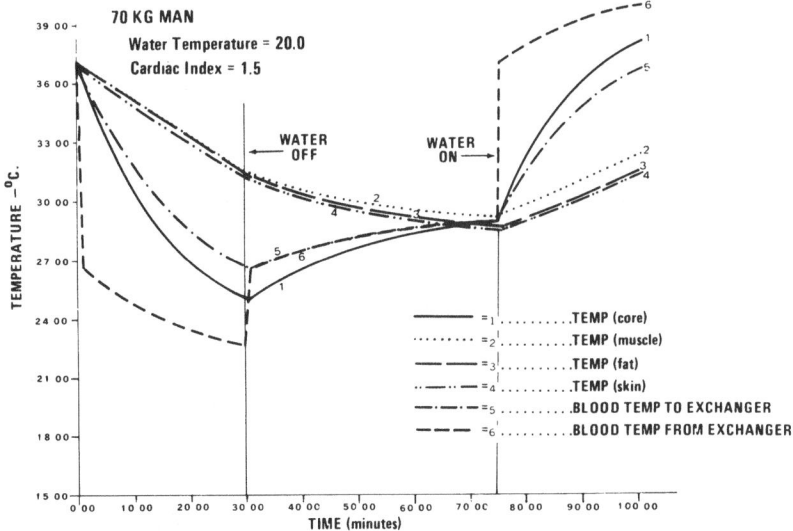

FIGURE 14

Combined cooling, hypothermia maintenance, and warming profiles for the compartments of the multiple-pool-patient model. Blood is cooled with $T_w = 20°C$ until core temperature = 25°C (approx.). Water flow is then turned off. After 45 min, water flow is turned on, with $T_w = 42°C$. Cardiac index = 1.5 l/min m².

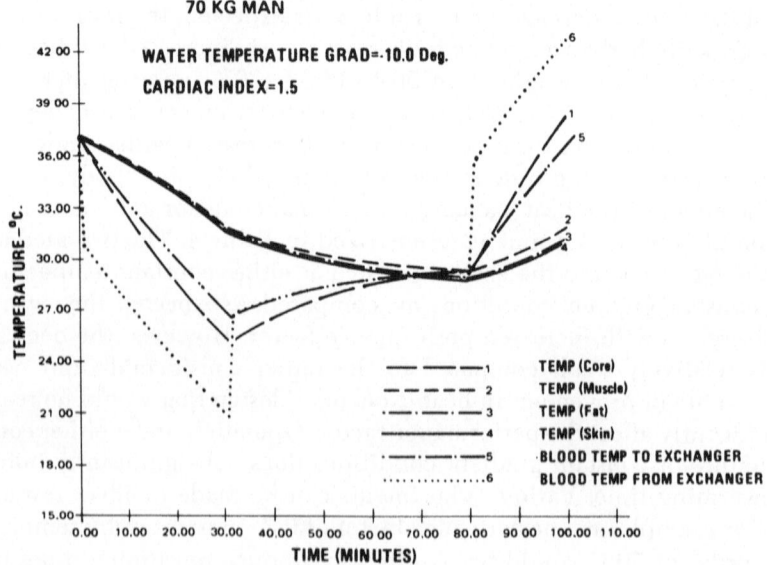

FIGURE 15

Combined cooling, hypothermia maintenance, and warming profiles for the compartments of the multiple-pool patient model. Blood is cooled, maintaining $T_w - T_b = 10°C$ until core temperature is 25°C. Water flow is then turned off for approximately 45 min. Water flow is then turned on with $T_w - T_b = 10°C$.

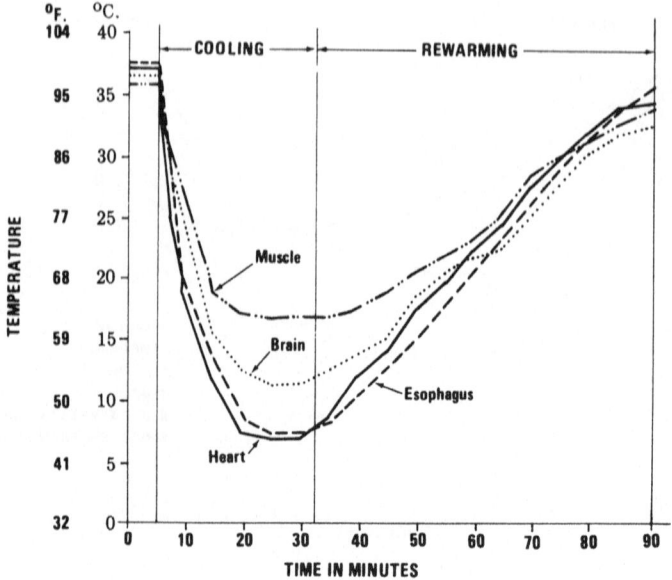

FIGURE 16

Temperature profiles in a dog during cardiopulmonary bypass with blood heat exchange. Note that the brain is considerably warmer than the core.

temperatures as the water temperature under constant temperature gradient conditions reaches the constant temperature case. For example, for a performance factor of 0.595, 8 min elapse before the water inlet conditions under a constant gradient of 10°C reach 42°C. At this point in the rewarming cycle, the core temperatures are essentially the same at 33°C. Thus, when rewarming from 29°C, either technique yields the same results as far as core temperature is considered. The analytical model was also used to simulate a cooling, hold, and rewarming cycle. Examples of these patterns are shown in Figs. 14 and 15 for a 70-kg person with a cardiac index of 1.5 for the case of constant temperature or constant gradient, respectively. In each case, as cooling proceeds, the core temperature is rapidly decreased until it reaches approximately 25°C. The cooling water temperatures used here are higher than those normally used in a clinical situation (1–5°C) but serve to better illustrate the time course of the intercompartmental temperature gradients. The other compartments lag, so that at the end of the cooling phase, a significant temperature difference exists. During the hold phase, heat is transferred from the warmer muscle, fat, and skin compartments to the core, causing its temperature to rise. At the end of the hold period, sufficient heat transfer has taken place, so that the direction of the intercompartmental thermal gradient is reversed, that is, the skin is cooler than the core. The rewarming cycle follows that previously described. These patterns are similar to those previously reported for monkeys in Chap. 12 and others shown for dogs in Fig. 16. The data indicate that large gradients develop between components during cooling,

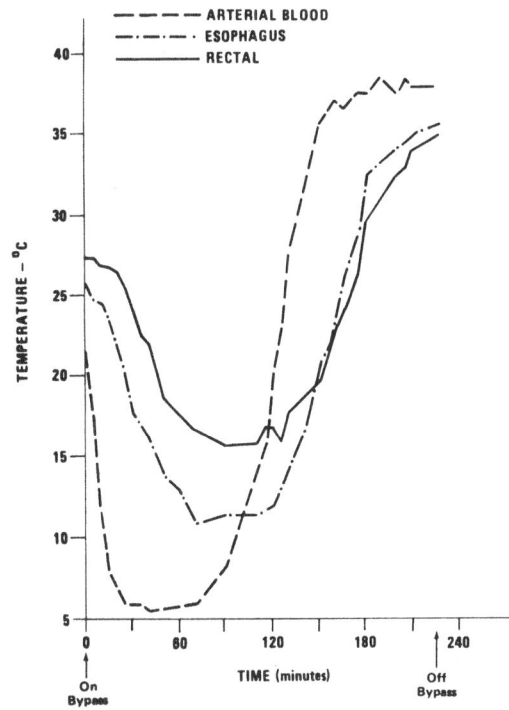

FIGURE 17

Temperature profiles in a patient on cardiopulmonary bypass with blood heat exchange, subjected to deep hypothermia.

with thermal crossovers between some sections during rewarming. This effect was also demonstrated by the model studies. The human data, taken for a deep hypothermia procedure (Fig. 17) show that temperature differences on the order of 5°C can exist between the esophagus and the rectum, which are indicative of core and muscle in the model. It is also interesting to note thermal crossover patterns between esophageal and rectal temperatures during rewarming.

4. SUMMARY AND RECOMMENDATIONS FOR FUTURE WORK

Whole-body hypothermia for metabolic protection of the surgical patient is an effective, widely used procedure that has only recently been quantified from a modern engineering standpoint. Experimental studies of heat transfer for blood and two-phase blood–oxygen mixtures in a modern blood-heat exchanger design are reported. Heat transfer effectiveness is strongly influenced by blood flow and gas bubbles which govern the blood-side's thermal resistance. The bubbles decrease the boundary layer thickness by a scrubbing action at the wall. Effectiveness is less strongly influenced by water flow rate, within a clinically relevant range, and the thermal properties of blood. Correlations of experimental data allow predictions of heat exchanger performance for this design. Lumped parameter models of the patient interfaced with the extracorporeal heat exchange provide reasonable predictions of compartmental cooling and rewarming profiles during extracorporeal hypothermia.

REFERENCES

1. Mohri, H., and Merendino, K. A., Hypothermia with or without a pump oxygenator, in *Surgery of the Chest*, 2d ed., J. Gibbon, D. Sabiston, F. Spencer, ed. (W. B. Saunders, Philadelphia, 1969).
2. Olsen, R. W. (Ph.D. dissertation, University of Texas Health Science Center at Dallas, 1983).
3. Eberhart, R. C., and Trezek, G. J., Central and peripheral rewarming patterns in postoperative cardiac patients, *Crit. Care Med.* 1, 239–251, 1973.
4. Curtis, R. M., *J. Extracorp. Tech.* 10, 179–186, 1978.
5. Riley, J. B., Winn, B. A., *In vitro* analysis of extracorporeal blood–heat exchange devices, *J. Extracorp. Tech.* 9, 134–144, 1977.
6. Nagieh, H. R., Curtis, R. M., Trezek, G. J., Heat transfer design and analysis of a bubble oxygenator, ASME preprint 82–WA/HT-81.
7. Holman, J. R., *Heat Transfer*, 4th ed. McGraw-Hill, New York, 1976.

Chapter 23

HEAT AND WATER TRANSPORT IN THE HUMAN RESPIRATORY SYSTEM

P. W. Scherer and L. M. Hanna

1. INTRODUCTION

The warming and humidification of inspired air in the human respiratory tract, often called respiratory air conditioning, is a well-controlled and remarkably stable process necessary for maintaining life. An understanding of heat and water movement from the underlying blood vessels to and from the air flowing in the respiratory passageways, the problem addressed in this chapter, requires the application of knowledge and techniques from both the biological and physical sciences. Complete understanding of this process and its relationship to other functions of the respiratory tract is still a distant goal.

The health and metabolic activities of the various types of cells lining the respiratory tract is intimately related to the heat and water transport from the airway surfaces. Serous and mucus secretion from these cells is very sensitive to changes in temperature and humidity in the ambient air, as shown by the almost immediate change in mucus flow rate with changing environment.[1-4] It has also been suggested by Litt[5] that the rheological properties of the mucus secretions are related to the net water loss or gain of the respiratory system. In the upper and lower respiratory tracts, cilia propel the mucus toward the epiglottis by either moving the serous layer in which they are embedded or by making direct contact with the more viscous mucus lying above, the point of contact being at the distal end of the cilia. Evaporation has been suggested to decrease the volume of the serous layer, resulting in a large resistance to ciliary motion by exposing more of the cilia length to the upper viscous mucus or gel layer. Conversely, an increase in the serous layer depth could result in decoupling the gel layer from cilia contact.[6-8] Critical temperatures at which cilia cease to beat have been observed and are dependent on humidity.[8,9] Humidification of the delivered gas mixture with a water droplet aerosol is common practice during anesthesia to prevent the dilaterious effects of drying on the mucociliary defense system. Knowledge of the rate of heat loss and evaporation of water from the mucosal surface

---

P. W. Scherer and L. M. Hanna • Department of Bioengineering, University of Pennsylvania, Philadelphia, Pennsylvania 19104.

under varying ambient air conditions is therefore essential to describe the mucociliary system fully and understand its dysfunction, which contributes to or causes disease. Several recent reviews are available[2,10] that discuss physiological aspects of the mucociliary defense system. An excellent overview of the efficiency of the respiratory air-conditioning process and obstacles to the solution of the transport problem is also presented by Proctor and Swift.[11,64]

We briefly review anatomical and physiological data relevant to the respiratory air-conditioning process as well as recent fluid mechanical measurements made in models and casts of various regions of the respiratory tract. Finally, we develop a new quasi-steady distributed parameter mathematical model of the coupled respiratory heat and water vapor transport process, summarize predictions of the model, and compare those predictions with some experimental measurements.

2. ANATOMY OF THE RESPIRATORY SYSTEM

The general anatomical features of the human respiratory tract are well covered in standard texts.[12,13] In this section, we comment on those features that are particularly important in the air-conditioning process. Figure 1 shows the main anatomical features of the upper and lower respiratory tract. The boundary between the two is customarily taken as the larynx. The lower tract is usually called the bronchial tree and is often idealized according to the model proposed by Weibel[14] as a symmetric network of tubes branching sequentially by dichotomy and containing 23 generations of branches. Particularly important to the air-conditioning process are (1) the physical

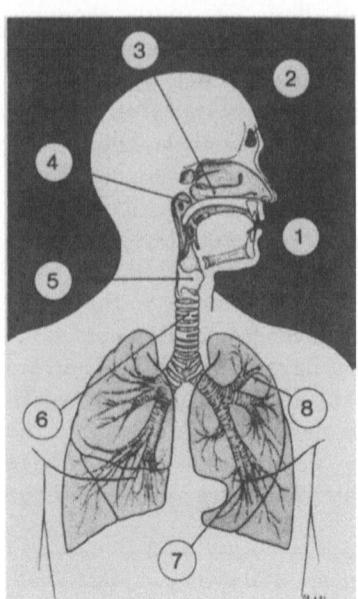

FIGURE 1

Diagram of the entire respiratory tract. The division between the upper and lower tract is usually taken as the larynx. Numbers indicate: (a) nasal vestibule, (2) frontal sinus, (3) middle meatus, (4) nasopharynx, (5) larynx, (6) trachea, (7) main bronchi, (8) peripheral bronchi. Reprinted with permission from *Respiratory Defense Mechanisms*, Marcel Dekker, vol. 5, part 1, 1977.

dimensions of the various airway passages, (2) the type and structure of the mucus membrane lining the passages at various levels, and (3) the blood supply and innervation of the airway passage walls. Figure 2 shows typical cross-sectional areas and average airstream temperature and humidity as a function of distance during normal inspiration. The channels of air flow in the nose are provided beneath the superior, middle, and inferior turbinates. The cross-sectional areas of the upper airways, especially the nasal passages, are variable and can be greatly reduced by blood filling the capillaries within the mucous membranes. The four sets of sinuses drain into these channels, providing an additional source of mucus and humidified air.

During inspiration, most of the air traversing the nasal cavity passes between the middle and the inferior turbinates and is never more than about 1 mm from a mucosal surface. This passageway is highly curved in both the anterior–posterior and medial–lateral directions. The result is that strong secondary air motions are induced by inertial forces in the curving airstream. These secondary motions, which may develop into full turbulence, cause intra-airway mixing of heat and water vapor and greatly increase the convective flux of heat and water vapor from the mucosal surface into the core of the flow. The constriction produced by the vocal chords in the larynx provides another source of strong secondary motion or turbulence in the airstream, which can persist for several generations into the bronchial tree.[15,16] The configuration and dimensions of the tracheobronchial tree are also available in the literature.[14,17] The lower respiratory tract has been generally considered, however, to be less important to the air-conditioning process than the upper tract, at least during normal, nasal breathing in a moderate environment.[18–20]

The types of mucous membrane lining the various regions of the respiratory system differ in different regions. Squamous epithelium lines the nasal

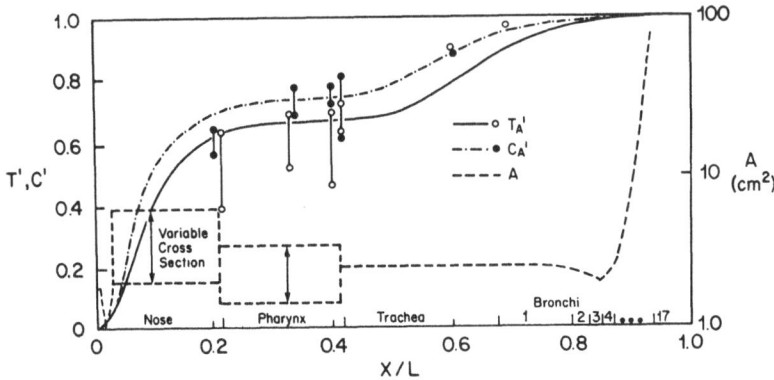

FIGURE 2

Total airway cross-sectional area *A* as a function of distance into the respiratory tract. Superimposed are the calculated inspiratory temperature and water vapor concentration for normal respiration in a moderate environment using the model developed on p. 299. Airway temperature and water vapor concentration are depicted as a fraction of the fully conditioned values and compared to experimental values that represent data taken by several investigators (see Refs. 18, 19, 32, 33 and 37).

entrance. A transition to a ciliated, columnar, mucus-secreting epithelium, which also lines the para-nasal sinuses, occurs just anterior to the nasal turbinates. Squamous epithelium again occurs between the nasopharynx and the upper trachea, with a transition back to ciliated columnar distal to the middle trachea. Cilia persist down to the level of the respiratory bronchioles[3,7] where the so-called mucociliary escalator begins.

The mucous membrane is of great importance in the air conditioning process. The mucus covering the membrane loses water through evaporation into the air flowing through the respiratory tract and gains water through condensation from the air phase. The rheological properties of respiratory mucus, the ease with which it is transported by the ciliary escalator and its ability to serve as a defense barrier to the underlying cells are all strongly connected to mucosal water content and temperature.[1,4,8,9] There is thus a direct connection between the respiratory air-conditioning process and the protective capabilities of the mucous layer.

The mucous membrane is a complicated structure, comprising several layers and containing many kinds of cells.[3,8,21] The submucosal layer contains glands with both serous and mucous cell types, which, together with goblet cells in the mucosa, contribute most of the respiratory mucus. Additional fluid passes through adjacent cell membranes[3,4,22] into the mucous layer. Control of respiratory mucous secretion is an area of active current research.[2,3] It is known that many irritants and infectious agents produce excess secretion and increased numbers of mucus-secreting glands over a period of several days.[3,8,23] Much less is known, however, about local parasympathetic nervous or chemical control of the volume and physicochemical properties of respiratory secretion in response to mucus evaporation over a period of seconds, minutes, or hours. The exact means by which mucus protects the underlying cells is also unknown. These are areas where further research is required before the interaction between respiratory air conditioning and mucous layer production can be understood and the structures thus protected and controlled.

Another important feature of respiratory tract anatomy and physiology that is not yet fully understood is the distribution and control of upper airway and bronchial blood flow. Ultimately, heat and water vapor carried away from the respiratory tract in the expired air must come from the blood. Blood flow to the upper airways is provided largely by branches of the external carotid artery. Branches of this artery supply a plexus beneath the mucous membrane of the main nasal passages that consists of erectile tissue. The blood volume of this plexus can change quickly in response to a change in ambient air conditions, among other stimuli, and in turn can change the cross-sectional area of the nasal air passages as indicated in Fig. 2. A delay in the response of the blood vessels in the nasal submucosa to sudden changes in ambient air conditions or a response different from normal has been observed in disease[24,25] and may have important implications for the system's ability to recover from insult.

Blood flow to the trachea and lower bronchial tree is supplied by branches of the inferior thyroid artery, the bronchial arteries, and distally by peripheral branches of the pulmonary arteries. A plexus of capillaries and venules exists in the layers of the mucosa, from the nasopharynx to the smallest peripheral bronchial tubes. The geometry of these plexae appears to be variable, with several layers in some regions and few in others.[13,26] There is little data available on the blood volume, blood flow rate, temperature, or the factors that control the blood flow rate in these plexae.[27–29] Recent measurements done on anesthetized sheep[30] showed total bronchial blood flow to be about 0.4% of cardiac output. Until there is a better understanding of the volume and local control of blood flow to various areas of the respiratory mucosa, it will be difficult to do accurate calculations of respiratory heat and water vapor transport. This represents another area in which progress toward quantitative understanding must await progress in experimental measurement.

3. PHYSIOLOGICAL MEASUREMENTS

Physiological observations, measurements, and speculation concerning respiratory heat and water vapor transport have been made more or less continuously since the beginning of the nineteenth century,[11,19,31] and a qualitative overview of this process has emerged.[32–34] Ambient air, which is normally below body temperature and complete water vapor saturation, is inspired air reaching body temperature and 100% water saturation around nose. The inspired air temperature and water vapor pressure are usually lower than that at the air–mucosal interface; consequently, convective transport of heat and water vapor occurs from the mucosal surface into the airstream. Because the velocity field in the flowing air is complex, the details of this convective transport process are not well understood. After passing the nasal turbinates and entering the nasopharynx, the inspired air has normally come to within about 70% of body temperature and water saturation.[18,32,35,36] Additional uptake of heat and moisture occurs beyond the nose, with the inspired air reaching body temperature and 100% water saturation around the third bronchial generation.[33,37] Evaporation of water vapor from the nasal and upper respiratory tract mucosa cools the mucous surface during inspiration. During expiration, warmer air from deeper segments of the bronchial tree and alveoli is cooled by heat exchange with the upper airway mucous membranes. The result is a reversal of convective heat and water vapor transport from the air phase to the liquid mucous phase, with a subsequent warming of the mucous layer, largely by the gain from the latent heat of vaporization on condensation of the water vapor. The overall result of this process is the net loss from the respiratory tract per day of about 350 kcal of heat (about 17% of the basal metabolic rate) and 250–400 ml of water.[32,34] The total heat loss includes both the so-called "sensible" heat loss, due to the change in inspired and expired air temperature (15% of total), and the latent or insensible heat loss (85% of total), due to the heat of vaporization carried out of the system by the added water vapor.

A large amount of experimental measurement has been done to describe the overall respiratory heat and water vapor exchange.[19,32,33,37,38] The respiratory losses depend on many factors, such as temperature and humidity of the incoming air, minute volume, and degree of exertion.[18,37,39,40] Temperature and humidity measurements taken simultaneously at the nostril entrance and in the nasopharynx during inspiration and expiration indicate the astonishing capacity of the nasal mucosa to condition air.[18–20,32] For example, ambient air inhaled at about −25°C has been found to be warmed to about 25°C when it reaches the nasopharynx,[20] a linear distance of about 9 cm! Air temperature increases of more than 70°C have been recorded between mouth and trachea during oral inspiration in dogs.[41]

Mouth breathing generally results in less warming and humidification of the inspired air in the upper airways than does nose breathing,[18,19] although measurements are highly variable, depending on the position of the tongue. The temperature of the nasal submucosa is normally found to be below body core temperature[25,33,42] and changes in response to changes in body skin temperature as well as ambient air temperature. A longitudinal gradient in tracheal and bronchial mucosal wall temperature sensitive to inspiratory conditions has also been recently measured during ventilation in anesthetized dogs.[43] The temperature of the mucosal surface was found to gradually increase with depth. Furthermore, phasic changes in tracheal submucosal temperature were not observed even though oscillatory changes in air temperature were.[42,43] Similar steady temperatures have also been observed in the nasal submucosa during oscillatory air flow in room air breathing.[42] Debate still exists over exactly what are the temperature and relative humidity of deep bronchiolar and alveolar air.[44–46] Expired alveolar air, when measured at the mouth or nostril, is usually found to be at less than body temperature (32°C for room air breathing) and at full water saturation at the expired air temperature. However, it must be remembered that this air has lost heat and water vapor to the upper airway walls on expiration.

One important physiological question is to what extent the air-conditioning process can be considered to operate in a quasi-steady-state mode. That is, are physical quantities, such as mucosal temperature and water vapor flux changing rapidly and continuously in time at all locations along the respiratory tract during inspiration and expiration? Alternatively, is a quasi steady state quickly reached, after which changes are small? Since a quasi-steady-state process is far easier to analyze mathematically than an unsteady process, it is desirable to know if this approximation is reasonable. There is some experimental evidence that suggests that nearly steady-state conditions are set up fairly rapidly during inspiration and expiration.[37,45,47] Measurements of airway air temperature[20,34,42] with thermocouples suggest that a quasi steady state exists to a good degree of approximation during most of inspiration and expiration over much of the upper and lower respiratory tract (Fig. 3). The quasi steady state appears to be set up within the first 20% of inspiration and even faster during expiration. A characteristic length into the airway beyond which air temperature and humidity do not fluctuate over the respiratory cycle has been found and called the isothermal saturation boundary (ISB).[37] The location of the ISB, at the second or third bronchial generation for normal respiration, is observed to shift with changes in ventilation rate, ambient air temperature, and humidity.

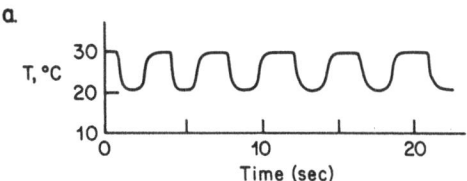

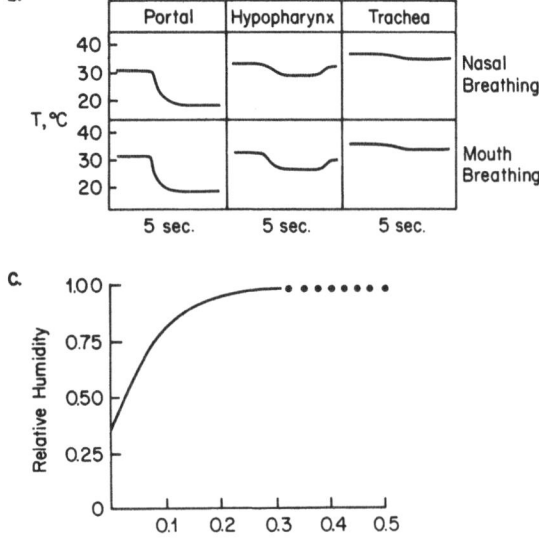

FIGURE 3

(a) Air temperature recording at the anterior nares of a subject breathing moderate room air. Time interval is 5 sec. The recording has not been corrected for the delayed time response of the thermistor. When this correction is included, temperature rise time will be even faster than shown (see Ref. 18). (b) Recordings at different positions in the air passages, again not corrected for the delayed time response of the thermistor (see Ref. 18). (c) Relative humidity of expired gas as a function of expired volume measured at the mouth during quiet breathing in a moderate environment. The data was graphed after delay and time constant corrections were made. From Refs. 18 and 45 with permission.

There is clearly a need for more local measurements of temperature and humidity along the respiratory tract during normal respiration. These measurements should be made not only in the air phase but also at the surface and within the mucosa and submucosa. The time response of the measurements must be sufficiently fast in order to accurately determine whether or not quasi steady states are reached. To date, rapid time response has been difficult to obtain in airway humidity measurement. Also, as previously noted, measurements of blood flow rates and volumes, and blood temperature within the upper respiratory tract would be valuable. At present, values for most of these quantities have to be guessed, or loosely inferred from very limited data.

4. MODELS

Because of the experimental difficulties associated with making detailed local measurements in the respiratory tract of animals and humans, several studies have been done that use physical models. The earliest systematic series of model experiments was done by Proetz,[48] who visualized air streamline patterns in steady unidirectional flow through a cast replica of the nose and upper airways. Using smoke to visualize the airflow, he observed a difference in streamline patterns between steady inspiratory and expiratory flow. From

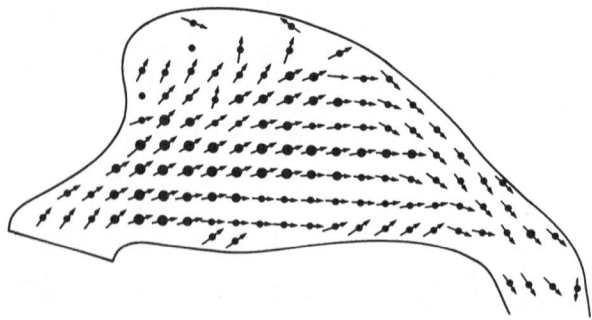

FIGURE 4
Velocity vector field for inspiratory nasal airflow as measured in a cast replica. The nostril is on
the left. The size of the dot indicates the magnitude of the velocity vector and the arrow the
direction (see Ref. 64). Reprinted with permission from *Respiratory Defense Mechanisms*, Marcel
Dekker, vol. 5, part 1, 1979.

the observed dispersion of the flowing smoke streams, Proetz concluded that
turbulence was present in the flow. Similar steady flow visualization measurements have been made more recently[64] on a clear plastic model of the nostril,
nasal passages, and nasopharynx. In addition to photographs of streamlines,
local velocities were measured (Fig. 4) using a miniature Pitot tube. A standing
eddy was observed in the olfactory area at the top of the main nasal passages.
Steady inspiratory and expiratory flow, visualized in casts of the human
trachea[15] led to the conclusion that flow is probably turbulent or highly
disturbed there during most of normal breathing. More recent measurements
have yielded the same finding.

The main limitation of these previous studies is that they have dealt with
only steady unidirectional flow. Recent flow visualization experiments by the
authors, using oscillatory flow through an accurate upper and lower airway
model (see Fig. 5) have demonstrated the important effects of asymmetry in

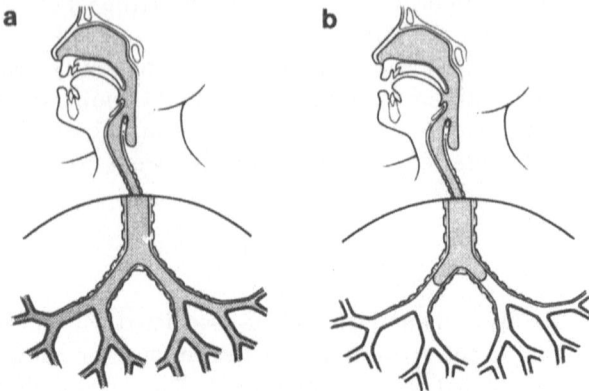

FIGURE 5
Schematic diagram of flow visualization during oscillatory flow in a respiratory tract model
showing trapping of inspired particles by convection at the end of expiration. Shown are (a) end
of first inspiration and (b) end of first expiration. Upper bronchi are magnified.

the flow chamber geometry during inspiration and expiration. A fraction (10–20%) of inspired gas is trapped within the upper airways at the end of expiration largely because of differences in forward and reverse airway velocity profiles. Typical examples for the lower airways are shown in Fig. 6. This gas, which was present at body temperature and 100% saturation in the bronchial tree at the end of inspiration, fills the upper airways at the end of expiration.

There is a need for more detailed flow measurements on models of the human upper airways. Specific information is needed on the boundary layer thicknesses and complete velocity profiles along the upper respiratory tract during oscillatory flow, at physiological values of Reynolds number (cf. App. 1) and respiratory frequency. In addition, local measurements of heat and water vapor flux rates from the model walls into the flowing airstream need to be made.

More flow visualization and velocity profile measurements have been made on models and casts of the lower tracheobronchial tree than on models of the upper respiratory tract.[50,51] These models have also been made primarily for steady unidirectional flow; however, oscillatory flow measurements have also been recently reported.[52] Though important, these lower airway model studies are likely to be less significant for the air conditioning process than upper airway studies would be, since, as previously noted, most humidification and warming of inspired air normally occurs in the upper airways. However, in mouth breathing, breathing through an endotracheal tube, or in stressful environments, conditioning of inspired air is not completed until the air passes further into the bronchial tree. Also, even though the transmucosal heat and water fluxes in the lower airways are normally small, they may become large enough under stressful conditions to cause serious medical consequences. Among these would be triggering smooth muscle contraction and airway constriction in asthma.[40,53] It has been found that the degree of postexertional bronchial obstruction in asthmatics correlates well with the rate of respiratory heat exchange. Furthermore, clinical improvement is obtained in these patients if inspired air temperature and humidity

FIGURE 6
A schematic representation of velocity profiles for the steady inspiratory flow (arrows to right) and the steady expiratory flow (arrows to left) through a model of a bronchial bifurcation. Longitudinal dashed lines show a non-reversible fluid particle trajectory over one oscillatory cycle. *A* represents fluid particle starting point, and *B* represents fluid particle end point for a single cycle (see Ref. 52).

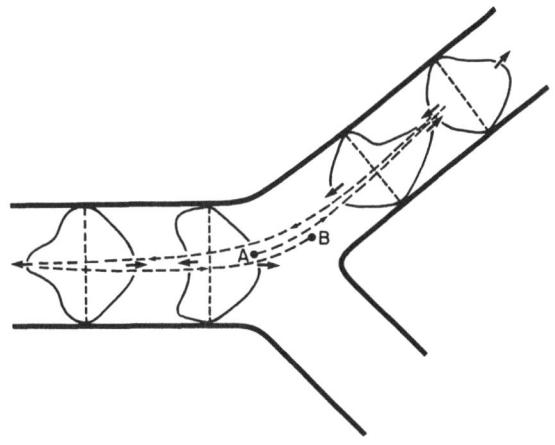

are raised toward expired values, to reduce sensible and insensible heat losses.[40] Also, prolonged breathing through endotracheal tubes has resulted in the death of ciliated cells in the tracheal mucosa and their replacement by a nonciliated squamous epithelium.[54]

Most analytical models reported on transport in the human respiratory tract involve overall heat and water vapor balances taken between inspiratory and expiratory air. Fanger et al.[55] note that differences in water content of expired and inspired air approximately follow the relationship

$$m_e = 0.029 + 0.20 m_i \tag{1}$$

where m is air–water content in kg H_2O/kg dry air. McCutchan and Taylor[56] proposed the equation

$$T_e = 32.6 + 0.066 T_i + 32 m_i \tag{2}$$

relating expired and inspired air temperatures and inspired air–water content.

Those few theoretical models that focus on local transport conditions in the respiratory system are limited to certain regions of the lower respiratory tract,[57] or the models' applicability is further limited by the neglect of wall blood flow and the assumption of constant mucous membrane temperature.[11] More useful theoretical models of respiratory air conditioning in animals have been reported[58–61]; Collins et al.[62] developed an overall steady-state model of heat and water transfer in the upper respiratory tract of the kangaroo rat. Although the model did not predict local temperatures and fluxes along the airway, it included respiratory blood flow and was successful in predicting expired air temperatures, rates of overall respiratory water loss, and efficiency of water vapor recovery.

Johnson et al.[43,47] succeeded in obtaining a convective heat transfer coefficient h_c averaged over three bronchial generations and expressed as the correlation (cf. App. 1)

$$\overline{\text{Nu}} = \frac{h_c D}{k} = 0.302 \, (\text{Re Pr})^{0.786} \tag{3}$$

This correlation is similar to one they obtained earlier in experiments on an iron pipe model. Their studies, however, encompassed only the dog bronchial tree and not the upper airways. Furthermore, Johnson et al. were unsuccessful in an attempt to derive inspiratory convective heat transfer coefficients for each individual generation of the upper bronchial airways. Recently, Nuckols[63] has reported the results of convective heat transfer experiments carried out on a cast replica of the human upper airways extending from mouth and nose to the trachea. He expressed his results in terms of a Nusselt number averaged over the entire cast for inspiration and expiration as follows: For inspiration,

$$\overline{\text{Nu}}_i = 0.028 \, (\text{Re Pr})^{0.854} \tag{4}$$

for expiration,

$$\overline{\mathrm{Nu}}_e = 0.0045 \, (\mathrm{Re} \, \mathrm{Pr})^{1.083} \qquad (\mathrm{Re} \leqslant 7{,}800) \tag{5a}$$

and

$$\overline{\mathrm{Nu}}_e = 0.310 \, (\mathrm{Re} \, \mathrm{Pr})^{0.585} \qquad (\mathrm{Re} > 7{,}800) \tag{5b}$$

In these expressions for $\overline{\mathrm{Nu}} = h_c D / k$, the characteristic length D is taken to be the diameter of the trachea, the Reynolds number Re is that in the trachea, and the Prandtl number Pr is averaged over the region.

It might appear that with the present lack of physiological data, constructing a mathematical model would be a fruitless exercise. However, it is precisely the present lack of data that increases the value of a good theoretical model. Once such a model has been constructed and adjusted to agree with the physiological data that is available, a prediction will be possible of data that has not yet been measured. An iteration process, between theory and experiment, can then be set up where both approaches are gradually refined, leading to a better understanding of local airway heat and water vapor transport.

4.1. Theoretical Model of Local Airway Heat and Water Vapor Transport

A complete, unsteady three-dimensional analysis of the respiratory air-conditioning process has not been done at the present time, due primarily to the geometrical complexity of the respiratory tract. One major limitation, which we have previously noted, is that the air velocity field in the upper and lower respiratory passages is not yet known in sufficient detail. Without this information, the convective fluxes of heat and water vapor across the mucosal walls cannot be reliably calculated. At present, use must be made instead of local, experimentally determined convective heat and mass transfer coefficients. These can be found from model replicas by using heat and mass transfer principles or estimated from literature values obtained for somewhat similar geometries. Once estimates are available for these coefficients, a complete one-dimensional local model of the respiratory air conditioning process can be developed.

It is then necessary to decide which physical parameters must be included in the model to obtain an adequate degree of realism, flexibility, and predictive capability. It appears that at least five interrelated variables must be included: the axially varying mix-mean temperature and water vapor content in the flowing airstream T_A, C_A; the mass flux rate of water vapor N from the mucosal surface into the flowing airstream; and the temperature and water vapor content at the mucus–air interface T_M, C_M. These five quantities, together with a locally specified blood flow rate and/or blood temperature in the airway wall, determine the convective and conductive fluxes of heat and water at all points along the respiratory tract.

A simplification in the theory is possible if heat and water transport during inhalation and exhalation can be considered to occur in a quasi steady state. Human airway temperature measurements suggesting that this assumption is justified have been previously noted (see Fig. 3).[18,20,32,33,42,43,45,47] Further support for this assumption is found by estimating the time required to set up steady-state thermal or mass concentration gradients in the airway walls. These characteristic times are of the order l^2/α, where l is the linear distance from the capillary or venous bed to the air–mucus interface and α is the thermal or mass diffusivity in the airway wall. The computed times are very short ($\approx 10^{-3}$ sec) compared to the time of normal inspiration (≈ 2.0 sec). The results suggest that wall temperature would quickly adjust to sudden changes in air, blood, or mucous surface temperature. Furthermore, if all of the latter parameters were steady in time at a given position, so would be the wall temperature. Conversely, if T_B, T_M, or T_A vary in time, then internal wall temperatures would not lag behind. As previously noted, steady wall and air temperatures have been observed during most of inspiration and expiration. Therefore, it appears that a steady-state, one-dimensional airway heat and water vapor transport model might be expected to have reasonable agreement with experimental measurements.

Along these lines, the authors have developed a theoretical schematic, shown in Fig. 7. Anatomical details of the airway wall are neglected, and the distance between the capillary and venous plexae and the air–mucus interface (Δy) is considered to be occupied by a homogeneous medium of constant thermal conductivity k_{BM}. Across this space, heat is assumed to be transported by steady-state conduction from the blood, at temperature $T_B(x)$, to the mucus–air interface, at temperature $T_M(x)$: The heat flux across this layer is given by

$$q_B = (-k_{BM}/\Delta y)(T_M - T_B) \tag{6}$$

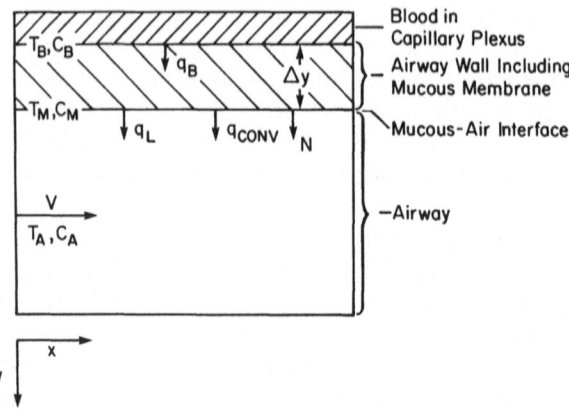

FIGURE 7

Schematic diagram of the airway model and the control volume used to determine heat and mass balances on a given region of the respiratory system and to derive Eqs. (6)–(14). All quantities are functions of distance x along the airways.

At the mucus–air interface, for all positions x along the airways, a steady-state heat balance exists and is expressed as

$$q_B = q_{CONV} + q_L \tag{7}$$

where q_{CONV} represents the rate of heat removal by convective flux into the airstream and q_L represents the rate of removal of latent heat due to surface water evaporation. In the model, we assume that the heat loss carried by the water flux across the mucus–air interface into the airstream is exactly balanced by a heat gain from water flowing in from the blood, so that these terms cancel one another in Eq. (7).

We define a mucosal surface–air heat transfer coefficient $h_M(x)$, such that q_{CONV} is given by

$$q_{CONV} = -h_M(x)(T_A - T_M) \tag{8}$$

The rate of latent heat loss is given in terms of water loss, NM_W and latent heat of vaporization, h_{fg}

$$q_L = NM_W h_{fg} \tag{9}$$

so the balance in Eq. (7) becomes

$$(k_{BM}/\Delta y)(T_B - T_M) = h_M(x)(T_M - T_A) + NM_W h_{fg} \tag{10}$$

The convective water flux from the mucus into the airstream, in moles of H_2O/cm^2 sec is given by

$$N = K_M(x)(C_M - C_A) \tag{11}$$

where $K_M(x)$ is the local convective water vapor mass transfer coefficient.

In a perfect steady state, the mucus layer thickness would have to remain constant in time at every point x along the respiratory tract. In the model proposed here, we assume that the change in thickness due to water evaporation is small and at least partially compensated for by a locally inward flux of water from the cells. Even if a significant change in mucus layer thickness does occur, the resulting local change in Δy would be small and the effect on Eq. (10), negligible. We also assume that there is no frank drainage of water in the x direction and the transport properties of the mucus layer are independent of the degree of hydration of the mucus.

The equilibrium relationship between the air temperature and water vapor content at the mucus–air interface is given from thermodynamics by an equation of the form

$$C_M = 22.4 \exp{[(-4.97 \times 10^3)/T_M]} \tag{12}$$

A water vapor mass balance on a "slice" of airway cross section (see Fig. 7), neglecting axial vapor diffusion, gives

$$v(x)\frac{dC_A}{dx} = \frac{P(x)}{A(x)}K_M(x)(C_M - C_A) \tag{13}$$

where $v(x)$ is local mean airstream velocity, $P(x)$ is local total airway perimeter, and $A(x)$ is local total airway cross-sectional area (see Fig. 3). Similarly, an energy balance on an airway slice neglecting axial diffusion gives

$$v(x)\frac{dT_A}{dx} = \frac{P(x)(T_M - T_A)}{\rho_A c_A A(x)}[h_c(x) + c_W M_W N] \tag{14}$$

Equation (14) is nonlinear due to the dependence of N on T_M and T_A through C_M and C_A. It is also possible to formulate the model in terms of local blood flow rate and inlet and outlet blood temperatures through the capillary and venous plexus. However, since local blood flow rate is not known at present, the simpler formulation using a spatially varying average wall blood temperature $T_B(x)$ is a convenient and flexible alternative.

Equations (10)–(14) constitute a system of five coupled nonlinear ordinary differential and algebraic equations for the five unknown functions $N(x)$, $T_M(x)$, $T_A(x)$, $C_M(x)$, and $C_A(x)$. Assuming appropriate values for $T_B(x)$, $A(x)$, $P(x)$, $h_M(x)$, $K_M(x)$ and $v(x)$, the system can be solved numerically to determine all the steady-state temperatures, humidities, and heat and water vapor fluxes along the human respiratory tract during inspiration or expiration. The convective heat transfer coefficient $h_M(x)$ can be estimated from the data of Johnson et al. for the lower airways (Eq. 3) and from Nuckols's data (Eqs. 4 and 5) for the upper airways. Mass transfer coefficients $K_M(x)$ can be estimated from the heat transfer coefficients, assuming a complete analogy between heat and mass transfer, such that the Sherwood (Sh) and Schmidt (Sc) numbers become analogous to the Nusselt (Nu) and Prandtl (Pr) numbers. The blood temperature distribution along the airways $T_B(x)$ can be estimated from the fact that expired air temperature is usually observed to be very close to nasal mucous membrane temperature, which is shown from Eqs. (10)–(12) to be very close to nasal blood temperature. Nasal blood temperature can thus be estimated from a given set of experimental data to be close to either nasal mucous membrane temperature or expired air temperature. As previously mentioned,[42,43] there is evidence that airway wall blood temperature varies little in time over a breath cycle but gradually rises from its value in the nose to body core temperature toward the peripheral bronchial generations. In the calculations that follow, we have assumed that blood temperature is at a constant plateau value in the nose and rises linearly at a constant rate of 0.3°C/cm to reach a higher plateau at body temperature toward the bronchial periphery.

With these assumptions used to determine $h_M(x)$, $K_M(x)$, and $T_B(x)$, Eqs. (10)–(14) were solved numerically and separately for inspiration and expiration as an initial-value problem, using a fourth-order predictor–corrector finite difference scheme, and linearizing Eq. (12) by a Taylor series expansion within each length increment Δx. Integration proceeded from the nose toward the alveoli for inspiration and from the alveoli toward the nose for expiration. The results for breathing at normal flow rate in a normal room environment and under maximal flow rate in a cold and dry environment are shown in Figs. 8 and 9.

Examination of these figures reveals several important points.

1. During inspiration, there are two major regions of air conditioning. The first is the main nasal cavity, where air temperature and humidity reach about 70% of their final values deep in the lung. The second occurs over the middle of the trachea and the upper generations of the bronchial tree. The position of the ISB for normal room air breathing is calculated to be at about the third bronchial generation, in agreement with experiment.

2. During inspiration and expiration in a moderate environment, airstream warming lags behind air humidification. The reverse is true when breathing cold, dry air.

3. During expiration, all water vapor and heat recovery occur in the main nasal cavity, with air exiting from the nose at very near blood temperature

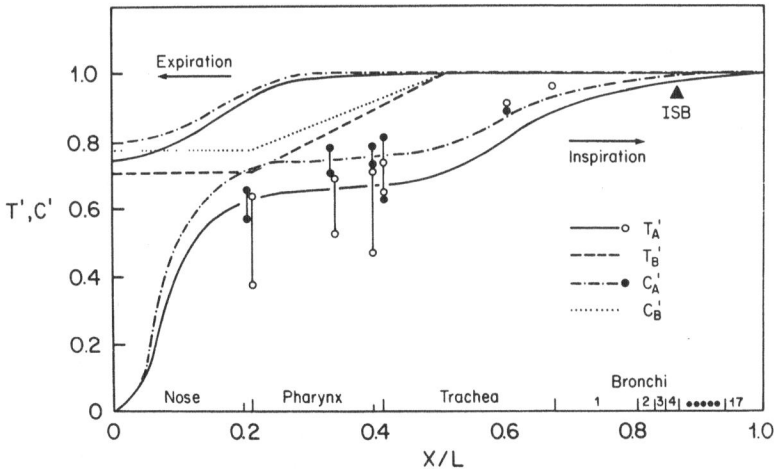

FIGURE 8

Calculated nondimensional temperature and water vapor concentration profiles predicted from Eqs. (10)–(14). for normal respiration ($\dot{V}$ = 300 cc/sec, inspiration air temperature and relative humidity are 23°C, and 30% RH) plotted against the nondimensional distance x into the airway measured from the nose. Nondimensional airway temperature T'_A, is defined as

$$T'_A(x) = \frac{T_A(x) - T_A(\text{insp})}{T_A(\text{core}) - T_A(\text{insp})}$$

The nondimensional airway water vapor concentration C'_A is similarly defined. T'_B is the value of T'_A in equilibrium with $T_B(x)$. C'_B is similarly defined. Based on experimental data (see Ref. 42), the blood temperature throughout the nasal cavity was chosen to be 33°C. The longitudinal blood temperature gradient was assumed to be 0.3°C/cm, such that the blood in the walls of the respiratory tract reaches body temperature in the trachea. Distance into the airways is nondimensionalized by the total length from nose to alveoli in Weibel's model (see Ref. 14).

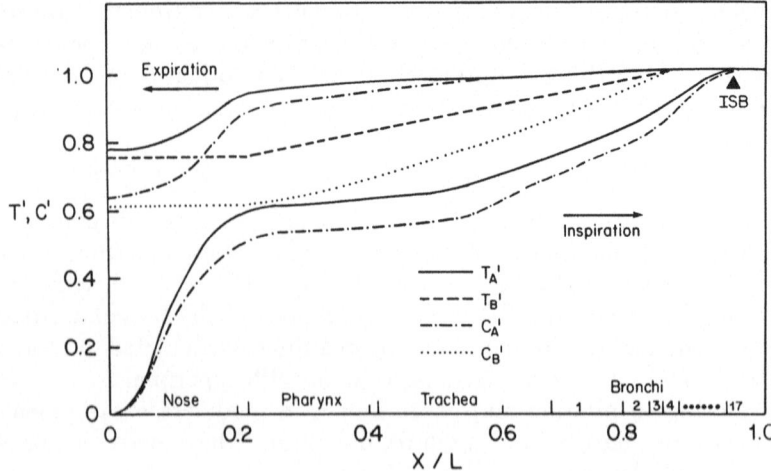

FIGURE 9

Calculated nondimensional temperature and water vapor concentration profiles using Eqs. (10)–(14) for maximal respiration ($\dot{V}$ = 2,000 cc/sec) of inspired cold dry air (0°C, 0% RH) plotted against the nondimensional distance x in from the nose. The blood temperature gradient was again assumed to be 0.3°C/cm, however the nasal blood temperature was assumed to be 28°C, the expired air temperature as suggested by experimental data (see Refs. 32 and 34). Here, the blood in the walls of the respiratory tract did not reach body temperature until the third bronchial generation.

and full water vapor saturation at that temperature. The total calculated heat and water vapor loss from the system is also in close agreement with experiment.

4. The position of the ISB shifts six or seven bronchial generations deeper into the lung with the maximal breathing rate of dry, cold air, in agreement with experiment.

5. The major parameter controlling the shape of the air-stream temperature and water vapor concentration curves is the blood temperature distribution along the airway walls $T_B(x)$. In order to obtain heat and water vapor recovery on expiration, T_B in the nose must be below body temperature, as it normally is. Further experimentation is required to determine how $T_B(x)$ is controlled and how it adjusts to changes in ambient air conditions.

6. The uptake of water over a respiratory cycle in the pharyngeal region is seen to be sensitive to both the wall blood temperature and the nasal airway cross-sectional area. In particular, a net gain in the laryngeal–oropharyngeal region can be converted to a net loss by (1) allowing airway wall blood temperature to increase to body core temperature at a more rapid rate or; (2) breathing through one nostril instead of two while keeping all other parameters constant.

7. From integration of the local surface water vapor loss and gain during inspiration and expiration in a normal room environment, it can be estimated that the mucus layer in the main nasal cavity would completely evaporate in

about nine breaths if no water influx occurred from the cells beneath the layer. The water volume influx necessary to maintain the nasal mucus layer at a depth of $10 \, \mu$m is calculated to be about $8.95 \times 10^{-5} \, cm^3/\text{sec cm}^2$. In this calculation, the mucus layer is assumed to be pure water. Assuming mucus to be a polymer solution will lower the water vapor pressure at the air interface and prolong the time for complete evaporation.

8. As previously noted, the model predicts that the difference between mucous membrane surface temperature $T_M(x)$ and wall blood temperature $T_B(x)$ is small and on the order of 0.1°C or less from the nasopharynx to the alveoli.

There are many possible applications for a theoretical respiratory air-conditioning model such as the one just discussed. The demonstration of the secondary region of conditioning in the upper generations of the bronchial tree and its relation to exercise-induced asthma[40] has already been mentioned. The model can also aid in elucidating the important relationship between the temperature and water content of the bronchial mucus and that of the airstream. The model can be used to predict the viscosity, elasticity, and depth of the mucus layer at different airway locations in response to different ambient air conditions. These properties of the mucus layer in turn affect its flow and transport by the cilia and its protective effect on the cells below. Other applications include analysis of respiratory heat and water transport in extreme environments, such as deep diving heat transfer (high-conductivity He/O_2 breathing) and breathing through adsorption systems (gas masks) and regenerative systems.

5. SUMMARY

Except for the skin, no other body system interacts with the environment so directly as the respiratory system. Sudden and significant changes in temperature and humidity must be accommodated, so that the alveolar environment is stabilized. Physiological and epidemiological studies suggest that susceptibility to, or exacerbation of, many respiratory diseases is strongly connected with ambient air conditions. Although the exact details of this connection are not yet clearly known for any disease, it is obvious that an improved quantitative understanding of respiratory air conditioning will help clarify the process of pathogen uptake in the airway and bring us closer to the day when many of these diseases can be prevented or their effects mitigated.

Research directed at uncovering more of the local details of the respiratory air-conditioning process provides an ideal opportunity for the fruitful collaboration of biological, physical, and clinical scientists. In the preceding discussion, we have attempted to survey the present state of knowledge and indicate some important gaps to be filled and possible avenues for further investigation. It is our hope that this discussion will stimulate the attainment of greater and more useful knowledge in this important biomedical field.

REFERENCES

1. Dalhamn, T., Mucous flow and ciliary activity in the trachea of healthy rats and rats exposed to respiratory irritant gases, *Acta Physiol. Scand. Suppl.* **36**, 123, 1956.
2. Keal, E. E., Physiological and pharmacological control of airway secretions, in *Respiratory Defense Mechanisms*, vol. 5, pt. 1, J. D. Brain, D. F. Proctor, and L. M. Reid, eds. (New York: Marcel Dekker, 1979).
3. Sturgess, J. M., Mucous secretions in the respiratory tract, *Pediatr. Clin. North Am.* **26**, 481–501, 1979.
4. Yaeger, H., Jr., Tracheobronchial secretions, *Am. J. Med.* **50**, 493–509, 1971.
5. Litt, M., Mucus rheology, *Arch. Intern. Med.* **126**, 417–423, 1970.
6. Litt, M., Mucus rheology and mucociliary clearance, *Mod. Prob. Pediatr.* **19**, 175–181, 1977.
7. Sleigh, M. A., The nature and action of respiratory tract cilia, in *Respiratory Defense Mechanisms*, vol. 5, pt. 1, J. D. Brain, D. F. Proctor, and L. M. Reid, eds. (New York: Marcel Dekker, 1979).
8. Widdicombe, J. G., Defense mechanisms of the respiratory system, *Resp. Physiol. II* **14**, 292–354, 1977.
9. Horstmann, G., Iravani, J., Melville, G. N., and Richter, H. G., Influence of temperature and decreasing water content of inspired air on the ciliated bronchial epithelium, *Acta Oto-Laryngol.* **84**, 124–131, 1977.
10. Widdicombe, J. G., Respiratory reflexes and defense, in *Respiratory Defense Mechanisms*, vol. 5, pt. 2, J. D. Brain, D. F. Proctor, and L. M. Reid, eds. (New York: Marcel Dekker, 1979).
11. Proctor, D. F., and Swift, D. L., Temperature and water vapor adjustment, in *Respiratory Defense Mechanisms*, vol. 5, pt. 1, J. D. Brain, D. F. Proctor, and L. M. Reid, eds. (New York: Marcel Dekker, 1979).
12. Hollinshead, W. H., *Textbook of Anatomy*, 3d ed. (New York: Harper and Row, 1974).
13. Warwick, R., and Williams, P. L., eds., *Gray's Anatomy*, 3d ed. (Philadelphia. W. B. Saunders, 1973).
14. Weibel, E. R., *Morphometry of the Human Lung* (Berlin: Springer-Verlag, 1963).
15. Dekker, E., Transition between laminar and turbulent flow in the human trachea, *J. Appl. Physiol.* **16**, 1060–1064, 1961.
16. Olson, D. E., Iliff, L. D., and Sudlow, M. F., Some aspects of the physics of flow in the central airways, *Bull. Physiol. Path.* **8**, 391–408, 1972.
17. Horsfield, K., and Cumming, G., Morphology of the bronchial tree in man, *J. Appl. Physiol.* **24**, 229–231, 1968.
18. Cole, P., Recordings of respiratory air temperature, *J. Laryng.* **68**, 295–307, 1954.
19. Ingelstedt, S., Studies on the conditioning of air in the respiratory tract, *Acta Oto-Laryngol. Suppl.* **131**, 1956.
20. Webb, P., Air temperature in respiratory tracts of resting subjects in cold, *J. Appl. Physiol.* **4**, 378–382, 1951.
21. Jeffrey, P. K., and Reid, L. M., The respiratory mucous membrane, in *Respiratory Defense Mechanisms*, vol. 5, pt. 1, J. D. Brain, D. F. Proctor, and L. M. Reid, eds. (New York: Marcel Dekker, 1979).
22. Lopez-Vidriero, M. T., Das, T., and Reid, L. M., Airway secretion: source, biochemical and rheological properties, in *Respiratory Defense Mechanisms*, vol. 5, pt. 1, J. D. Brain, D. F. Proctor, and L. M. Reid, eds. (New York: Marcel Dekker, 1979).
23. Reid, L., An experimental study of hypersecretion of mucus in the bronchial tree, *Br. J. Exp. Pathol.* **44**, 437–445, 1963.
24. Speisman, I. G., Vasomotor responses of the mucosa of the upper respiratory tract to thermal stimuli, *Am. J. Physiol.* **115**, 181–187, 1936.
25. Mudd, S., Grant, S. B., and Goldman, A., The etiology of acute inflammations of the nose, pharynx, and tonsils, *J. Lab. Clin. Med.* **6**, 253–275, 1921.
26. Dawes, J. D. K., and Prichard, M. M. L., Studies of the vascular arrangements of the nose, *J. Anat.* **87**, 311–322, 1953.
27. Drettner, B., Blood vessel reactions in the nasal mucosa, *Int. Rhinol.* **1**, 40, 1963.

28. Malcolmson, K. G., The vasomotor activities of the nasal mucous membranes, *J. Laryngol. Proc.* **73**, 73, 1959.
29. Rubenstein, E., Pardee, R. C., and Eldridge, F., Alveolar-capillary temperature, *J. Appl. Physiol.* **15**, 10–12, 1969.
30. Magno, M., and Fishman, A. P., private communication.
31. Proctor, D. F., Historical background, in *Respiratory Defense Mechanisms*, vol. 5, pt. 1, J. D. Brain, D. F. Proctor, and L. M. Reid, eds. (New York: Marcel Dekker, 1979).
32. Cole, P., Further observations on the conditioning of respiratory air, *J. Laryngol.* **67**, 669–681, 1953.
33. Dery, R., Pelletier, J., Jacques, A., Clavet, M., and Houde, J. J., Humidity in anesthesiology, III: Heat and moisture patterns in the respiratory tract during anesthesia with the semiclosed system, *Can. Anaesth. Soc. J.* **14**, 287–298, 1967.
34. Walker, J. E. C., Wells, R. E., Jr., and Merrill, E. W., Heat and water exchange in the respiratory tract, *Am. J. Med.* **30**, 259–267, 1961.
35. Cramer, I. I., Heat and moisture exchange of respiratory mucous membrane, *Ann. Otol. Rhin. Laryng.* **66**, 327–343, 1957.
36. Seeley, F. E., Study of changes in temperature and water vapor content of respiratory air in the nasal cavity, *Heat./Piping/Air Cond.* **12**, 377–383, 1940.
37. Dery, R., The evolution of heat and moisture in the respiratory tract during anesthesia with a nonrebreathing system, *Can. Anaesth. Soc. J.* **20**, 296–309, 1973.
38. Schmidt-Nielsen, K., Hainsworth, F. R., and Murrish, D. E., Counter-current heat exchange in the respiratory passages: Effect on water and heat balance, *Resp. Physiol.* **9**, 263–273, 1970.
39. Deal, E. C., Jr., McFadden, E. R., Jr., Ingram, R. H., Jr., Straus, R. H., and Jaeger, J. J., Esophageal temperature during exercise in asthmatic and nonasthmatic subjects, *J. Appl. Physiol.* **46**, 484–490, 1979.
40. Deal, E. C., Jr., McFadden, E. R., Jr., Ingram, R. H., Jr., Straus, R. H., and Jaeger, J. J., Role of respiratory heat exchange in production of exercise-induced asthma, *J. Appl. Physiol.* **46**, 467–475, 1979.
41. Moritz, A. R., and Weislager, J. R., Effects of cold air on the air passages and lungs, *Arch. Int. Med.* **75**, 233–240, 1945.
42. Cole, P., Respiratory mucosal vascular responses, air conditioning, and thermoregulation, *J. Laryngol.* **68**, 613–622, 1954.
43. Johnson, C. E., Linderoth, L. S., Jr., and Nuckols, M. L., An analysis of sensible respiratory heat exchange during inspiration under environmental conditions of deep diving, *ASME Trans. J. Biomech. Eng.* **99**, 45–53, 1977.
44. Dery, R., Humidity in anesthesiology, IV: Determination of the alveolar humidity and temperature in the dog, *Can. Anaesth. Soc. J.* **18**, 145–151, 1971.
45. Ferrus, L., Guenard, H., Vardon, G., and Varene, P., Respiratory water loss, *Resp. Physiol.* **39**, 367–381, 1980.
46. Green, I. D., and Nesarajah, M. S., Water vapor pressure of end-tidal air of normals and chronic bronchitics, *J. Appl. Physiol.* **24**, 229–231, 1968.
47. Johnson, C. E., and Linderoth, L. S., Jr., Deep diving respiratory heat and mass transfer (Durham: NC Duke University, 1976).
48. Proetz, A. W., Air currents in the upper respiratory tract and their clinical importance, *Ann. Otol. Rhinol. Laryngol.* **60**, 439–467, 1951.
49. Proctor, D. F., Airborne disease and the upper respiratory tract, *Bac. Rev.* **30**, 498–513, 1966.
50. Schreck, R. M., and Mockros, L. F., Fluid dynamics in the upper pulmonary airways, in *3d AIAA Fluid and Plasma Dynamics Conf., Los Angeles, 1970.*
51. Schroter, R. C., and Sudlow, M. F., Flow patterns in models of the human bronchial airways, *Resp. Physiol.* **7**, 341–355, 1969.
52. Haselton, F. R., and Scherer, P. W., Bronchial bifurcations and respiratory mass transport, *Science* **208**, 69–71, 1980.
53. Chen, W. Y., and Horton, D. J., Heat and water loss from the airway and exercise-induced asthma, *Respiration* **8**, 305–313, 1977.
54. Hilding, A. C., Laryngotracheal damage during intratracheal anesthesia, *Ann. Otol. Rhinol. Laryngol.* **80**, 565, 1971.

55. Fanger, P. O., McNall, P. E., and Nevins, R. G., Predicted and measured heat losses and thermal comfort conditions for human beings, *Symposium on Thermal Problems in Biotechnology* (New York: ASME, 1968).
56. McCutchan, J. W., and Taylor, C. L., Respiratory heat exchange with varying temperature and humidity of inspired air, *J. Appl. Physiol.* 4, 121–135, 1951.
57. Saidel, G. M., Kruse, K. L., and Primiano, F. P., Jr., Heat and water transport in the trachea, presented at the 73d Annual Meeting of AICHE, Chicago, Nov. 20, 1980.
58. Cole, G. W., and Scott, N. R., A Mathematical model of the dynamic heat transfer from the respiratory tract of a chicken, *Bull. Math. Biol.* 39, 415–433, 1977.
59. Hutchinson, J. C. D., Evaporative cooling in fowls, *J. Agric. Sci.* 45, 48–59, 1955.
60. Roper, W. E., Heat and moisture transfer in the avian respiratory system (Ph.D. dissertation, Michigan State University, East Lansing, 1969).
61. Seymour, R. S., Convective heat transfer in the respiratory system of panting animals, *J. Theor. Biol.* 35, 199–127, 1972.
62. Collins, J. C., Pilkington, T. C., and Schmidt-Nielsen, K., A model of respiratory heat transfer in a small animal, *Biophys. J.* 11, 886–914, 1971.
63. Nuckols, M. L., Heat and water transfer in the human respiratory system at hyperbaric conditions (Ph.D. dissertation, Duke University, Durham, NC, 1981).
64. Swift, D. L., and Proctor, D. F., Access of air to the upper respiration system, in *Respiratory Defense Mechanisms*, vol. 5, pt. 1, J. D. Brain, D. F. Proctor, and L. M. Reid, eds. (New York: Marcel Dekker, 1979).

Chapter 24

HEAT TRANSFER IN TEETH

R. F. Boehm

1. INTRODUCTION

Concern about heat transfer effects in teeth began in a formal way in the 1930s and 1940s. This coincided with the development of higher and higher speed dental drills that aided quick removal of carious formations. As drill speeds increased and bur-cooling options become more numerous, specific evaluations were made of the resulting effects on teeth.

In the 1960s and the 1970s, this activity reached a high point. In this era, fairly sophisticated experiments supplemented with numerical models were widely used. With these approaches, workers were able to address a variety of questions regarding the thermal environment in teeth under numerous circumstances. This period also witnessed the limited use of the laser as a tool in dental practice.

In what follows, the discussion of heat transfer in teeth will be divided into five categories, viz.: the teeth, their form and properties; "normal" thermal response and possible damage; effects of restorative processes or materials; preventive applications in dentistry, including the use of lasers; and formulation of heat transfer analysis in teeth.

As might be anticipated, almost all of the heat transfer processes in teeth take place in the conductive mode. Hence, virtually all studies reported here analyze some aspect of this phenomenon. As is the case with all conduction studies, boundary conditions have a profound effect on the results.

2. THE TOOTH

2.1. Its Form and Major Disease

While a detailed treatise on the various physical, chemical, and physiological aspects of teeth is not the object of this work, some qualitative description is in order. In general, the tooth is composed of three parts: enamel, dentin, and pulp.

Enamel is the most inorganic (less than 2% organic) and physically the hardest of any of the body tissues. It is composed mainly of hydroxyapatite

---

R. F. Boehm • Department of Mechanical and Industrial Engineering, University of Utah, Salt Lake City, Utah 84112.

(HAP) crystals that are covered by a very thin layer of an acid-resistant polymer known as cutile. The HAP crystals are in the form of prisms that are generally at right angles to the surface. This structure is traversed by small inorganic pathways extending from the surface toward the dentin–enamel junction. In any location, tooth enamel is less than a few millimeters thick.

The main structure of the tooth is formed by dentin, which lies just under the enamel. Dentin essentially forms the complete tooth root except for a thin overcovering of cementum. The dentin is about 15 times more organic than enamel, being about 30% overall. Fluid-filled tubules form the basic structure of the dentin. Each of these tubules contains a protoplasmic process extending from the dental pulp to the tooth's dentin–enamel junction.

The pulp region of the tooth is much like vital tissue elsewhere in the body, consisting of numerous blood vessels, nerves, and cells. These cells are capable of participating in defense against injury and in some self-repair. The pulp also consists of numerous connective tissue cells and fibers, including the highly specialized odontoblasts, which are related to dentine formation. An approximate representation of two human teeth is given in Fig. 1, where both an incisor and molar are shown.

Much of the work will be addressed later in this chapter is related to the repair of carious lesions. "Dental caries" is a widespread malady that is particularly active in young people who partake of excess sugar and do not use good dental hygenic techniques.[1]

The early carious lesion in enamel is characterized by significant subsurface demineralization. The pit and fissure surfaces of the molars are especially susceptible, as are all interproximal surfaces. While some effort has been expended in the dental community on preventive techniques (e.g., fluoride treatment), by far the major effort is directed toward restorative processes. In

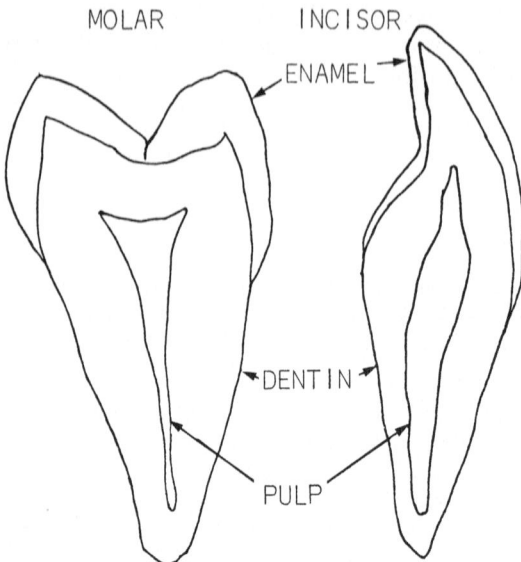

FIGURE 1
Diagrams of two typical teeth showing enamel, dentin, and pulp regions.

the latter, a small drill is used to remove the carious material (i.e., it forms a cavity preparation), and then some restorative material is used to fill the cavity thus prepared. As will be discussed later, the energy produced in cutting can have a very negative effect on tooth vitality. Because of this, a great deal of attention has been given to thermal aspects of the restorative processes.

2.2. Tooth Properties

The properties of primary interest for many thermal analyses in teeth include density, specific heat, and thermal conductivity. Of secondary importance are the mechanical (strength) properties, such as Young's modulus and the ultimate strengths, as well as the thermal expansion coefficient, because of the role they play in thermal stress calculations. The next category of importance is that of optical and infrared properties of the tooth, which are used in the much less proven dental techniques involving lasers.

A caution is in order: As in all human tissue, variations in properties can easily occur between individuals and even between teeth of the same individual. In some cases, a specific property for a large number of samples is reported. However, because of the complexity involved in determining many of these properties, a relatively small number of samples is usually used. Hence, absolute ranges of variability are often difficult to bracket. Also,

TABLE 1
Thermal Properties of Teeth

Property	Enamel values	Ref.	Dentin values	Ref.
Density, g/cm^3	2.8	2	1.96 ± 0.01	3
Specific heat, cal/g°C	0.17 ± 0.01	3	0.38 ± 0.01	3
Thermal conductivity, cal/sec cm°C	$(2.23 \pm 0.11) \times 10^{-3\,a}$	4	$(1.39 \pm 0.12) \times 10^{-3\,b,k}$	4
	$1.50 - 1.62 \times 10^{-3\,d}$	5	$(1.36 \pm 0.07) \times 10^{-3\,b,k}$	4
	$1.51 - 1.58 \times 10^{-3\,e}$	5	$1.07 \times 10^{-3\,f}$	5
			$0.96 \times 10^{-3\,g}$	5
			$(2.38 \pm 0.2) \times 10^{-3\,h}$	6
			$\left(2.29\,^{+0.55}_{-0.42}\right) \times 10^{-3\,i}$	6
			$\left(2.21\,^{+0.20}_{-0.27}\right) \times 10^{-3\,j}$	6
Thermal diffusivity, cm^2/sec	$4.69 \times 10^{-3\,a}$	3	$1.87 \times 10^{-3\,b}$	3
			$1.36 \times 10^{-3\,c}$	3

[a] Longitudinal, 50°C.
[b] Perpendicular to tubules, 50°C.
[c] Parallel to tubules, 50°C.
[d] Varying angles, 26–29°C.
[e] Cut parallel to rods, 26–29°C.
[f] Crown, cut parallel to tubules, 26–29°C.
[g] Root, cut parallel to tubules, 26–29°C.
[h] 39.5–45.2°C.
[i] 73.5–79.0°C.
[j] 86.3–91.0°C.
[k] Deemed to be most appropriate values for dentin by Brown *et al.*[3]

determining a particular property may be especially difficult with samples typical of those from teeth, possibly yielding a degree of uncertainty with values cited. Keep all of these factors in mind when referring to the properties that follow.

2.2.1. Density, Specific Heat, Thermal Conductivity, and Thermal Diffusivity

Values for the various properties so critical to even elementary heat transfer analyses are given in Table 1. The wide range of values is apparent, particularly on the thermal conductivity of dentin.

One also finds in the literature some studies of a less basic nature from which thermal properties might be inferred. An example of this type of situation is the work of Heithersay and Brannstrom[7] where time–temperature

TABLE 2
Mechanical Properties of Human Teeth

Property	Enamel	Ref.	Dentin	Ref.	Whole tooth	Ref.
Elastic modulus	$3.7 - 5.6^a$	8	$1.0 - 1.5$	8	$1.7 - 2.5$	8
($\times 10^{-4}$), MPa	$2.6 - 4.1^b$	8	$0.38 - 2.2$	9		
	$0.83 - 1.2^c$	8	$1.7–1.9$	11		
	$6.6 - 9.1^a$	10				
	$7.1 - 8.4^b$	10				
Proportional limit	2.36^a	8	1.73	8	1.65	8
($\times 10^{-2}$), MPa	1.46^b	8	$1.53 - 2.08$	11		
	1.06^c	8				
	$2.82 - 4.47^a$	10				
	$2.16 - 4.16^b$	10				
Ultimate compressive	2.77^a	8	3.47	8	2.36	8
strength ($\times 10^{-2}$), MPa	1.94^b	8	$2.63 - 3.1$	11		
	1.25^c	8				
	$2.29 - 5.13^a$	10				
	$2.50 - 4.43^b$	10				
	3.8^d	12				
Ultimate tensile	8.3^e	13	40^f	13		
strength, MPa	$30 - 35^g$	14	$22.4 - 50+$	9		
			$32 - 65^g$	14		
Shear strength, MPa	90 ± 19^h	14	138 ± 30^h	15	152 ± 30^h	15
Thermal expansion			7.5^k	16	8.3^i	16
($\times 10^6$)/°C			12^l	16	11.4^j	16

^{a} Cusp area.
^{b} Side area.
^{c} Occlusial area.
^{d} Bending test.
^{e} Standard deviation = 2.4 MPa.
^{f} Standard deviation = 6.2 MPa.
^{g} Used diametral compression to calculate.
^{h} 100 μ punch-calculated shear.
^{i} Root area.
^{j} Crown area.
^{k} 0.5°C/min. heating rate.
^{l} 5°C/min. heating rate.

plots are given for conduction across enamel slabs. Other studies of a related nature are given in Section 3.

2.2.2. Mechanical Properties

Calculations of thermal stress behavior in teeth requires the knowledge of mechanical property values. During the 1970s, these types of studies were of interest in analyzing tooth response to hot and cold food, restorative processes, and laser applications. More details will be given on these aspects later, but several of the pertinent mechanical properties are given in Table 2; values for thermal expansion coefficients are also given there.

2.2.3. Optical Properties

Optical properties of crystalline materials are of great interest for the structural analysis of these materials. Details of these types of studies are beyond the scope of the present work. Instead, we focus more on data that may be pertinent to describing tooth enamel interaction with thermal radiation, both visible and infrared. However, since the main application of this type of data is in the use of the laser, which is relatively recent and limited, there is a paucity of data in this area. Almost all data lie in the visible and near infrared regions.

The index of refraction for human enamel was measured in a very careful study by Spitzer and Ten Bosch.[17] They noted that there was little variation between enamel samples, including human and bovine.

The data for monochromatic, hemispherical reflectivity from the enamel of tooth slabs are shown on Fig. 2. Compared there are data taken using an integrating sphere[18] as well as data inferred from a transmission technique.[19]

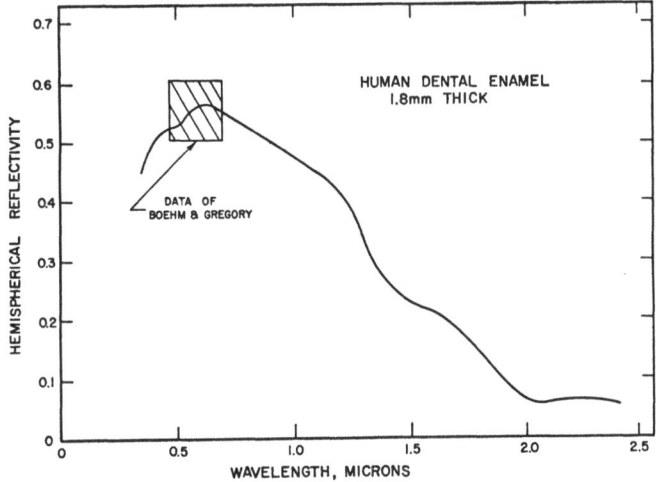

FIGURE 2
Monochromatic hemispherical reflectivity for human tooth enamel surfaces (see Refs. 16 and 19).

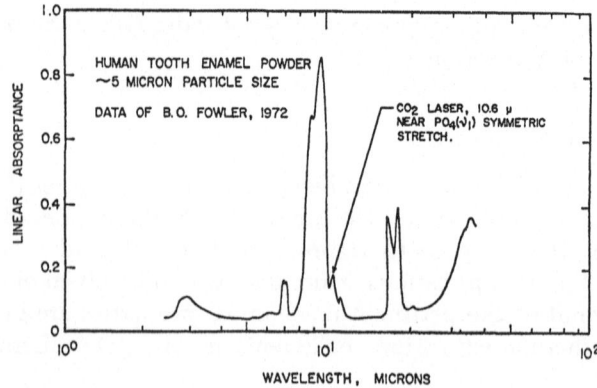

FIGURE 3
Absorption data for enamel powder. From Ref. 20 with permission.

Data above 2.5 μm for tooth surfaces appears to be lacking. However, data for transmission measurements for enamel powder are shown in Fig. 3. Note the large absorption near 10 μm.

3. THERMAL RESPONSE IN "NORMAL" TEETH

There are many physical insults that occur to teeth through restorative and preventive processes that have thermal ramifications. However, in this section, we deal with only those processes that occur normally (e.g., breathing, eating, and drinking) and review basic thermal damage studies that have been performed.

The single most important factor for any thermal treatment or heat-related study of teeth is: will the thermal effects harm the teeth? At least two key studies have been performed on this crucial question, each reaching different conclusions.[21,22]

In 1952, Lisanti and Zander reported thermal studies on dog teeth *in vivo*.[21] A thermode heating element was mounted in a buccal cavity preparation to place the heat source near the pulp area. A small opening through

TABLE 3
Temperature and Time of Heat Application on Dentin[a,b]

Thermode temperature °C	Application time (sec)
52	5
66	10, 20, 60
93	60
149	5, 10, 20
316	10, 20

[a] From Ref. 21.
[b] Nominally 1.0-mm thick dentin.

the opposite (lingual) side was formed, where a fine wire thermocouple was inserted into the pulp chamber. The thermode temperature and time of application were set in each test (168 anterior teeth were used in all, both test and control) according to Table 3.

Temperatures at the junction between the pulp and the dentin near the point of heat application were taken every 5 sec for 1 min. Temperatures achieved in the pulp chamber after 5 and 60 sec are shown in Table 4.

Teeth were extracted at intervals of 4 hr, 24 hr, 48 hr, 1 week, 1 month, and 2 months after thermal injury. In each case, teeth in which cavities were prepared, but to which no heat was applied, served as a control. Extracted teeth were then prepared for celloidin sections after being ground in a longitudinal plane and stained with hemotoxylin and eosin.

Four hours after introduction of temperatures of 52–316°C, a separation of the odontoblastic layer from the dentin was seen, the degree of separation being proportional to the temperature applied. Some minor separation was also seen in the control teeth examined after 4 hr, due only to the cavity preparation procedure. Teeth extracted after 1 month to which heat was applied showed no irritation. In all heated samples, even those in contact with 316°C, pulpal repair was complete after a 2-month period. Teeth heated by the lower temperatures healed more quickly. The authors conclude their

TABLE 4
Results of Lisanti and Zander[a]

Thermode temperature, °C	Dentin thickness,[b] mm	Pulp temperature rise after	
		5 sec, °C	60 sec, °C
52° ± 2°	1.10	1.7	7.7
	1.20	1.7	7.8
	0.90	1.9	7.9
	0.88	1.9	7.8
66 ± 2°	1.33	3.0	10.9
	1.21	3.2	11.2
	0.92	2.8	11.1
	0.81	2.9	11.2
93° ± 3°	1.52	5.6	14.0
	1.24	6.3	14.5
	0.95	6.3	15.2
	0.91	6.2	14.9
149° ± 6°	1.14	12.6	31.2
	0.99	12.7	31.7
	1.23	12.1	31.0
	1.43	11.7	30.0
316° ± 6°	1.82	24.1	46.2
	1.34	24.8	47.8
	1.12	24.2	48.5
	0.93	23.7	50.8

[a] From Ref. 21.
[b] Dimension between thermode tip and pulp.

paper with the statement, "All pulps showed healing after thermal injury regardless of the temperature applied."

Zach and Cohen[22] performed a related experiment. Major differences between the Zach and Cohen work and that of Lisanti and Zander[21] included the following characteristics in the Zach and Cohen work: Heat was applied to the outside of the tooth—no excavation was performed. A soldering iron was used as the heat source. Also, while temperatures in the pulp region were measured via a thermistor (a thermocouple was used by Lisanti and Zander) inserted through the side of the tooth opposite the point of heat application, this was done only to determine temperature response of the tooth. Actual thermal treatment and histologic studies were performed on the tooth opposite the one where temperatures were measured. In addition, the *Macaca rhesus* monkey was used for this study, while dogs were used in the earlier study. These differences should be kept in mind when results from the two studies are compared.

The temperature of the soldering iron was 275°C. Contact with the tooth was maintained for periods from 5–20 sec to produce the range of thermal changes studied. Intrapulpal temperature changes of 2.2–16.7°C were produced in the thermally measured teeth and reproduced on the opposite side of the jaw. Specimens were obtained at intervals of 2, 7, 14, 56, and 91 days after heat application.

Studies of teeth where intrapulpal temperature increases of 2.2°C occurred were identifiable from control (unheated) specimens; no noticeable damage was done. At an increase of 5.6°C in the pulp chamber, response was apparent, both immediately and after prolonged intervals. After a period of 1–2 weeks, repair processes had begun, while after 56 days, most pulps had overcome their trauma, with little change occurring at longer periods. It should be noted, however, that several pulps of small teeth failed to survive a 5.6°C heat rise and became necrotic. This number amounted to approximately 15% of the samples heated 5.6°C above normal.

At a 11.1°C pulp temperature rise, the critical range in pulpal response was frequently exceeded; some, however, did repair. Approximately 60% of teeth in this category failed to recover.

At a 16.7°C increase, all teeth so treated showed an overwhelming, irreversible necrotic response. No histologically recognizable pulps were recovered after heating to this temperature level. The results of this study are summarized in Fig. 4.

Comparisons of the work of Lisanti and Zander with that of Zach and Cohen lead to contrary conclusions. While all of the teeth treated by Lisanti and Zander repaired, even with measured pulp temperature rises over 50°C, Zach and Cohen reported that 100% of the teeth whose pulp temperatures where elevated 16.7°C became necrotic.

An explanation of this basic difference in conclusions is not easily accomplished. Had pulp temperature been inferred rather than measured, the explanation might have been simple. The best possible explanations are as follows. If the pulp volume is significantly larger in one case, that case may be able to withstand a point source of heat more readily. Also, if one case cooled off

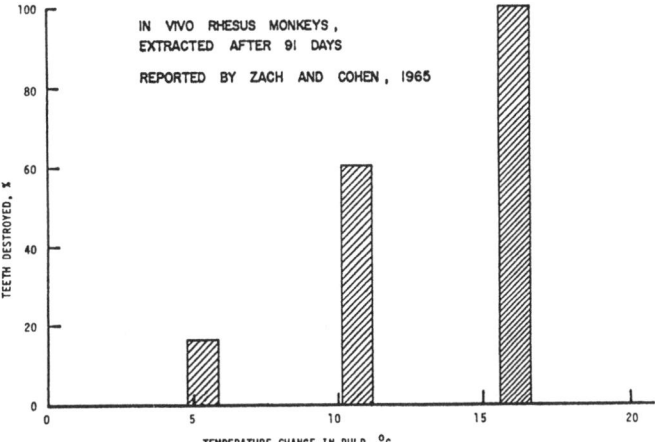

FIGURE 4
Percentage of teeth destroyed due to elevation of pulpal temperature. From Ref. 22 with permission.

more quickly after removing the heating object, that case would also be more likely to repair, all other aspects being equal. The harmful reactions are undoubtedly proportional to the product of the elevated temperature level and the time elapsed at that level. Certainly the Zach and Cohen criterion is safer, and it has been used for predicting tooth vitality with treatment by a thermal process.[23]

Other studies of basic tooth physiological responses are quite limited. One of the more fundamental of these is the work by Delbalso and Todd.[24] They found in their study of the effects of thermal injury on pulpal hydrolases that thermal injury resulted in approximately a 15% reduction in the activities of N-acetyl-D-glucosaminidase, β-galactosidase, and β-glucuronidase, and acid and alkaline phosphatases.

Pain sensation reaction to thermal gradients has been of interest. Matthews[25] performed a very careful study where 55 pulpal nerves were tested at temperature environments of 5, 37, and 55°C. Of these, five responded to heat, while 17 responded to cold. Cold-sensitive nerves gave a smaller response with shorter latency than heat-sensitive ones. The response of the heat-sensitive ones did not stop when heat was removed. This work and the work of Naylor[26] give many interesting insights into the mechanism of thermally induced pain.

Temperatures and pressures in unaltered teeth have been of interest in a variety of studies. For example, Selden[27] measured pulp temperatures near the tooth apex and correlated them with the oral cavity temperature. He found that as the oral cavity temperature rose, the periapical temperature increased more. The temperature variation in the gingiva sulci was reported by Mukherjee.[28]

Van Hassel and Brown[29] studied the pressure response to temperature in the pulp. They used a cannula (a device to measure simultaneously the pressure and temperature) in canine teeth of dogs. The teeth were wrapped with Nichrome wire to effect the heating. Findings included the fact that pressure increased linearly with temperature. Van Haseel and Brown attributed the pressure rise to two components—one reversible and one irreversible. The reversible component was hypothesized to be due to vaso-dilatation, while the irreversible one was thought to be due to the penetration technique.

One of the more interesting studies of thermal effects in teeth—both the generation and movement of heat—was performed by Brown and Goldberg.[30] They used fine-wire thermocouples to measure surface temperatures at several locations on several teeth *in vivo*. Their studies led them to conclude that conduction from the peridontal tissue is a major source of heat influencing the tooth's surface temperature. Circulation and pulp metabolism were found to have relatively little effect. Other authors have explored general thermal effects and their means of measurement.[31,32]

Eating and open-mouth breathing are two processes that are encountered regularly by teeth. Depending on the food, drink, or air temperature and the heat transfer coefficient involved, these processes could result in severe impacts on the teeth. Jacobs *et al.*[33] measured the heat transfer coefficients for various foods; a sample of some of their values is given in Table 5. These values can be used with Newton's law of cooling to calculate heat transfer, viz.,

$$q = hA(T_{\text{ambient material}} - T_{\text{tooth surface}}) \tag{1}$$

In the open-mouth breathing process, the situation is complicated because of the variable nature of the air flow and its temperature. Boehm[34] has given the following expresssion for the exhalation air temperature:

$$T_{\text{ex}} \cong T_{\text{body}} - 0.2(T_{\text{body}} - T_{\text{dry bulb}}) \tag{2}$$

where $T_{\text{dry bulb}}$ is the dry bulb temperature of the air. During inhalation

$$T_{\text{in}} = T_{\text{dry bulb}} \quad \text{(usual situation)} \tag{3}$$

or

$$T_{\text{in}} = T_{\text{wet bulb}} \quad \text{(extreme situations)} \tag{4}$$

TABLE 5
Heat Transfer Coefficients for Various Foodstuffs[a]

Food	Heat transfer coefficient (cal/cm^2 sec °C)		
	Immediate	Intermediate	Long-term
Yogurt from refrigerator	2.66×10^{-2}	2.66×10^{-2}	2.66×10^{-2}
Carbonated beverage, poured	5.86×10^{-2}	5.86×10^{-2}	5.86×10^{-2}
Ice cream from freezer	1.42×10^{-2}	3.39×10^{-2}	1.21×10^{-2}
Instant potatoes from stove	1.02×10^{-2}		6.51×10^{-3}

[a] From Ref. 33.

Heat transfer coefficients for teeth during breathing varied considerably—from 340–1,135 W/m² °C. Most values fell in the 568–850 W/m² °C range.

Analytical approaches to predict temperature in teeth were quite popular in the 1960s, see, for example, Ref. 35. During the 1970s, several papers addressed the numerical analyses of heat transfer in teeth, see, for example, Ref. 33 and 36–38. The relatively low-thermal conductivity of tooth material is, as has been shown earlier, a dominant factor in the conduction phenomena. This has led some workers to investigate the effects of thermal stress and possible cracking in transient condution situations.

Since the tooth is made up of materials that have low-thermal conductivity and ultimate strength, the effects of thermal stress cracking have been studied. Related discussion on this topic will be given in the sections on restorative and preventive processes. Studies have been reported of thermal cycling of teeth under extremes of temperatures, with an examination following for thermal-cracking phenomena.[33,38–40] One set of results from a study of this type is shown in Table 6.

Peultier and Frank[39] also studied thermal cooling of teeth for short time periods; young teeth were found to be more sensitive than mature teeth to thermal shocks. No difference was noted between carious and healthy teeth in their propensity to cracking. Also, no significant difference was found in the reaction of teeth from different parts of the mouth. It should be noted, however, that the tests that indicated significant cracking shown in Table 6 required relatively high temperature differences.

A more detailed study of cracking due to thermal cycling was reported by Brown and co-workers,[38,40] typical results are shown in Fig. 5. In Figure 5, each cycle is defined as 30 sec at 52°C and then 30 sec at 24°C, for a differential of 28°C impressed every minute. In general, the cracking is complete at about 2,000 cycles. The authors indicated that common foods or beverages could cause the types of cracking indicated, but they were uncertain if the cracks always result in deleterious effects. Interesting correlations between average crack length and patient age are also given.

TABLE 6
Thermal Cycling and Cracking[a]

Temperature limits, °C	Time at each temperature, min	Total test time, min	Percentage with cracks at completion
0–100	0.5	10	100
0–80	1.0	10	100
0–65	1.0	10	100
0–55	1.0	10	100
0–45	1.0	10	100
0–35	1.0	10	50
0–30	1.0	10	0
30–65	1.0	10	0
21–65	1.0	10	50

[a] From Ref. 39.

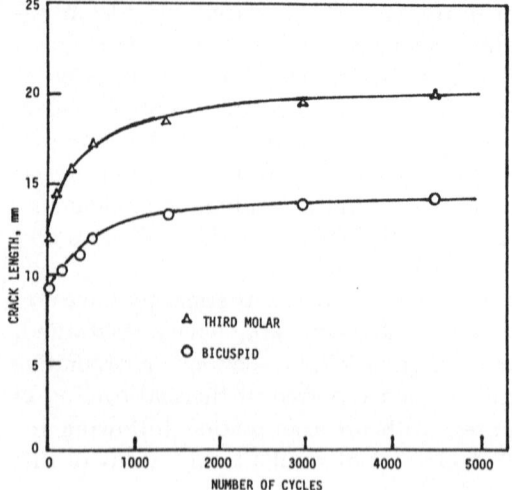

FIGURE 5
Results of crack growth due to thermal
cycling. Each cycle is defined as 30 sec
at 52°C, 30 sec at 24°C. From Ref. 38
with permission.

A few studies have been reported that attempt to calculate stresses in
teeth associated with some kinds of transient thermal behavior.[36,38] Both
finite-difference and finite-element programs are used for predicting temperatures and stresses. Generally, these studies indicate that tooth enamel can
be quite susceptible to cracking during rapid cooling; see further discussion
on the formulation of models for temperature prediction in Section 6.

4. THERMAL FACTORS DUE TO TOOTH REPAIR

As most readers are aware, dental caries are a widespread fact of modern
life.[1] The typical restorative procedure to arrest carious action is to remove
sufficient portions of the tooth both to eliminate the carious formation and
hold a filling in the cavity thus formed. Over the years, the drill used to
perform the cutting has progressed from a foot-driven belt drive to a motor-
driven belt drive, until now air turbines are used almost exclusively. These
latter instruments cause the dental drill (bur) to turn at very high speeds. This,
in turn, allows the tooth material to be cut very rapidly, resulting in a very
high rate of energy deposition in the tooth or in the cuttings, depending on
the technique of the dentist.

Many dental researchers have addressed questions related to the thermal
environment in teeth during high-speed cutting. Perhaps more papers have
been written on this thermal aspect in dentistry than on any other. In what
follows, a large number of these reports will be noted, but only a few will
be reviewed. The reason for this is that many of the papers are qualitative.
In many of the more quantitative ones, agreement may be lacking with the
work of others, and insufficient detail may be given so as to render comparison
almost impossible. However, because of the summary nature of the present
work, only a few details can be given.

4.1. Studies of Temperature Rise in Cutting

"Dry cutting" may mean that no coolant is used on the bur. Many times, this is done in a research project to bracket maximum temperatures, quantify energy inputs, or for some other reason. On the other hand, there are times the reported work tends to support or deny the hypothesis that temperature levels in the pulp become harmful as a result of cutting processes. Those who believe that the resulting temperature from cutting without a coolant are harmful tend to divide on which type of coolant is most effective.

The two main coolants that are considered for use in cutting procedures are air and water. Air is easier to handle and to work around (it does not obscure the view when cutting), but the better heat transfer characteristics of water are generally recognized. The question is usually whether air is a satisfactory coolant.

Reports related to cutting studies are quite numerous and include Refs. 32, 41–70. Several general articles of a qualitative nature have appeared[41,42] as well as reviews of testing techniques.[32,43]

Sorenson *et al.*[44] performed a study of cutting where they investigated several aspects related to energy deposition. The researchers observed that mechanical hardness had no effect on the energy interaction. They also observed the obvious characteristic that the applied load greatly affects the cutting speed of the air turbine. However, they found that the magnitudes of applied load under which the heat transfer rate and the rate of material removal were maximized were not the same. Specifically, Sorenson and co-workers found that maximum heat transfer occurred at a lighter cutting load than did the maximum rate of material removal, but both aspects were related to the load through a parabolic function. They gave several plots that substantiated this observation. Sorenson and co-workers also designed a clever hand piece that combined the cutting and temperature measurement functions.[45]

Several other studies have explored many of the variables during cutting,[46–53] included are the effects of bur rotational speed,[46,49] type of bur,[46,48,51–53] and force on hand piece.[46,49,52,53] Temperature increases to 95°C above the initial tooth temperature were reported,[47,49] but some numerical models indicate temperatures probably approach 370°C.[52] Some have indicated relatively low temperature increases in the pulp, about 10°C.[50]

Perhaps the most careful engineering studies of energy dissipation during cutting have been performed by Brown and co-workers.[52,53] Not only were many variables measured during cutting experiments, but numerical analyses of temperatures and stresses generated during cutting were also performed. They concluded that temperatures sufficiently high to cause biological damage to the pulp are generated. They were also able to show graphically that surface cracking of the enamel did occur and thermal stresses sufficient to fracture the tooth, generated during cutting, were probably present. Their research certainly ranks among the most quantitative and reliable studies performed in this area.

A fundamental study of cell destruction (supplementing the vitality studies reported earlier by Lisanti and Zander, and Zach and Cohen) was reported by Kawahara and Yamagami.[54] Their investigation of possible cell responses to heat and vibration were oriented toward understanding cutting-related damage.

They tested a two-day culture of "strain *L* fibroblast" treated cells and found that cellular responses were reversible to the following time–temperatures pairs: 20 min at 50°C, 10 min at 60°C, and 5 sec at 80°C. At temperatures greater than 45°C, vibration aided the destruction of cells. Note that these temperatures are considerably higher than the 100% kill limit described by Zach and Cohen.[22] Hence, it must be said that the mechanism by which temperature affects pulp vitality is not well understood.

With potentially high temperatures being generated during cutting without coolants, it is not too surprising that several workers have investigated the effectiveness of coolants on the cutting process. Because an air spray does not cloud eye contact with the cut during the process, some dentists prefer that approach; there are several studies in this area.[55–57] However, the better the heat transfer characteristics of an air–water or water-only spray have also been of interest.[58–67] Although cutting with some form of water cooling appears to be much more effective for removing heat, several of those who support the use of air-only cooling cite the following rationale:

1. Better view of the cut is obtained while using air as coolant.
2. Major aspects of thermal control can be accomplished with correct procedure (i.e., control of the force used in cutting).
3. Water may overcool the pulp, causing pulp discomfort.

Typical arguments of this sort are expressed by Schuchard and Watkins.[62] However, as was mentioned earlier, the actual mechanism of pulp damage has not been quantified. Hence, severance of the odontoblastic processes, dehydration, and, perhaps, cooling have all been cited as problems in this area.[56] Also, some experimental techniques (see, for example, the work of Schuchard[57]) may be subject to errors, possibly leading to incorrect conclusions.

One of the more convincing studies in this area compared air spray, air–water spray, hollow bur water spray, and water syringe.[67] The study compared, among other things, the temperature rise at the dentin–enamel junction as a function of time for the various cooling modes. After 5 sec, the following approximate temperature increases were noted: air, 110°C; air–water, 38°C; hollow bur water, 12°C; water syringe, 5°C. These differences became more distinct with time. Note that the water flow was essentially the same except for the air-only case. Over one order of magnitude range of temperature increases was found for any given water flow rate when applied in the different forms. This study underlines the fact that, not only is the type of coolant important, but so is its mode of application. Primarily, this is so because high-speed burs tend to impede coolant flow into the cavity area during cutting.[62]

Thermal stresses initiated during cutting procedures are severe enough to cause cracking in teeth.[53,68–70] Both experimental[68–70] as well as combined numerical–experimental studies have been performed on this phenomenon.[53]

The completed cavity and filling material have also been of interest from a thermal point of view. Temperature rises during full-crown preparations,[71] polystyrene and gutta percha restorations,[72] and other materials[73] have been reported.

As one would expect, the filling material has a significant interaction with the tooth structure, both thermally and mechanically. In general, these restorative materials (gold is one example) have a much higher thermal conductivity, thermal expansion coefficient, and mechanical strength than does the tooth structure. (See typical results in reference 4.) Since the engineering properties of these materials are quite common, they will not be dealt with here. The effects of lining materials for cavities are dealt with by Braden;[74,75] thermal effects of polishing processes have been treated by several workers,[76–78] while the temperature changes that occur during removal of amalgam restorations have also been addressed.[79]

The dimensions of a restorative cavity are affected by the thermal environment; actual measurements of these movements during heating and cooling were measured by Roydhouse and Paxon.[80] Thermal cycling was performed on restorations of various ages, and the resulting marginal leakage was measured;[81] some sealing techniques showed good inhibition of leakage.

5. IMPLICATIONS OF PREVENTIVE PROCESSES

5.1. Overview

It is generally thought that an ounce of prevention is worth a pound of cure. This is certainly true in the case of the most prevalent dental disease—dental caries. As we have discussed, the cure might possibly kill the tooth organism, depending on how it is treated; at the very minimum, the tooth structure is weakened.

The problem with preventive dentistry is that, as in many other health applications, the cure has been elusive. Typically, three strategies are accepted as major impactors on preventing tooth decay: (1) combatting the cariogenic microflora, (2) modifying the diet, and (3) increasing tooth resistance through multiple use of fluorides.[1] Another technique, perhaps less well accepted, is the use of pit and fissure sealants. (See, for example, Ref. 82).

Of the various techniques considered, the only serious thermal implications are found in some fluoride application methods and the use of pit and fissure sealants. These two topics are dealt with briefly in Section 5.2.

5.2. Pit and Fissure Sealants

By far, the majority of the work in this area has involved using a polymeric material to seal the chewing surfaces of the molars. (See reference 82.) In this case, temperature increases in the sealant material could reach 30°C,[83] a very large increase. However, this is probably not a significant factor in tooth health, since there is not much material involved, and hence little energy deposition. The material is applied to the outside of the tooth where the effects of convection and radiation losses could play important roles compared to conduction to the pulp.

Greater thermal implications are possible from the proposed use of lasers to install sealants made of toothlike materials.[18] In this application, a ceramic material is essentially fused (although at moderate temperatures) to the tooth crown. Again, small amounts of materials are involved, so that they do not affect occlusion and mastication processes.

The laser has been proposed for dental applications almost since its invention. For an excellent review of the bulk of the early work, the reader is referred to the article by Stern.[84] Many of the early studies were oriented toward using the laser as a cutting tool. However, the thermal insult to the pulp was almost always too great.

In dealing with safe applications of lasers to teeth, the following factors are clear: If small energy beams of large intensity are used, high surface temperatures can be achieved without significant harm to the pulp.[23] Input energies below 1.5 J over any short period of time, such that conduction into the tooth interior dominates surface losses appears to assure tooth safety, based on Zach and Cohen's work on pulp vitality.[22] However, if not applied appropriately, this procedure could cause cracking of the dental enamel due to thermal stresses.[85] Using typical pulses on the order of 1 sec or less, a "cracking threshold," was defined in terms of an energy flux.[85] A plot of typical data based on more information than was used in that paper is shown in Fig. 6.[86] Depending on the specific application, either the pulp vitality limitation or surface cracking limitation may dominate.[85]

A photograph of a portion of a tooth under moderate magnification is shown in Fig. 7. This photograph encompasses two laser irradiation

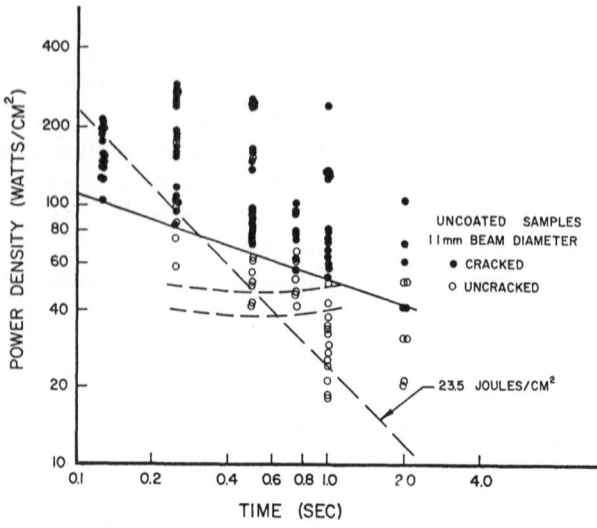

FIGURE 6

Damage threshold for surface cracking in teeth using laser pulses. The dashed pair of curves represent theoretical cracking predictions (see Ref. 86).

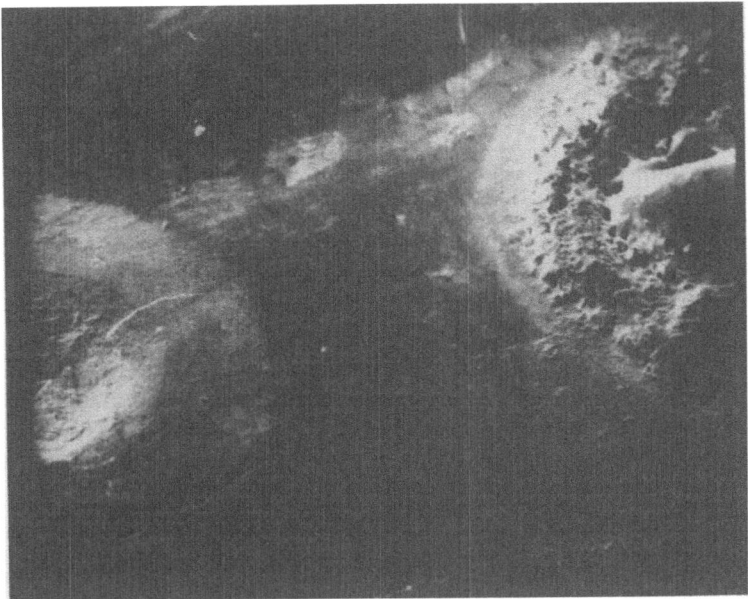

FIGURE 7

A photomicrograph (×53) showing two sites of safe laser irradiation on a tooth. In one, an apatite material has been fused to the tooth. No material was fused at the other location (see Ref. 86).

locations—one where material was fused safely, and one where no material was fused.

5.3. Fluoride Uptake Enhancement

Fluoride is usually applied to drinking water,[87] dentifrices, or locally applied jells. Although the mechanism of fluoride uptake in tooth enamel has not been well understood, it is hypothesized that a chemical reaction takes place with the apatite base of tooth enamel.[18] An excellent review and theoretical model of this situation has been put forth by Baechler,[88] who hypothesized that an increase in temperature should increase the fluoride uptake. At the time of Baechler's thesis, conflicting data were available in the literature on this point. The more recent data of Putt *et al.*[89] seem to substantiate an increased fluoride uptake with an increase in temperature.

6. FUNDAMENTALS OF THERMAL ANALYSIS

6.1. Theoretical Basis

Compared to heat transfer in many biological systems, which can be a combination of conduction, convection, and, perhaps, radiation, heat transfer in teeth takes place only by conduction. Some circulation and metabolic processes do occur in the pulp material, but as was discussed in Section 3, this accounts for a much smaller energy flow than the flow axially along the tooth from the peridontal tissue.[30] A detailed derivation of conduction heat transfer is given in Chap. 6. With reference to the typical tooth profile in Fig. 1, the transient three-dimensional conduction equation is

$$\rho c \frac{\partial T}{\partial t} = \nabla k \nabla T + q \tag{5}$$

In general, k will vary with location and orientation within the tooth, while ρ and c will vary only with location. Since q has been shown to be very small, it will be omitted from further consideration. To complete the problem formulation, an initial condition and two boundary conditions for each spatial coordinate are required. The initial condition is almost always set by the fact that the tooth is at uniform temperature prior to energy input.

The boundary conditions must be tailored to the particular problem. Analyses of heat transfer in teeth have been carried out with specific wall temperature, specified wall heat flux, or a specified convection to ambient (Newton's law of cooling). Specified wall heat flux can be applied to the laser-cutting problem; away from the zone of irradiation, the surface is modeled as a heat insulator. Many times, heat transfer in the tooth enamel and dentin is of interest; in this case, a separate analysis for both materials at the enamel–dentin junction is carried out, matching both temperatures and heat fluxes at the enamel–dentin interface.

The governing equation and boundary conditions are usually taken to be linear: in practice, this may not be the case. In the analysis of heat transfer in teeth, three nonlinearities may typically arise: two in the boundary conditions and one in the governing equation. The heat balance expressed in the governing equation may be rendered nonlinear by a thermal conductivity that varies with temperature (see Table 1). Natural convection may occur at the tooth surface, making the boundary condition nonlinear, because the heat transfer coefficient in Newton's law of cooling depends on temperature. Also, in a situation that occurs much less frequently, thermal radiation from the tooth may be important; this may occur because of a high surface temperature, such as that caused by laser irradiation. While these nonlinearities cause complications in analytical solutions, the numerical approaches discussed in Section 6.2 can be used.

Analytical solutions to many systems are available. With few exceptions, the geometry, nonhomogeneous material, and boundary conditions on teeth render analytical solutions very difficult or impossible. Hence, most practical problems will require numerical analysis and computer solutions.

6.2. Results from Tooth Modeling

In a few simple situations, heat transfer in teeth can be modeled analytically. One has to be careful, however, for while gross assumptions about the geometry and makeup of the tooth can always be made, they may not result in a very good model of the physical situation. For example, a molar may be modeled as an infinitely long cylinder of a single substance, but the finite length and nonhomogeneous nature of a tooth may render predictions that deviate quite widely from the actual situation.

Perhaps the most successful analytical model is the use of semi-infinite slab results for short-time descriptions of situations where the thermal stimuli occur in a very short period of time. For example, consider the instantaneous response of a tooth surface to a msec (or shorter) laser pulse. The initial response of the tooth are well represented by a semi-infinite model.[85] The temperature variation on the surface of a tooth, following a Gaussian beam irradiation profile is shown in Fig. 8. This model should not be used to represent long-term behavior, as the accuracy will be quite poor.

Specific solutions to the heat conduction equation for teeth are obtained with the aid of a computer. The numerical approach allows a variety of complexities to be incorporated if desired. Computer codes used for the solution of bioheat transfer problems typically use finite-difference solutions, but recently, some workers have begun to use finite-element techniques (see Chap. 18, for example). The latter approach has the advantage that irregular shapes can easily be accommodated; it does require the availability of a computer with a large storage capacity.

Typical situations analyzed with numerical models include the investigation of the effects of low-order heating or cooling of the teeth as might occur during eating or drinking[33,36] or the effects of a specified heat flux at the boundary or in the interior of a tooth as might occur in cutting[52,53] or laser irradiation.[23] While it is not possible to give a large number of typical results, a few representative situations will be noted.

The influence of the surface heat transfer coefficient on the temperature distribution at the dentin–enamel junction of a central incisor is shown in

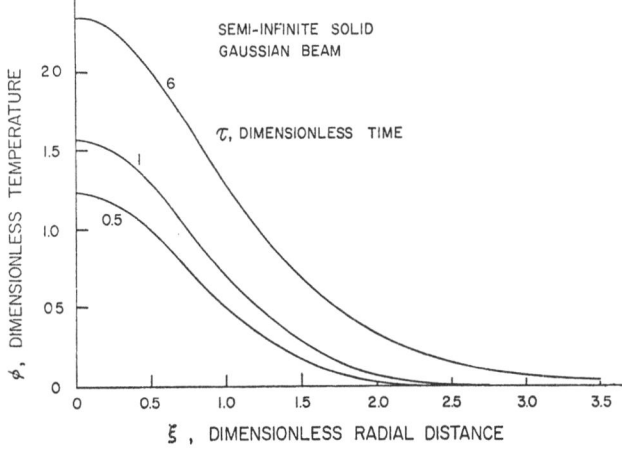

FIGURE 8

Dimensionless temperature variation on the surface of a semi-infinite solid due to absorption of laser energy from a Gaussian distributed beam (see Ref. 85).

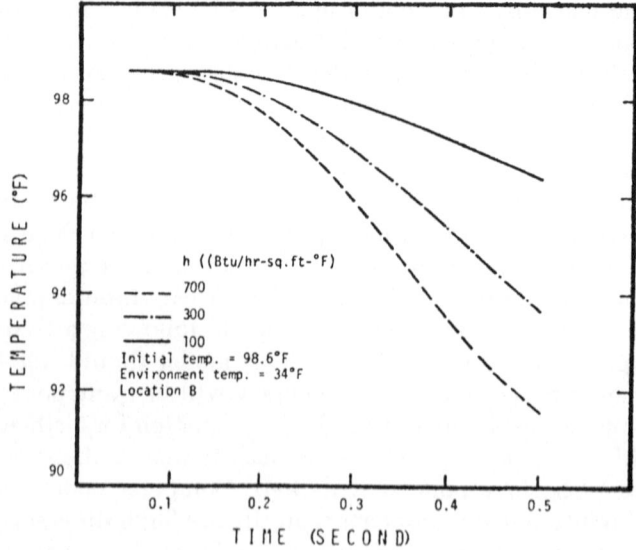

FIGURE 9

Temperature–time plot for a point on the dentin–enamel junction of a central environmental subjected to a sudden decrease in environmental temperature for various values of the heat transfer coefficient (see Ref. 36).

Fig. 9.[36] This represents the temperature variation after simulated contact with a substance at a cold temperature. Pertinent heat transfer coefficients (see Table 5) can be used to determine the appropriate parametric curve.

Study of an imposed heat flux on a tooth has been the most prevalent application of numerical modeling to date. For example, the effect of the energy liberated in the tooth during the cutting processes is of much practical concern. By use of a finite-element routine, Brown and co-workers[53] estimated the steady temperature distribution within the tooth as a function of the amount of heat input; an example of this is shown in Fig. 10. Relatively high temperatures are predicted within 1 mm of the cut surface; however, neglect of time dependence causes an over estimation of the tooth temperature. The value of the numerical analysis is that it can be used to predict temperatures in locations where measurements would be nearly impossible.

Some calculated temperature–time traces that were simulated for laser irradiation are shown in Fig. 11.[23] Comparisons made between the calculations and some experiments on the resulting correlation are fairly good. One aspect should be noted relating to this figure and the results shown in Fig. 8. Figure 11 should be more accurate for long times, while the results shown in Fig. 8 should be most valid for short times, as was previously noted. The numerical technique used to predict the temperatures in Fig. 11 could be somewhat inaccurate in the first one or two times steps used in the calculation if very short laser pulses are considered. Hence, it is recommended that an analytical solution be used to predict the maximum surface temperatures and the times at which they occur. The numerical solution can then be used to predict long time as well as spatially removed point temperatures.

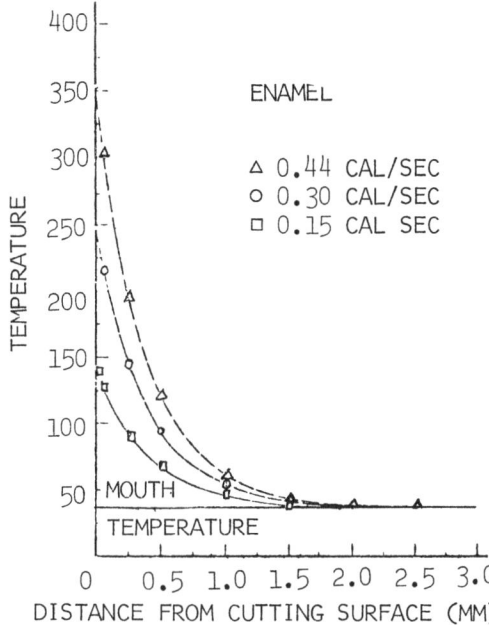

FIGURE 10
Time averaged temperature distribution in enamel, measured from the surface being cut. Various energy deposition rates with no coolant are analyzed. From Ref. 53 with permission.

7. FUTURE DIRECTIONS

As was summarized, most of the naturally occurring and typical restorative processes have been analyzed in some depth. Very good correspondence has been demonstrated between the measurement of thermal effects and the prediction of the same. Heat transfer theory can be used very effectively in

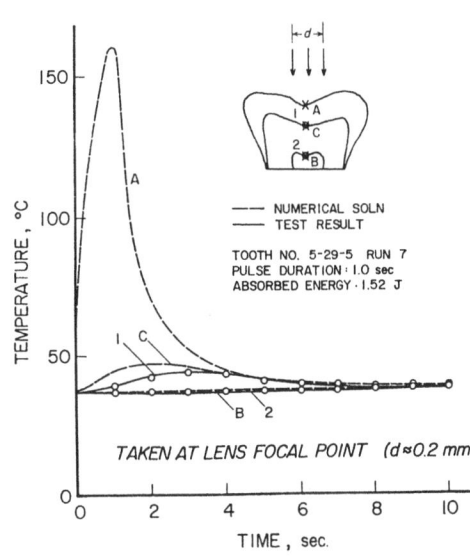

FIGURE 11
Comparison of numerical results at three positions in the tooth with experimental results at two positions for a CW (manually pulsed) laser of 1.0 sec pulse duration (see Ref. 23).

estimating thermal impacts. While other creative and definitive work in this area will undoubtedly be done in years to come, this is not the fertile ground that it was in the early 1960s. Some interest will probably linger in the development of cutting techniques and topics that minimize thermal insults.

Probably the most potential lies in analysis associated with new directions in dentistry. How much emphasis will occur here is uncertain, because dentistry has been a field characterized by small refinements in old techniques with very little interest in bold new approaches. Sooner or later, some of the aspects described in Section 5, as well as some not anticipated there, will attract more widespread interest. Thermal analyses of these processes will be required. It is anticipated that heat transfer theory will be used very effectively, as it has been used for other processes in the past, to assess these new processes.

ACKNOWLEDGMENTS. I wish to express my appreciation to the following students: T. Baechler, C. K. Blair, M. J. Chen, R. Gregory, S. Janke, J. Rich, L. Stewart, and J. Webster as well as to several others who aided me in developing my understanding of thermal processes in teeth and to the National Institute of Dental Research for furnishing financial support in this area. I also wish to acknowledge the less direct but extremely valuable assistance from those in the University of Utah's Dental Engineering Laboratory, including Drs. W. S. Brown and B. Lloyd.

REFERENCES

1. Mandel, I. E., Dental caries, *Am. Sci.* **67**, 680–688, 1979.
2. Sicher, H. *Orban's Oral Histology and Embryology*, 6th ed. (C. V. Mosby, St. Louis, 1966).
3. Brown, W. S., Dewey, W. A., and Jacobs, H. R., Thermal properties of teeth, *J. Dent. Res.* **49**, 752–755, 1970.
4. Craig, R. G., and Peyton, F. A., Thermal conductivity of tooth structure, dental cements, and amalgam, *J. Dent. Res.* **40**, 411–418, 1961.
5. Soyenkoff, B. C., and Okum, J. H., Thermal conductivity measurements of dental tissues with the aid of thermistors, *J. Am. Dent. Assoc.* **57**, 23–30, 1958.
6. Lisanti, V. F., and Zander, H. A., Thermal conductivity of dentin, *J. Dent. Res.* **29**, 493–497, 1950.
7. Heithersay, G. S., and Brannstrom, M., Observations on heat-transmission experiments with dentin, 1. Laboratory study, *J. Dent. Res.* **42**, 1140–1145, 1963.
8. Stanford, J. W., Paffenbarger, G. C., Kumpula, J. W., and Sweeney, W. T., Determination of some compressive properties of human enamel and dentin, *J. Am. Dent. Assoc.* **57**, 487–492, 1958.
9. Lehman, M. L., Tensile strength of human dentin, *J. Dent. Res.* **46**, 197–201, 1967.
10. Craig, R. G., Peyton, F. A., and Johnson, D. W., Compressive properties of enamel, dental cements, and gold, *J. Dent. Res.* **40**, 936–945, 1961.
11. Craig, R. G., and Peyton, F. A., Elastic and mechanical properties of human dentin, *J. Dent. Res.* **37**, 710–718, 1958.
12. McGinley, M. B., Lloyd, B. A., Despain, R. R., and Brown, W. S., Tensile strength of enamel, *IADR Abstracts 1972*, no. 871.
13. Bowen, R. L., and Rodriguez, M. M., Tensile strength, testing method, and values for some dental materials, *J. Dent. Res.* **39**, 768–769, 1960.
14. Hannah, C. M., Tensile properties of human enamel and dentin, *J. Dent. Res.* **50**, abstract 113, 1971.

15. Cooper, W. E. G., and Smith, D. C., Determination of the shear strength of enamel and dentin, *J. Dent. Res.* **47**, 997, 1968.

16. Barton, J. A., and Dickson, G., Thermal and hygroscopic dimensional changes of tooth structure, *IADR Abstracts 1970*, p. 217.

17. Spitzer, D., and Ten Bosch, J. J., The absorption and scattering of light in bovine and human dental enamel, *Calcif. Tissue Res.* **17**, 129–137, 1975.

18. Boehm, R., Baechler, T., Webster, J., and Janke, S., Laser processes in preventive dentistry, *Opt. Eng.* **16**, 493–496, 1977.

19. Boehm, R. F., and Gregory, R. W., An estimation of the optical properties of human tooth enamel in the visible wavelength region, AIAA paper 75-714, 1975.

20. Fowler, B. O., National Institute for Dental Research, private communication, 1972.

21. Lisanti, V., and Zander, H., Thermal injury to normal dog teeth: *in vivo* measurements of pulp temperature increases and their effect on the pulp tissue, *J. Dent. Res.* **31**, 548–558, 1952.

22. Zach, L., and Cohen, G., Pulp response to externally applied heat, *Oral Surg. Oral Med. Oral Pathol.* **19**, 515–530, 1965.

23. Boehm, R. F., Chen, M. J., and Blair, C. K., Temperatures in human teeth due to laser heating, ASME paper 75-WA/Bio-8, 1975.

24. Delbalso, A. M., and Todd, M. J., The effects of thermal injury on pulpal hydrolases, *Oral Surg.* **40**, 801–803, 1975.

25. Matthews, B., Cold-sensitive and heat-sensitive nerves in teeth, *J. Dent. Res.* **47**, 974–975, 1968.

26. Naylor, M. N., Studies on the mechanism of sensation to cold stimulation of human dentin, in *Sensory Mechanisms in Dentin*, D. J. Anderson, ed. (Pergamon, Oxford, Eng., 1963).

27. Seldon, H. S., Periapical temperatures of teeth undergoing endodontic therapy, *J. Dent. Res.* **45**, 1424–1429, 1966.

28. Mukherjee, S., The temperature of the gingival sulci, *J. Periodontol.* **49**, 580–584, 1978.

29. Van Hassel, H. J., and Brown, A. C., Effect of temperature changes on intrapulpal pressure and hydraulic permeability in dogs, *Arch. Oral Biol.* **14**, 301–315, 1969.

30. Brown, A. C., and Goldberg, M. P., Surface temperature and temperature gradients of human teeth *in situ*, *Arch. Oral Biol.* **11**, 973–983, 1966.

31. Newton, A. V., and Mumford, J. M., Thermal flow through human teeth, *Dent. Practit.* 84–86, Nov. 1970.

32. Jarby, S., On temperature measurements in teeth, *Odonto. Tids.* **66**, 421–471, 1958.

33. Jacobs, H. R., Thompson, R. E., and Brown, W. S., Heat transfer in teeth, *J. Dent. Res.* **52**, 248–252, 1973.

34. Boehm, R. F., Thermal environment of teeth during open-mouth respiration, *J. Dent. Res.* **52**, 75–78, 1972.

35. Braden, M., Heat conduction in normal human teeth, *Arch. Oral Biol.* **9**, 479–486, 1964.

36. Jacobs, H. R., Brown, W. S., and Ting, V. C., Influences of heat transfer on teeth, ASME paper 71-WA/HT-30, 1971.

37. Takahashi, N., Kitagami, T., and Komori, T., Evaluation of thermal change in pulp chamber, *J. Dent. Res.* **56**, 1480, 1977.

38. Lloyd, B. A., McGinley, M. G., and Brown, W. S., Thermal stress in teeth, *J. Dent. Res.* **57**, 571–582, 1978.

39. Peultier, N., and Frank, R. M., Effects of thermal shocks on human dental enamel, *J. Dent. Res.* **46**, 1249, 1967.

40. Brown, W. S., Thompson, R. E., and Jacobs, H. R., Thermal fatigue in teeth, *J. Dent. Res.* **51**, 461–467, 1972.

41. Robinson, H. B. G., and Lefkowitz, W., Operative dentistry and the pulp, *J. Prosthet. Dent.* **12**, 985–1001, 1962.

42. Stanley, H. R., Pulpal response to dental techniques and materials, *Dent. Clin. North Am.* **15**, 115–126, 1971.

43. Aplin, A. W., Sorenson, F. M., and Cantwell, K. R., Method for measuring temperature changes in the tooth during restorative procedures, *J. Dent. Res.* **42**, 925–933, 1963.

44. Sorenson, F. M., Cantwell, K. R., and Aplin, A. W., Thermogenics in cavity preparation using air turbine handpieces: the relationship of heat transferred to rate of tooth structure removal, *J. Prosthet. Dent.* **14**, 524–532, 1964.

45. Sorenson, F. M., Phatak, N. M., and Everett, F. G., Thermal pulp tester: a new instrument, *J. Dent. Res.* **41**, 961–965, 1962.

46. Walsh, J. P., and Symmons, H. F., A comparison of the heat production and mechanical efficiency of diamond instruments, stones, and burs at 3,000 and 60,000 rpm, *N. Z. Dent. J.* **45**, 28, 1949.

47. Hudson, D., and Sweeney, W., Temperatures developed in rotating dental-cutting instruments, *J. Am. Dent. Assoc.* **47**, 127–133, 1954.

48. Peyton, F. A., Temperature rise in teeth developed by rotating instruments, *J. Am. Dent. Assoc.* **50**, 629–632, 1955.

49. Vaughn, R. C., and Peyton, F. A., The influence of rotational speed on temperature rise during cavity preparation, *J. Dent. Res.* **30**, 737–744, 1951.

50. Wheatcroft, M. G., Harnett, J. E., and Smith, W. F., Tooth pulp temperature changes produced by cutting with air-turbine handpieces, *J. Dent. Res.* **39**, 753, 1960.

51. Eames, W. B., and Nale, J. L., A comparison of cutting efficiency of air-driven fissure burs, *J. Am. Dent. Assoc.* **86**, 412–415, 1973.

52. Lloyd, B. A., Christensen, D. O., Jacobs, H. R., and Brown, W. S., Heat transfer in teeth during restoration, Annual meeting, American Institute of Chemical Engineers, Philadelphia, Nov. 1973.

53. Brown, W. S., Christensen, D. O., and Lloyd, B. A., Numerical and experimental evaluation of energy inputs, temperature gradients, and thermal stresses during restorative procedures, *J. Am. Dent. Assoc.* **96**, 451–458, 1978.

54. Kawahara, H., and Yamagami, A., *In vitro* studies of cellular responses to heat and vibration in cavity preparation, *J. Dent. Res.* **49**, 829–835, 1970.

55. Aplin, A. W., Sorenson, F. M., and Cantwell, K. R., Thermogenics in cavity preparation using the air-turbine handpiece, *J. Dent. Res.* **40**, 769, 1961.

56. Bhaskar, S. N., and Lilly, G. E., Intrapulpal temperature during cavity preparation, *J. Dent. Res.* **44**, 644–647, 1965.

57. Schuchard, A., Surface temperature response by use of air coolant in restorative procedures, *J. Am. Dent. Assoc.* **75**, 1188–1193, 1967.

58. Crawford, W. H., Thermodynamics in cavity preparations, *N.W. Dentistry* 296–302, Sept. 1957.

59. Peyton, F. A., Effectiveness of water coolants with rotary-cutting instruments, *J. Am. Dent. Assoc.* **56**, 664–675, 1958.

60. Schuchard, A., and Watkins, C., Temperature response to increased rotational speeds, *J. Dent. Res.* **39**, 738, 1960.

61. Zach, L., and Cohen, G., Thermogenesis in operative technics: comparison of four methods, *J. Prosthet. Dent.* **12**, 977, 1962.

62. Schuchard, A., and Watkins, C. E., Thermal and histologic response to high-speed and ultrahigh-speed cutting in tooth structure, *J. Am. Dent. Assoc.* **71**, 1451–1458, 1965.

63. Woods, R. M., and Dilts, W. E., Temperature changes associated with various dental-cutting procedures, *J. Can. Dent. Assoc.* **35**, 311–315, 1969.

64. Hamilton, I. A., and Kramer, I. R., Cavity preparation with and without waterspray, *Br. Dent. J.* **123**, 281–285, 1967.

65. Carlton, M. L., Jr., and Dorman, H. L., Comparison of dentin and pulp temperatures during cavity preparation, *Tex. Dent. J.* **87**, 7–8, 1969.

66. Pinsky, L. D., The coolant controversy in operative dentistry, *Wis. State Dent. Soc.* **47**, 279–282, 1971.

67. Lloyd, B. A., Rich, J. A., and Brown, W. S., Effect of cooling techniques on temperature control and cutting rate for high-speed dental drills, *J. Dent. Res.* **57**, 675–684, 1978.

68. Kasloff, A., Cracks in tooth structure associated with rotary-cutting instruments, *J. Dent. Res.* **40**, 769, 1961.

69. Kasloff, A., A continuing study of cracks in teeth associated with various rotary-cutting instruments, *J. Can. Dent. Assoc.* **28**, 244, 1962.

70. Kasloff, Z., Enamel cracks caused by rotary instruments, *J. Prosthet. Dent.* **14**, 109–116, 1964.

71. Grajower, R., Kaufman, E., and Stern, N., Temperature of full-crown preparations with modelling compound, *J. Dent. Res.* **54**, 212–217, 1975.

72. Jarby, S., and Dansgaard, W., The thermal effects of different operations on teeth, *Odonto, Tids.* **67**, 207–240, 1959.

73. Plant, C. G., Jones, D. W., and Darvell, B. W., The heat evolved and temperatures attained during setting of restorative materials, *Br. Dent. J.* **137**, 233–238, 1974.

74. Braden, M., Heat conduction in teeth and efficiency of lining materials, *J. Dent. Res.* **42**, 1084, 1963.

75. Braden, M., Heat conduction in teeth and the effects of lining materials, *J. Dent. Res.* **43**, 315–316, 1964.

76. Sorenson, F. M., Aplin, A. W., and Cantwell, K. R., Preliminary studies on thermal production by some commonly used polishing procedures in dentistry, *J. Dent. Res.* **40**, 769, 1961.

77. Christensen, G. J., and Dilts, W. E., Thermal change during dental polishing, *J. Dent. Res.* **47**, 690–693, 1968.

78. Grajower, R., Kaufman, E., and Rajstein, J., Temperature in the pulp chamber during polishing of amalgam restorations, *J. Dent. Res.* **53**, 1189–1195, 1974.

79. Wittrock, J. W., Morrant, G. A., and Davies, E. H., A study of temperature changes during removal of amalgam restorations, *J. Prosthet. Dent.* **34**, 179–186, 1975.

80. Roydhouse, R. H., and Paxon, P. R., Thermal changes in dimension of restorative cavities, *J. Dent. Res.* **49**, 567–571, 1970.

81. Guzman, H. J., Swartz, M. L., and Phillips, R. W., Marginal leakage of dental restorations subjected to thermal stress, *J. Prosthet. Dent.* **21**, 166–175, 1969.

82. Bounocore, M. G., Caries prevention in pits and fissures sealed with an adhesive resin polymerized by ultraviolet light: a two-year study of a single adhesive application, *J. Am. Dent. Assoc.* **82**, 1091–1093, 1971.

83. Von Fraunhofer, J. A., and Williams, B., Heat liberation during the setting of four fissure sealants, *Br. Dent. J.* **136**, 498–499, 1974.

84. Stern, R., Dentistry and the laser, *Laser Applications in Medicine and Biology*, vol. 2, M. L. Wolbarsht, ed. (Plenum, New York, 1974).

85. Boehm, R., Rich, J., Webster, J., and Janke, S. Thermal stress effects and surface cracking associated with laser use on human teeth, *J. Biomech. Eng.* **99**, 189–194, 1977.

86. Boehm, R. F., Development of new processes for preventive dentistry, final report, grant DE0554, National Institute of Dental Research, Oct. 1, 1978.

87. McClure, Frank, *Water Fluoridation: The Search and The Victory* (Washington, D.C.: U.S. Department of Health, Education and Welfare, 1970).

88. Baechler, T. K., Diffusion with chemical reaction: fluoride incorporation in tooth enamel (Ph.D. diss., University of Utah, 1976).

89. Putt, M. S., Beltz, J. F., and Muhler, J. C., Effect of temperature of SnF_2 solution on tin and fluoride uptake by bovine enamel, *J. Dent. Res.* **57**, 772–776, 1978.

90. Ozisik, M. N., *Heat Conduction* (Wiley, New York, 1980).

91. Arpaci, V. S., *Conduction Heat Transfer* (Addison-Wesley, Reading, MA, 1966).

92. Carslaw, H. S., and Jaeger, J. S., *Conduction of Heat in Solids*, 2nd ed. (Oxford, London, 1959).

Chapter 25

ANALYSIS AND APPLICATION OF THERMOGRAPHY IN MEDICAL DIAGNOSIS

T. J. Love

1. INTRODUCTION

Thermography is the term used to describe the technique of mapping skin temperature patterns and the interpretation of that information to assess physiologic status. The temperature map is typically obtained with a scanning infrared radiometer; a CRT typically provides a visual image of the map. Temperature levels may be displayed by various shades of grey or by colors representing discrete temperature levels. Photographs of the displayed image provide a record of the temperature patterns. Liquid crystals applied directly to the skin or encapsulated in sheets of plastic have also been used to portray skin temperature. Skin temperature maps have been used in a wide range of medical applications. These applications include, but are not limited to, breast cancer detection, assessment of peripheral vascular disease, soft tissue trauma, infection, rheumatology, detection of thyroid cancer, viability of skin flaps, monitoring densitized tissues, diagnosis of occlusive disease of the carotid artery, and diagnosis and treatment-monitoring in headache. Applications will be treated in detail later in this chapter.

Thermography has the obvious advantage of being a noninvasive technique that provides the physician with a spatially coherent map of surface temperature. It will be demonstrated that this map provides an indication of local cutaneous blood perfusion, which may be related to certain physiological states.

Since the thermal map reflects the effects of perfusion, metabolism, and environmental temperature, it cannot be viewed in the same context as the radiograph. However, like the radiograph, the thermal modality provides a painless and completely noninvasive test that may be performed with relative ease.

The primary disadvantages of thermography are related to the lack of specificity of the examination and the careful control of environmental conditions and instrument calibration that are required. It should also be

T. J. Love • School of Aerospace, Mechanical, and Nuclear Engineering, University of Oklahoma, Norman, Oklahoma 73019.

emphasized that thermography is not in competition with X ray or ultrasound, since it is only an image of surface temperature and primarily a physiologic measurement, as opposed to the anatomic measurement provided by the other imaging modalities.

Skin temperature is a very labile quantity, which depends on the environment as well as the physiologic activity in the tissues, thus, the quality of the thermogram depends on operator technique. For best results, the instrumentation should be calibrated and the patient's environment carefully controlled. In most applications, the quality of the thermograms has depended on the technician's adjustment of the contrast and brightness. As a result, physicians working with their own technique have been more successful[1] than those who used thermograms taken independently at another clinic.[2]

Much of the early medical literature has described thermography in terms of an observation of rather mysterious infrared rays emanating from the diseased region of the anatomy. In breast cancer studies, many investigators mistakenly considered the elevation of skin temperature associated with malignancy to be the result of heat conduction from the hot tumor to the surface.[3–6] The fallacy of this model will be demonstrated later in this chapter. The point that must be made is that if skin temperatures are to be used in diagnosing disease and monitoring therapeutic treatment, it is essential to understand the heat transfer processes that contribute to those patterns. The present work examines those factors contributing to thermographic patterns and provides a basis for estimating physiologic activity on the basis of those patterns.

2. INSTRUMENTATION

The scanning infrared camera operates on the principle of focusing the long wavelength infrared radiation emitted from a small area of skin onto a liquid nitrogen-cooled semiconductor detector. The thermal radiation reaches the detector via a rotating mirror that moves the focal spot over the skin in a manner much as a television camera scans a visible scene. The response of the detector is amplified and displayed on a cathode ray tube with a scan synchronized to the scanning mirror (Fig. 1). Thus, the image on the cathode ray tube represents a map of the infrared radiation leaving the skin. This radiation is principally the thermal emission of the skin and is related to the skin's temperature. The transfer function between thermal emission and temperature is considered in detail later in this chapter.

Early thermographs with a thermopile detector[7] had scan rates of 15 min. per image. Modern instruments use scan rates ranging from 30 frames per sec to a frame each 4 sec. These instruments use either indium–antimonide, with detection primarily between 3–5 μm or mercury cadmium telluride, with detection in the 8–15-μm wavelength range. These detectors are normally operated at liquid nitrogen temperatures. There are a number of instruments based on this technology that are commercially available in the United States, Western Europe and Japan.

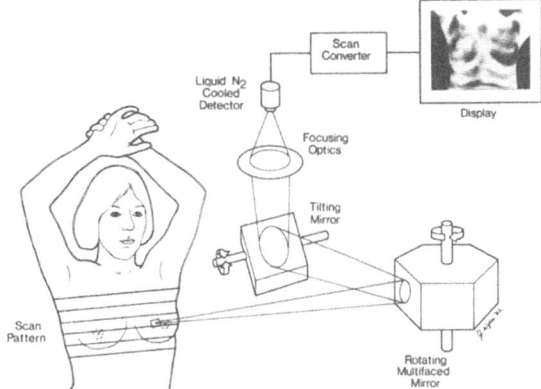

FIGURE 1
Schematic of an infrared thermography measurement system.

A number of factors must be considered when evaluating data from such instruments. In general, experience has shown that the accuracy of the temperature determination is much lower than the temperature resolution reported in the literature. Before using an instrument for quantitative temperature measurements, it should be carefully calibrated. The following items should be checked using an accurate hohlraum (isothermal cavity):

1. Linearity of response over the temperature range.
2. Uniformity of output over the field of view.
3. Modulation transfer function of the temperature pattern.

With regard to the last item, it is possible to qualitatively view spatial temperature variations on the cathode ray tube viewer. However, the accuracy with which the temperature of a hot spot may be determined is a function of its size and the temperature of the surrounding space. This may be checked experimentally by observing the temperature reading of a hohlraum with different apertures. This is a very important consideration when attempting to measure skin temperatures over superficial veins. Vein temperatures are an important indicator of regional blood flow, as will be noted later.

Liquid crystal thermography has been used as an alternative method to the scanning infrared camera and is receiving increasing attention, owing to its low cost.[8,9] Liquid crystals are composed of long chain cholesteric molecules that exhibit a short range order, similar to that of crystalline polymers. Light is selectively reflected from the liquid crystal, giving rise to colors. The color reflected is sensitive to the molecular state, which changes with temperature. The liquid crystal response may be calibrated, so that the color reflected is related to temperature. Thus, skin temperature patterns may be obtained by applying a thin layer of the material to the skin and observing the color change. The method requires that a black paint be first applied to the skin and the liquid crystal applied over the paint. A technique using liquid crystals encapsulated in flexible, reusable plastic sheets is much easier to use.

A record of the temperature patterns is retained by making a color photograph of the color pattern. A drawback of this method is the fact that the skin temperature is altered during the application process. In addition, the plastic sheet may be difficult to contour to the body, with resulting errors in observed temperature if the contact, and thus the thermal conductance, are not uniform.

The recent development of a pyroelectric vidicon tube for thermal infrared imaging offers the possibility of an infrared television system.[10] The pyroelectric vidicon tube is similar in operation to a standard vidicon, except that the target is a dielectric material that, on irradiation, undergoes a pyroelectric conversion from a temperature pattern to a charge pattern. This conversion, which results from a change in polarization, produces a charge-induced field that is probed by an electron beam that scans the target in the vidicon. The discharge current generated by the interaction of the beam and the thermal image is amplified and transmitted in a manner similar to that for a television signal. A major advantage of the pyroelectric vidicon is that it is compatible with commercial television-signal-processing electronics and does not require liquid nitrogen cooling of the detector. Unfortunately, the pyroelectric vidicon has not yet been developed with sufficient image quality and temperature discrimination, so that it is suitable for medical thermography. There have been numerous improvements in the design in recent years, and the possibility for useful thermograms obtained with this instrument will most likely occur in the future.

For the commercially available, single element indium antimonide or mercury cadmium telluride detector-based instruments, the detector response is a function of the total incident energy. It is therefore necessary to make design trade-offs. Some instruments have a faster scan rate, with only 100 lines of spatial resolution, while other instruments provide a standard, television format spatial resolution, with slower frame rates. These slower frame rates create some difficulty in focusing the image but provide much improved image quality. A recent experimental unit uses a serial scan, multi-element detectors.[11] It employs circuit time delays and a signal-summing circuit, which permits thermal infrared imaging with the standard television format of 60 frames per sec. The signals are interlaced to provide improved spatial and radiometric accuracy.

In all commercially available scanning infrared camera designs, the output of the detector and associated electronics is proportional to the temperature of an equivalent black body. In the general case, the output is proportional to the radiative energy input from a blackbody, where T is the absolute temperature, i.e., σT^4. However, when small temperature differences are to be measured, with reference to the radiant flux from a surface at known reference temperatures T_{ref}, i.e., $T - T_{ref}$, the radiative flux difference can be approximated by the temperature difference

$$\sigma(T^4 - T_{ref}^4) \approx 4\sigma(T_{ref})^3(T - T_{ref}) \qquad (1)$$

Thus, within the limits of this approximation, radiative flux can be estimated. As shown in Chap. 27 this temperature measurement can be quite sensitive and accurate.

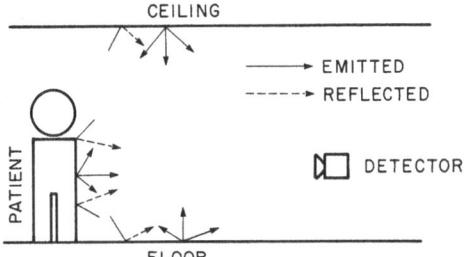

FIGURE 2
Radiative exchange between patient,
room, and camera.

The radiation sensed by the instrument will include thermal energy
emitted from all surfaces in the field of view, by virtue of the temperatures
of those surfaces, and the reflection of the infrared radiation incident on the
surface from the surroundings (Fig. 2). The physical nature of human skin is
such that the reflection of radiation from skin in the detector's spectrum is
less than 1%. Thus, the energy sensed by the thermograph is essentially that
emitted by the skin surface, with a flux proportional to the fourth power of
the absolute temperature of the skin. The range of measured temperatures
observed in a clinic, with an ambient temperature of 22°C, will lie between
that temperature and the core temperature of the subject; most observations
will fall between 26 and 36°C. If the instrument is used to monitor hyperthermia therapy, this upper limit must be raised approximately 10°C.

For most instruments and applications, the technician conducting the
examination adjusts the brightness and contrast controls in order to obtain a
suitable image, the observed thermal details depend on these adjustments.
Most physicians who achieve some degree of success with thermographic
diagnostics spend considerable time with the instrument and the technician
in order to develop a reliable technique for qualitative interpretation. The
standard procedure for developing a diagnostic examination is to have the
physician observe a number of normal subjects along with a group of patients
with known pathology. Variations in the thermal patterns are noted, usually
by comparison with symmetric contralateral regions, local temperature elevations and prominent vascular patterns are also noted.

Skin temperature distributions may also be obtained from negative films
by using densitometric measurements, calibrated against reference hohlraums
in the field of view. Recent development of high-speed analog-to-digital
converters permits the direct digital storage of the signal. Anliker[12] processed
the digital image but did not calibrate it in terms of absolute temperature.
Haberman[11] included digital images of four reference temperatures on each
sequence, which permitted actual temperature measurements of the thermal
image. These images were processed by digital computer to provide various
parameters, based on the skin distribution. These parameters were used in
various correlations with the incidence of breast tumors.

3. OPTICAL PROPERTIES OF SKIN

An important consideration in relating the output of the scanning infrared
camera to the temperature of the skin is the skin emissivity. The emittance

ratio of the observed radiative emission at given wavelength and skin temperature to the ideal value is defined by Planck's radiation law (Appendix 1). Experimental results are reported for the spectral emissivity of skin between 3 and 4 μm by Steketee.[13] In this range, the emittance is independent of wavelength and also the pigmentation of the skin observed in the visible spectrum. The value Steketee gives in this range is 0.98 ± 0.01 for both *in vivo* and *in vitro* measurements, which is in good agreement with other works.[14,15]

These values were obtained in the direction normal to the skin surface. Although the emittance may vary in nonnormal directions, the diffuse nature of emission from the skin permits use of the normal emissivity value up to about 45°C from the normal without introducing significant error. By performing an energy balance on the sensed area, it can be shown that temperature measurements by thermography should be valid up to an angle of 75°C.[16]

Since the clinical environment involves a room temperature of approximately 20°C, the energy reflected by the skin compensates for the reduction of net emission (emissivity less than unity). A simple relationship may be derived that provides an error estimate.[17]

$$\text{Sensed temperature} = \left[\frac{\varepsilon T_s^4 + (1 - \varepsilon) T_\infty^4}{T_s^4} \right] T_s \qquad (2)$$

where ε is the skin emissivity, T_s is the actual skin temperature, and T_∞ is the room temperature. For a 303 K (30°C) skin temperature and a 293 K (20°C) clinic temperature, the ± 0.01 variation of emissivity reported in the literature will result in only a ± 0.01°C variation in observed temperature. This value is less than the error range of most instruments.

One other point to be made is that the skin is opaque to thermal radiation. The temperature recorded by radiometric methods is a measurement of thermal radiation that originates from within approximately 45 μm of the skin surface. Thus, the observed temperature is essentially that of the surface.

In many cases, quantitative temperature data are required rather than relative qualitative images, in these cases, it is important to use care in evaluating temperature. As previously shown, the radiative flux leaving the skin is characterized by the nominal black body flux, which is proportional to the fourth power of the absolute temperature. We must, however, recognize that the optical path of the system, response of the liquid-nitrogen-cooled detector, and the electronic amplification will modify the signal significantly. Further, if a photographic image of the cathode ray tube is the final output, it is easy to see that many factors, including the photographic emulsion and development may influence the accuracy of the temperature determination.

Because of these and other factors that can influence the final determination, it is important to include a series of reference targets in the instrument field of view. A minimum of four reference temperatures should be included. One reference should be at a temperature corresponding to the minimum and one to the maximum temperature in the field of view, the other two should be equally spaced along the thermal scale.

These references can be conveniently constructed from blocks of aluminum, which should be drilled to a depth of about 3 cm with a 2-cm diam. drill. The interior of the block is painted with a flat black paint. The reference is heated by an electric resistor controlled by a small solid-state controller with a thermistor transducer. A separate thermistor should be mounted on each block and the four blocks mounted in styrofoam insulation and an appropriate case. The temperatures of the references may be monitored by the implanted thermistors and a digital temperature display.

With these four reference temperatures in the field of view, a linear scale or a curve may be fitted throughout the four temperatures and the temperatures of the image determined by interpolation. For "on line" digitized images, this interpolation may be accomplished by the computer, for film images, the negative is evaluated using a transmission photomicro densitometer. In such a system, the interpolation is normally accomplished by hand. Some thermographic cameras have a color CRT display. With such systems the colors are adjustable to represent discrete temperature intervals, and the reference temperatures are displayed in color and the image evaluated accordingly. For digital systems and color level systems, the cameras may be stable enough, so that an image of the four reference temperatures can be recorded at the beginning and end of a study. If care is taken not to adjust the instrument during the study, the necessity of having the reference in each picture is avoided. It must be reemphasized that the stability of the instrument should be carefully checked.

4. CONTROL OF THE CLINIC ENVIRONMENT

For medical diagnostics, the objective of thermography is to assess physiologic function from skin temperature patterns. In order to properly use the skin temperature measurement for this purpose, it is important to recognize and control the influence of external heat transfer effects. The principal modes of heat transfer that must be considered are convective and radiative exchanges with the room. The energy balance with the environment includes the net exchange by radiation between the skin and the room's walls and the convective heat transfer to the air moving over the skin. In a thermographic clinic with controlled temperature and humidity, the evaporative loss from dry skin can be shown to be a second-order effect. The physician who understands these factors is in a position to use the modality successfully in a clinical setting.

The radiative contribution from an isothermal environment to the energy balance of the skin is a straightforward computation. The skin absorptivity is near unity, and therefore the skin absorbs nearly all of the incident energy. Thus, maintaining the clinic at as nearly a uniform temperature as possible permits the control of the absorbed energy. This is done not only from the standpoint of improving the accuracy of quantitative measurements, but it is also very important in qualitative thermography. The effect of variation in the room's wall or adjacent skin temperature on the net energy flux at the skin surface may be seen by noting the effect of a nonisothermal environment on the radiative transfer. Since most wall paints, like skin, have an emittance near unity, the following expression will approximate the net radiative flux from the skin to its surroundings:

$$q_{net} = \sigma\left(T_s^4 - \sum_{j=1}^{N} F_{sj}T_j^4 \right) \tag{3}$$

where F_{sj} is the geometric view factor from the skin to the jth surface at temperature T_j. For isothermal surroundings, T_j is constant and $F_{sj} = 1$ if the skin views only these surroundings. In the situation where the skin views other parts of the body (such as the medial aspect of the limbs), the net radiative flux is decreased by the fraction of the hemispherical solid angle that is subtended by the other parts of the body. Windows, cold exterior walls, or other objects, such as hot lights, may cause significant skin temperature variations and thus produce artifacts that may lead to errors in interpretation.

Uniformity of convective heat transfer at the skin surface is necessary for reliable comparison of temperature profiles in contralateral, symmetric regions of the body. Furthermore, if determination of other physiologic parameters, such as perfusion rate is required, the environmental conditions must also be uniform. In our practice, the convective heat exchange with the skin has been regulated by having a larger than normal airflow through the room. Our experience has suggested that the air temperature should be controlled by an automatic temperature controller with a band width of ±0.1°C. The airflow rate should be approximately 0.5 m³/sec for each occupant in the room, including both patients and technicians, an additional 1.0 m³/sec should be provided to account for miscellaneous heat gains. The clinic should be constructed in the interior portion of a building away from exterior walls. The airflow should be introduced through a perforated ceiling from a plenum that extends over the entire room. Recommended airflow rates should provide a uniform downward flow of air with a velocity between 0.05 and 0.08 m/sec. The return air should be collected at the floor level through registers, well spaced around the room's perimeter.

Such an arrangement will provide uniform airflow and avoid skin temperature artifacts, which will inevitably occur in a clinic with normal air-conditioning design. A calculation of the Reynolds number/Grashof number ratio,† which compares forced convection with thermally induced free convection flows, demonstrates that the convective heat transport from the skin will otherwise be dominated by (variable) free convection. The recommended down draft configuration avoids other difficulties in providing forced airflow over each patient in a uniform manner.

It is necessary for the patient to remain supine for at least 10 min in the clinic environment before the thermogram is taken. This provides a reproducible basal metabolic level and also a repeatable steady-state environmental heat exchange condition.

† See Appendix 1.

5. APPLICATIONS OF THERMOGRAPHY

5.1. Breast Cancer Diagnosis

The death toll from breast cancer is about 50,000 per year in the United States. The prognosis for long-term survival is dramatically increased if this cancer is identified and treated at an early stage. The possibility of developing improved methods for early detection was responsible for much of the early development of thermography. A study by Lilienfeld[18] demonstrated that physical examination, mammography, and thermography each provided approximately the same sensitivity and specificity. However, these modalities selected different populations. The combination of all three modalities provides the best probability for early detection of breast cancer.

It should be pointed out that the interpretation of breast thermograms is not a simple pattern recognition problem. There is a significant overlap between normal patterns and those found in a patient with cancer. In general, the physician must compare the symmetry of the superficial vascular patterns, the areolar temperatures, and the temperature profiles between contralateral regions. Since most cameras in current use are not calibrated, the physician must also compare the degree of asymmetry with that observed in the entire clinic population. Figure 3 is an illustration of a breast thermogram. The pattern of vascularity in this patient represents a suspicious finding. In this case, the patient had benign fibrocystic disease, and the pattern remained unchanged over a period of years. Incidentally, one of the problems with all three detection modalities is that differential diagnosis of fibrocystic disease versus cancer is difficult.

It should also be pointed out that thermography is not a substitute for x-ray mammography. Thermography may be used in conjunction with physical examination as a harmless and inexpensive method of screening women for breast cancer. One study conducted with mobile clinics in a rural area indicated that continued thermography/physical examination produced a significant improvement in detection over physical examination alone.[19] X radiation examination carries a small but finite risk of tissue destruction by ionization; exposure intensity and duration must therefore be limited. The radiation exposure and expense of x-ray mammography may be significantly reduced by thermography.

Most of the published references dealing with breast thermography rely on the results of a visual interpretation, by a physician, of a photograph of the thermographic display. In an extensive study, Haberman[11] studied a large population (10,000) of normal women, as well as symptomatic patients. The study was conducted in a carefully controlled environment with an experimental scanning infrared camera that recorded calibrated absolute digital images. These images were studied with a large digital computer, and statistical pattern recognition schemes were developed during the course of the analysis. The study demonstrated that the specificity and sensitivity of computer-evaluated absolute temperature patterns may provide a viable screening method for early breast cancer. The study also indicated that qualitative

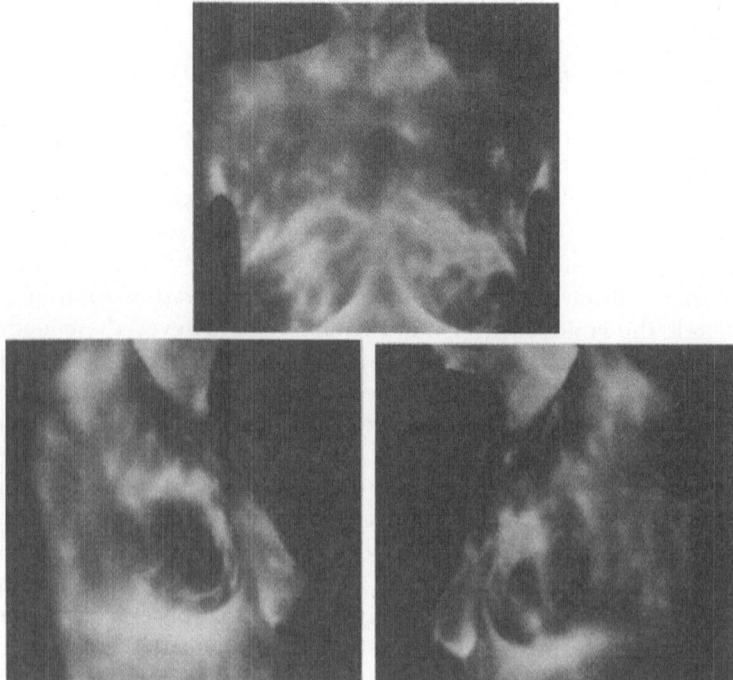

FIGURE 3

Thermogram of female breast. The anterior view and right and left oblique are normally taken for diagnostic studies. Note the superficial venous patterns indicating increased blood flow. This patient had fibrocystic disease and has been followed at 6-month intervals for 8 years with minimal changes noted.

thermograms, properly taken in a controlled environment and analyzed by experienced physicians, can make an important contribution to the early detection of breast cancer.

The thermal pattern of breast cancer appears to be the result of a regional stimulus of the tissues by the early development of the malignant tumor. Blood perfusion increases in response to this stimulus, which in turn causes some increase and asymmetry in regional temperature and also increases superficial vascularity. The thermographic signal *should not* be interpreted as the conduction of heat from a "hot" tumor to the skin. This misunderstanding, which has often appeared in the literature, contributes to a misinterpretation of the breast thermogram.

The chief problem with breast cancer detection by thermography, assuming a reliable technique is employed, results from the variability of the "normal" female's breast temperature pattern. Breast cancer is very difficult to detect by any modality in the early stage. The incidence of 1–2 per 1,000 makes clinical studies very expensive. However, the death rate from the disease, combined with the concern for the damaging effects of x-ray, makes thermography a viable alternative methodology for screening and diagnosis.

5.2. Peripheral Vascular Disease

This application of thermography allows the indirect observation of the peripheral circulation in terms of blood-borne heat that is conducted to the skin surface. With proper instrument calibration and environmental control, a quantitative assessment can be made of the superficial peripheral circulation.[20] The detection of more deeply seated abnormalities may also be accomplished by thermography.[21,22] Thermographic imaging of the extremities provides a detailed map of temperature variations. Since ischemic conditions produce a decrease in temperature from normal values, these maps may be quite dramatic. While gross changes in temperature are usually clinically obvious, small progressive changes and borderline cases require a careful instrumented study that is facilitated by thermography. Ring[21] proposes using a thermal stress test, with serial thermographic studies of the skin temperature recovery following stress. The test he used involved cooling the extremity in water at 20°C and observing the subsequent hyperemic reaction over a 10-min period. The test appears to be capable of differentiating scleroderma and Raynaud's phenomena. The test may have additional value in assessing the level for amputation and monitoring the effectiveness of nonsurgical therapy.

A note of caution must be raised in thermographic examinations of the extremities. The effects of smoking, drug administration, and other vasomotor effects must be considered. Variation in central temperature will also produce peripheral effects. In addition, the emotional stress of the patient can also cause significant changes in the temperatures of the extremities. All of these factors must be carefully monitored and controlled in order to produce reliable results.

5.3. Soft Tissue Trauma

The assessment of muscle strain as well as bruised tissues may be made by thermography, based on the blood perfusion to the region. Qualitative comparisons of contralateral symmetry as well as controlled serial examinations are of value.[22,23]

5.4. Assessment of Infection

Thermography has proven valuable in identifying appendicitis.[24] It has also been successfully used to identify abdominal abscess in the post–surgical care of patients who have undergone abdominal surgery.[25]

5.5. Monitoring Desensitized Tissues

Thermography has been used successfully to monitor the soft tissues in limbs of people suffering from such diseases as diabetes, in which nerve damage causes a loss of feeling. Monitoring such areas helps avoid the trauma of repeated activities that may lead to ulceration and infection.[26,27]

5.6. Rheumatology

Thermography has demonstrated potential in assessing arthritic conditions and treatment in rheumatology.[31,32]

5.7. Plastic Surgery

Thermographic methods have been successfully used in predicting the viability of skin flaps and other skin grafts and in assessing burns.[31,32]

5.8. Thyroid Cancer

Thermography has been used to identify and assess thyroid nodules. In general, nodules that are inactive on nuclear medicine examination and show an increased thermal pattern by thermography have a very high probability of being malignant.[33,34] In the thermographic examination of the thyroid, a cooling technique may be employed to enhance the thermal image. Figure 4 illustrates the result of such a procedure. The throat is initially warm compared to the face and thorax, because it is a protected area of the anatomy (Fig. 4a). A plastic bag of water that has equilibrated to the 20°C clinic temperature is placed gently over the neck of the supine patient for 60 sec. A series of thermograms are then taken as the skin rewarms from the cooled condition (Fig. 4b). The malignant nodule will ordinarily be characterized by an asymmetrical increase in temperature and warm superficial vascularity, as shown at 1 min. post-cool (Fig. 4c).

The preceding sections give an indication of some of the possible uses of thermography. There have been many other applications reported in the literature. The interested reader is cautioned to read these reports critically and develop applications of interest under carefully controlled conditions. In any application, it is extremely important to recognize the environmental influences on the skin temperature patterns. It is likewise important to study the normal variations of thermal patterns in control subjects as well as symptomatic patients.

6. RELATIONSHIP OF BLOOD FLOW TO TEMPERATURE PATTERN

An additional purpose of this work is to present an heuristic method for estimating blood flow rate from the thermographic skin temperature pattern. The method is based on several critical assumptions, which may not all be met in practice. Nevertheless, the significance of obtaining a spatially coherent map of blood flow provides sufficient rationale for proceeding with this approach.

We begin with the observation that in the thermographic clinic, the skin temperature over the trunk of the body will normally range between 27 and 30°C. The higher values represent temperatures in protected crevices in the

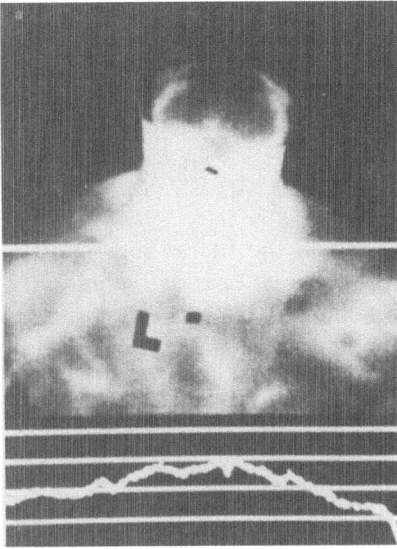
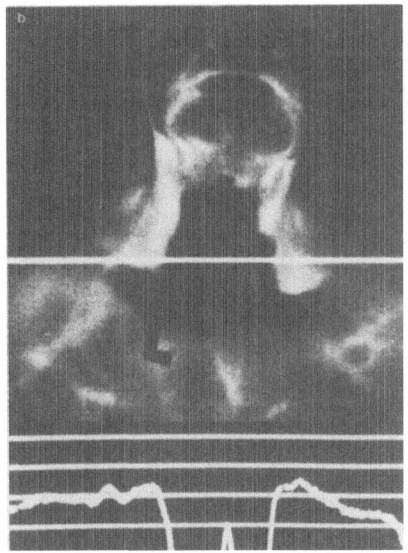
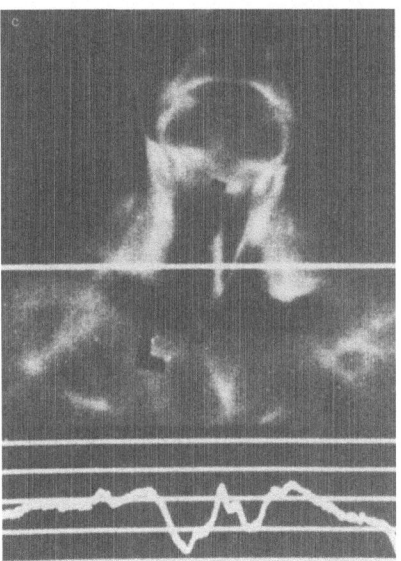

FIGURE 4
Thermogram of thyroid. (a) Throat prior to cooling; (b) immediately after cooling; (c) one-minute postcooling.

skin or over active superficial veins. The lower values are normally seen over avascular, fatty tissue. Tissue temperature increases to core temperature 2–3 cm below the skin surface, and it is assumed that the heat conducted to the surface is far outweighed by the convective contribution. It can also be shown (see the following) that heat production, such as from an actively metabolizing tumor, is negligible in comparison with blood-borne heat. Thus, the heat transport to the skin surface q may be grossly represented by the convection of heat in macroscopic vessels q_c. If we assume that the blood perfusing a region supplied by macroscopic vessels is entirely collected in superficial veins and that the blood in these veins is at the average temperature $T_{\bar{v}}$ of the tissues drained by the same veins, then the total heat flux to the skin surface may be approximated as

$$q = q_c = \rho_b c_b w_b (T_a - T_{\bar{v}}) \tag{4}$$

We have ignored the detailed processes by which the blood-borne heat is distributed to the tissue at the capillary level. The heat thus delivered to the skin surface must be transported to the environment. Under the competing influences of perfusion, environmental temperature T_0, and surface heat transfer coefficient h, the skin temperature "floats" to a level that provides the necessary thermal driving potential $T_s - T_0$ to remove heat from the skin. The surface heat loss may be expressed as q_s, where

$$q_s = h(T_s - T_0) \tag{5}$$

Under the simplifying assumptions just given and assuming in addition that the heat transfer coefficient is uniform and constant over the surfaces of interest and thermal radiation may be ignored, the expressions for delivered and rejected heat Eqs. (4) and (5) may be equated. Carrying this out and rearranging, we obtain for the perfusion rate w_b

$$w_b = \frac{h(T_s - T_0)}{\rho_b c_b (T_a - T_{\bar{v}})} \tag{6}$$

Now if we can obtain expressions for T_a and $T_{\bar{v}}$, we are in a position to provide at least a relative assessment of w_b. We may obtain T_a within approximately 0.2°C by assuming it to be equal to core temperature, obtained at rectal, oropharyngeal, tympanic membrane, or other site. We assume the skin temperature directly over macroscopic superficial veins will normally be 0.2°C cooler than the blood in those veins, in the thermographic clinic environment. Thus, we use the infrared thermogram to measure both T_s and $T_{\bar{v}}$. In order to compare perfusion rates serially, or in adjacent or contralateral regions, expressions for w_b (Eq. 6) are obtained at the two sites in terms of the relevant temperatures. Again, assuming that h, ρ_b, c_b, T_a, and T_0 do not vary between these sites, we may obtain the ratio of the superficial (tissue) perfusion rates as

$$\frac{w_{b2}}{w_{b1}} = \frac{T_{s2} - T_0}{T_{s1} - T_0} \frac{T_a - T_{\bar{v}1}}{T_a - T_{\bar{v}2}} \tag{7}$$

where subscripts 1 and 2 refer to the appropriate serial values or regions. Thus, it is possible to obtain an estimate of the superficial perfusion distribution directly from the thermographic image.

Note that the ratio of blood perfusion at the two sites [i.e., Eq. (7)] is a function of only the measured temperatures. Unfortunately, the ratio is very sensitive to the arterial supply temperature and the temperatures of the veins. These temperatures are difficult to measure with accuracy. For the thorax or abdomen, it may be possible to correlate an oral or esophageal temperature with the arterial temperature, such a correlation may be more difficult in the extremities. The skin temperature superficial to the veins is also difficult to measure accurately. Most scanning infrared cameras that are now commercially available cannot provide an accurate vein temperature. In general, the quantitative reading is influenced by the size of the vein and the temperature of the surrounding skin.

Nonetheless, the preceding analysis provides a basis for assessing physiologic activity from the skin temperature patterns. Even when quantitative values are impractical because of instrumentation limits or lack of proper environmental control, the analysis permits a qualitative assessment of thermal images.

Qualitative thermographers consider symmetry of both avascular tissue temperature and superficial vein temperatures in their evaluations. The number of superficial veins in the thermographic image may also be indicative of regional involvement. It should also be noted that the superficial venous plexus is collateral with a deep venous system. Several factors control the partitioning of blood flow in the deep and superficial venous systems, some of which are discussed in Chaps. 3, 5, and 7. Contributions of the superficial veins seem to increase in importance when there is an "excess" blood flow to an area. Examples that are particularly noted in the case of the female breast include pregnancy and lactation. At these times, the metabolic activity in the breast requires an increased blood flow. There is a visible change in the thermal image: a dense network of superficial veins appears over the surface of the breast, which are engorged with warm venous blood. After ceasing lactation, these venous patterns disappear, and the breast thermal image returns to that pattern observed prior to pregnancy. A second example of the appearance of increased superficial venous temperature occurs with exercise. Strenuous activity of the limbs will result in an increased skin temperature over the involved muscles, with an increase in the temperature of the superficial veins drained by these muscles. These patterns also disappear after a period of rest.

It must be emphasized that the increased skin temperature noted in thermography is a result of increased regional blood flow and is only a secondary indicator of metabolic activity. It is unfortunate that in the case of breast cancer, there is a general misconception that the thermographic signal is a result of a tumor acting as a heat source. A simple energy balance considering the oxygen concentration in the arterial blood supply establishes a limit to the increased temperature in normal tissue. The normal concentration in arterial blood is 20 ml oxygen/100 ml of blood. The normal concentration in the venous blood is approximately 16 ml oxygen/100 ml of blood. Based on well-documented metabolic studies, the average heat release in the tissues is 4.8 Kcal/liter of oxygen. For flow in an adiabatic system, the heat removed from normal tissue may be estimated by rewriting Eq. (4)

$$q = \rho_b c_b w_b (T - T_a) \tag{8}$$

where T is the average normal tissue temperature. Now q can be related to the oxygen consumption rate, which, by the Fick relationship, is the product of the blood flow rate and the difference in O_2 concentration between the arterial and venous blood supply. Equating Eq. (8) and the Fick relationship, the temperature excess of normal tissue $(T - T_a)$ can be estimated. For normal oxygen extraction, the elevation of tissue temperature due to metabolic activity is 0.2°C. If the improbable situation existed where all of the arterial oxygen were used, the temperature elevation would be only 1.0°C.

It should be noted that there are two instances in which the heat generation in the tissues might lead to skin temperatures in excess of the value previously calculated. The first is the case of transient intensive exercise, where the muscles incur an oxygen debt. Even in this case, the blood perfusion will be maximal, persisting for some time after the exercise while supplying oxygen and removing lactic acid and other metabolites. The second case of interest is the situation where there is dissipative heat generation in the tissues from applying an external radiation field, such as microwave or ultrasound. Based on this analysis, it must be recognized that excessive metabolic heating from tumors is not responsible for temperature variations of several degrees noted in thermographic examinations. In chronic infection or tissue injury, as in malignancy, the metabolic heat generation is generally limited by the oxygen supply of the blood and thus the temperature will be dominated by the blood flow as previously demonstrated. Thus, the presence of any stimulus that causes increased metabolic activity will require increased blood flow to the region. This blood is supplied to the small arteries and arterioles at essentially deep-body temperature. This increases the temperature of the subcutaneous tissues and provides a region of increased skin temperature. The stimulus may be a malignant tumor; the result of a tissue injury, such as an infection or a bruise; or merely the result of massaging the area.

It is impractical to improve the analysis of superficial perfusion without knowledge of the detailed structures, thermal properties, and perfusion networks in the cutaneous and subcutaneous regions. Such an improvement would entail a tremendous increase in mathematical complexity. For example, the presence of bony structures will undoubtedly cause some change in the values of the skin temperatures, fat layers of locally variable thickness will do the same. Nevertheless, the experimental evaluation of this approximate analysis demonstrates the potential of the method.

Figure 5 illustrates the correlation of the thermographic determination of blood perfusion (Eq. 7) with independent measurements obtained with venous occlusion plethysmography.[36] These measurements were obtained in the upper extremities. Since measuring blood perfusion in the thorax is difficult

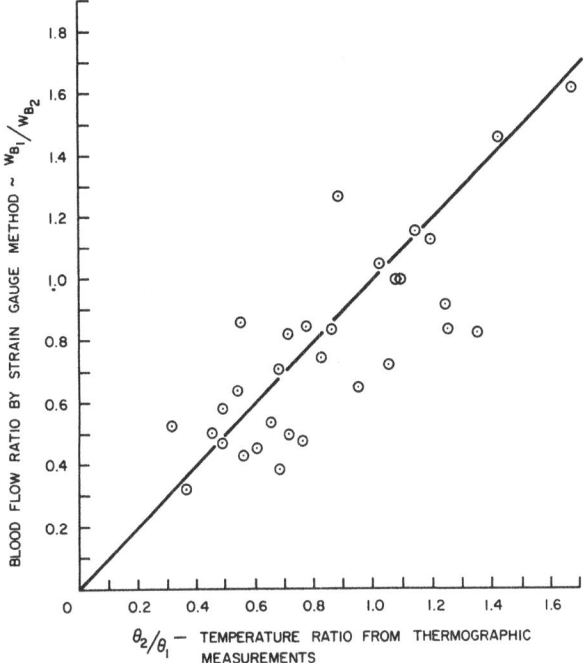

FIGURE 5
Correlation of cutaneous regional blood flow with regional dimensionless temperature ratio for the forearm. This graph compares the changes in blood flow in the forearm pre-exercise, immediately postexercise, and later after exercise. The strain gage is a venous occlusion plethysmographic measurement.

if not impossible with noninvasive techniques, there are no data available for comparison in the case of the breast.

The experiment involved a group of subjects in whom blood flow measurements were made before and after exercise. The exercise consisted of a variable number of actions with a calibrated handgrip. A blood pressure cuff rapidly filled with compressed air was used to occlude the veins in the biceps region of the arm. A mercury strain gage encircling the proximal forearm measured the rate of blood accumulation at the time of occlusion. This strain gage was calibrated by using a volume plethysmograph on the forearm. A calibrated thermogram of the forearm was made prior to the venous occlusion, and blood flow rate was measured before and immediately after exercise. In each subject, an avascular area over the medial aspect of the distal portion of the flexor muscles and the superficial basilic vein was identified. The temperatures of these regions were measured using a densitometer on the negative photograph of the thermogram CRT image. The ratio of blood perfusion in the forearm before and after exercise was computed from the thermographic temperatures, using Eq. (7) and compared with a similar ratio computed from the slope of the strain gage resistance change immediately after occlusion. Although there appears to be some scatter in the data, the correlation is about as good as is ordinarily obtained in plethysmographic measurements. In fact, the scatter of data comparing the strain gage and the volume measurements themselves was only slightly less.

Computed blood perfusion measurements in the breast, based on quantitative thermographic measurements [Eq. (7)] of 407 normal female subjects aged 35–80[35] indicated a range of blood perfusion rates from 0.008 kg/m^3 sec–0.024 kg/m^3 sec. In the report on the investigation of computer-evaluated digital temperature images as a method of screening for breast cancer, the perfusion ratio (Eq. 7), was computed and found to be a highly correlated parameter with tumors or tumorlike regions.[11]

7. SUMMARY

The wide range of published applications for thermography indicate the potential of this modality. Thermography is still an underdeveloped technique. There still exists a great need for improvement in quantitative methods, with careful control of heat transfer to the skin by controlling environmental temperatures and airflow. The expected range of normal temperature patterns should be established and standardized in order to be carefully compared with temperature patterns over corresponding anatomical regions in diseased patients. The correlation of these measurements with pathology is still a subject for clinical research. The present work demonstrates that with careful clinic control and calibration of the scanning infrared camera, regional blood perfusion estimates may be made. Additional correlative studies should be made to provide better relationships between superficial perfusion and skin temperature measurements. These analyses can be used in connection with controlled superficial heating or cooling of the tissues and the observance of thermal recovery profiles in order to gain additional information from the dynamic response. Many of the studies reported in the literature should be extended to include calibrated quantitative observations under controlled environmental conditions. Such controls should eliminate the variability of instrument adjustments, help provide better reproducibility of results between clinics, and thus aid the judgement of the analyst in estimating the significance of cutaneous temperature differences.

REFERENCES

1. Haberman, J. D., *Image Analyses of Medical Thermograms, Critical Reviews in Radiological Sciences*, vol. 2, (Chemical Rubber Publishing Company, Cleveland 1971).
2. Moscowitz, M., Milbrath, J., Gartside, P., Zermino, A., and Mandel, D., The lack of efficacy of thermography as a screening tool for minimal and stage I breast cancer, *N. Eng. J. Med.* **295**, 219–252, 1976.
3. Lawson, R. N., and Chughtai, M. B., Breast cancer and body temperature, *Can. Med. Assoc. J.* **88**, 68–70, 1963.
4. Draper, J. W., and Boag, J. W., The calculation of skin temperature distribution in thermography, *Phys. Med. Biol.* **16**, 2, 201–211, 1971.
5. Draper, J. W., and Boag, J. W., Skin temperature distributions over veins and tumors, *Phys. Med. Biol.* **16**, 4, 645–654, 1971.

6. Gautherie, M., Bourjat, P., Quenneville, Y., and Gros, C., Heat production capacity in breast epitheliomas, *Rev. Europ. Etudes Clin. Biol.* **17**, 776, 1972.
7. Barnes, R. B., Thermography of the human body, *Science* **240**, 1963, 870.
8. Lelik, F., Kezy, G., and Solymossy, O., The diagnosis of locomotive disorders of domestic animals by contact thermography, *ACTA Thermographica* **2**(1), 13, 1977.
9. Pochaczevsky, R., and Meyers, P., The value of vacuum-contoured, liquid crystal, dynamic breast thermoangiography, *Acta Therm.* **4**(1), 8–16, 1979.
10. Watton, R., Harper, B., and Burgess, D., Infrared T.V.: An inexpensive portable pyroelectric camera, *Infrared Physics*, vol. 16 (Pergamon, New York 1976) pp. 81–85.
11. Haberman, J. D., Phase II final report: mass screening for breast cancer by automatic infrared pattern recognition, University of Oklahoma Health Sciences Center, NCI contract number N01-CB-43869, 1980.
12. Anliker, M., and Friedli, P., Evaluation of high-resolution thermograms by on-line digital mapping and color coding, *Appl. Rad. Nucl. Med.* **5**(3), 114–115, May/June 1976.
13. Steketee, J., Spectral emissivity of skin and pericardium, *Phys. Med. Biol.* **18**(5), 686–694, 1973.
14. Watmough, D., and Oliver, R., Some physical factors relevant to infrared thermography, *Phys. Med. Biol.* **15**(1), 178, 1970.
15. Hardy, J. D., The radiating power of human skin in the infrared, *Am. J. Physiol.* **127**, 454, 1939.
16. Martin, C. J., and Watmough, D., Thermal scanning of curved surfaces, *Acta Thermo.* **2**(18), 1977.
17. Love, T. J., and Haberman, J. D., Heat transfer in medical thermography, *Therm. Quarterly,* **11**(1), spring 1977.
18. Lilienfeld, A., Barnes, J., Brasfield, R., Connell, J., Diamond, E., Gershon-Cohen, J., Haberman, J., Isard, H., Lane, W., Lattes, R., Miller, J., Seamon, W., and Sherman, R., An evaluation of thermography in the detection of breast cancer, *Cancer* **24**, 1206, 1969.
19. Haberman, J., Love, T., and Francis, J., Screening a rural population for breast cancer using thermography and physical examination techniques: Methods and results, *Ann. N. Y. Acad. Sci.* **335**, 492–500, 1980.
20. Cooke, E. D., Thermography diagnosis of deep venous thrombosis, *Acta Therm. Suppl.* **1**, 7, 55, 1978.
21. Ring, E. F. G., A thermographic index for the assessment of ischemia, *Acta Therm.* **5**(1), 1980.
22. Karpman, H., Knebel, A., Semel, C., and Cooper, J., Clinical studies in thermography application of thermography in evaluating musculoligamentous injuries of the spine, *Arch. Environ. Health* **20**, 112, 1970.
23. Connell, J., Morgan, E., and Rousselot, L., Thermography in trauma, *Ann. N. Y. Acad. Sci.* **121**, 171, 1964.
24. Buwalda, G., Possible applications of thermography in abdominal conditions, *Bibl. Radiol.* **5**, 198–200, 1969.
25. Farrell, C., and O'Hara, A. E., The use of thermography in the pediatric patient, *Clin. Pediatr.* (*Philadelphia*) **11**, 673, 1973.
26. Sandrow, R., Torg, J., Lapayowker, M., and Resnick, E., The use of thermography in the early diagnosis of neuropathic arthropathy in the feet of diabetics, *Clin. Orthop.* **88**, 31, 1972.
27. Verhonick, P., Lewis, D., and Goller, H., Thermography in the study of decubitus ulcers, *Nurs. Res.* **21**(3), 233, 1972.
28. Cosh, J., Collins, A., and Ring, E., Infrared radiometry and thermography: Their applications in rheumatology, *Proc. Roy. Soc. Med.* **65**, 890, 1972.
29. Ring, E. F., Thermography and rheumatic diseases, *Bibl. Radiol.* (6), 97–106, 1975.
30. Zysno, E. A., and Rusch, D., Thermographic methods in rheumatology, *Verh. Dtsch. Ges. Rheumatol.* **2** suppl., 2, 231, 1972.
31. Bloomenstein, R. B., Viability prediction in pedicle flaps by infrared thermography, *Plast. Reconstr. Surg.* **52** 185, 1975.
32. Watson, A. C., and Vasilescu, C., Thermography in plastic surgery, *J.R. Coll. Surg.* (*Edinburgh*) **17**, 247, 1972.
33. Planiol, T., Floyrac, G., and DeMeyer, J., Thermography in malignant thyroid nodules, *Acta Therm.* **1**(3), 126–137, 1976.

34. Samuels, B. I., Thermography: A valuable tool in detection of thyroid disease, *Radiology* **102**, 59, 1972.
35. Goin, J., Love, T., and Haberman, J., Absolute temperature thermography: Observations of skin temperature variations of the human breast, ASME paper no. 79 WA/HT-70, 1979.
36. Francis, J., Roggli, R., Love, T., and Robinson, C., Thermography as a means of blood perfusion measurement, *Trans. ASME, J. Biomech. Eng.* **101**(4), 246, 1979.

Chapter 26

COMPUTER-AIDED TOMOGRAPHIC THERMOGRAPHY

Michael M. Chen

1. INTRODUCTION

The use of thermography as a medical diagnostic tool has been described in Chap. 25. In contrast to x-ray and ultrasound-imaging techniques, which detect structural differences, thermography is sensitive to functional parameters, such as circulation and metabolic rate, thus providing information not obtainable with other methods.

Despite the promise of such features and apart from the cost of equipment, thermography does have shortcomings that have thus far limited its acceptance. Although the surface temperature distribution is a manifestation of the distribution of interior thermal parameters, such as lesions, the usually observed thermal pattern is neither a direct nor an unique one. This is due to two reasons: First, heat transfer proceeds both longitudinally and transversely, the latter having the net effect of "smearing" the image, so that it appears diffuse and without a sharp outline. Second, since the interior tissue is three-dimensional but the observed steady-state thermographic image is only two-dimensional, some of the information is lost, and it is not possible to ascertain the depth or structure of the tissue nonuniformity leading to a given surface temperature nonuniformity. It is likely that these factors have contributed to some of the negative assessment of thermography as a breast cancer screening technique.[1,2]

It is thus useful to inquire whether some modified procedure of thermographic measurements can contain three-dimensional information and whether such information can be extracted and made available to the diagnostician.

It is clear that in the protocols currently employed—that of taking a single picture of steady-state skin temperature—thermography cannot possibly contain more than two dimensions of information. However, if a series of pictures were taken under varying conditions, it at least seems possible that some information about the third dimension could be retained. With the aid of a

Michael M. Chen • Department of Mechanical and Industrial Engineering, University of Illinois, Urbana, Illinois.

digital computer, such information can be extracted and reconstituted into suitable three-dimensional images to be used for diagnostic purposes. The computer-based, image-enhancing techniques developed for image transmission in space exploration may also be called on to improve the quality of the reconstituted image.

The class of problems in determining thermophysical parameters from temperature measurements is called the inverse heat conduction problem, since the usual, or "direct," problem in heat conduction is to determine the temperature from known distributions of thermophysical properties. Such inverse problems have been investigated for a number of applications and physical configurations.[3,4] Although most of these previous studies have dealt with one-dimensional problems, recent analytical and experimental results have shown that three-dimensioal problems relevant to thermography can also be treated.[5-7]

2. THE GOVERNING EQUATION AND RELEVANT PARAMETERS

The observable data in the thermogram is the experimental surface temperature distribution $T_e = T_e(y, z)$, determined by infrared scanning or some other technique. Here, we have chosen x to denote the depth coordinate. In order for the observed data to contain three-dimensional information, we must perform the observation repeatedly, as a function of an adjustable parameter P, such that an analytical distribution T_s may be obtained, where

$$T_e = T_s(y, z, P) \equiv T(0, y, z, P) \tag{1}$$

We shall later identify P with either the heat transfer coefficient h or time t. The expression $T_s(y, z, P)$ is the consequence of internally distributed heat sources, thermophysical properties, and the applied boundary conditions, subject to the heat conduction equation

$$\rho c \frac{\partial T}{\partial t} = \nabla k \nabla T + S \tag{2}$$

and the boundary condition

$$-k \frac{\partial T}{\partial x} = h(T - T_0) \tag{3}$$

where T_0 is ambient temperature. The important biological parameters are the sweat rate, which affects the heat transfer coefficient h, the metabolic rate, which constitutes part of the volume heat source S; and the blood perfusion rate, whose effects have been variously included in the thermal conductivity k or a temperature-dependent heat source term.[5] For the present discussion, we shall follow the more common practice of including perfusion effects in the heat source

$$S = \rho_b c_b w_b (T_b - T) + q \tag{4}$$

where w_b is the volumetric rate of blood perfusion per unit volume of tissue, T_b is the blood temperature, and q denotes the metabolic heat source per unit volume. The derivation of this equation has been discussed in detail in Chaps. 6 and 7. As a rule, the variation of k is considerably less than the variation of w_b in Eq. (4). Therefore, we shall assume that T_0, h, and k are uniform and that nonuniformities in w_b and q indicate the presence of pathology.

$$w_b = w_b(x, y, z) \tag{5}$$

$$q = q(x, y, z) \tag{6}$$

Thus, the bioheat equation becomes

$$\rho c \frac{\partial T}{\partial t} = k \nabla^2 T + \rho_b c_b w_b (T_b - T) + q \tag{7}$$

Our task is to solve the inverse problem, which determines the unknown functions $w_b(x, y, z)$ and $q(x, y, z)$, from the observed function $T_s(y, z)$. In the following discussions, we shall find it more convenient to examine solutions for the direct problem before arriving at a technique for solving the inverse problem.

3. SEPARATING THE PERFUSION AND METABOLIC-HEATING TERMS

For simplicity and without loss of generality, we shall choose the blood temperature T_b as the datum for temperature measurement. Thus,

$$T_b = 0 \tag{8}$$

There now remains one inhomogeneous term $q(x, y, z)$ in differential equation (Eq. 7) and an inhomogeneous term T_0 in the boundary condition, Eq. (3). These two inhomogeneous terms can easily be separated. Let T_p denote the solution to the system

$$\rho c \frac{\partial T_p}{\partial t} = k \nabla^2 T_p - \rho_b c_b w_b T_p \tag{9}$$

$$- k \frac{\partial T_p}{\partial x} = h(T_p - T_0) \tag{10}$$

and let T_m denote the solution to the system

$$\rho c \frac{\partial T_m}{\partial x} = k\nabla^2 T_m - \rho_b c_b w_b T_m + q(x, y, z) \tag{11}$$

$$-k\frac{\partial T_m}{\partial x} = hT_m \tag{12}$$

It is clear that the sum of T_p and T_m satisfies the original equation (Eq. 7) and boundary condition (Eq. 3)

$$T = T_p + T_m \tag{13}$$

Note that T_p is unaffected by the metabolic-heating term q. Thus, T_p reflects only variations of $w_b(x, y, z)$ and represents the perfusion component of the solution. On the other hand, T_m is influenced by the inhomogeneous term $q(x, y, z)$ and thus includes the consequence of metabolic heating. Also note that T_m is independent of the ambient temperature T_0 and that $T_p = 0$ everywhere if $T_0 = 0$. In other words, the amplitude of the p component of the observed surface temperature $T_{ps}(y, z, p)$ is proportional to T_0 (actually, $T_0 - T_b$). This separation of T_m from T_p is not only mathematically convenient, but also physically realizable. The procedure would be as follows.

1. Set ambient temperature equal to blood temperature, and obtain an analytical surface temperature distribution $T_s(x, y)$. Since $T_p = 0$ for this ambient temperature, $T_{ms}(x, y) = T_s(x, y)$.
2. Set ambient temperature to any desired T_0, and obtain another $T_s(x, y)$.
3. Since T_m is independent of T_0, the difference between this second $T_0(x, y)$ and $T_{m0}(x, y)$ is $T_{ps}(x, y)$.

4. EFFECTS OF METABOLIC HEATING

In this section, we inquire whether the unknown source function $q(x, y, z)$ can be determined by observations of $T_{m0}(y, z, P)$. For this exercise, we shall assume that the perfusion rate w_b is uniform. the effects of nonuniform $w_b(x, y, z)$ will be examined in a forthcoming section.

4.1. Reduction to One Dimension

Instead of working with the three-dimensional equations, it is both theoretically and practically convenenient to work with the Fourier components of q and T. In other words, q and T will be expanded in terms of appropriate sine and cosine series satisfying the transverse boundary conditions.

$$q(x, y, z) = \sum_{i=0}^{\infty} \sum_{j=0}^{\infty} Q_{ij}(x) \, {}^{\cos}_{\sin}\chi_i y \, {}^{\cos}_{\sin}\chi_j z \tag{14}$$

$$T_m(x,y,z,P) = \sum_{i=0}^{\infty} \sum_{j=0}^{\infty} T_{m,ij}(x,P) \, {}^{\cos}_{\sin}\chi_i y \, {}^{\cos}_{\sin}\chi_j z \tag{15}$$

The differential equation (Eq. 11) and boundary condition (Eq. 12) thus become the set of one-dimensional equations

$$\rho c \frac{\partial T_{m,ij}}{\partial t} = k \frac{\partial^2 T_{m,ij}}{\partial x^2} - [k(\chi_i^2 + \chi_j^2) + \rho_b c_b w_b]T_{m,ij} + Q_{ij} \tag{16}$$

$$-k \frac{\partial T_{m,ij}}{\partial x} = hT_{m,ij}, \qquad i,j = 0,1,2,\ldots \tag{17}$$

Our task is to find a scheme to determine $q(x)$ from measurements of the $(i,j)^{\text{th}}$ components of the surface temperatures $T_{ms,i,j}(P) = T_{m,i,j}(0,P)$ for a continuous range of an independently variable parameter P. Two procedures are possible:

(a) Vary the heat transfer coefficient h, and determine $T_{ms,i,j}(h)$ from steady-state measurement of the surface temperature. In this case, h serves as the continuously varying parameter P.

(b) Vary h impulsively, and determine $T_{ms,i,j}(t)$ from the transient measurement of the surface temperature. In this case, $P = t$.

On closer examination, the two methods are in fact analogous. In transient heat conduction problems, the depth of the thermal layer δ_T, affected by the transient is[6]

$$\delta_T = \left(\frac{k}{\rho c} t\right)^{1/2} \tag{18}$$

The time-dependent temperature distribution for $x \ll \delta_T$ can be well represented by a quasi-steady solution with an appropriate boundary placed at $x \cong \delta_T$. Therefore, the foregoing scheme (b) is, in fact, equivalent to varying the Biot number by varying the effective slab thickness. In scheme (a), the Biot number is varied by varying the heat transfer coefficient.

Since the solutions to steady-state heat transfer is arithmetically more compact than those for transient solutions, we shall examine these equations on the basis of scheme (a). It is noted, however, that scheme (b) is far more practical experimentally, since a large number of steady-state measurements is essentially impossible.

As an example, we shall consider a slab of infinite thickness d with an insulated rear surface. Rather than dealing with a continuum of sources $q_{ij}(x)$, it is more instructive to examine the equations for a discrete source of strength q located at a point $x = x_\lambda$

$$Q_{ij}(x) = q\delta(x - x_\lambda) \tag{19}$$

where δ is the Dirac delta function

$$\delta(x) = 0, \qquad x \neq 0$$

$$\delta(x) = \infty, \qquad x = 0 \tag{20}$$

$$\int_{0-\varepsilon}^{0+\varepsilon} \delta(x)\, dx = 1$$

This in essence follows methods employing Green's function, from which the solution for a continuum of sources can be constructed by integration (cf. Chap. 11). Introducing the dimensionless variables,

$$x^* = x/d, \qquad \lambda = x_\lambda/d \tag{21}$$

$$T^* = T_{m,i,j} k/qd \tag{22}$$

$$\text{Bi} = hd/k \qquad \text{(Biot number)} \tag{23}$$

$$\chi^2 = \chi_i^2 d^2 + \chi_j^2 d^2 + \rho_b c_b w_b d^2/k \tag{24}$$

the system of equations (Eqs. 16 and 17), subject to the steady-state assumption, becomes

$$\frac{\partial^2 T^*}{\partial x^{*2}} - \chi^2 T^{*2} + \delta(x^* - \lambda) = 0 \tag{25}$$

$$\frac{\partial T^*}{\partial x^*}(0) = \text{Bi}\ T^*(0) \tag{26}$$

$$\frac{\partial T^*}{\partial x^*}(1) = 0 \tag{27}$$

This can be solved to yield the surface temperature

$$T_s^* \equiv T_{ms,i,j} k/qd$$

$$T_s^* = \frac{e^{-\chi\lambda} + e^{-\chi(2-\lambda)}}{\chi + \text{Bi} - (\chi - \text{Bi})\, e^{-2\chi}} \tag{28}$$

which shows, as expected, the surface temperature $T_{ms,i,j}$, henceforth to be abbreviated as T_s, a function of the source strength q, the location λ, and the Biot number Bi.

The inverse problem for this discrete source is determining q and λ from surface temperature measurements, which on casual observation would appear to be quite easy. All we seem to have to do is make two surface temperature observations using two values of the Biot number and then solve for two unknowns q and λ. For example, if observations T_{s1} and T_{s2} are associated with Biot numbers Bi_1 and Bi_2, respectively, we could eliminate q immediately by taking the ratio of the two temperatures.

However, this ratio is independent of λ, i.e., the source position. Therefore, it is not possible to determine the position of a discrete source by two observations using two different Biot numbers.

The generalization of this conclusion is that it would be extremely difficult, if at all possible, to determine the unknown source function $q(x, y, z)$ from a series of steady-state surface temperature measurements $T_s(y, z, h)$ with variable h or a series of transient surface temperature measurements $T_s(y, z, t)$.

The latter possibility was investigated numerically for a one-dimensional geometry using a Kiefer–Wolfowitz (K–W) identification algorithm.[7] A direct numerical calculation was used to generate a surface temperature transient $T_s(t)$ for a specific q and λ. The physical situation that was simulated was a heating transient after severe surface cooling to steady-state conditions. Using the results of this numerical calculation as the "correct" solution, the K–W algorithm was used to adjust q and λ, starting from some arbitrary values, until the surface temperature transient agreed with the correct transient. It was found that the values of q and λ that could produce satisfactory convergence depended on the starting values and were not unique. Satisfactory convergence could be obtained for several combinations of q and λ for a given correct solution. This numerical experiment was done without considering the presence of noise in the measurements, which would tend to confound results even further.

5. EFFECTS OF BLOOD PERFUSION

We shall now consider the relationship between the perfusion rate $w_b(x, y, z)$ and $T_p(x, y, z, t)$. Since w_b is a coefficient of one of the homogeneous terms of the differential equation, this problem is directly analogous to one of the one-dimensional inverse problems that have been treated in the literature.[3,4] Therefore, our only task is to seek a scheme to reduce the problem to one-dimensional form. Following the standard procedure of inversion, using transient measurements, we shall assume that a steady-state condition is first established as an initial condition and then the boundary conditions are changed and the transient observed in order to yield interior information. The initial condition $T_{pi}(x, y, z)$ thus satisfies the system of equations

$$k\nabla^2 T_{pi} - \rho_b c_b w_b T_{pi} = 0 \tag{29}$$

$$-k\frac{\partial T_{pi}}{\partial x} = h_i(T_{pi} - T_0) \tag{30}$$

The transient component of the temperature $T_{pt} = T_p - T_{pi}$ satisfies the system

$$\rho c \frac{\partial T_{pt}}{\partial t} = k\nabla^2 T_{pt} - \rho_b c_b w_b T_{pt} \tag{31}$$

$$-k \frac{\partial T_{pt}}{\partial x} = h(T_{pt} - \theta) \tag{32}$$

$$T_{pt} = 0, \quad \text{at } t = 0 \tag{33}$$

where

$$\theta = [1 - (h_0/h)](T_a - T_{pi}) \tag{34}$$

Note that T_{pt} has a homogeneous initial condition, a considerable simplification.

We shall consider a finite slab of thickness d with an insulated rear surface. It will be convenient to normalize the variables with respect to d and θ. Since we shall be dealing exclusively with T_{pt} in the remainder of this section, the subscript pt will not be retained.

$$T^+ = T_{pt}/\theta \tag{35}$$

$$x^+ = x/d, \qquad y^+ = y/d, \qquad z^+ = z/d \tag{36}$$

$$w_b^+ = \rho_b c_b w_b d^2 / k \tag{37}$$

$$t^+ = kt/\rho c d^2 \tag{38}$$

Separating the Uniform and Nonuniform Components

Let w_b be separated into a uniform component $\bar{w}$ and a nonuniform component w'. Let $\bar{T}$ denote solutions to Eqs. (31)–(33) with uniform perfusion $\bar{w}$. In dimensionless form, the equation for $\bar{T}$ is

$$\frac{\partial \bar{T}^+}{\partial t^+} = \frac{\partial^2 \bar{T}^+}{\partial x^{+2}} - \bar{w}^+ \bar{T}^+ \tag{39}$$

$$-\frac{\partial \bar{T}^+}{\partial x^+} = \text{Bi}\,(T^+ - 1), \qquad \text{at } x^+ = 0 \tag{40}$$

$$\frac{\partial \bar{T}^+}{\partial x^+} = 0, \qquad \text{at } x^+ = 1 \tag{41}$$

$$\bar{T}^+ = 0, \qquad \text{at } t^+ = 0 \tag{42}$$

where

$$w_b^+ = \bar{w}^+ + w'^+ \tag{43}$$

The solution for this system of equations is easily accomplished, and $T^+(x^+, t^+)$ will be considered a known function in subsequent discussions. We now define

$$T^+ = \bar{T}^+ + T'^+ \tag{44}$$

Substituting Eqs. (43) and (44) into the dimensionless form of Eqs. (31)–(34) and neglecting second- and higher-order terms in w^+ and T'^+, the equation, boundary and initial conditions for $T'^+(x)$ are

$$\frac{\partial T'^+}{\partial t^+} = \frac{\partial^2 T'^+}{\partial x^{+2}} - \bar{w}^+ T'^+ - w'^+ \bar{T}^+ \tag{45}$$

$$-\frac{\partial T'^+}{\partial x^+} = \text{Bi } T'^+, \quad \text{at } x^+ = 0 \tag{46}$$

$$\frac{\partial T'^+}{\partial x^+} = 0, \qquad \text{at } x^+ = 1 \tag{47}$$

$$T'^+ = 0, \qquad \text{at } t^+ = 0 \tag{48}$$

It is now possible to decompose w^+ and T'^+ into transverse Fourier components satisfying appropriate transverse boundary conditions, which lead to mutually independent equations for each Fourier mode.

$$w^+(x^+, y^+, z^+) = \sum_{i=1}^{\infty} \sum_{j=1}^{\infty} w_{ij}^+(x^+) \, {}^{\cos}_{\sin}\chi_i y^+ \, {}^{\cos}_{\sin}\chi_j z^+ \tag{49}$$

$$T'^+(x^+, y^+, z^+) = \sum_{i=1}^{\infty} \sum_{j=1}^{\infty} T_{ij}'^+(x^+) \, {}^{\cos}_{\sin}\chi_i y^+ \, {}^{\cos}_{\sin}\chi_j z^+ \tag{50}$$

$$\frac{\partial T_{ij}'^+}{\partial t^+} = \frac{\partial^2 T_{ij}'^+}{\partial x^{+2}} - \chi^2 T_{ij}'^+ + \bar{T}^+ w_{ij}^+ \tag{51}$$

$$-\frac{\partial T_{ij}'^+}{\partial x^+} = \text{Bi } T_{ij}'^+, \quad \text{at } x^+ = 0 \tag{52}$$

$$\frac{\partial T_{ij}'^+}{\partial x^+} = 0, \qquad \text{at } x^+ = 1 \tag{53}$$

$$T_{ij}'^+ = 0, \qquad \text{at } t^+ = 0 \tag{54}$$

where χ denotes a combination of the transverse wave numbers and the mean perfusion parameter $\bar{w}^+$

$$\chi^2 = \chi_i^2 d^2 + \chi_j^2 d^2 + \bar{w}^+ \tag{55}$$

Equations (51)–(54) constitute a one-dimensional transient problem. Note that $T_{ij}'^+(0, t^+)$, or the $(i, j)^{\text{th}}$ component of the surface temperature $T_s^+(y^+, z^+, t^+)$, can be easily determined by Fourier analysis of the thermographic measurements.

In order to determine $w_{ij}^+(x^+, y^+, z^+)$ from $T_{ij}'^+(0, t^+)$, a number of schemes can be followed. A frequently employed method for the inversion of an one-dimensional heat conduction equation is to express the unknown function, in this case w_{ij}^+, as a function of a suitable form with a number of undetermined parameters

$$w_{ij}^+(x^+) = w_{ij}^+(x^+, P_1, P_2, \ldots, P_n) \tag{56}$$

Computations of surface temperatures using the assumed function are then compared with the observations, while the parameters are adjusted to minimize the error. A survey of such methods can be found in reference 4.

If desired, an explicit technique can also be derived by expanding w_{ij}^+, $\bar{T}_{ij}^+$, and $T_{ij}'^+$ as a finite Fourier series, which would lead to an explicit solution of the surface temperature $T_{ij}'^+(0, t^+)$ in terms of coefficients of the series representation of w_{ij}^+. The determination of these coefficients from experimental data would then involve a linear algebraic procedure.

6. A NUMERICAL SIMULATION

To further examine the practicality of the tomographic thermography procedure, a numerical simulation was carried out in which simulated thermographic measurements were obtained by numerical computation followed by tomographic inversion. The resulting perfusion rate distribution was then compared with the originally assumed perfusion pattern. A simulated noise was also included to accurately reflect realistic clinical conditions. A description of these calculations and results follow.

The initial and boundary conditions consistent with the transient protocol described earlier can be expressed by the following relations:

$$-\frac{\partial T}{\partial x} = \frac{h}{k}(T - T_h), \quad \text{at } x = 0 \tag{57}$$

$$\frac{\partial T}{\partial x} = 0, \qquad \text{at } x = d \tag{58}$$

$$T = 0, \qquad \text{at } t = 0 \tag{59}$$

Equation 57 describes the boundary condition at the interface of the skin and the ambient fluid, while Eq. (58) refers to an insulated back surface. Equation (59) simply states that a steady state has been reached prior to introducing the transient.

6.1. Simulation of Thermographic Data

To simulate the thermographic data, the surface temperatures $T(0, y, z, t)$ are computed from Eq. (3) in conjunction with the initial and boundary conditions described earlier. The domain of computation was a rectangular volume of depth $d = 1.80$ cm (x coordinate), width $W = 2.0$ cm (y coordinate), and height $H = 2.0$ cm (z coordinate), with insulated side and back surfaces. The front surface was exposed to a constant temperature medium through a heat transfer coefficient h (Fig. 1). The "pathology" was represented as a variation in w_b, while the thermal conductivity k was taken as constant over the region of interest. The latter assumption was made partly because we wish to restrict this study to a relatively simple model, partly because existing data appear to show that the thermal conductivity for pathologic tissues is not significantly different from normal tissues. The consequences when both blood perfusion and thermal conductivity are nonuniform will be treated in future investigations. To conserve computer time, the nonuniformities were

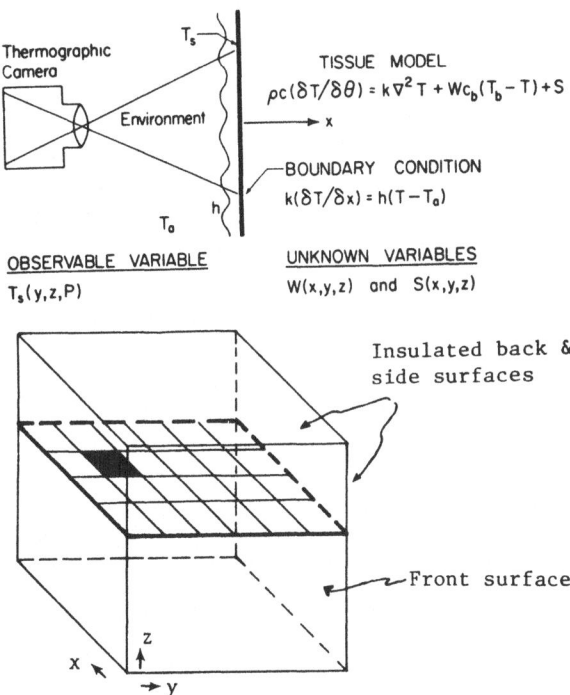

FIGURE 1
Schematic representation of the three-dimensional thermographic problem (see Ref. 5).

assumed to be two dimensional. In other words, there was no variation in the z direction and only the variation in the x–y plane need be considered.

The elemental thermal parameters characterizing the hypothetical bio-material are given in Table 1, where w_b and w_{b1} refer to normal and pathologic tissues, respectively. It should be pointed out that in the present simulation, the ratio of these w_b values is 1.2. This results in approximately a 0.4–0.5°C difference between hot and cold surface spots in the steady state after the initial perturbation.

Initial and boundary conditions are defined as follows:

$t < 0$ biomaterial at uniform temperature of 37°C

$t \geq 0$ front surface brought into contact with an ambient fluid of 17°C. Convection to ambient is characterized by a heat transfer coefficient h.

Bearing in mind these assumptions and conditions, Eq. (3) is solved numerically for $T(0, y, t)$, assuming a given $w_b(x, y, z)$ distribution and using the explicit form of the infinite-difference approximation. Thus, the compilations are carried out using standard finite-difference procedures.

6.2. Algorithm for Tomographic Reconstruction

The procedure in tomographic reconstruction is to use the surface temperatures $T(0, y, z, t)$ in conjunction with the bioheat equation and initial and boundary conditions in determining the three-dimensional distribution of the blood perfusion rate $w_b(x, y, z)$. The present algorithm follows the method described earlier based on linearization and Fourier decomposition of the nonuniform temperature and perfusion parameters into transverse modes. In this way, the equation governing each Fourier mode is only one-dimensional in x and solved iteratively to determine the correct $w_{ij}(x)$. The different Fourier components of the perfusion distribution are recombined to form the desired three-dimensional distribution $w_b(x, y, z)$

$$w_b(x, y, z) = \bar{w} + \sum_{i=0}^{\infty} \sum_{j=0}^{\infty} w'(x) \, {}^{\cos}_{\sin}\chi_i y \, {}^{\cos}_{\sin}\chi_j z + \bar{w}_u \qquad (60)$$

It is noted here that an additional uniform component $\bar{w}_u(x)$ has been included to indicate that in reality, w_b should be a function of x. This contribution,

TABLE 1
Parameters Characterizing the Test Biomaterial

k (W/m°C)	h (W/m²°C)	c [(W sec)/(m³°C)]	x (cm)	t (sec)	w_b w_{b1} (dimensionless)		c_b (J/kg°C)
0.36	30.0	3×10^6	0.40	10	2.19	2.63	4.0×10^3

TABLE 2

1. Establish initial conditions.
2. Impulse change of boundary conditions.
3. Transient surface temperature measurements (thermography).
4. Harmonic analysis.
5. Inversion: determine $w_{ij}(x)$ from $T_{sij}(t)$.
6. Synthesis of $w_b(x, y, z)$.
7. Display.

however, is very small, as evidenced by the close agreement between the simulated mean temperatures and the calculated ones obtained from Eqs. (17) and (18) using an x-independent uniform component $\bar{w}$. The computation process is schematically shown in the flow diagram in Table 2.

6.3. Discussion

The results of the computations are shown in Figs. (2)–(8). In Fig. 2, the simulated steady-state skin temperature distribution, as might be measured by conventional thermography, is shown along with the assumed inhomogeneity of the perfusion parameter. It is shown that while the temperature does show a peak in the center, corresponding to the high perfusion nodes in the front center, there is no indication whatsoever of the two high perfusion regions at about 1.6 cm from the surface. In addition, the observed temperatures contain no information on either the depth or the amplitude of the inhomogeneity.

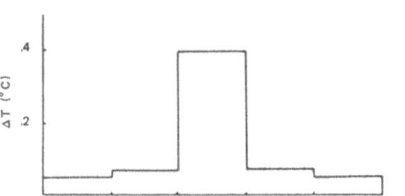

Front surface temperature profile T(0, y, ∞) along any horizontal line. Base level was 27.5 C and temperature range was .4 C.

FIGURE 2

Hypothetical perfusion pattern and corresponding front surface steady-state temperature profile. The numbers in each element denote the dimensionless blood perfusion rates. To convert to conventional units kg/(m³ sec), the numbers are multiplied by $k/\rho_b c_b d^2$. (a) Front surface temperature profile $T(0, y, z)$ along any horizontal line. Base level was 27.5°C and temperature range was 0.4°C. (b) Assumed perfusion pattern $w_b(x, y, z)$ at midplane ($z = 0$).

		FRONT		
	2.19	2.63	2.19	
		2.63		
		2.19		
		2.19		
2.19	2.63	2.19	2.63	2.19

Assumed perfusion pattern W(x, y, z) at midplane (z = 0).

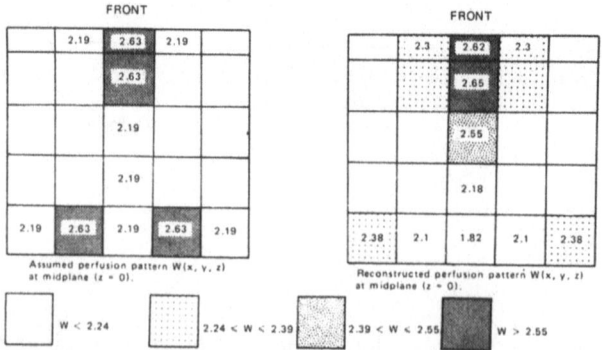

FIGURE 3

Inhomogeneities near the surface and at two deeply imbedded sites. (left) Assumed perfusion pattern $w_b(x, y, z)$ at midplane ($z = 0$). (right) Reconstructed perfusion pattern $w_b(x, y, z)$ at midplane ($z = 0$). Reprinted in part from Ref. 7 with permission.

The tomographical reconstruction of the same inhomogeneities is shown in Fig. 3, where it is seen that the reconstructed distribution of the perfusion parameter is qualitatively a faithful reproduction of the original assumed pattern. Quantitatively, the reconstruction correctly determined the magnitude and the location of the front-center inhomogeneity, as well as the depth of the two inhomogeneities 1.6 cm from the surface. Although neither the magnitude nor the lateral position of the deeply buried inhomogeneities were accurately computed, it should be noted that the inaccuracy in the lateral position corresponded only to one node of the finite-difference formulation. In other words, the error could conceivably be associated with the truncation error of the finite difference scheme. Considering the depth of the inhomogeneities, the accuracy attained was quite gratifying.

Figure 4 shows the assumed and reconstructed perfusion patterns for two inhomogeneities located about 1.4 cm below the surface. Again, a relatively faithful reconstruction was obtained.

In Fig. 5, two inhomogeneities relatively close to the surface were considered. Although the patterns were qualitatively correct, the depth of the

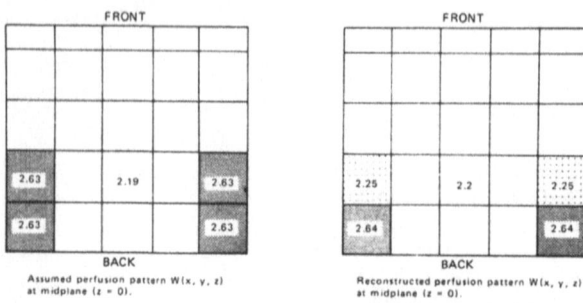

FIGURE 4

Two deeply imbedded inhomogeneities. (left) Assumed perfusion pattern $w_b(x, y, z)$ at midplane ($z = 0$). (right) Reconstructed perfusion pattern $w_b(x, y, z)$ at midplane ($z = 0$). Reprinted in part from Ref. 7 with permission.

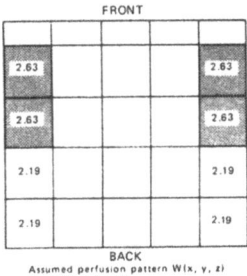

FIGURE 5

Two superficially located inhomogeneities. (left) Assumed perfusion pattern $w_b(x, y, z)$ at midplane ($z = 0$). (right) Reconstructed perfusion pattern at midplane ($z = 0$).

inhomogeneities in the reconstructed pattern was rather inaccurate. The cause for this is not known and should be further investigated.

A fact of life in multiple parameter estimation calculations is that often the solution is not unique. To minimize this problem, it is usually desirable to use any additional information we might have on the unknown parameters. For our case, it seems reasonable that on the basis of known blood flow data, we could establish lower and upper bounds for the perfusion parameter. The effects of such bounds on the accuracy of the tomographical computations are shown in Figs. 6 and 7, which have the same assumed perfusion patterns. In Figure 6, the bounds for w_{ij} are ± 0.3 throughout; in Fig. 7, they are ± 0.3 for $i = 1$, $j = 1, 2, 3$, and ± 0.1 for $i = 3$, $j = 1, 2, 3$. It is seen that Fig. 6b is more accurate than Fig. 7b in terms of the magnitude of the perfusion parameter for the rear center node.

Finally, in Fig. 8, we show a larger 9×9 matrix. The essential features are present, although the extent of the inhomogeneity is somewhat exaggerated. It should also be mentioned that the duration of the transient (exposure time) is approximately 12 min., while for the smaller matrices, the exposure time is close to 16 min.

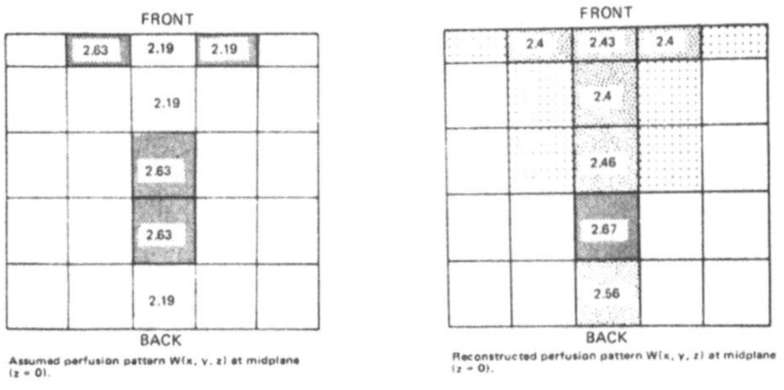

FIGURE 6

Superficial and subcutaneous inhomogeneities. Same bounds for all modes. (left) Assumed perfusion pattern $w_b(x, y, z)$ at midplane ($z = 0$). (right) Reconstructed perfusion pattern $w_b(x, y, z)$ at midplane ($z = 0$).

FRONT

2.63	2.19	2.63		
	2.19			
		2.63		
		2.63		
	2.19			

BACK
Assumed perfusion pattern W(x, y, z)
at midplane (z = 0).

FRONT

2.4	2.43	2.4		
	2.4			
	2.5			
	2.29			
	2.16			

BACK
Reconstructed perfusion pattern W(x, y, z)
at midplane (z = 0).

FIGURE 7
Superficial and subcutaneous inhomogeneities. Different bounds for different modes. (left) Assumed perfusion pattern $w_b(x, y, z)$ at midplane ($z = 0$). (right) Reconstructed perfusion pattern $w_b(x, y, z)$ at midplane ($z = 0$).

7. CONCLUSIONS

The foregoing discussion shows that tomographical thermography is indeed feasible. However, considerable research remains to be done before the technique can be made a practical one.

REFERENCES

1. Island, H. J., Thermography in mass screening of cancer: successes and failures, in *Thermal Characteristics of Tumors: Applications in Detection and Treatment, Ann. N. Y. Acad. Sci.* **335**, 489–491, 1980.
2. Threatt, B., Norbeck, J. M., Ullman, N. S., Kummer, R., and Roselle, P. F., Thermography and breast cancer: an analysis of a blind reading, in Thermal Characteristics of Tumors: Applications in Detection and Treatment, *Ann. N. Y. Acad. Sci.* **335**, 501–519, 1980.

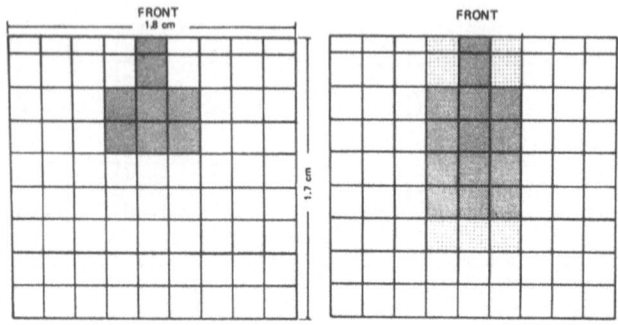

FIGURE 8
True (left) and reconstructed (right) perfusion matrices. $X = Y = 0.2$ cm, $\Delta t = 6$ sec, exposure time $\cong 12$ min. The pattern corresponds to the midplane ($z = 0$) and the slab's dimensions are $1.7 \times 1.8 \times 1.8$ cm. Reprinted from *Ann. N. Y. Acad. Sci.* with permission.

3. Beck, J. V., Nonlinear estimation applied to the nonlinear inverse heat conduction problem, *Int. J. Heat Mass Transfer* **13**, 703–716, 1970.

4. Rubinsky, B., Shitzer, A., Analytic solutions to the heat equation involving a moving boundary with application to the change of phase problem (the inverse Stefan problem), *ASME Trans. J. Heat Transfer* **100**, 300–303, 1979.

5. Chen, M. M., Pedersen, C. O., and Chato, J. C., On the feasibility of obtaining three-dimensional information from thermographic measurements, *J. Biomech. Eng.* **99K**, 58–64, 1977.

6. Pantazatos, P., and Chen, M. M., Computer-aided tomographic thermography: a numerical simulation, *J. Bioeng.* **2**, 397–410, 1978.

7. Chen, M. M., and Pantazatos, P., Tomographical thermography, in *Thermal Characteristics of Tumors: Applications in Detection and Treatment, Ann. N.Y. Acad. Sci.* **335**, 438–442, 1980.

Part V

SELECTED TOPICS

Chapter 27

TEMPERATURE MEASUREMENT

Thomas C. Cetas

1. INTRODUCTION

Thermometry began with the development of the air thermometer in the early seventeenth century by Santorio Santorre (Sanctorius).[1] Sanctorius attached a scale to the measurement of heat and cold and went on to record seasonal changes as well as body temperatures obtained by putting the patient's mouth or hand around the air bulb of the thermometer. An early observation was that body temperature was quite stable, although it tended to rise quickly and to a high level if the patients were not feeling well. It took another 150 years to develop a practical, calibrated thermometer that would provide repeatable measurements in people. The historical development of the thermometer has been given by Middleton[1]; a review of modern developments, especially as they relate to heat therapy has recently been published by Cetas.[2] A useful source of thermometry information, encompassing a broad spectrum from fundamental scale definition to practical problems, is the series *Temperature*,[3] which appears approximately once each decade. The fundamentals of thermometry are reviewed by Quinn and Compton[4] and more recently by Hudson.[5]

The historical development of the thermometer shows that thermometry has been related to medicine since its beginning. In addition, the development of the science of thermometry and its contribution to medicine have depended on technical advances, and the converse is just as true. Finally, practical temperature scales, as opposed to the theoretical thermodynamic scale (Kelvin), have been based from the beginning on defined fixed temperature points and material properties. All three of these points are true today, and the remainder of this chapter is, to some extent, an elaboration of these points.

2. TEMPERATURE SCALES

There are two types of temperature scales. The first is based on the theoretical concept of temperature and was developed by thermodynamic reasoning as man began to understand the nature of heat and temperature. The absolute temperature scale was established in terms of the energy and

---

Thomas C. Cetas • Division of Radiation Oncology, University of Arizona, Tucson, Arizona 85724.

work involved when a heat engine operated between two temperatures. When defined this way, temperatures are independent of such material properties as mercury or gas expansions, which resolves the perplexity of early thermometer makers due to the fact that different materials do not expand proportionately. An absolute zero exists in the thermodynamic postulation of a temperature scale, and although it cannot be reached, it can be approached asymptotically. Ratios of temperatures measured on this thermodynamic scale are more significant physically than additions of temperatures.

The second type of temperature scale is the practical or working scale. Currently, the entire world, by agreement, refers all temperature measurements to the International Practical Temperature Scale of 1968 (IPTS-68).[6] This scale, based on careful thermodynamic experiments, was established by an international committee and represents the current state of the art as an approximation to the Kelvin thermodynamic scale. The IPTS-68 is believed to be accurate to within about 0.01 K over most of its range below 904 K. For comparison, the precision, but not necessarily the accuracy, of most fundamental thermometry standards below the steam point is 10^{-4} K. The procedure of the IPTS-68 is to define the temperature of several natural fixed points, such as the triple point of water, and provide a protocol for interpolation and extrapolation.

A list of the standard fixed temperature points is given in Table 1. Between these fixed points, a precisely constructed platinum resistance thermometer

TABLE 1
Fixed points of the IPTS-68[a]

Equilibrium state	Assigned value of IPTS	
	T_{68}(K)	t_{68}(°C)
Triple point of equilibrium hydrogen	13.81	−259.34
Liquid and vapor phase equilibrium of hydrogen at a pressure of 33,330.6 N/m² (25/76 standard atmosphere)	17.042	−256.108
Boiling point of equilibrium hydrogen	20.28	−252.87
Boiling point of neon	27.102	−246.048
Triple point of oxygen	54.361	−218.789
Boiling point of oxygen	90.188	−182.962
Triple point of water[b]	273.16	0.01
Boiling point of water[b,c]	373.15	100.
Freezing point of zinc	692.73	419.58
Freezing point of silver	1,235.08	961.93
Freezing point of gold	1,337.58	1,064.43

[a] Except for the triple points and one equilibrium hydrogen point (17.042 K), the assigned values of temperature are for equilibrium states at a pressure $p_0 = 1$ standard atmosphere (101,325 N/m²). In the realization of the fixed points, small departures from the assigned temperatures will occur as a result of differing immersion depths of thermometers or the failure to realize the required pressure exactly. If allowance is made for these small temperature differences, they will not affect the accuracy of realization of the scale.
[b] The water used should have the isotopic composition of ocean water.
[c] The equilibrium state between the solid and liquid phases of tin (freezing point of tin) has the assigned value of $t_{68} = 231.9681$°C and may be used as an alternative to the boiling point of water.

is used for interpolation. The interpolation function involves a polynomial with 21 terms carried to 16 significant figures. A given platinum thermometer is then calibrated at the fixed points to determine the corrections necessary to make it reproduce the scale of the ideal thermometer. In principle, the entire scale can be set up by following rules given in the text of the IPTS-68. Calibration of one platinum thermometer against another is not necessary. In practice, only a few of the national standards laboratories in various countries have established the IPTS-68-defined fixed points and defined the scale in this way. The national laboratories have calibrated many platinum resistance thermometers against the IPTS-68 and use them to calibrate thermometer probes sent to them.

3. THERMOMETER PROBES

A thermometer is any instrument that measures temperature, and any temperature-dependent parameter can serve as the basis for a thermometer. Thermometer designers are concerned with producing the maximum dependence of their instruments on temperature and eliminating any dependence on extraneous or environmental parameters. It is not a trivial task to construct a stable, accurate, and convenient instrument that measures only temperature. Consequently, we will discuss some of the sources of error for each type of thermometer.

Even if a thermometer behaves ideally, errors in measurement can arise from poor technique. If a few general concepts are kept in mind, however, these errors can usually be reduced, or at least their presence and magnitude can be noted. Probe thermometers sense their own temperature, which is determined by establishing equilibrium with the medium to be measured. Consequently, the temperature sensed is an average over a finite volume that depends on the presence of gradients, the thermal properties of the medium, and those of the probe. Heat is exchanged by conduction, convection, and radiation, each of which must be considered when measuring temperatures.

Conduction is described by Fourier's law, which states that heat flow is proportional to the temperature gradient, the thermal conductivity of the medium, and the cross-sectional area through which heat is flowing (cf. Appendix 1). Temperature measurements in the presence of steep gradients lose significance, because the thermometer will sense an average value. Furthermore, thermal conduction down the thermometer leads or shaft may affect the temperature sensed and change gradients and temperature profiles drastically. If thermal equilibrium is desired, for example, when calibrating one thermometer against another, the system should be constructed so that the paths of heat flow do not cross the region where equilibrium is desired (that is, where the thermometers are located). Convection can induce steep thermal gradients. Major blood vessels are not necessarily in thermal equilibrium with the surrounding tissue. Similarly, if a calibration bath is filled with a medium, such as an unstirred, viscous oil, the temperature will vary significantly throughout the bath.

Surface temperatures are also difficult to monitor because of radiant heat exchange as well as convective and conductive exchange. Probes placed on the surface will alter all three mechanisms and sense not only the temperature of the altered surface, but a temperature influenced by the radiative, convective, and conductive exchange of the probe with the environment. Generally, radiometric techniques, such as thermography, are preferable for surface measurements (cf. Chap. 25) if the surface has an emittance close to unity. Tissues, for example, have an emittance of the order of 0.98.

Thermometers can be characterized by several features. The choice of a particular type of thermometer for a given task will depend on the relative merits of these features as well as personal preferences and the availability of equipment:

Sensitivity: Some parameter, such as resistance, electromotive force, or reflected light intensity, changes monotonically with temperature; and a means is available to measure this change with sufficient precision. In medical applications, 0.1°C sensitivity is usually adequate, but higher precision is often helpful for determining small temperature differences of importance in calibrating perfusion estimation and studying thermoregulation.

Accuracy: The thermometer must reproduce absolute temperatures within the limits necessary for the task. In most biomedical situations, 0.1°C accuracy is sufficient. However, certain biochemical studies require higher accuracy if experimenters are to compare results properly.

Stability: Most thermometers drift with age and so must be checked periodically and recalibrated when necessary.

Response time: Thermometer readings must keep pace with the actual temperature fluctuations of interest. The response time is usually characterized by a time constant that can be defined in the following way. A thermometer is subjected to a step change in temperature T_0 by plunging the probe into a water bath. The indicated change in temperature T will approach T_0 exponentially. The time constant is the time required for the difference between T and T_0 to reduce $1/e$, or 37% of T_0. Equilibrium, a residual difference within 5% of T_0, is assumed to be reached after an interval corresponding to about three time constants.

Passivity: The thermometer should not significantly perturb the medium to be measured, either through thermal conduction, heating artifact, or tissue reactions (e.g., coagulation, cell adhesion). Induction heating of electrically conducting sensors and leads has been a principal limitation of thermometry in electromagnetic diathermy fields.

Size: The sensor should be small compared to the size of the object to be measured and small enough to resolve significant temperature gradients. Its heat capacity (or more properly, its thermal inertia, which is the product of density and specific heat; and thermal resistivity) should be low, so that it will not produce a cold spot. The size of the probe is linked intimately to the two preceding items and the problem of durability.

Temperature indication: This refers to how temperatures are obtained from the thermometric parameter. For example, on a clinical thermometer, the height of mercury column is read against a scale etched on a glass tube. Electronic thermometers provide an analog scale or digital reading. High precision, accuracy, and stability of the order of 0.1°C require, however, the individual calibration of the thermometer.

These features are not all inclusive nor entirely independent. A discussion of several common types of thermometers follows; an attempt is made to describe their features, drawbacks, instrumentation, and sources of error.

3.1. Mercury-in-Glass

The mercury-in-glass thermometer is one of the oldest thermometers; it is essentially unchanged in form and accuracy from the time of Fahrenheit.[1] Laboratory quality thermometers, accurate to about 0.1°C, can be obtained. If calibrated individually, thermometers with 0.1°C divisions can be used to maintain a scale to approximately 0.02°C. Careful, well-trained technicians at national standards laboratories routinely and reproducibly read them to 0.005°C!

The major deficiency of the mercury-in-glass thermometer is related to its large mass: It cannot be used for point measurements in regions with steep thermal gradients; it has a very long time constant; and it is easily broken. Occasionally, the mercury column becomes separated, especially during shipment, but carefully heating the thermometer to well above its range, so that the mercury expands into the enlarged cavity at the top of the column, followed by slow cooling will usually correct the problem. Alternatively, the thermometer can be cooled, so that all of the mercury is drawn into the bulb at the bottom. This is easily accomplished by lowering the thermometer into the neck of a flask of liquid nitrogen. The thermometer may break, however, if it contacts the liquid or is cooled or warmed too quickly.

Glass and mercury do not expand proportionately, and glass exhibits some hysteresis in its thermal properties. Consequently, it is preferable in precision work to always work with increasing temperatures unless very long equilibrium times are possible. Gently tapping the thermometer before reading will help relieve strains in the mercury column due to surface tension. It is preferable to use mercury-in-glass thermometers in a totally immersed fashion. If this is not possible, some corrections may be necessary, because the stem and column are at temperatures different from the medium under study. Glass has a tendency to flow over a period of time, which will cause a change in the bulb's volume and consequently a shift in the thermometer's calibration. An occasional check of the reading of the thermometer at the ice point will reveal any changes that may have occurred. The shift at the ice point can be added to the entire calibration, so that a complete recalibration is not necessary. If precise work with mercury-in-glass thermometers is planned, a copy of the NBS report by Wise[7] is useful.

Recently, we tested 22 new clinical thermometers at random: they were all found to be accurate to ± 0.15°C at 37°C. Abbey *et al.* [8] noted in a sample of nearly 300 thermometers that 97.7% were accurate to 0.1°C initially, but only 76% of the thermometers retained that accuracy figure after 10 months. They noted, however, that one brand of the four they tested was much less reliable than the other three.

3.2. Platinum Resistance Thermometer

In ideal metals at a temperature of absolute zero, electrons originating from the outer shells of the constituent atoms are free to travel about the well-organized lattice structure without hindrance. At temperatures above absolute zero, the electrons are scattered by collisions with vibrating atoms in the lattice structure. The higher the temperature, the greater the degree of vibration and consequently the greater the resistance to electron transport. A temperature-dependent electrical resistance, which is linear within the biologically significant range of room temperature, results from this phenomenon.

The standard platinum resistance thermometer (SPRT) is now specified as the interpolating instrument on the International Practical Temperature Scale of 1968 (IPTS-68) for determining temperatures between the defined fixed points of 13.81–903.89 K.[6] Riddle *et al.*[9] have written an excellent monograph on the use of these instruments. Most medical laboratories will find that the SPRT is more than they require. Many industrial grade (non-SPRT) platinum resistance thermometers are available commercially; they typically have a resistance of about 100 Ω at 0°C, compared to 25.5 Ω for the SPRT. The temperature coefficient is approximately 0.4%/°C. The individual grade thermometers exist in a variety of configurations but cannot be made small enough to fit into hypodermic needles or fine catheters; thus, they are primarily useful for calibrating other biomedical thermometers.

3.3. Thermistor

Thermistors also measure temperature by a resistance principle; however, unlike platinum thermometers, they are semiconductors and have different thermometric characteristics. Electrons in a semiconductor at low temperatures are bound to the atoms that make up the solid. In contrast to insulators, the energy ΔE required to excite the electrons into the conduction band and make them available to carry electrical current is of the order of thermal energies at ordinary temperatures (0.025 eV). The number of conduction electrons, and hence the conductivity σ, is determined by the Boltzman distribution function

$$\sigma = \sigma_0 \exp\left(-\Delta E / kT\right) \tag{1}$$

For this application, σ_0 and ΔE are constant, k is the Boltzman constant, T is the absolute temperature in K. The reciprocal of this expression, the resistivity $(1/\sigma)$, decreases exponentially with temperature. Taking the natural logarithm of the expression for resistivity and writing in terms for the resistance

R of a specific device, we obtain

$$1/T = A_0 + A_1 \ln R \qquad (2)$$

where A_0 and A_1 are constants. Finally, if we follow Steinhart[10] and allow one additional term to correct for nonideal semiconductors, we have

$$1/T = A_0 + A_1 \ln R + A_2 (\ln R)^3 \qquad (3)$$

A_0, A_1, and A_2 are constants to be determined from the calibration of the probe. This expression represents the temperature versus resistance characteristic for thermistors to better than 0.01°C, from 0–55°C.[11] A feature of this method is that only three constants are required to characterize the thermometer. Accurate temperatures can be computed directly by this expression from the measured resistance and displayed, tabulated, or used in computer analysis.

Thermistors are constructed from a variety of sintered oxides, such as MgO; a common form is a small bead molded around two fine platinum leads. The beads are fired, encapsulated in glass, and probes of various styles are constructed from them. They are made in a range of sizes from 0.3 mm or less to perhaps 2 mm in diameter; the fabrication of thermistors is still an art. Stability varies from probe to probe, even if made from the same basic material, and glass-encapsulated beads tend to be more stable than disks and those encapsulated in epoxy.

Thermistors have several advantages for biomedical applications; for instance, they can be made quite small. In our laboratory, we use commercial probes mounted in 25 AWG hypodermic needles or Teflon catheters. These probes have a high sensitivity of about 4%/°C, which is approximately an order of magnitude greater than that for platinum resistors. Thermistors have high precision and when calibrated, high accuracy with relatively modest instrumentation. The stability of thermistors varies, but if care is taken in selection, it can be very good; our standard thermistor has not drifted in six years within our measurement uncertainty of ± 0.003°C at the ice point. Two identical needle probes with sequential serial numbers were purchased at the same time, and one of them has not drifted, while the other has required recalibration several times to keep it accurate to within 0.05°C.

The most annoying deficiency in thermistors is that they are not really interchangeable; that is, they cannot be made to have the same R vs. T characteristic to within 0.1°C. Schemes have been designed to make them interchangeable within 0.3–0.5°C, especially if larger sensors are permissible and compound units can be used. Generally, a choice must be made between such larger, direct reading probes and smaller (needle-sized) probes. Electronic circuits to convert resistance measurements to directly reading temperature units are sensitive to the specific thermistor bead (cf. Section 5). Many of the problems associated with producing interchangeable probes and circuits for directly reading temperature units can be alleviated by using a microprocessor system. In such a scheme, coefficients A_0, A_1, and A_2 of Eq.

3 for a specific probe can be stored and the temperature computed from the specific resistance readings. Thus, accuracy, small size, and direct reading all become possible in one design.

Direct current potentiometric methods or Wheatstone bridges are used for the resistance measurement when high precision is required. We use the former in a "four-lead" configuration for our standard thermistor thermometers. Two leads carry probe current and two measure the voltage difference, thus eliminating effects of lead resistance. Current-reversal techniques, which eliminate effects of contact potentials in the voltage measurement should be used for precise thermometry (better than ±0.01°C).

Probes used in medical or biological work are often calibrated as two-lead devices. For these less precise applications, a digital multimeter (DMM) with a resistance measurement mode can be used. Most DMMs use currents about an order of magnitude too large for great precision, causing self-heating of the thermistor on the order of 0.5–2°C. In our laboratory, we use a standard current of 10 μA for reading thermistors of a few kΩ. In our clinic, our DMMs use a reading current of about 100 μA for a 1,000-Ω thermistor. The self-heating error in a medium of high thermal conductivity is less than 0.01°C; however, it is much greater in a thermally resistive medium, such as air. Heat conduction down the shaft of a hypodermic needle probe can be a troublesome source of error. The needle should be inserted more than 4 mm into a thermally conductive media in order ensure that its reading will be accurate to within 0.1°C.[12] If the medium is poorly conductive, such as a fat layer, the error will be worse. Finally, the presence of strong electromagnetic fields, such as those used in hyperthermic work, can cause serious errors.

3.4. Thermocouples

A current is observed to flow in a circuit composed of two different metals when the junctions are at two different temperatures. The current is derived from the dissimilar thermal diffusion rates of electrons in the two metals between the hot and cold junctions.

A thermocouple thermometer is constructed by placing one junction of the two metals at the point where the temperature T_1 is to be measured and a second junction, where lead wires to the voltmeter are connected, at a reference temperature T_0, such as in an ice point cell. Frequently, the reference junction is simulated electrically and incorporated into the reading microvoltmeter. Because the thermoelectric force dE/dT is nearly constant near mammalian temperatures, linearization circuits are relatively simple. Units that are capable of reading directly in temperature units are common.

A copper versus constantan (a copper–nickel alloy) thermocouple is frequently used for biomedical purposes; it has a relatively high sensitivity of about 40μV/°C. Probes are constructed by joining wires of pure copper with constantan by a weld joint. With care, the variation from probe to probe can be kept to less than 0.1°C. For reasons discussed later, this is usually the limit on the resolution as well.

Thermocouples can be made very small and thus are quite attractive for many biomedical applications. Probes in 29-gauge needles are readily fabricated, and a technique for constructing probes of 10 μm diam. for intracellular applications has been developed. The probe consists of a copper–nickel junction fabricated on a quartz substrate.[13,14] A hydrogen flame, glass probe puller is used to obtain 5–30 μm tip diam. Thin layers of copper and nickel are vapor deposited on the quartz substrate to form a single thermocouple at the tip of the probe, which is less than 0.5 μm thick. The thermal properties of the quartz are such that $k\rho c_p$ and α approximate values for biological media.

The theoretical rise time has been modeled as a function of tip diameter and measured experimentally.[15] Experimental measurements indicate a slightly longer rise time than the calculated values for all probe sizes. The thermoelectric EMF for these probes is 21 μV/°C, which is constant from ambient to 90°C. These probes have been used for laser photocoagulation studies. The vacuum-deposited copper has a relatively high reflectance, so that a laser-irradiated probe absorbs less than 20% of the incident energy. The response of the probe to directly absorbed laser energy is so much faster than the response to conducted heat that the two components can usually be separated. The direct absorption response appears as a step-temperature increase at the beginning of a light pulse and a step-temperature decrease at the end as shown in Fig. 1. These probes have proven to be extremely durable in experiments that involve their insertion through the sclera of the eye and contact with the biological tissue for periods of 4–6 hr. Their linearity, thermal EMF, and fast response time have allowed recordings to be made of the response to laser irradiation for temperature rises from 1–45°C and from 1 msec to 60 sec.

A vapor-deposited platinum–tellurium thermocouple has recently been developed by Guilbeau and Mayall,[16] using fabrication techniques similar

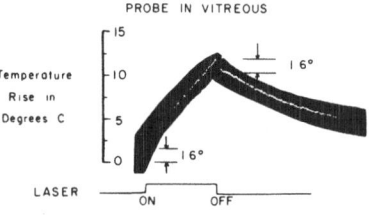

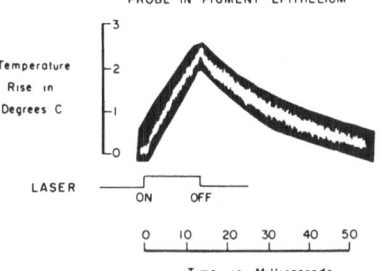

FIGURE 1
Temperature rise in the vitreous (top) and pigment epithelium (bottom) of the rabbit eye due to Ar laser irradiation. Laser power was reduced by approximately a factor of 5 for the pigment epithelium measurement. From Ref. 15 with permission.

to the one described earlier. A 1 μm sensing junction was achieved, with a sensitivity of 300 μV/°C and a 90% response time of less than 50 msec.

Sources of error in thermocouple systems arise from contact potentials, which can easily exceed 10 μV; mechanical working of the wire, which causes strain-induced voltage artifact; inhomogeneity in the alloy; or impurity in the wires. Errors from these last two effects can be kept below 0.1°C for measurements in biomedical environments. Electronic biases and drifting in the reading instrumentation also appear as temperature shifts. For these reasons, thermocouples are less satisfactory as thermometer secondary standards. Finally, since they are electrical conductors, thermocouples will be affected by electromagnetic fields in terms of noise, self-heating, and field perturbation. Some precautions can be taken to reduce these problems, such as filtering the input to the electronic reading unit, inserting probes at normal incidence to electrical fields, electrically insulating the leads from current-induced fields, and, when possible, electromagnetically shielding the thermocouples. These measures, although satisfactory in some cases, may not suffice in others. Since one lead of a thermocouple is typically a ferromagnetic alloy, it will be significantly affected (heated) by a strong radiofrequency magnetic field. Consequently, such a thermocouple may be unsuitable for measurements in the fields produced by magnetic induction high-frequency diathermy.

3.5. Other Electronic Devices

Nearly any circuit parameter that varies with temperature can be used to provide a thermometer, provided variations in that parameter due to other effects can be controlled. Circuits have been designed that use both diodes and transistors as active elements in thermometer systems. Recently, integrated circuits have been developed specifically as thermometers; they can be incorporated into larger electronic systems for control and monitoring. Integrated circuits appear to be quite linear, which relieves the need for complicated linearization circuits or nonlinear calibration algorithms.

3.6. Quartz Thermometer

The quartz thermometer relies on the temperature dependence of the resonant frequency of a quartz transducer.[17] It is an elegant device with a direct-reading accuracy of 10^{-4} K and is used in many laboratories as a secondary standard. It is subject to vibration, shock, and drift, and so its calibration must be verified periodically, as must all other systems described in this chapter.

3.7. Bimetallic Strips

These thermometers are mentioned because they are so common in the medical environment and are frequently used for routine measurements with low-accuracy requirements. They are constructed by bonding together two metals with different thermal expansion coefficients and then coiling the bimetallic strip into a flat spiral. As the strip is heated, the coil expands according to the different expansion rates of the two elements in the strip; a pointer attached to the strip indicates the temperature on a scale. Such thermometers are cheap, rugged, and do not require expensive electronics for readout. Measurement errors are caused by the large thermal mass of the probe and by thermal conduction down the shaft. These thermometers are often inaccurate; we tested three units used for monitoring temperatures in premature infant incubators, and two read low by 8°F and one by 2°F!

3.8. Liquid Crystal Indicators

Cholesteric liquid crystals show a strong color play with temperature. As the liquid crystal is heated, the twisted stack of long molecules will twist more (or unwind). Light at a wavelength commensurate with the spacing of these layers is preferentially reflected. Consequently, the apparent color of the liquid crystal will pass through the visible spectrum as it is heated and so serve as a thermometer. These indicators have several applications in medicine. In one, they are used to measure oral or rectal temperatures. We tested seven of these units and found them to be accurate within $0.2 \pm 0.2°C$, in agreement with a more extensive evaluation reported by Besley *et al.*[18] Other uses of liquid crystals in thermometry have been for monitoring critical care patients[19] and for providing an alternative means of obtaining breast thermograms.[20] In another application, they have been applied to optical fibers for constructing thermometers that can be used in strong electromagnetic fields.[21]

3.9. Fiber Optic Thermometers

Recently, a new class of thermometers has been developed that uses light propagating through optical fibers as a means of transmitting thermal information. The impetus for this scheme has come from the necessity of determining temperatures in the presence of strong electromagnetic fields,[12,20–22] such as those used in studies of biological effects or in electromagnetically induced hyperthermia for cancer therapy. These thermometers may also be useful in other applications, such as surgery, with the most stringent requirements for electrical leakage. Such thermometers may ultimately be interchangeable, disposable, and competitive with current electronic thermometers.

Several different types of optical thermometric sensors have been proposed. The first probe used a liquid crystal sensor.[21] As previously described, a cholesteric liquid crystal undergoes a color change with temperature within a certain finite range. If a red light is incident on the crystal, the intensity of the reflected light will be a function of the sensor temperature. Unfortunately, liquid crystals do not appear to have the stability necessary for accurate thermometry. They drift in terms of color play with temperature and in the magnitude of the reflectance; furthermore, they appear to show some hysteresis.

The birefringence of a single crystal of $LiTaO_3$ cleaved along certain directions is temperature dependent and can be used as the sensitive element in a thermometer. This optical thermometer[25,26] exploits the better stability of single-crystal systems. A practical probe can be made by coating one side of the crystal with a dielectric mirror and the other side with a polarizing film that serves as polarizer of the incident light and analyzer of the light transmitted both ways through the crystal. The coated crystal is attached with transparent epoxy to the end of the optical fibers: probes have been made less than 0.5 mm in diameter. The temperature response for this thermometer is sinusoidal, but the period can be made sufficiently broad, typically from 20–55°C, that the thermometer is single valued over the temperature range of interest. The use of an optical reference channel removes the effects of drift in the light source, photodetector, and signal processing electronics.

One problem with many optical thermometers has been the drift of the monochromatic emitted light with temperature. Christensen[27,28] exploited this effect for the GaAs light-emitting diode by recognizing that its reciprocal, absorption of light, would also be temperature dependent. He constructed a thermometer by placing a small chip of GaAs on the end of the fiber bundle, irradiated by an external LED. This technique is based on the band-edge absorption of infrared light as it excites valence-band electrons across the forbidden energy gap into the conduction band as the light passes through the GaAs chip. The variation of band-gap energy with temperature provides a varying absorption of emitted, narrow band light from the LED as a function of sensor (GaAs chip) temperature. The temperature characteristic of this probe is approximately linear over a range from 20–50°C, with a short-term accuracy of 0.1°C. The probes range from 0.3–0.6 mm in diam. Christensen[27] has also developed a sophisticated electronic system, including a built-in calibrator for the probe.

Several optical thermometers have been designed that use the fluorescence decay time of crystals or phosphors as the temperature-sensitive parameter.[29–31] The temperature-sensitive substance is excited with a light source. The stimulated emission decay rate is monitored, either by integrating the total light emitted during two fixed time intervals of the decay,[29,30] or by monitoring the decay ratio at two emission wavelengths.[31]

Another interesting technique uses the peak wavelength of the Raman scattering from a ruby crystal for thermometry in lipid membranes.[32] This is an excellent example of using available systems to accomplish specific needs. The expansion with temperature increase of a liquid contained in a small capsule has also been used in a fiber optic thermometer application.[33] The reflectance of light from the meniscus varies with the liquid expansion, which can be monitored in a fashion similar to other systems. An infrared detector has been coupled to optical fibers to measure the thermal radiation at the end of the fiber and hence determine the temperature in another application.[34] These systems, however, appear to be more useful at temperatures higher than the range of interest in biology and medicine.

Perhaps the most useful characteristic of fiber optic thermometers is that they are insensitive to the strong electromagnetic fields used to induce hyperthermia in cancer therapy and to warm tissues in physical medicine and research on the biological effects of these fields. But the optical probes have other useful features as well. Most probes are very small: 1-mm diam. now is considered large. Glass has a relatively low thermal conductance, so that the thermometers cause little thermal perturbation and, in addition, measure the temperature of only a very small region near the sensor. As a consequence of their small size, the probes have very fast time constants, which are increased somewhat when protective teflon jackets are used; the jackets can be removed in special circumstances. Finally, these probes have been developed with modern electronic reading units and so have convenience and accuracy features that may not be available in more conventional thermometer systems.

The major hindrance to the broader application of fiber optic thermometers is that they are still very expensive. Handling the optical fibers during construction is tedious. Optical components, which are analogous to electronic analog or integrated circuits, switches, directional couplers, and even suitable connectors are still under development. Consequently, fabricating these thermometers is currently quite laborious, and they are less amenable to mass production techniques. However, commercial sources are now available, since private corporations are beginning to recognize the size of the potential market as hyperthermia therapy in cancer treatment becomes accepted.

3.10. Other Thermometers for Use in Electromagnetic Fields

Another approach to constructing thermometers that can be used in the presence of strong electromagnetic fields is to modify a conventional type of thermometer so that it is less sensitive to the field. Bowman's device[35] uses a small high-resistance thermistor as the sensor. The leads are constructed from very high-resistance, carbon-impregnated teflon; the measuring current is carried by one pair of leads, while the potential is measured across the other pair. The high resistivity reduces the dipole currents induced by the electric field; close proximity of the leads minimizes the encompassed area and so minimizes magnetically induced loop currents; the four-terminal configuration eliminates errors associated with lead resistances. Present thermometers of this design are less than 1 mm in diam. Some electromagnetic losses still occur in this thermometer, but they are small and of the same order of magnitude as the tissue displaced by the probe; Larsen and colleagues[36] have worked on related types of thermometers.

Olsen and Molina[37] have constructed thermocouples from high-resistivity, nonmetallic conductors. The high resistivity reduces interference from the electric field as previously described. Further, the materials are not magnetic, so they are less sensitive to magnetically induced noise and heating.

4. THERMOGRAPHY

In recent years, scanning infrared thermographic cameras have been used for observing and measuring thermal variations across exposed surfaces. Most applications infer structural flow and metabolic information from the intensity of emitted infrared radiation. It is possible, however, to also measure temperature with modified thermographic equipment; this subject is treated in Chap. 25.

5. CALIBRATION

Temperature measurement accuracy to 0.1°C cannot be guarenteed in off-the-shelf thermometers without individual calibration. Probe thermometers mounted in either flexible catheters or hypodermic needles are not direct reading, interchangeable, and stable at this level of accuracy. Thus, a facility with calibration media temperatures known to 0.02°C should be a part of every laboratory or clinic in which 0.1°C temperature precision is required. When secondary, working thermometers are calibrated against the standard thermometer, uncertainties should be well below the levels of biological interest. Furthermore, calibration drift in the secondary probes can be detected before it becomes significant.

Two approaches to a calibration facility are possible. The first and most common is to maintain a standard thermometer that has a calibration that can be referred to the IPTS-68.[6] Some type of temperature-controlled medium, such as a water bath or a copper block, is used to maintain the unknown and standard thermometers at the same temperature for comparing readings. The other approach is to use established cells of natural fixed point temperatures, such as the ice point, 0°C, or gallium point, 29.772°C, and observe the thermometer reading at these points.

Precision mercury-in-glass thermometers are available from a number of scientific suppliers and can be read to this accuracy.[7] In addition, NBS has developed special thermometers (SRM 933 and 934) for their Standard Reference Materials program that have precise calibrations at four specific temperatures—0°C, 25°C, 30°C, and 37°C.

Thermometric fixed points also can be used for calibration. Table 2 shows several fixed points that are of interest for studies requiring precise thermometry between −30–100°C. The melting point of very pure (99.9999% pure) gallium has been shown to be useful as a reference temperature for

TABLE 2
Fixed-Point Temperatures of Biomedical Interest

		Reference
H_2O (freezing point)	0°C (273.15 K)	6
H_2O (triple point)	0.01 (exact)	6
Phenoxybenzene (triple point)	26.87	
Ga (Melting point)	29.772	38, 51
$Na_2SO_4 \cdot 10\ H_2O$ (salt hydrate transition)	32.373	39
$KF \cdot 2\ H_2O$ (salt hydrate transition)	41.421	39
$Na_2\ HPO_4 \cdot 7H_2O$ (salt hydrate transition)	48.222	39
H_2O (boiling point)	100.00	6

calibration.[38,51] Complete cells, ready for use, can be obtained from the National Bureau of Standards and Yellow Springs Instruments Company.

Magin and co-workers[39] devised a technique for using the hydration–dehydration point of certain salts in solution as temperature fixed points (see Table 2). Fresh reagent grade chemicals ($Na_2SO_4 \cdot 10H_2O$, $Na_2HPO_4 \cdot 7H_2O$, $KF \cdot 2H_2O$) are heated a few degrees above the salt hydration transition temperature. The salt is poured into a vacuum flask and stirred mechanically. The thermometers are placed in oil-filled capillary wells and calibrated as with gallium cells. Three calibration points are adequate to determine the three coefficients that describe the thermistor resistance–temperature characteristic. Designs for baths and details of measurement technique in reference media are discussed at length in Refs. 2 and 12.

Two other precepts of thermometry are that no thermometer is necessarily stable and none is indestructible. Hence, for accurate work, three standard quality thermometers must be maintained to determine which one may have drifted, and all thermometers must be checked regularly, especially the standards. We have found that operator carelessness and thermometer element deterioration contribute to the thermometer's finite lifetime. In our laboratory, the average lifetime of thermometers is several months except for our well-guarded standards.

6. THERMOMETRY IN THERAPEUTIC ELECTROMAGNETIC FIELD-INDUCED HEATING

As previously described, measuring temperature in the presence of strong electromagnetic fields is a difficult task. Fields emanating from the power mains are an aggravation to nearly all measurements. Precision measurements at many sites are plagued by radiofrequency (rf) interference from nearby radio and television transmitters. Thus carefully shielding and grounding equipment, or even entire rooms, may be required to eliminate the problems caused by environmental rf fields.

The nature and significance of the perturbing effect of thermometers in strong fields, especially in relation to electromagnetic heating, has been demonstrated graphically with a thermographic camera by Johnson and Guy.[22] Some workers have claimed that their data were only minimally affected by the presence of electromagnetic fields. In one test that we performed, using 3-MHz current fields to heat a man's arm, no measurable artifact appeared in temperatures monitored with a needle-mounted thermistor. However, when the same treatment was applied to an eleven-year-old girl with osteogenic sarcoma, much poorer electromagnetic coupling to the arm was achieved, and a thermometry artifact of about 1°C was observed.

The discrepancy between these observations may be traced to the many factors that influence the magnitude of the perturbation. They include the field distribution; the orientation and position of the thermometer leads with respect to field orientation and phase; the degree of shielding, including that due to tissue and the geometry of the subject and the radiator. Three types of interaction between the thermometer and the field contribute to the artifact. The first is electromagnetic pickup by the electronic measurement system. The second is self-heating of the thermometer element by direct absorption of energy from the electromagnetic field. The third is perturbation of the electromagnetic field, caused by reflection or shunting of the electric field by the thermometer or its sheath.

Several techniques can be used to reduce the magnitude of these effects. The instruments should be well shielded and grounded, with special care taken to avoid ground loops, including those involving capacitive coupling; radiofrequency filters in the sensing circuitry and the power lines help as well. The thermometer leads should run perpendicular to the electric field vector and be twisted tightly to reduce magnetic induction pickup. When possible, extraneous fields radiating from the source should be eliminated. With respect to interference in the sensor itself, the thermometer should be electrically isolated from the electromagnetic field. For example, thermistors mounted in hypodermic needles must not be shorted to the needle, or excessive rf noise will be fed to the instrumentation. More accurate temperature information may be obtained by periodically shutting off the radiative power source in order. For some experiments, re-entrant wells can be prepared in the subject prior to heating; low thermal mass thermometer probes are inserted immediately after the field is extinguished. The maximum temperature attained can be determined by extrapolating the cooling curve to the instant that heating ceased.

In other cases, the sensor can be inserted so that the tissues electrically shield most of the sensor leads. In most of these cases the magnitude of any temperature artifact is proportional to the power levels that produce the artifact. At frequencies below 5 MHz (wavelength > 60 m), the thermometer leads tend to make relatively poor antennas, which reduces the problem with artifacts. At higher frequencies, these lead antennas become more efficient, and the problem with noise increases. As microwave frequencies are approached, near 1 GHz, the fields can be more easily confined, and noise is again less of a problem. The magnitude of an artifact can sometimes be estimated by quickly switching off the field and watching the thermometer's cooling rate. A fast change implies electromagnetic noise effects or the cooling of a small mass is confined to the probe itself and so is artifactual; a slower rate suggests the cooling of a larger tissue volume. Some investigators are using microcomputers to shut off the radiation source and switch in the thermometers for monitoring in a regular fashion.

These solutions are not entirely satisfactory for a number of reasons. First of all, they frequently are not adequate; significant interference may remain. Shielding the sensor may reduce electromagnetic interference but may also produce a substantial reflection or shunting of the electromagnetic field. The near field of an antenna can have electric field components in all directions, and thus it may be impossible to place the leads normal to the field. The fields are especially complex in regions of curved interfaces that can focus the fields. Hot spots caused by the presence of the probe may occur somewhat removed from the probe. Thus, the common technique of momentarily switching off the field and distinguishing between the fast jump in the sensor reading that is attributed to electromagnetic interference and the slower change attributed to a real temperature change will not necessarily reveal localized tissue heating caused by the presence of the thermometer. Many situations exist where continuously monitoring the temperature is necessary for precisely controlling heating fields. Even switching the field off for a few seconds will result in a substantial temperature drop if a large blood flow rate exists in the region being heated. Finally, in a given experiment or clinical treatment, ethical or physiological constraints take precedence over physical considerations. Nevertheless, these techniques must be used until new thermometers specifically designed for use in strong electromagnetic fields become generally available.

7. THERMOMETRY IN THERAPEUTIC ULTRASOUND-INDUCED HEATING

Thermometry in conjunction with ultrasonic heating is employed two ways: as a dosimetric technique to guide hyperthermia,[40] and evaluate the properties of tissue exposed to ultrasound.[41] The latter technique is discussed in Chapter 8; we will consider only the former in this chapter.

Thermometry in the presence of the intense ultrasonic fields required for therapy is less of a problem than it is for electromagnetic heating. This is because the ultrasonic field is confined to dense media, electromagnetic fields do not occur, and the signal processing systems are electrically based rather than mechanical measurements. Hence, interference in the temperature measurement must involve electroacoustic coupling, which is relatively weak.

Nevertheless, thermometry is not free of problems peculiar to this environment. Two principal concerns are to avoid preferential heating of the thermometer probe and to avoid perturbing the ultrasonic field by the probe. Both of these heating artifacts are reduced by using probes with dimensions that are small compared to the ultrasound wavelength. For example, the wavelength of a 1 MHz ultrasound wave in soft tissue or water is about 1.5 mm. Lele[40] uses thermocouples that are 25 and 50 μm in diameter. In addition to reducing the scattering and direct absorption of the ultrasonic wave, the small probe size also minimizes thermal perturbations from heat conduction down the leads. If the probes can be oriented so that the cross-sectional area exposed to the propagating wave is minimized, the interference is reduced further. Probes that have been sheathed in soft polyethylene or any other highly absorbing material should be avoided, since the ultrasonic energy absorption coefficient (in dB/cm) of soft polyethylene is approximately five times greater than that of most soft tissues.[41] One further source of direct probe interference arises from pulsed fields. The propagating ultrasonic wave can produce relative motion between the probe[42] and the medium. This leads to dissipative (viscous) heating near the probe, which is a transient problem, however, with a short relaxation time and so is not of concern in the steady state.

The most difficult aspect of applying thermometry to ultrasound thermal dosimetry is common to all forms of heating by wave propagation. Heterogeneous tissues and interfaces between tissue planes cause wave scattering. Consequently, hot spots or shadows are very difficult to predict. Bone–soft tissue interfaces and air–soft tissue interfaces are particularly troublesome in ultrasound applications. Reflections from bone will cause hot spots in the soft tissue just in front of the bone. Conversely, if the coupling to the bone is good, preferential heating of the bone will occur. Thus, single-point temperature measurements may completely miss the occurrence of thermally significant events at a point a few millimeters away.

8. SUMMARY

Some of the basic principles and underlying assumptions of thermometry have been presented to clarify the science. Conventional thermometry is not arcane or difficult, but care in application and interpretation is necessary in order to avoid errors. On the other hand, in thermal dosimetry, it is quite easy to obtain utterly nonsensical results. Clinical thermometry has been especially important in physical medicine, clinical hyperthermia for cancer treatment, managing heat stroke and controlled hypothermia, such diagnostic procedures as screening for breast cancer or observing cardiovascular circulatory problems. Numerous clinical applications of thermometry are found in other chapters of this book.

ACKNOWLEDGMENTS. Support for this effort has come from NIH grant CA17343. I am indebted to several people for assistance. In particular, Mr. Bruce Herman of the Ultrasonics Branch of the Bureau of Radiological Health and Dr. Padmaker Lele of MIT assisted on the section on thermometry in ultrasonic fields. Dr. R. Nasoni provided insight on ultrasonic tomographic thermometry. The extent of the discussion on each subject reflects my interest and awareness of the subject, and neither an endorsement nor criticism of any device or system is intended.

REFERENCES

1. Middleton, W. E. K., *A History of the Thermometer* (Johns Hopkins Press, Baltimore, 1966).
2. Cetas, T. C., Thermometry, in *Therapeutic Heat and Cold*, 3rd ed., J. F. Lehmann, ed. (Williams and Wilkins, Baltimore, 1982).
3. Plumb, H. H., ed., *Temperature*, vol. 4 (Instrument Society of America, Pittsburgh, 1972).
4. Quinn, T. J., and Compton, J. P., The foundations of thermometry, *Rep. Prog. Phys.* **38**, 151, 1975.
5. Hudson, R. P., Measurement of temperature, *Rev. Sci. Instrum.* **51**, 871, 1980.
6. Preston-Thomas, H., International practical temperature scale of 1968, *Metrologia* **5**, 35, 1969, amended ed. of 1975, *Metrologia* **12**, 7, 1976.
7. Wise, J. A., Liquid-in-glass thermometry, *NBS Monograph* **150**, (U.S. Government Printing Office, Washington, D.C., 1976).

8. Abbey, J. C., Anderson, A. S., Close, E. L., Hertwig, E. P., Scott, J., Sears, R., Willens, R. M., and Packer, A. G., How long is that thermometer accurate? *Am. J. Nurs.* **78**, 1375–1376, 1978.

9. Riddle, S. L., Furukawa, G. T., and Plumb, H. H., Platinum resistance thermometry. *NBS Monograph 126*, (U.S. Government Printing Office, Washington, D.C., 1972).

10. Steinhardt, J. S., and Hart, S. R., Calibration curves for thermistors, *Deep Sea Res.* **15**, 497, 1968.

11. Trolander, H. W., Case, D. A., and Harruff, R. W., Reproducibility, stability, and linearization of thermistor resistance thermometers, *Temperature*, vol. 4, H. H. Plumb, ed. (Instrument Society of America, Pittsburgh, 1972), p. 997.

12. Cetas, T. C., and Connor, W. G., Thermometry considerations in localized hyperthermia, *Med. Phys.* **5**, 79, 1978.

13. Cain, C. P., and Welch, A. J., Thin-film temperature sensors for biological measurements, *IEEE Trans. Biomed. Eng.* **BME-21**, 421, 1974.

14. Battist, L., Goldner, F., and Todreas, N., Construction of a fine wire thermocouple capable of repeated insertions into, and accurate positioning within, a controlled environmental chamber, *Med. Biol. Eng.* **7**, 445, 1969.

15. Priebe, L. A., Cain, C. P., and Welch, A. J., Temperature rise required for production of minimal lesions in *macaca mulatta* retina, *Am. J. Ophthal.* **79**, 405, 1975.

16. Guilbeau, E. J., and Mayall, B. I., Microthermocouple for soft tissue temperature determination, *IEEE Trans. Biomed. Eng.* **28**, 301, 1981.

17. Benjaminson, A., and Rowland, F., The development of the quartz resonator as a digital temperature sensor with a precision of 1×10^{-4}, in *Temperature*, vol. 4, H. H. Plumb, ed. (Instrument Society of America, Pittsburgh, 1972).

18. Besley, L. M., and Kemp, R. C., *In vitro* evaluation of accuracy of a single-use clinical thermometer, *Med. J. Aust.* **2**, 337, 1978.

19. Lees, D. E., Schuette, W., Bull, J. M., Whang-Peng, J., Atkinson, E. R., and MacNamara, T. E., An evaluation of liquid crystal thermometry as a screening device for intraoperative hyperthermia, *Anesth. Analg.* **57**, 669, 1978.

20. Logan, W. W., and Lind, B., Improved liquid cholesterol ester crystal thermography of the breast, *J. Surg. Oncol.* **8**, 363, 1976.

21. Rozzell, T. C., Johnson, C. C., Durney, C. H., Lords, J. L., and Olsen, R. G., A nonperturbing temperature sensor for measurements in electromagnetic fields, *J. Microwave Power* **9**, 241–249, 1974.

22. Johnson, C. C., and Guy, A. W., Nonionizing electromagnetic wave effects in biological materials and systems, *Proc. IEEE* **60**, 1972.

23. Cetas, T. C., Thermometry in strong electromagnetic fields, in *The Physical Basis of Electromagnetic Interactions with Biological Systems*, L. S. Taylor and A. Y. Cheung, eds. (University of Maryland, College Park, 1978), p. 261.

24. Christensen, D. A., Thermal dosimetry and temperature measurements, *Cancer Res.* **39**, 2325, 1979.

25. Cetas, T. C., Hefner, R. D., Snedaker, C., and Swindell, W., Further developments of the birefringent crystal optical thermometer, abstract presented at USNC/URSI Series on Biological Effects of Electromagnetic Waves, Amherst, MA, 1976.

26. Cetas, T. C., A birefringent crystal optical thermometer for measurements in electromagnetically induced heating, in *Proceedings 1975 USNC/URSI Symposium*, C. C. Johnson and M. L. Shore, eds. (HEW Publication 77-8011, vol. 11, 1976), p. 338.

27. Christensen, D. A., A new nonperturbing temperature probe using semiconductor band edge shift, *J. Bioeng.* **1**, 541, 1977.

28. Christensen, D. A. Experience with a four-probe nonperturbing temperature-monitoring system, presented at 3d Int. Symp. on Cancer Therapy by Hyperthermia, Drugs, and Radiation, Fort Collins, 1980, Abstract Book, p. 57.

29. Sholes, R. R., and Small, J. G., Fluorescent decay thermometer with biological applications, *Rev. Sci. Instrum.* **51**, 882, 1980.

30. Samulski, T., and Shrivastava, P. N., Photoluminescent thermometer probes: temperature measurements in microwave fields, *Science* **208**, 193, 1980.

31. Wickersheim, K. A., and Alves, R. B., Recent advances in optical temperature measurement, *Indust. Res. Dev.* **21**, 82 1979.
32. Cavatorta, F., Schoen, P. E., and Sheridan, J. P., An optical nonperturbing probe for temperature measurements in biological materials exposed to microwave radiation, presented at Bio-electromagnetics Symposium, Seattle, 1980, *Abstract B-34*, p. 491.
33. Deficis, A., and Priou, A., Nonperturbing microprobes for measurement in electromagnetic fields, *Microwave J.* **20**, 55, 1977.
34. Dakin, J. P., and Kahn, D. A., A novel fiber optics temperature probe, *Opt. Quant. Elec.* **9**, 540, 1977.
35. Bowman, R., A probe for measuring temperature in radiofrequency-heated material, *IEEE Trans. Microwave Theory Tech.* **MTT-24**, 43, 1876.
36. Larsen, L. E., Moore, R. A., Jacobi, J. H., Halgas, F. A., and Brown, P. V., A microwave compatible MIC temperature electrode for use in biological dielectrics, *IEEE Trans. Microwave Theory Tech.* **MTT-27**, 673, 1979.
37. Olsen, R. G., and Molina, E. A., The nonmetallic thermocouple: a differential temperature probe for use in microwave fields, *Radio Sci.* **14**, 81, 1979.
38. Sostrum, H. E., Melting point of gallium as a temperature calibration standard, *Rev. Sci. Instrum.* **48**, 127, 1977.
39. Magin, R. L., Statler, J. A., and Thornton, D. D., Inorganic salt hydrate transitions as temperature-fixed points in biomedical thermometry, *Advances in Biomedical Engineering*, M. K. Wells, ed. (ASME, New York, 1979), pp. 151–154.
40. Lele, P. P., Induction of deep, local hyperthermia by ultrasound and electromagnetic fields, *Rad. Environ. Biophys.* **17**, 205, 1980.
41. Wells, P. N. T., *Biomedical Ultrasonics* (Academic, New York, 1977).
42. Fry, W. J., and Fry, R. B., Determination of absolute sound levels and acoustic absorption coefficients by thermocouple probes—theory, *J. Acoust. Soc. Am.* **26**, 294–310, 1954.

Chapter 28

SENSITIVITY ANALYSIS OF ERRORS INDUCED IN THE DETERMINATION OF TISSUE PERFUSION

Avraham Shitzer and Robert C. Eberhart

1. INTRODUCTION

One of the important tasks in physiology and medicine is estimating tissue blood perfusion rate and heat generation rate due to metabolism and other sources. This information is required for experimental studies as well as for clinical purposes and has been the subject of numerous ingenious efforts. The difficulties in performing these measurements are mainly due to (1) the complex anatomical structure of the tissue and interwoven blood supply network of vessels of various sizes and (2) the lack of noninvasive techniques that are capable of directly measuring these quantities. As a result, indirect methods have been devised in which the required information is inferred from measurements of related physical and physiological parameters. Due to these difficulties, the time and spatial resolution of the estimated parameters is compromised and one must accept lower accuracy and precision with these methods.

Since estimating the tissue heat production rate by the heat clearance method has been used less frequently than estimating the tissue blood perfusion rate, consideration of this problem will be limited to some remarks later on. Techniques for estimating blood flow rates, discussed in Chap. 9, may be classified according to the size of the blood vessels. For instance, cardiac output is routinely estimated by transient washout in major vessel or stepwise or bolus perturbation of indicator concentration. One may also find detailed hydrodynamic studies of the velocity profiles and pressure drops in the large vessels and even in arterioles and venules.[1,2] These studies may be invasive, involving the introduction of a flowmeter into the vessel;[3] noninvasive techniques may also be employed, such as laser–Doppler anemometry.[4]

Flow studies involving the smaller vessels are more complicated and less amenable to direct measurement, and results are usually presented as the

Avraham Shitzer • Department of Mechanical Engineering, Technion, Israel Institute of Technology, Haifa, Israel.   *Robert C. Eberhart* • Department of Surgery, University of Texas Health Science Center, Dallas, Texas 75235.

volume of averaged blood perfusion rate, i.e., nondirectional rate of blood flow through the capillary bed. Among the methods commonly employed to obtain these average values are: (1) the measurement of injected radioactive or other diffusible gases,[5] (2) the measurement of locally generated hydrogen ions,[6] (3) the dissipation of heat in the tissue (thermal dilution or heat clearance),[7] and (4) the measurement of radio-labeled microspheres of aggregated albumin that are introduced locally into the circulation and eventually trapped in the tissue bed.[8] Many of these techniques have been reviewed by Paradise and Fox.[9] Thermal dilution techniques have recently been reviewed by Eberhart *et al.*[10]

All of these methods may be categorized as indirect; average tissue blood perfusion rates are obtained indirectly from continuously or intermittently monitoring the specific indicator. The monitoring sites may be inside the tissue volume of interest, or by employing a Fick relation, they may be in the blood vessels supplying and draining that volume. Thus, the process of estimating blood perfusion rate is based on mass and/or energy balances that define the phenomenological relationship between indicator concentration and blood perfusion rate that exists in the physical model. Estimation also depends on the accuracy of the collected data and the method of parameter estimation employed. All of these steps are prone to various degrees of error and may yield quantitative results that are considerably in error.

The purpose of this chapter is to present an analysis of the sensitivity of the estimated tissue blood perfusion rate to errors in the pertinent parameters and variables. The analysis is presented in terms of a generalized indicator, e.g., injected heat or mass. Particular attention is given to the error due to the concentration of the diffusible indicator. The analysis is presented in terms of normalized partial derivatives (sensitivity coefficients). They are obtained from a general diffusion equation involving a blood-perfusion-dependent convection term. Subsequently, the diffusion term is removed, and the equation is allowed to degenerate into simple one- and two-compartment washout models.

2. ANALYSIS

Measurement errors (observation, round off, systematic, calibration, drift, etc.) are unavoidable in any experimental setup and may cause considerable deviations of the calculated values from the "true" ones. These factors may offset or augment one another. The objective of the experimenter in this context is to first determine the magnitude of anticipated measurement errors and then design the appropriate experiments and parameter estimation techniques in order to minimize errors in the calculated parameters. Such an approach allows attention to be focused on accurately measuring those quantities that may make large contributions, or have sensitivity coefficients, to the final results.

Much effort has already been devoted to developing error analysis theories, e.g., reference 11, and parameter estimation techniques, e.g., reference 12. In the following sections, a parameter sensitivity analysis is presented based on total and partial derivatives obtained from the equation modeling the physical phenomenon.

Suppose there exists a phenomenological relationship, based on first principles, among several parameters, $x_1, x_2, \ldots, x_m$. The relationship for the jth parameter is assumed to be described by

$$x_j = x_j(x_i), \qquad i \neq j \tag{1}$$

It follows that the normalized sensitivity coefficient (NSC) of x_i with respect to x_j is given by

$$A_i = \frac{x_i}{x_j} \frac{\partial x_j}{\partial x_i} \tag{2}$$

A_i determines the relative change of x_j, i.e., dx_j/x_j, should dx_i/x_i change by unity. This coefficient has long been employed in engineering analysis.[13,14]

In some cases, it may be desirable to calculate the NSC of a certain parameter that is stated implicitly, e.g., x_n with respect to x_i. By chain differentiation,

$$\frac{\partial x_j}{\partial x_n} = \frac{\partial x_j / \partial x_i}{\partial x_i / \partial x_n} \tag{3}$$

and employing the reciprocity principle among partial derivatives

$$\frac{\partial x_i}{\partial x_j} = \left(\frac{\partial x_j}{\partial x_i}\right)^{-1} \tag{4}$$

then the desired NSC is given by

$$\frac{x_i}{x_n} \frac{\partial x_n}{\partial x_i} = \frac{(x_i/x_j)(\partial x_j/\partial x_i)}{(x_n/x_j)(\partial x_j/\partial x_n)} = A_i A_n^{-1} \tag{5}$$

These expressions can now be used to study the propagation of errors in parameter estimation.

3. RESULTS

The application of sensitivity analysis is discussed in the following sections.

3.1. Heat Transfer in a Tumor Model

As a first illustration, a steady-state tissue heat balance is considered. A simplified model of a tumor surrounded by normal tissue is studied. The two regions are assumed to have different thermophysical and physiological properties. Matching conditions for both temperature and heat flux are specified at the neoplastic–normal tissue interface. The equation describing the heat balance for the two regions is the one-dimensional bioheat equation given by[15]

$$\frac{d^2 T_i}{dy_i^2} + \left(\frac{\beta_i}{b_i}\right)^2 (T_i - T_a) + \frac{\rho_i q_i}{k_i} = 0 \tag{6}$$

where $i = 1$ indicates the neoplastic tissue region $0 < y_1 < b_1$, $i = 2$ indicates the normal tissue region $0 < y_2 < b_2$, and w_i is the blood perfusion parameter given by

$$\beta_i^2 = \frac{\rho_{bi} c_{bi} w_{bi} b_i^2}{k_i} \tag{7}$$

Equation (6) is solved subject to the following boundary conditions:

$$\text{at } y_1 = y_2 = 0, \quad T_1 = T_2, \text{ and } \quad k_1 \frac{\partial T_1}{\partial y_1} = -k_2 \frac{\partial T_2}{\partial y_2} \tag{8}$$

$$\text{at } y_1 = b_1, \quad \frac{\partial T_1}{\partial y_1} = 0 \tag{9}$$

$$\text{at } y_2 = b_2, \quad T_2 = T_2^0 \text{ (const)} \tag{10}$$

The dimensionless solution to this set of equations is represented by

$$\theta_1 = \gamma_1 + (\Omega - \gamma_1 + \gamma_2) \frac{\cosh[\sqrt{\beta_1}(1 - \xi_1)]}{\cosh \beta_1} \tag{11}$$

$$\theta_2 = \gamma_2 + \frac{\Omega \sinh[\sqrt{\beta_2}(1 - \xi_2)] + (1 - \gamma_2) \sinh \sqrt{\beta_2} \xi_2}{\sinh \sqrt{\beta_2}} \tag{12}$$

where

$$\theta = \frac{T_i - T_a}{T_2^0 - T_a}; \quad \gamma_i = \frac{q_i}{\rho_{bi} c_{bi} w_{bi}(T_2^0 - T_a)}$$

$$\Omega = \frac{(\gamma_1 - \gamma_2)\tanh\sqrt{\beta_1} + \kappa(1 - \gamma_2)}{\tanh\sqrt{\beta_1} + \kappa \coth\sqrt{\beta_2}}$$

$$\kappa = \frac{k_2 w_2}{k_1 w_1}\left(\frac{b_1}{b_2}\right)^2, \quad \xi_i = \frac{y_i}{b_i} \tag{13}$$

Equations (11)–(13) define the relationship between the temperature fields in both the neoplastic (Eq. 11) and the normal (Eq. 12) tissues in terms of the various physical and physiological parameters. Now suppose the temperature fields and other parameters are measured and the data are used to estimate the blood perfusion parameters. This inverse problem can be stated in a form similar to Eq. (1), namely,

$$w_{bi} = w_{bi}(\rho_{bi}, c_{bi}, k_i, b_i, q_i, T_i, T_a, T_2^0) \tag{14}$$

The sensitivity coefficients of w_{b_i} with respect to any of the other parameters may now be calculated according to Eqs. (4) and (5). To avoid lengthy expressions, details are omitted. However, the simple case is treated in Section 3.4.

Results pertaining to tumor blood perfusion w_{bi} are shown in Fig. 1. The sensitivity of this parameter to errors in five variables, i.e., tumor and normal tissues temperatures (T_1, T_2), normal tissue blood perfusion rate (w_{b_2}) and neoplastic tissue metabolic rate (q_1), thermal conductivity (k_1) and tumor half-thickness (b_1), are plotted vs. the distance from the tumor's center. It is observed that the sensitivity of w_{b_1} to errors in temperature measurement is dominant; it is at least two orders of magnitude higher than that attributed to any of the other parameters. It is further observed in Fig. 1 that the sensitivity of w_{b_1} to errors in T depends on the location of the temperature probe. Consequently, the farther away from the center of the tumor the placement of this probe, the greater the sensitivity will be. Of all the parameters studied, the neoplastic tissue thermal conductivity k_1 has the smallest sensitivity coefficient, which decreases with distance from the center of the tumor until

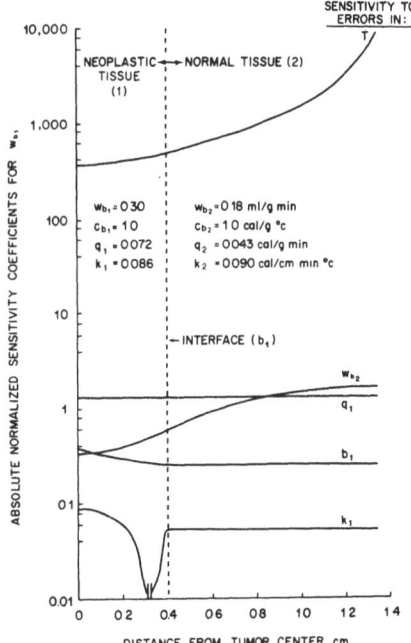

FIGURE 1
Steady-state absolute values of normalized sensitivity coefficients for neoplastic tissue blood flow w_{bi} due to errors in the indicated parameters. Results are plotted vs. distance from the tumor center (see Ref. 20).

a minimal value is attained. Thereafter, k_1 increases and becomes almost constant in the normal tissue region.

The most interesting result from this example is in relation to temperature measurement, since this example implies that in predicting blood perfusion, extreme care should be exercised in measuring temperature. It is interesting to note that due to the reciprocity principle, the opposite statement also applies; tissue temperature is rather insensitive to errors in blood perfusion rate under the conditions of this example. This conclusion holds with regard to the precision and accuracy of the measuring system, the selection of the placement site for the probe, and the accuracy of the parameter estimation routine. Another result of this analysis suggests that much larger errors in any of the other parameters studied may be tolerated due to their lower sensitivity coefficients. This result should not be construed as a recommendation for removing the accuracy requirements for these parameters. It merely indicates that should difficulties arise in securing these requirements, they may be somewhat compromised without overly affecting the accuracy of estimating w_{b_1}.

3.2. Heat Clearance in Myocardium

As a second illustration, the prediction of blood perfusion rate in a contracting myocardium is chosen.[16] It is assumed that the dynamic heat balance in this tissue is given by the transient bioheat equation. In this case, the walls of the contracting myocardium are assumed to be maintained at a constant temperature equal to the arterial level T_a. The initial condition is given by

$$T = T_a - U[1 - (2y/L - 1)^4] \tag{15}$$

and the response of the tissue to a thermal perturbation provided by a sudden return of the arterial inflow temperature to its steady-state value is monitored.[16] The solution to this problem is given by

$$T = T_a + \frac{\gamma}{\sqrt{\beta}}\left\{1 - \frac{\sinh\left[\sqrt{\beta}(1-\zeta)\right] + \sinh\left(\sqrt{\beta}\zeta\right)}{\sinh\beta}\right\}$$

$$- \sum_{n=1}^{\infty} B_n \sin\left(\lambda_n\zeta\right)\exp\left[-\alpha(\beta + \lambda_n^2)t/L^2\right] \tag{16}$$

where

$$\gamma = \rho q/k, \qquad \zeta = y/L, \qquad \lambda_n = n\pi$$

and

$$B_n = \frac{4}{(2n-1)^3\pi^3}\left\{48U\left[1 - \frac{8}{(2n-1)^2\pi^2}\right] + \frac{\gamma L^2}{1 + \beta/(2n-1)^2\pi^2}\right\} \tag{17}$$

Results for the normalized sensitivity coefficients of the blood perfusion rate with respect to temperature, thermal conductivity, and myocardial heat generation rate are shown in Fig. 2. Curves are given for two tissue depths—0.1 cm (near the epicardial surface) and 0.5 cm (centerline) and are plotted versus time elapsed from removal of the thermal perturbation. It is again seen that estimation of blood perfusion rate is extremely sensitive to temperature errors, which by far exceed the contribution of any of the other parameters studied.

Another interesting observation is the appearance of lines of discontinuity, as is seen in Fig. 2. The values of the NSC in the vicinity of these lines go up to infinity and then through a change of sign return to finite values. The mathematical reason for the appearance of these discontinuities is the presence of w_b, i.e., β, in the time-dependent argument of the exponential terms in Eq. (16). Consequently, an incremental change in w_b will cause a crossover of the temperature profiles at a certain point in time, i.e., $\partial T/\partial w_b \to 0$ at this point. Thus, the sensitivity coefficient, which is determined by the reciprocal of this value, i.e., $\partial w_b/\partial T$, would go to infinity, yielding the observed lines of discontinuity. In this figure, as is the case for all other figures in this chapter, absolute values of the sensitivity coefficients are shown. Therefore, a line of discontinuity, as discussed, would not indicate the appropriate sign of the sensitivity coefficient. This information is not required, however, for the purposes of the present study; one is interested in the absolute values of the sensitivity coefficient and the relative error so that their contribution to maximum error may be calculated.

Results shown in Fig. 2 clearly indicate that there exist certain time intervals for which the sensitivity of w_b to errors in the measurement of the other parameters is minimal. These curves also indicate that the error in w_b due to temperature measurement error is reduced for measurements in the

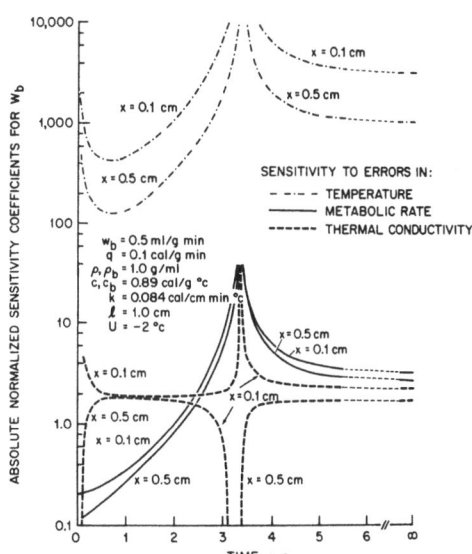

FIGURE 2
Temporal variations or normalized sensitivity coefficients of myocardial blood perfusion w_b due to errors in T, q, and k. Results are shown for two locations: subepicardial, $x = 0.1$ cm, and midmyocardial, $x = 0.5$ cm (see Ref. 20).

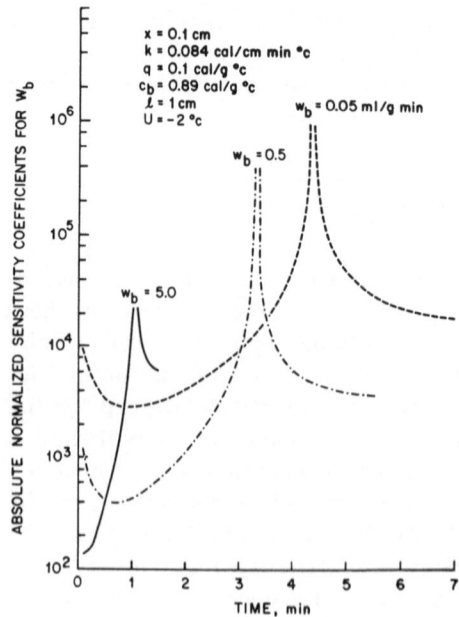

FIGURE 3
Effect of the value of myocardial blood perfusion rate w_b on its normalized sensitivity coefficients due to errors in T (see Ref. 20).

deeper layers, e.g., temperature at $x = 0.5$ cm. In other cases, the position dependence of the NSC is more complex, as in the case of the error in w_b due to the error in thermal conductivity.

Since it is evident from Fig. 2 that the sensitivity of w_b to errors in temperature measurement is highest and clearly dominates the others, the value of w_b for which these coefficients are calculated was varied 100-fold. Results are shown in Fig. 3. Curves are plotted for a position fixed near the epicardial surface $x = 0.1$ cm. It is noted that as a general trend, the lower the value of w_b, the higher its sensitivity to temperature errors. Minimal values of temperature sensitivity coefficients are obtained at about 150, 400, and 2,800 for $w_b = 5.0$, 0.5, and 0.05 ml/g min., respectively. The occurrences of these minima also seem to be inversely dependent on w_b, i.e., the lower the value of w_b, the longer the time required to reach minimum sensitivity to temperature errors.

Additional data on the behavior of the sensitivity indicate that the lowest values are obtained for errors in the estimation of heating rate q. However, sensitivities with respect to both q and k seem to decrease as the value of w_b increases. It is also noted that the lowest sensitivity to errors in q is obtained shortly following removal of the thermal perturbation. Data also indicate that this sensitivity rapidly increases at later times. The sensitivity to errors in k is almost constant in time, except near the interval of discontinuity; this result is qualitatively the same as that shown in Fig. 2.

3.3. One-Compartment Model with Internal Sources

The next illustration involves a well-stirred, one-compartment model with a single inlet and a single outlet through which blood is flowing at a constant rate w_b without any recirculation. In addition, heat is being generated inside the compartment at a constant rate q. At time $t = 0$, the temperature of the incoming blood stream is suddenly changed as a sharp step, and the response of the system to this perturbation is monitored by measuring the temperature of the outgoing stream. This problem, also referred to as a one-compartment "washout" with internal sources, is quite common in physiology, medicine, and engineering.[17] The differential equation describing this system may be obtained by eliminating the heat conduction term from the bioheat equation to yield,

$$\rho c \frac{dT}{dt} = \rho_b c_b w_b (T_a - T) + q \tag{18}$$

The initial condition specified for Eq. (18) is

$$\text{at } t = 0, \qquad T = T_a + U \tag{19}$$

for which the solution is

$$T = T_a + \frac{q}{\rho_b c_b w_b} + \left(U - \frac{q}{\rho_b c_b w_b} \right) \exp \left(-\frac{\rho_b c_b w_b}{\rho c} t \right) \tag{20}$$

The normalized sensitivity coefficients of w_b to temperature T and heating rate q are calculated from Eq. (20) and plotted in Fig. 4. Two cases are

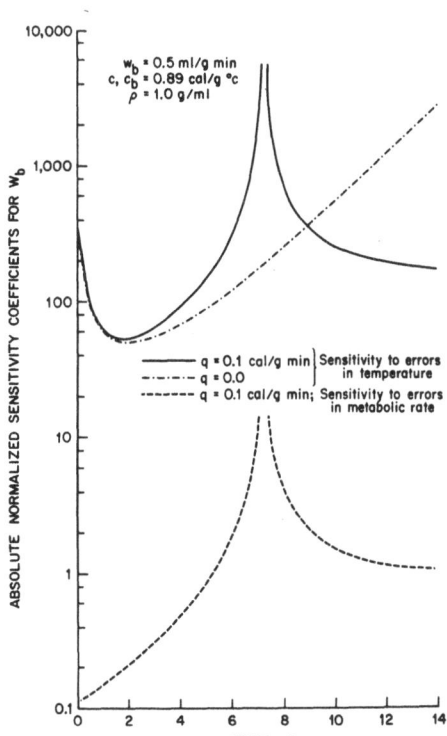

FIGURE 4
Temporal variations of the normalized sensitivity coefficients of blood perfusion w_b due to errors in T and q in a washout model with internal heat sources (see Ref. 20).

presented in this figure: one, similar to the previous illustration in Section 3.2, involves a 0.1 cal/g min heat generation term. The other assumes no heat to be generated inside this compartment, i.e., a simple washout response. As in Section 3.2, the sensitivity of w_b to errors in temperature measurement is dominant. It is noteworthy that for the simple washout problem, i.e., $q = 0$, the discontinuity in the NSC of w_b with respect to T is eliminated. However, large values are observed at both very short and very long times following the introduction of the perturbation. In addition, the interval of the minimal NSC for this case seems to be extended in comparison to that obtained for the case with internal heat sources.

For the examples in Sections 3.2 and 3.3 (Figs. 2–4), the NSC of w_b to errors in T exhibits a similar behavior in that minimal values and lines of discontinuity are obtained with respect to both time and position as long as $q \neq 0$. The times and positions of occurrence of these two extreme values may be calculated from the NSCs obtained from Eqs. (16) and (20), respectively, by differentiating with respect to time and equating with zero. A similar derivation can be employed with regard to position. To simplify the calculations, the simpler case without heat conduction, i.e., Eq. (20) was chosen. The time of occurrence of the minimum NSC, t^*, is given by the implicit expression

$$
\left(T_a + \frac{q}{\rho_b c_b w_b}\right)\frac{\rho_b c_b w_b}{\rho c}t^* - \left(U - \frac{q}{\rho_b c_b w_b}\right)\exp\left(-\frac{\rho_b c_b w_b}{\rho c}t^*\right)
$$
$$
= \frac{UT_a - (q/\rho_b c_b w_b)^2}{U - (q/\rho_b c_b w_b)} \tag{21}
$$

where the time of occurrence of the line of discontinuity, t^{**}, is given by

$$
t^{**} + \left[1 - \exp\left(\frac{\rho_b c_b w_b}{\rho c}t^{**}\right)\right]\frac{\rho c q}{(\rho_b c_b w_b)^2}\left[\frac{1}{U - (q/\rho_b c_b w_b)}\right] = 0 \tag{22}
$$

It is obvious that in order to minimize the contribution due to errors in measuring T, the estimation of w_b should be performed at or around t^*. By the same reasoning, data collected around t^{**} should be excluded to achieve the same purpose. Equation (21) may also be employed to find the minimal sensitivity coefficients for the simple one-compartment model without heat or other indicator sources by setting q equal to zero. A discussion of this case follows.

3.4. Simple One-Compartment Washout Model

This one-compartment washout model is widely used in many disciplines to indicate the response to diffusible indicators besides heat, such as inspired Xe, injected dye, or H_2 evolved at an electrode. Therefore, Eq. (20) was modified to

$$
X(t) = X(0)\exp(-\phi t) + C[1 - \exp(-\phi t)] \tag{23}
$$

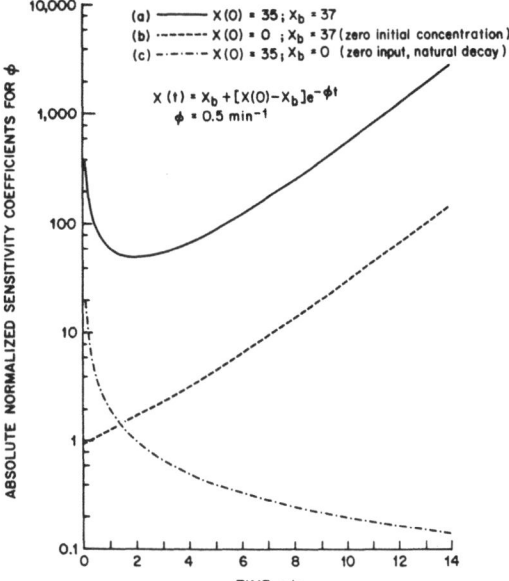

FIGURE 5

Temporal variations of the normalized sensitivity coefficients of the flow parameter ϕ due to errors in X for a linear, one-compartment washout model with constant input concentration. Results are shown for three sets of initial tissue and indicator concentrations.

where X represents the concentration of the diffusible indicator in the compartment that may be assumed equal to tissue concentration; C indicates the concentration of the diffusible indicator in the inflowing stream; $\phi = f/V$ is the flowrate parameter, which is the reciprocal of the system time constant $\tau = 1/\phi$. Three cases may be identified for this problem:

(a) nonzero initial and inlet concentrations $X(0)$, $C \neq 0$;
(b) zero initial with nonzero inlet concentrations, $X(0) = 0$, $C \neq 0$; and
(c) nonzero initial with zero inlet concentrations, $X(0) \neq 0$, $C = 0$ (representing, e.g., radioactive decay)

Normalized sensitivity coefficients of ϕ with respect to X and t for these three cases are derived in the appendix to this chapter, and results are plotted in Fig. 5. As in the previous illustrations, it is seen that for both cases of nonzero inlet concentration, cases (a) and (b), the NSC of ϕ to errors in measured concentration of the diffusible indicator is extremely high. Only for case (a) are low values and a minimum obtained shortly following the introduction of perturbation. Case (c), representing, for example, the washout of an injected radioactive indicator or natural decay, seems to behave in the opposite manner; it starts off with a high sensitivity to indicator concentration error, which continuously diminishes as time progresses. As a matter of interest, it can be shown that case (a) is the linear sum of cases (b) and (c), although this is not seen in Fig. 5, due to the different normalizing factors employed.

3.5. Parallel, Two-Compartment Washout Model

To further demonstrate the sensitivity of compartmental analysis to errors in measuring indicator concentration, a two-compartment, parallel system was analyzed. The general equation describing the outlet concentration of this system in response to a constant input concentration of a diffusible indicator is given by

$$Y(t) = C + \sum_{i=1}^{2} v_i [Y_i(0) - C] \exp(-\phi_i t) \tag{24}$$

where v_i is the normalized flow distribution function, i.e., $v_i = f_i / \Sigma f_i$, and $i = 1, 2$ indicate the first and second compartments, respectively. Normalized sensitivity coefficients for ϕ with respect to the outlet concentration as obtained from Eq. (24) are shown in Fig. 6. For the set of parameters chosen, qualitatively similar results to the one-compartment system, case (a), are obtained. It is observed that for identical initial conditions, the faster flow compartment has the lower sensitivity coefficients for about the first 6.5 min. following the onset of the perturbation. Thereafter, the sensitivity values of this compartment increase at a much higher rate than do those of the slower flow compartment.

Figure 6 also indicates the existence of minimum sensitivity values for both compartments. This phenomenon prompted the calculation of the time of occurrence of these minima, as carried out in Section 3.3. The evaluation was performed by differentiating the expressions obtained for $(Y/\phi_i)(\partial \phi_i / \partial Y)$ with respect to time and equating to zero. A Newton–Raphson scheme was employed to calculate the times of occurrence of these minimum sensitivity coefficients. The two curves plotted in Fig. 6 show the variations of the NSCs for both ϕ_1 and ϕ_2 for one set of parameters. Subsequently, the time of occurrence of the minimal NSCs and their values are calculated for a tenfold

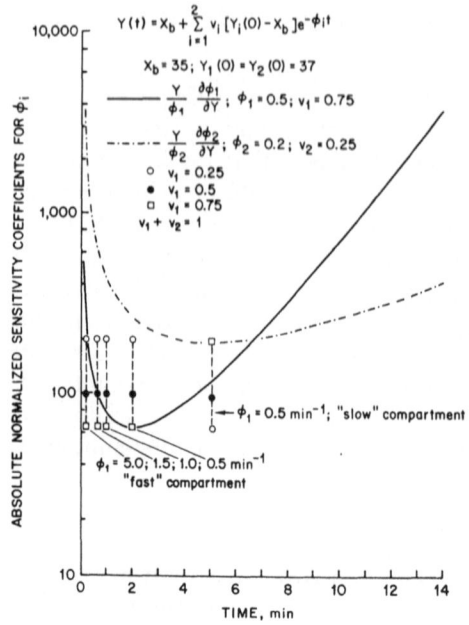

FIGURE 6

Temporal variations of the normalized sensitivity coefficients of the flow parameters ϕ_1 and ϕ_2 due to errors in Y for a two-compartment, parallel washout model with constant input concentration.

range of ϕ_1 and a threefold range of v_1. The results are indicated by symbols and connecting broken lines. It is observed that for a given value of ϕ_1, the time of occurrence of the minimum NSC is essentially constant for a range of v_1, decreasing with increases in ϕ_1. The minimum value of the NSC for ϕ_1 is seen to change considerably with the value of the flow distribution function v_1 for any given value of ϕ_1, exhibiting an inverse relationship.

3.6. Embedded Heated Temperature Probe

As a final illustration of the application of this technique, one of the works by Bowman, Balasubramaniam, and Woods is analyzed (cf. Ref. 18 and Chap. 9). These investigators employed the steady-state bioheat equation to derive an expression for the temperature field developed in a tissue into which a heated thermistor bead was inserted. The temperature of the bead was calculated by the steady-state heat equation and subsequently averaged over the volume of the bead. The expression for the bead's temperature is given by (Eq. 14 in Ref. 18)

$$\Delta T_p = \frac{\Gamma a^2}{3k_p}\left\{0.2 + \frac{k_p}{k}\left[1 + a\left(\frac{\rho\rho_b c_b w_b}{k}\right)^{1/2}\right]^{-1}\right\}$$

(25)

where k_p and k are bead and tissue intrinsic thermal conductivities, respectively; Γ is the volumetric electrical power required to maintain ΔT_p above a certain given temperature; and a is the apparent radius of the thermistor bead. Equation (25) is further manipulated by Bowman *et al.*, who define an "effective thermal conductivity" (Eq. 16 in Ref. 18)

$$k_{\text{eff}} = k\left\{1 + a\left(\frac{\rho\rho_b c_b w_b}{k}\right)^{1/2}\right\}$$

(26)

from which an explicit equation for blood perfusion rate can be obtained.

Bowman *et al.* were concerned with uncertainty in the measured value of perfusion and analyzed the sensitivity of perfusion estimate from Eq. (26) for errors in estimating the effective thermal conductivity $dk_{\text{eff}}/k_{\text{eff}}$. Their results, obtained by differentiating Eq. (26), are plotted in their Fig. 1.[18] These results indicate very large sensitivity coefficients of w_b to errors in k_{eff} for $k_{\text{eff}}/k < 1.1$. It can also be shown, based on Eq. (26), that the absolute normalized sensitivities of w_b to errors in a and c_b are constant, being equal to 2 and 1, respectively.

It was felt, however, that since the effective thermal conductivity is calculated from Eq. (25), which involves temperature measurements, the sensitivity of w_b to errors in measuring ΔT_p should also be calculated. This was done by applying the methods presented herein to Eq. (25); results are shown in Fig. 7. Solid curves show the sensitivity coefficients of w_b due to errors in ΔT_p for two values of k/k_p—unity and 4.13, which is the value given by Bowman *et al.* for their studies.[18] It is seen that for low values of the dimensionless blood perfusion parameter β, the sensitivity coefficients are

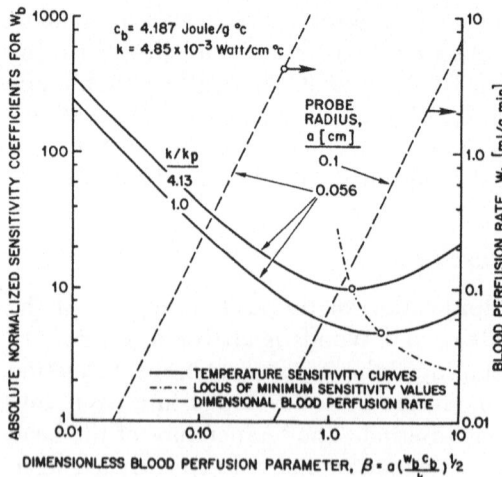

FIGURE 7

Normalized sensitivity coefficients for the blood perfusion rate w_b due to errors in ΔT_p for the self-heated thermistor probe. Also plotted is the locus of the minima of these coefficients as a function of β and the blood perfusion rate for two values of the probe radius a (see Ref. 20).

rather high, exceeding 20 for $\sqrt{\beta} < 0.10$. Thereafter, these values continue to decrease to a local minimum and then resume the increasing trend. As a matter of interest, the locus of the minima, given by

$$\beta_{min} = 1 + 5k_p/k \tag{27}$$

is also shown in this figure. This locus indicates a decrease in the minimum sensitivity coefficient as β increases and k/k_p decreases.

To further explore the practical implication of this information, the dimensional blood perfusion rate w_b was also plotted as a function of β in Fig. 7. It is observed that since β varies linearly with the radius of the bead, in order to minimize the sensitivity coefficient, this radius a should be increased. For example, for $\beta = 1.0$, which is close to the region of minimum sensitivity error, with $c_b = 4.187$ J/g °C and $k = 4.85 \times 10^{-3}$ W/cm °C, the radius of the bead probe should be 0.2 cm in order to optimally measure a blood flow of 1.74 g/cm^3 min. For the same data but a smaller radius of the bead at 0.056 cm, the blood flow would become much higher, at about 22 g/cm^3 min around the region of minimum sensitivity. It follows that in order to decrease the sensitivity of w_b to errors in measuring ΔT_p, the radius of the thermistor bead should be increased, thereby increasing the degree of insult to the tissue.

4. DISCUSSION

The examples chosen to illustrate the applicability of the sensitivity analysis are all taken from a class of problems dealing with the indirect, or inverse, estimation of blood perfusion rate. These problems involve introducing a diffusible indicator into the tissue while monitoring its response to the perturbation. As formulated here, these problems are all modeled by variants of a second-order, partial differential equation including storage, diffusion, generation and convection terms, which we have generally referred to in this book as the bioheat equation. The convection term employed in this work represents one plausible method of describing the effects of blood perfusion on the tissue indicator balance that is particularly applicable in the capillary bed. Other expressions for modeling blood convection effects in tissue have been proposed by Chen and Holmes (cf. Ref. 19 and Chap. 7). Boundary and initial conditions used in this study are not general but nevertheless are considered adequate for the purposes of the present study.

It should be noted that this chapter considers only the sensitivity of one parameter to changes in other parameters. In certain situations, useful information may be obtained from the time rate of change as well. For example, the slope of a curve may be used rather than the measured value itself. The extension of the sensitivity analysis presented here to gradients is straightforward and will not be discussed.

Solutions for the temperature and concentration fields of the examples considered in this study all involve an exponential term in one form or another. The argument of this exponent invariably includes the blood perfusion parameter; e.g., Eqs. (12), (16), (20), (23), and (24). When the sensitivity of this parameter to measurement errors in either temperature or concentration is calculated, the exponential term remains. This exponential term includes a positive argument that dominates the value of the sensitivity parameter and may attain excessively high values, particularly for long time periods. This phenomenon is not typical with regard to any of the other parameters and may partially account for the lower sensitivity values obtained for them.

In all cases studied, the sensitivity of the flow parameter ϕ or its reciprocal τ is highest with respect to measurement errors in the concentration of the specific indicator passing through the system. This sensitivity may vary with time (e.g., Figs. 2 and 5) and/or position (e.g., Fig. 1), and may also exhibit regions of minimal values (e.g., Fig. 3). The implication of these observations is twofold: (1) When designing an experimental protocol, a sensitivity analysis of the type presented here should be performed in order to identify parameters possessing the highest sensitivity values and also the most favorable sites for placing the measuring probes; and (2) data processing and parameter evaluation procedures and techniques must consider the appropriate time and space domains, so that minimum possible error propagation is obtained.

A typical question may be posed regarding the most profitable investment in equipment from an experimental accuracy standpoint. Thus, an investigator may be able to justify the need for a more refined and accurate system for measuring a certain critical parameter. As pertains to the present study, this single parameter seems to be the temperature (or concentration) of the diffusible indicator. The sensitivity coefficients calculated for this parameter are very high and would inevitably cause a large error to be propagated into the estimation of the flow parameter. Fortunately for the heat clearance method, certain temperature measurement techniques are currently capable of indicating temperatures to ±0.05°C or even better. Thus, at normal body temperatures of, say, 37°C, the absolute relative error in temperature measurement dT/T

would be about 0.00135. Thus, even with a sensitivity coefficient of 100, the value of the absolute error in estimating flow $d\phi/\phi$ would be 13.5%. It is clearly seen that the combined effort of both accurately measuring temperature, or concentration, and estimating the flow parameter at the time periods and locations where the sensitivity parameter is at or around its minimal value, is warranted.

Although this chapter has primarily focused on the estimation of blood perfusion, the method presented may also be applied to other parameters. As already mentioned, the estimation of tissue heat generation rate *in situ* is an important problem in physiology. Whenever the heat generation term is included in the modeling equation, its sensitivity to temperature measurement errors can be calculated in the same manner as for blood perfusion. It can be shown that the ratio of the sensitivity coefficients of w_b to T and w_b to q represents the ratio of the sensitivity coefficient of q to T. Sensitivity coefficients for any of the parameters with respect to the others may be obtained in a similar manner.

Although somewhat restricted to the examples treated here, results of this study nevertheless indicate the following conclusions:

(a) Consideration should be given early in the experimental design stage to identifying critical variables that will yield the highest sensitivity coefficients with respect to the sought parameter.

(b) The experimental protocol should be designed to secure the least contribution of measurement errors both in terms of equipment accuracy and probe placement.

(c) During the parameter estimation process, the appropriate time intervals for which the minimum sensitivity coefficients occur should be identified and employed to reduce the probability of error propagation. Sufficient computational accuracy should be retained to avoid unnecessary loss of measurement accuracy during this process.

(d) Estimating the flow parameters, or time constants, in one- and parallel two-compartment models appears to be extremely sensitive to concentration measurement errors, which vary with time.

The method developed here is general and in no way restricted to the class of problems presented in this chapter. This method does not, however, preclude or supersede other techniques usually applied to reduce the uncertainty associated with experimental results; it rather complements them and should be used to improve the final outcome.

ACKNOWLEDGMENT. Supported in part by the American Heart Association (Texas affiliate).

REFERENCES

1. Anliker, M., Towards a nontraumatic study of the circulating system, in *Biomechanics, Its Foundations and Objectives*, Y. C. Fung *et al.*, eds. (Prentice Hall, Englewood Cliffs, NJ, 1971), pp. 337–379.

2. Bugliarello, G., Hung, T. K., and Quevedo, C. E., Fluid flow, in *Biomedical Engineering*, J. H. U. Brown *et al.*, eds. (F. A. Davis, Philadelphia, 1971), pp. 43–69.
3. Wyatt, D. G., Theory, design, and use of electromagnetic flow meters, in *Cardiovascular Flow Dynamics and Measurements*, N. H. C. Hwang and N. A. Norman, eds. (University Park Press, Baltimore, 1977), pp. 89–150.
4. Baker, D. W., and Daingle, R. E., Noninvasive ultrasonic flowmetry, in *Cardiovascular Flow Dynamics and Measurements*, N. H. C. Hwang and N. A. Normann, eds. (University Park Press, Baltimore, 1977), pp. 151–190.
5. Lassen, N. A., On the theory of the local clearance method for measurement of blood flow, including a discussion of its application to various tissues, *Acta Med. Scand. Suppl.* **474**, 136–145, 1967.
6. Koyamata, T., Local myocardial blood flow measured by the use of a needle-type Pt-H_2 electrode, in *Recent Advances in Studies on Cardiac Structure and Metabolism: The Metabolism of Contraction*, P. E. Roy and G. Rona, eds. (University Park Press, Baltimore, 1975), pp. 522–538.
7. Perl, W., Heat and matter distribution in body tissues and the determination of tissue blood flow by local clearance methods, *J. Theor. Biol.* **2**, 201–235, 1962.
8. Rudolph, A. M., and Heymann, M. A., Circulation of the fetus *in utero*: methods for studying distribution of blood flow, cardiac output, and organ blood flow, *Circ. Res.* **21**, 163–184, 1967.
9. Paradise, N. F., and Fox, I. J., Regional blood flow measurement, in *Dye Curves: The Theory and Practice of Indicator Dilution*, D. A. Bloomfield, ed. (University Park Press, Baltimore, 1974), pp. 335–362.
10. Eberhart, R. C., Shitzer, A., and Hernandez, E. J., Thermal dilution methods: Estimation of tissue blood flow and metabolism, in *Thermal Characteristics of Tumors: Applications in Detection and Treatment*, R. K. Jain and P. M. Gullino, eds. (New York Academy of Sciences, New York, 1980), pp. 107–132, 335.
11. Clifford, A. A., *Multivariate Error Analysis* (Wiley, New York, 1973).
12. Beck, J. V., and Arnold, K. J., *Parameter Estimation in Engineering and Science* (Wiley, New York, 1977).
13. Mickley, H. S., Sherwood, T. K., and Reed, C. E., *Applied Mathematics in Chemical Engineering* (McGraw-Hill, New York, 1957).
14. Tomovic, R., *Sensitivity Analysis of Dynamic Systems* (McGraw-Hill, New York, 1963).
15. Pennes, H. H., Analysis of tissue and arterial blood temperatures in the resting human forearm, *J. Appl. Physiol.* **1**, 93–122, 1948.
16. Hernadez, E. J., Hoffman, J. K., Fabian, M., Siegel, J. H., and Eberhart, R. C., Thermal quantification of regional myocardial perfusion and heat generation, *Am. J. Physiol.* **236**, 345–355, 1979.
17. Jacquez, J. A., *Compartmental Analysis in Biology and Medicine* (Elsevier, New York, 1972).
18. Bowman, H. F., Balasubramaniam, T. A., and Woods, M., Determination of tissue perfusion from *in vivo* thermal conductivity measurements, ASME paper #77-WA/HT-40, 1977.
19. Chen, M. M., and Holmes, K. R., Microvascular contributions in tissue heat transfer, in *Thermal Characteristics of Tumors: Applications in Detection and Treatment*, R. K. Jain and P. M. Gullino, eds. (New York Academy of Sciences, New York, 1980), pp. 137–150, 335.
20. Shitzer, A., Eberhart, R. C., and Eisenfeld, J., Estimation of tissue blood perfusion rate from diffusible indicator methods: A sensitivity analysis. *Trans. ASME J. Biomech. Eng.* **102**, 258–269, 1980.

APPENDIX

As an example, the normalized sensitivity coefficient of the flow parameter ϕ with respect to the indicator concentration $X(t)$ is derived from Eq. (23). Note that ϕ is given implicitly in Eq. (23); thus,

$$\frac{\partial X(t)}{\partial \phi} = [C - X(0)]t \exp(-\phi t) \tag{A.1}$$

Inverting according to Eq. (4),

$$\frac{\partial \phi}{\partial X(t)} = \left[\frac{\partial X(t)}{\partial \phi}\right]^{-1} = \frac{1}{[C - X(0)]t \exp(-\phi t)} \tag{A.2}$$

and normalizing according to Eq. (2) yields

$$\frac{X(t)}{\phi} \frac{\partial \phi}{\partial X(t)} = \frac{1}{\phi t}\left\{\frac{C + [X(0) - C]\exp(-\phi t)}{[C - X(0)]\exp(-\phi t)}\right\}$$

$$= \frac{1}{\phi t}\left[\frac{C \exp(\phi t)}{C - X(0)} - 1\right] \tag{A.3}$$

Equation (A.3) gives the desired NSC of ϕ with respect to $X(t)$. In this particular example, ϕ may be expressed explicitly by manipulating Eq. (23)

$$\phi = \frac{1}{t}\ln\left[\frac{X(0) - C}{X(t) - C}\right] \tag{A.4}$$

from which

$$\frac{\partial \phi}{\partial X(t)} = \frac{1}{t[C - X(t)]} = \frac{1}{[C - X(0)]t \exp(-\phi t)} \tag{A.5}$$

Equation (A.5) is identical to Eq. (A.2) and on normalizing would yield the required NSC.

The normalization process involves dividing and multiplying the partial derivative by the appropriate parameters, as indicated by Eq. (2). The purpose of this manipulation is to present the final quantity in a nondimensional form, thereby avoiding the need to consider the units of the various parameters as long as they are consistent. Unfortunately, certain widely used scales of units do not have a common zero point of reference, e.g., the Celsius and Kelvin temperature scales. Consequently, values of the NSCs calculated with respect to one scale will have to be recalculated by appropriately considering the ratio of the two scales. All NSCs presented in this chapter are calculated by parameters whose units are specified in degrees Celsius.

To demonstrate this aspect of sensitivity calculation, assume that $X(t)$ is given in degrees Celsius. The systematic measurement error for $X(t)$ is estimated at $\pm0.05°C$. At a normal body temperature of 37°C, the relative temperature error in this scale of units would be $dX(t)/X(t) = 0.05/37 = 0.00135$. The minimal NSC for the flow parameter ϕ from Fig. 5 is about 50. Thus, the relative error to be expected in calculating the value of ϕ at this point in time would be $d\phi/\phi = 50 \times 0.00135 = 0.0675$, or about 6.8%.

Now suppose the temperature is expressed in degrees Kelvin. The absolute systematic error is again estimated at ±0.05 K. The relative error in this scale is

$$\frac{dX(t)}{X(t)} = \frac{0.05}{(37 + 273)} = 0.000161$$

The NSC of ϕ with respect to $X(t)$ expressed in degrees Kelvin may be obtained from the values shown in Fig. 5. This is done by equating the relative error of the flow parameter

$$\frac{d\phi}{\phi} = \text{NSC}_\text{C}\,\frac{dX(^\circ\text{C})}{X(^\circ\text{C})} = \text{NSC}_\text{K}\,\frac{dX(\text{K})}{X(\text{K})} \qquad (A.6)$$

where subscripts C or K indicate calculation in degrees Celsius or Kelvin, respectively. Equation (A.6) may be solved for NSC_K

$$\text{NSC}_\text{K} = \text{NSC}_\text{C}\,\frac{X(\text{K})}{X(^\circ\text{C})}\,\frac{dX(^\circ\text{C})}{dX(\text{K})} \qquad (A.7)$$

which would yield a value of about 419 for the NSC in this specific example and a relative error $d\phi/\phi = 419\,(0.000161) = 0.0675$, or about 6.8% as in the preceding example.

SELECTED THERMOPHYSICAL PROPERTIES OF BIOLOGICAL MATERIALS

John C. Chato

The data in this tabulation have been selected primarily from the comprehensive tables of Chato[1] and Bowman.[2] The emphasis is on biological and medical applications. A few food materials are included; for these materials, other references related to the food industry should be consulted (e.g., Ref. 3).

Since the thermophysical properties of tissues depend on composition (i.e., protein, fat, water) and structure, which vary considerably among specimens, most of the data presented should be considered to have limited accuracy in a specific application. The data represent good approximations for the properties of similarly structured tissues with similar composition.

Among the components of biological tissues, water has the highest thermal conductivity, 0.63 W/m °C. Thus, its value should be considered as an upper limit for *in vitro* tissue thermal conductivity. Blood circulation *in vivo*, however, can produce a higher "apparent" thermal conductivity. This can be observed by comparing *in vitro* and *in vivo* results for similar tissues.

Constitutive expressions for thermophysical properties for generalized tissues are given in Table 4. Generalized plots of thermal reflection and emittance of human skin are given in Figs. 1 and 2, respectively.

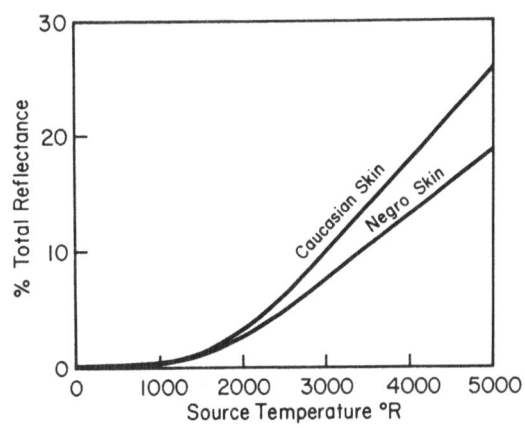

FIGURE 1
Variation with source temperature of the average, total, angular-hemispherical reflectance of the skin when irradiated by a blackbody or graybody source. From Boehm, R. F., and Tuft, D. B., Engineering radiation heat transfer properties of human skin, ASME Paper No. 71-WA/HT-37, 1971.

John C. Chato • Department of Mechanical and Industrial Engineering, University of Illinois, Urbana, Illinois 61801.

TABLE 1
Conversion Factors

To convert	from	to	multiply by
k	W/m °C	W/cm °C	10^{-2}
		cal/sec cm °C	0.2390×10^{-2}
		Btu/hr ft °F	0.5778
α	m^2/sec	cm^2/sec	10^4
		ft^2/hr	3.875×10^4
$k\rho c_p$	W^2 sec/m^4 °C^2	W^2 sec/cm^4 °C^2	10^{-8}
		cal^2/sec cm^4 °C^2	5.712×10^{-10}
		Btu^2/hr ft^4 °F^2	8.616×10^{-6}

TABLE 2
Thermal Properties of Solid Biological Materials above 0°C

Description	Conductivity k W/m °C	Diffusivity α $(m^2/sec) \times 10^7$	Inertia $k\rho c_p$ $(W^2 \ sec/m^4 \ °C^2) \times 10^{-6}$
A. Materials *in vivo* (values depend on blood perfusion)			
Bone, bovine and caprine	0.33–3.1		
Brain, cat	0.56–0.66	1.1–1.2	
Cartilage, scapula, bovine	1.8–2.8		
Cartilage, scapula, caprine	1.4–1.9		
Kidney, sheep	0.60–1.2	2.0–4.3	
Liver, canine	0.60–0.90	1.5–2.4	
Muscle, canine	0.70–1.0	0.7–1.3	
Skin, animal and human	0.48–2.8	0.4–1.6	1.0–7.0
B. Materials *in vitro* (room to body temperatures)			
Bone, fresh to several months postmortem	0.41–0.63[a]		0.87
Dry bone	0.22		
Bone marrow, bovine	0.22		
Brain, bovine, cat, and human	0.16–0.57	0.44–1.4	
Fat	0.094–0.37		0.28–0.51
Heart	0.48–0.59	1.4–1.5	
Kidney	0.49–0.63	1.3–1.8	1.4
Kidney[b], cortex, and medulla	$(-0.04 + 0.664w^c)$		
Liver	0.42–0.57[a]	1.1–2.0	1.0
Liver, parenchyma	0.32	1.7–2.0	
Lung		2.4–2.8	
Muscle	0.34–0.68	1.8	0.94–2.0
Skin, animal and human	0.21–0.41	0.82–1.2	1.2–2.2
Spleen	0.45–0.60	1.3–1.6	
Tumors, 37°C[d]			
General range	0.47–0.58		
Scirrhous carcinoma of the breast	0.40		
Squamous cell of the lung	0.67		

[a] Higher values were reported but are questionable, because they are above that of water.
[b] From Ref. 4.
[c] w is the mass fraction of water.
[d] From Ref. 5.

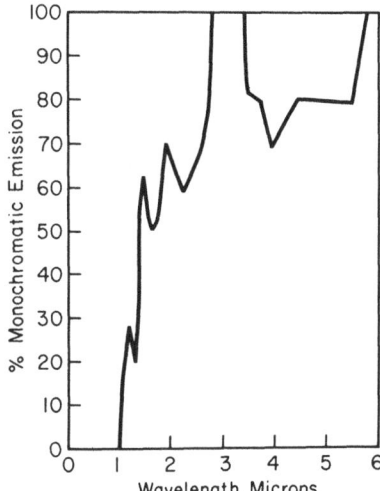

FIGURE 2
The spectral emittance of human skin at room temperature. The corresponding total emittance is 0.993. Reported values for total emittance range from 0.954 to 1.0; most authors use values between 0.98 and 0.99. From Elam, R., Goodwin, D. W., and Williams, K. L., Optical properties of the human epidermis. By permission of *Nature* 198 (4884), 1001–1002, 1963.

TABLE 3
Thermal Properties of Liquid Biological Materials above 0°C

Description	Conductivity k W/m °C	Diffusivity α (m²/sec) × 10⁷	Inertia $k\rho c_p$ (W² sec/m⁴ °C²) × 10⁻⁶
A. Biological fluids (at room to body temperatures)			
Agar gel, 1–1.75%	0.60–0.70	1.6	
Blood	0.48–0.60		
Blood, human	$[0.57-0.12H^a]$		
Blood, canine, 1% heparin	$[0.6\,(1-0.2V^b)$ $-2.26 \times 10^{-3}(37 - T \text{ °C})]$		
Blood plasma	0.57–0.60		
Humor, aqueous and vitreous	0.58–0.59		
Milk, regular and skimmed	0.53–0.59		
Cream (double Devon)	0.31		
Cod liver oil	0.17		
Egg white	0.56		
Egg yolk	0.34–0.42		
Gastric juice	0.44		
Urine	0.56		
Water	0.59–0.63		
B. Frozen, then thawed materials			
Brain, bovine, slow freeze–thaw	0.50		
Kidney, bovine, slow freeze– thaw	0.51–0.54		
Liver, bovine, slow freeze–thaw	0.50		
Muscle, bovine, sliced frozen	0.53		
Muscle, human, sliced frozen	0.44		

[a] H is the hematocrit fraction.
[b] V is the red cell volume fraction of blood.

TABLE 4
Thermal Properties of Frozen and Cold Materials

Description	Conductivity k W/m °C	Diffusivity α $(m^2/sec) \times 10^7$	Inertia $k\rho c_p$ $(W^2 sec/m^4 °C^2) \times 10^{-6}$
Beef, eye of loin, 69.5% water			
0°C	0.29		
−3 to −17°C	1.0		
Beef neck, fresh			
−7°C	0.41–0.47		
Beef			
−130°C	1.55		
Beef, 0.9% fat, 75% water, parallel to fiber			
0°C	0.48		
−10 to −25°C	1.36–1.54		
Beef, flank, lean, 3.4% fat, 74% water			
−10 to −25°C	1.07–1.21		
Beef, lean, 78.5% water, perpendicular to fiber			
0°C	0.48		
−5°C	1.06		
- 10°C	1.35		
−20°C	1.57		
Beef, fat, 74.5% water, perpendicular to fiber			
0°C	0.48		
−5°C	0.93		
−10°C	1.20		
−20°C	1.43		
Fat, bovine			
0 to −20°C	0.20–0.30		
Fat, porcine			
0 to −20°C	0.19–0.40		
Liver, bovine, slowly frozen			
−32°C	0.2[a]		
Liver, bovine, quickly frozen			
−195°C	0.1[a]		
Liver, rabbit	1.1		
Pork, leg, 6.1% fat, 72% water, −10 to −25°C			
parallel to fiber	1.4–1.6		
perpendicular to fiber	1.2–1.4		
Pork, lean neck			
−8°C	0.70–0.74		
Whole blood			
−10°C	1.6	8.7	
−20°C	1.7	10.4	
−40°C	1.9	13.6	
−60°C	2.1	16.9	
−80°C	2.4	20.4	
−100°C	2.7	23.7	

(continued)

TABLE 4 (cont.)

Description	Conductivity k W/m °C	Diffusivity α $(m^2/sec) \times 10^7$	Inertia $k\rho c_p$ $(W^2 sec/m^4 °C^2) \times 10^{-6}$
Plasma			
−10°C	2.0	9.7	
−20°C	2.1	11.4	
−40°C	2.3	15.1	
−60°C	2.6	18.8	
−80°C	2.9	22.9	
−100°C	3.2	26.9	
Packed cells			
−10°C	1.2	6.8	
−20°C	1.3	8.2	
−40°C	1.5	11.0	
−60°C	1.7	14.1	
−80°C	2.0	17.2	
−100°C	2.3	20.4	

[a] Questionably low value.

TABLE 5
General Expressions

k, W/m °C = 0.054 + 0.573 m_{water}[a] for m > 0.2[b]

k, W/m °C = $\rho \sum_n k_n m_n / \rho_n = \rho \times 10^{-3}$ (0.628 m_{water} + 0.117 $m_{protein}$ + 0.231 m_{fat})[c]

c_p, J/kg °C = 4.2×10^3 (0.4 + 0.6 m_{water}), approximate[c]

c_p, J/kg °C = $\sum_n c_{pn} m_n = 4.2 \times 10^3\, m_{water} + 1.09 \times 10^3\, m_{protein} + 2.3 \times 10^3\, m_{fat}$[c]

$1/\rho$, m³/kg = $\sum m_n / \rho_n = m_{water}/1{,}000 + m_{protein}/1{,}540 + m_{fat}/815$[c]

Fur and feathers
k, W/m °C = 0.025 + 0.0048 × (thickness in cm)

[a] m is mass fraction.
[b] From reference 6.
[c] From reference 7.

REFERENCES

1. Chato, J. C., Heat transfer in bioengineering, in *Advanced Heat Transfer*, B. T. Chao, ed. (University of Illinois Press, Urbana, 1969), pp. 404–412.
2. Bowman, H. F., Cravalho, E. G., and Woods, M., Theory measurement, and application of thermal properties of biomaterials, *Ann. Rev. Biophys. Bioeng.* 4, 58–69, 1975.
3. Qashou, S., Vachon, R. I., and Touloukian, Y. S., Thermal conductivity of foods, *ASHRAE Trans.* 78–1, 165–183, 1972.
4. Holmes, K. R., Pienta, C., Ryan, W., and Chen, M. M., *In vitro* thermal conductivity and water content of rabbit kidney cortex and medulla, *Physiologist* 24, 69, 1981.

5. Bowman, H. F., Thermodynamics of tissue heating: Modeling and measurements for temperature distributions, in *Physical Aspects of Hyperthermia*, G. H. Nussbaum, ed. (American Institute of Physics, New York, 1982).
6. Spells, K. E., The thermal conductivities of some biological fluids, *Phys. Med. Biol.* **5**, 150, 1960.
7. Cooper, T. E., and Trezek, G. J., Correlation of thermal properties of some human tissues with water content, *Aerospace Med.* **42**, 24–27, 1971.

FINITE-DIFFERENCE AND FINITE-ELEMENT METHODS OF SOLUTION

Avraham Shitzer, Linda J. Hayes, Robert W. Olsen, and Robert C. Eberhart

1. INTRODUCTION

The advent of digital computers has brought about the widespread application of numerical methods to solve boundary-value problems. These methods are particularly useful in handling problems that are not amenable to analytical solutions. Among these are problems involving variable properties, irregular boundary conditions, nonlinearities, and complex geometries. Bioheat transfer problems typically involve these and other matters that make them obvious candidates for numerical treatment.

A number of books have been published on numerical techniques, e.g., Refs. 1–3; application of these techniques to heat transfer problems has also been discussed.[4,5] The purpose of Appendix 3 is to briefly review some fundamental concepts in numerical techniques. In addition, two particularly useful techniques, i.e., finite difference and finite element, are presented and their application to the bioheat transfer equation is demonstrated.

2. FINITE DIFFERENCE

The basic concept underlying this numerical technique is the approximation of derivatives in the differential equation and boundary conditions in terms of differences. It will be recalled that in calculus, the derivative of the dependent variable $T(x)$ with respect to the independent variable x at x_i is defined by

$$\frac{dT}{dx}\bigg|_{x_i} \equiv \lim_{\Delta x \to 0} \frac{\Delta T}{\Delta x}\bigg|_{x_i} = \lim_{\Delta x \to 0} \frac{T(x_i + \Delta x) - T(x_i)}{\Delta x} \tag{1}$$

Avraham Shitzer • Department of Mechanical Engineering, Technion, Israel Institute of Technology, Haifa, Israel.   *Linda J. Hayes* • Department of Aerospace Engineering and Engineering Mechanics, University of Texas, Austin, Texas 78712.   *Robert W. Olsen and Robert C. Eberhart* • Department of Surgery, University of Texas Health Science Center, Dallas, Texas 75235.

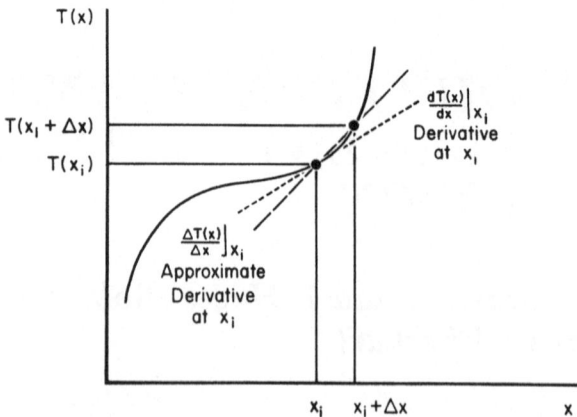

FIGURE 1
The derivative and its numerical approximation.

Reference to Fig. 1 will indicate that the derivative of $T(x)$ at x_i is actually the slope, or tangent, of this function at x_i. This exact derivative may be approximated by drawing the line that connects $T(x_i + \Delta x)$ with $T(x_i)$, as shown by the dotted line in Fig. 1. The slope of this line, which is the approximate derivative of the dependent variable $T(x)$ with respect to the independent variable x at $x_i + \Delta x$, is given by

$$\left.\frac{dT}{dx}\right|_{x_i} \doteq \left.\frac{\Delta T}{\Delta x}\right|_{x_i} = \frac{T(x_i + \Delta x) - T(x_i)}{\Delta x} \tag{2}$$

Although the slopes of these two lines are different, it is apparent that they may become closer as Δx is made smaller. Higher order derivatives are calculated in a similar manner, i.e., by taking the difference of the derivatives of the immediate lower order. An alternative technique for calculating the approximate derivatives is to expand the dependent function in terms of a Taylor series

$$T(x) = T(x_i) + \Delta x \left.\frac{dT}{dx}\right|_{x_i} + \frac{(\Delta x)^2}{2!} \left.\frac{d^2 T}{dx^2}\right|_{x_i} + \frac{(\Delta x)^3}{3!} \left.\frac{d^3 T}{dx^3}\right|_{x_i} + \cdots \tag{3}$$

Once the approximate derivatives have been obtained, they may be substituted into the differential equation that is to be solved and into the expressions of the boundary and initial conditions. These substitutions transform this set of differential equations into a series of algebraic equations. Solution of these algebraic equations, which may still be quite involved, yields an approximate solution to the original problem.

The degree of conformity of the approximate solution to the exact one depends on many factors; among these are the specific numerical formulation employed, the stability and convergence of the solution, the effect of truncation errors, etc. The final result of this powerful technique is always a compromise, however good it may be, between the required accuracy and computing expenses.

As an illustration, the finite-difference formulation of the nonsteady bioheat transfer problem in a tissue with a nonuniform blood flow distribution is presented; see Chap. 12 and Ref. 6 for a description of the problem and definitions of all terms in the model. Analytical solution of this problem is not possible in general, which thus warrants the application of a numerical technique. For the purpose of this illustration, it is assumed that the tissue may be described as a semi-infinite slab of half-thickness L. The biological tissue represented by this slab is a homogeneous medium with isotropic thermal properties. One side of the slab $X = L$ is exposed to an environment at temperature T_∞, with surface heat transfer coefficient h. The other side $X = 0$ is assumed to be insulated and an adiabatic condition is applied. The heat balance of this tissue is assumed to be given by the bioheat equation. Blood flow distribution in the tissue is given by the function $w_b(x)$.

Mathematically the problem may be stated in dimensionless form as

$$\frac{\partial \theta}{\partial \tau} = \frac{\partial^2 \theta}{\partial \xi^2} - \beta \theta \tag{4}$$

subject to the boundary and initial conditions

$$\text{at } \xi = 0, \qquad \frac{\partial \theta}{\partial \xi} = 0 \tag{5}$$

$$\text{at } \xi = 1, \qquad \frac{\partial \theta}{\partial \xi} = \text{Bi}(1 - \theta) \tag{6}$$

$$\text{at } \tau = 0, \qquad \theta = \theta_0(\xi) \tag{7}$$

where

$$\theta = \frac{T - T_a}{T_\infty - T_a}; \qquad \tau = \frac{\alpha t}{L^2}; \qquad \xi = \frac{x}{L}; \qquad \text{Bi} = \frac{hL}{k};$$

$$\beta = \frac{\rho_b c_b L^2 w_b(x)}{k} \tag{8}$$

To solve this set of equations numerically, the tissue model is divided into $M + 1$ equally spaced nodes as shown in Fig. 2. Here, node ϕ represents the surface, and node M represents the center. In order to increase the accuracy of the numerical results, the surface node is analyzed in a manner that more realistically weights the effect of blood perfusion in that region. As shown in Fig. 3, near the surface node, the driving potential for heat exchange between tissue and arterial temperatures is calculated not at this node, but rather at an arbitrary distance $\Delta x/4$ inward from it. In effect, the nominal difference between arterial and tissue nodal temperature $T_a - T_0$ is replaced by a weighted tissue temperature at the surface node. At all other nodes, the driving potential is taken as $T_a - T_m$. This modification considerably improves the accuracy of the results obtained by the numerical solution method.[6]

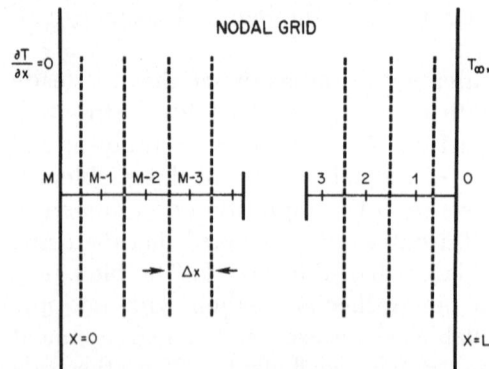

FIGURE 2
Nodal grid for a one-dimensional, finite-element model of an organ or tumor. A generalized Newtonian boundary condition is applied at $x = L$.

Equations (4)–(7) and the modification just discussed may now be converted into differential–difference equations. A typical interior nodal equation will have the form

$$\frac{\partial \theta_m}{\partial \tau} = \frac{1}{(\Delta \xi)^2} \{\theta_{m-1} - [2 + \beta_m(\Delta \xi)^2]\theta_m + \theta_{m+1}\} \tag{9}$$

while the equation for the center node, node M, can be written as

$$\frac{\partial \theta_M}{\partial \tau} = \frac{1}{(\Delta \xi)^2} \{2\theta_{M-1} - 2[1 + \beta_M(\Delta \xi)^2]\theta_M\} \tag{10}$$

The surface node, node 0, requires special treatment as previously described, and the equation for this node becomes

$$\frac{\partial \theta_0}{\partial \tau} = \frac{1}{(\Delta \xi)^2} \left\{ -2\left[1 + \mathrm{Bi}\, \Delta \xi + \frac{3\beta_0}{8}(\Delta \xi)^2 \right]\theta_0 \right.$$

$$\left. + \left[2 - \frac{\beta_0(\Delta \xi)^2}{4} \right]\theta_1 + 2\mathrm{Bi}\, \Delta \xi \right\} \tag{11}$$

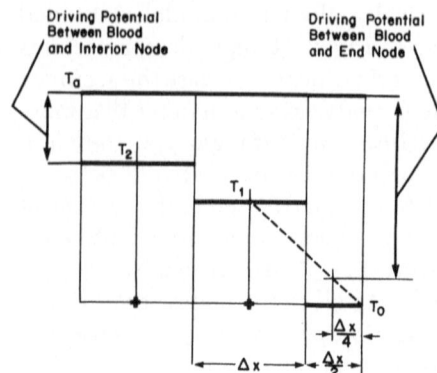

FIGURE 3
Model for the end node at the tissue surface. See text for details.

The problem may be formulated for solution by any of the suitable known techniques. In Appendix 3, we chose to employ the well known Crank–Nicholson solution in the time variable.[5] The set of algebraic equations is formulated in terms of an $(M + 1)$ by $(M + 1)$ matrix

$$
\begin{bmatrix}
(1+A) & -E_1 & 0 & 0 & \cdots & 0 & 0 & 0 \\
-p/2 & (1+c_1) & -p/2 & 0 & \cdots & 0 & 0 & 0 \\
0 & -p/2 & (1+c_2) & -p/2 & \cdots & 0 & 0 & 0 \\
\vdots & \vdots & \vdots & \vdots & \cdots & \vdots & \vdots & \vdots \\
0 & 0 & 0 & 0 & \cdots & p/2 & (1+c_{M-1}) & -p/2 \\
0 & 0 & 0 & 0 & \cdots & 0 & -E_{M-1} & (1-D)
\end{bmatrix}
\begin{bmatrix}
\theta_0^{n+1} \\
\theta_1^{n+1} \\
\theta_2^{n+1} \\
\vdots \\
\theta_{M-1}^{n+1} \\
\theta_M^{n+1}
\end{bmatrix}
$$

$$
+
\begin{bmatrix}
(1-A) & E_1 & 0 & 0 & \cdots & 0 & 0 & 0 \\
p/2 & (1-c_1) & p/2 & 0 & \cdots & 0 & 0 & 0 \\
0 & p/2 & (1-c_2) & p/2 & \cdots & 0 & 0 & 0 \\
\vdots & \vdots & \vdots & \vdots & \cdots & \vdots & \vdots & \vdots \\
0 & 0 & 0 & 0 & \cdots & p/2 & (1-c_{M-1}) & p/2 \\
0 & 0 & 0 & 0 & \cdots & 0 & E_{M-1} & (1-D)
\end{bmatrix}
\begin{bmatrix}
\theta_0^{n} \\
\theta_1^{n} \\
\theta_2^{n} \\
\vdots \\
\theta_{M-1}^{n} \\
\theta_M^{n}
\end{bmatrix}
$$

$$
+ 2p\mathrm{Bi}\,\Delta x
\begin{bmatrix}
1 \\
0 \\
0 \\
\vdots \\
0 \\
0
\end{bmatrix}
= 0
\tag{12}
$$

where

$$
A = p\left[1 + \mathrm{Bi}\,\Delta\xi + \frac{3\beta_0(\Delta\xi)^2}{8}\right]
\tag{13}
$$

$$
c_m = p\left[1 + \frac{\beta_m(\Delta\xi)^2}{2}\right]
\tag{14}
$$

$$
D = p\left[1 + \frac{3\beta_M(\Delta\xi)^2}{8}\right]
\tag{15}
$$

$$
E_m = p\left[1 - \frac{\beta_m(\Delta\xi)^2}{8}\right]
\tag{16}
$$

$$
p = \Delta\tau/(\Delta\xi)^2
\tag{17}
$$

n and $n + 1$ represent time steps such that

$$
\Delta\tau = \tau_{n+1} - \tau_n
\tag{18}
$$

Equation (12) was programmed and run for a set of problems involving nonuniform blood flow[6]; some solutions are given in Chapter 12. Partial verification of the computer code was done by comparing the results with those obtained from a closed form analytical solution. The comparison was necessarily limited to the simple case of uniform blood flow distribution. Normalized blood flow rate (β) in these calculations was allowed to range from 0 (no flow) to 1,000 (100-fold physiological). Time $\Delta \tau$ and spacing $\Delta \xi$ steps were also varied. The worst case percentage difference between the two techniques, defined as the greatest percentage difference in temperature between the analytical and numerical techniques at corresponding values of ξ and τ, was characterized over a wide range of $\Delta \xi$ and $\Delta \tau$ for values of ξ from 0–1 and all τ from 0 to equilibrium.

Results from these studies indicate that as β increases, smaller $\Delta \xi$ and $\Delta \tau$ are required to minimize error. For cases of physiological interest, β will range from 0.1 to 10. Additional calculations for $\beta = 2$, $\Delta \tau = 0.001$, and $\Delta \xi = 0.025$ showed that the maximum error in θ was less than 7.6×10^{-3}, while for $\beta = 10$, the maximum error was less than 7.6×10^{-2}.

Additional verification of the difference equations, computer codes, and numerical parameters $\Delta \xi$ and $\Delta \tau$ was obtained by diagonalization of the Crank–Nicholson propagation matrix. It is known that the Crank–Nicholson technique provides an unconditionally convergent solution for the parabolic-type heat conduction equation. Studies indicate that smaller values of $\Delta \xi$ and $\Delta \tau$ are necessary to minimize error as the perfusion parameter β increases.[6] If the matrix has been properly constructed, and the grid size and time step are adequately small, the absolute eigenvalues of the propagation matrix must all be real and less than unity. Further, if the solutions are to be nonoscillatory, the eigenvalues must all be positive. This was found to be the case for all of our Crank–Nicholson integrations, with either uniform or nonuniform blood flow distributions.

3. FINITE ELEMENT

One of the limitations of the finite-difference technique occurs when the problem to be solved has an irregular geometry or involves dissimilar materials within its boundaries. In these cases, the finite-element technique, which involves an element by element representation of approximate solutions, may be more useful.

It is recalled that in the finite-difference technique, the derivatives are approximated by differences to simplify the problem. In the finite-element technique, the problem to be solved is first restated in an integral or averaged formulation. The domain of the problem space is divided into a finite number of simple subdomains of specified shapes, the finite elements. The integral formulation is used to find a solution that satisfies, in a weighted average sense, the original differential equation and its boundary conditions.

The basic concept of this technique can be understood by reference to the calculus of variations.[7–9] This technique seeks functions that minimize a given integral over a region of interest, subject to certain boundary conditions[9-11]; however, a detailed discussion of this technique is beyond the scope of this Appendix. The interested reader is referred to the literature published on the subject, e.g., Refs. 9 and 11. Our purpose will be served by demonstrating the application of the finite-element technique to solve the bioheat equation.

We begin the demonstration of the finite-element method by introducing the bioheat equation with time-independent physical properties

$$\rho c \frac{\partial T}{\partial t} - \nabla(k\nabla T) + \rho_b c_b w_b T - (\rho_b c_b w_b T_a + q_m) = 0 \tag{19}$$

In Eq. (19), variables ρ and c are functions of space, whereas variables w_b, q_m, and T are functions of space and time. Equation (19) may be written in a simpler form by introducing

$$A = \rho c, \qquad B = \rho_b w_b c_b, \qquad D = -(\rho_b w_b c_b T_a + q_m) \tag{20}$$

Then we have

$$A \frac{\partial T}{\partial t} - \nabla(k\nabla T) + BT - D = 0 \tag{21}$$

To derive the weighted residual statement, we select an arbitrary, smooth weighting function $v = v(x, y)$ that is piecewise continuous and independent of time. Then we multiply Eq. (21) by v, integrate with respect to x and y and obtain

$$\int_\Omega \left[A \frac{\partial T}{\partial t} v - v\nabla(k\nabla T) + BTv - Dv \right] dx\, dy = 0 \tag{22}$$

$v\nabla \cdot (k\nabla T)$ can be integrated by parts using the relationship

$$v\nabla(k\nabla T) = \nabla(vk\nabla T) - k\nabla T\nabla v \tag{23}$$

Substituting Eq. (23) into Eq. (22) yields

$$\int_\Omega \left(A \frac{\partial T}{\partial t} v + k\nabla T\nabla v + BTv - Dv \right) dx\, dy - \int_\Omega \nabla(vk\nabla T)\, dx\, dy = 0 \tag{24}$$

The last integral on the right-hand side can be transformed into a boundary integral by using the divergence theorem

$$\int_\Omega \nabla(vk\nabla T)\, dx\, dy = \int_{\partial\Omega} v(k\nabla T \cdot \mathbf{n})\, dS \tag{25}$$

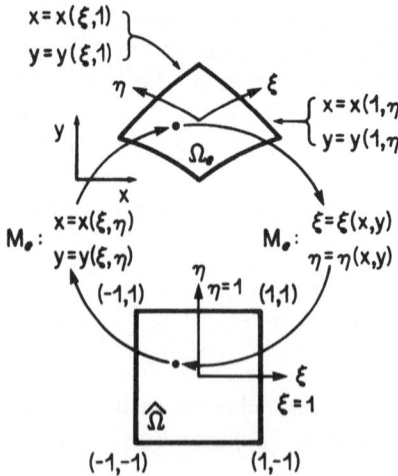

FIGURE 4
Coordinate maps for quadrilateral elements.

where $\partial\Omega$ is the boundary of domain Ω over which we integrate. This allows the incorporation of heat flux and convection boundary conditions into the variational statement. Substituting the boundary integral expression, Eq. (25), into Eq. (24), yields

$$\int_{\Omega} \left(A\frac{\partial T}{\partial t}v + k\nabla T\nabla v + BTv - Dv \right) dx\,dy = \int_{\partial\Omega} (k\nabla T \cdot \mathbf{n})v\,dS \qquad (26)$$

Now let us specify a convection boundary condition

$$-k\nabla T(x, y, t) \cdot \mathbf{n} = H(x, y)[T(x, y, t) - T_\infty(x, y, t)], \quad \text{for } (x, y) \text{ on } \partial\Omega \qquad (27)$$

Substituting Eq. (27) into the right-hand side of Eq. (26) yields

$$\int_{\Omega} \left(A\frac{\partial T}{\partial t}v + k\nabla T\nabla v + BTv - Dv \right) dx\,dy = \int_{\partial\Omega} HT_\infty v\,dS - \int_{\partial\Omega} HTv\,dS \qquad (28)$$

The continuous domain Ω is now subdivided into an approximation domain Ω_h containing the "finite elements" Ω_e (Fig. 4), and certain points in Ω_h will be called nodes or data points. For a two-dimensional problem, the shapes of the elements are normally triangular or quadrilateral, with straight or curved sides. The finite-element solution $T_h(x, y, t)$ is written in terms of basis functions $\phi_j(x, y)(x, y)$, which are simple polynomials (linear or quadratic) over each element

$$T_h(x, y, t) = \sum_{j=1}^{N} T_j(t)\phi_j(x, y)(x, y) \qquad (28a)$$

where the $T_j(t)$ are constants that multiply the basis functions and vary in time. The $T_j(t)$ is the value of temperature at the jth node at time t.

This leads to the following system of ordinary differential equations at time $t^{(2,7)}$

$$\sum_{j=1}^{N} C_{ij} \frac{\partial T_j}{\partial t} + \sum_{j=1}^{N} K_{ij} T_j = f_i, \quad i = 1, 2, \ldots, N \tag{29}$$

with

$$C_{ij} = \int_{\Omega_h} A \phi_i \phi_j \, dx \, dy \tag{30}$$

$$K_{ij} = \int_{\Omega_h} \left[k \left(\frac{\partial \phi_i}{\partial x} \frac{\partial \phi_j}{\partial x} + \frac{\partial \phi_i}{\partial y} \frac{\partial \phi_j}{\partial y} \right) + B \phi_i \phi_j \right] dx \, dy$$

$$+ \int_{\Omega_h} H \phi_i \phi_j \, dS \tag{31}$$

$$f_i = \int_{\Omega} D \phi_i \, dx \, dy + \int_{\partial \Omega_h} H T_\infty \phi_i \, dS \tag{32}$$

These coefficients are used to construct global matrices for the region of interest **C**, **K**, and **f**, which may be considered capacitive, conductive, and convective and surface exchange matrices, respectively. Thus, a matrix representation of Eq. (29) is given by

$$\mathbf{C\dot{T}}(t) + \mathbf{KT}(t) = \mathbf{f} \tag{33}$$

where $\dot{\mathbf{T}}$ and $\mathbf{T}$ are $N \times 1$ matrices, with $\dot{\mathbf{T}}$ being the time derivatives of the nodal temperatures **T**.

The temperatures T_j have been approximated in space at the nodes and must now be approximated in time as well. Several techniques are available to propagate the solution forward in time by regular increments.[8,13,14] The well-established Crank–Nicholson method has been shown to be suitable for solving the transient bioheat equation.[14] This method is based on a relationship of the form[8]

$$T^{(n+1)} = T^{(n)} + \frac{\Delta t}{2} \left[\frac{\partial T^{(n)}}{\partial t} + \frac{\partial T^{(n+1)}}{\partial t} \right] \tag{34}$$

The temperature at time $n + 1$ is a function of the temperature at the previous instant n and the arithmetic mean of the time derivatives of temperature at the beginning and end of time interval Δt. This procedure leads to a system of linear equations that must be solved in order to calculate the temperatures at each step in time.[15] Stability requirements dictate the maximum Δt that may be used.[7,14]

Both the finite-difference and the finite-element techniques result in a linear system of equations that must be solved at each time step. The matrices corresponding to each of these methods are sparse, symmetric, and banded, and they are about the same size. The finite-difference method is simple to understand and easy to implement in a model with a rectangular polar or spherical geometry. The finite-difference form of the bioheat equation requires special treatment at the boundary, as noted, to assure accuracy. Flux or convective heat transfer rate specifications at the boundary also require special difference equations to improve accuracy.[8] For problems involving non-homogeneous materials, special equations must be derived for interface nodes to ensure a balance of heat flux across the interface.

Finite-difference techniques do not work well when the bioheat transfer problem involves a complicated, curved geometry. For a single material, rectangular geometry case, both finite-difference and finite-element methods give comparable solutions. The finite-element method is much harder to comprehend than the finite-difference method; however, it offers several advantages over finite-difference methods. It can easily be applied to materials with nonconstant properties and temporally and spatially varying perfusion rates. Continuity of temperature and heat flux are automatically embedded in the finite-element equations, and special considerations do not have to be made at interface nodes. A prescribed temperature, heat flux, or convective heat transfer rate can be specified at the surface, and these boundary conditions are easily and accurately applied. Perhaps the most important feature of the finite-element method is its ability to treat geometrically complex structures, such as mammalian tissues, by dividing the problem into small, curved elements.

4. CONCLUSION

Appendix 3 presents the fundamental concepts underlying two of the most commonly used numerical techniques, namely, finite difference and finite element. We have attempted to present not only the basics of these techniques but also to demonstrate their application to analysis with the bioheat equation. Until recently, closed form solutions have been the primary result of bioheat transfer analysis. However, it now appears that a proliferation in the application of numerical methods will occur as the bioheat transfer problems under consideration become more complex.

REFERENCES

1. Forsythe, G. E., and Wasow, W. R., *Finite-Difference Methods for Partial Differential Equations* (Wiley, New York, 1960).
2. Zienkiewicz, O. C., *The Finite-Element Method in Engineering Science*, 3d ed. (McGraw-Hill, London, 1977).
3. Hildebrand, F. B., *Introduction to Numerical Analysis*, 2d ed. (Wiley, New York, 1974).

4. Isenberg, J., and de Vahl Davis, G., Finite-difference methods in heat and mass transfer, in *Topics in Transport Phenomena*, C. Gutfinger ed. (Halsted, New York, 1975), pp. 475–553.

5. Adams, J. A., and Rogers, D. F., *Computer-Aided Heat Transfer Analysis* (McGraw-Hill, New York, 1973).

6. Elkowitz, A. B., Shitzer, A., and Eberhart, R. C., Transient temperature profiles in tissues with nonuniform blood flow distributions, *ASME Trans. J. Biomech. Eng.* **104** (3), 202–208, 1982.

7. Becker, E. B., Cary, G. F., and Oden, J. T., *Finite Elements: an Introduction*, vol. 1 (Prentice Hall, Englewood Cliffs, NJ, 1981).

8. Myers, G. E., *Analytical Methods in Conduction Heat Transfer* (McGraw-Hill, New York, 1971).

9. Finlayson, B. A., *The Method of Weighted Residuals and Variational Principles* (Academic, New York, 1972).

10. Sokolnikoff, T. S., and Redheffer, R. M., *Mathematics of Physics and Modern Engineering*, 2d ed. (McGraw-Hill, New York, 1966).

11. Weinstock, R., *Calculus of Variations* (McGraw-Hill, New York, 1952).

12. Ergatoudis, I., Irons, B. M., and Zinkiewicz, O. C., Curved, isoparametric, "quadrilateral" elements for finite-element analysis, *Int. J. Solids Struct.* **4**, 31–42, 1968.

13. Hayes, L. J., A users' guide to PARAB—a two-dimensional, linear, time-dependent finite-element program, Texas Institute for Computational Mechanics, Report 80-10, 1980.

14. Nelson, M. (Master's thesis, University of Texas at Austin, 1982).

15. McCalla, T. R., *Introduction to Numerical Methods and Fortran Programming* (Wiley, New York, 1967).

INDEX